D0207550

# Photophysics of Polymers

ACS SYMPOSIUM SERIES **358**

# Photophysics of Polymers

**Charles E. Hoyle,** EDITOR
*University of Southern Mississippi*

**John M. Torkelson,** EDITOR
*Northwestern University*

Developed from a symposium sponsored
by the Division of Polymer Chemistry, Inc.,
at the 192nd Meeting
of the American Chemical Society,
Anaheim, California,
September 7-12, 1986

American Chemical Society, Washington, DC 1987

**Library of Congress Cataloging-in-Publication Data**

Photophysics of polymers/Charles E. Hoyle, John M. Torkelson, editors

"Developed from a symposium sponsored by the Division of Polymer Chemistry at the 192nd Meeting of the American Chemical Society, Anaheim, California, September 7-12, 1986."

p. cm.—(ACS symposium series, ISSN 0097-6156; 358)

Bibliography: p.
Includes index.
ISBN 0-8412-1439-5

1. Polymers and polymerization—Congresses.
2. Photochemistry—Congresses.

I. Hoyle, Charles E., 1948- . II. Torkelson, John M. III. American Chemical Society. Division of Polymer Chemistry. IV. American Chemical Society. Meeting (192nd: 1986: Anaheim, Calif.) V. Series.

QD381.8.P47 1987
547.7'0455—dc19 87-27307
CIP



PRINTED IN THE UNITED STATES OF AMERICA

# ACS Symposium Series

## M. Joan Comstock, *Series Editor*

# Foreword

The ACS SYMPOSIUM SERIES was founded in 1974 to provide a medium for publishing symposia quickly in book form. The format of the Series parallels that of the continuing ADVANCES IN CHEMISTRY SERIES except that, in order to save time, the papers are not typeset but are reproduced as they are submitted by the authors in camera-ready form. Papers are reviewed under the supervision of the Editors with the assistance of the Series Advisory Board and are selected to maintain the integrity of the symposia; however, verbatim reproductions of previously published papers are not accepted. Both reviews and reports of research are acceptable, because symposia may embrace both types of presentation.

# Contents

# Preface

SIGNIFICANT QUESTIONS ARE ARISING as the field of polymer science continues to develop. Many of these questions can be addressed only at the most fundamental levels.

Polymer photophysics has evolved over the past two decades, providing the polymer scientific community with a viable tool for probing polymer structure on a molecular scale. Herein lies its true value. By investigating photophysical phenomena of polymer systems, we can develop an accurate picture of how polymers exist in solution and in solid phases.

The chapters in this book are designed to provide scientists who are engaged in basic and applied polymer research with a clear understanding of the current status of polymer photophysics. The concepts developed in this volume should stimulate others to apply photophysical techniques in their own research.

We thank the authors for their outstanding response. The cooperation of each contributor has made assembling this book an unexpected pleasure. We also thank Patricia Linton and Susan Barge for their efforts above and beyond the call of duty.

I (CEH) thank Fred Lewis and James Guillet for introducing me to the world of photochemistry and photophysics. I also thank my family for their patience and understanding in allowing me to spend precious time preparing this book. This demonstration of love by Karen, Abbie, and Austin makes this book invaluable.

A special acknowledgment is extended to the Petroleum Research Fund, administered by the American Chemical Society, and to the ACS Division of Polymer Chemistry, Inc., for their generous support. Their financial contributions allowed several foreign contributors to participate in the ACS meeting upon which this book is based.

CHARLES E. HOYLE
University of Southern Mississippi
Hattiesburg, MS 39406-0076

JOHN M. TORKELSON
Northwestern University
Evanston, IL 60201

August 4, 1987

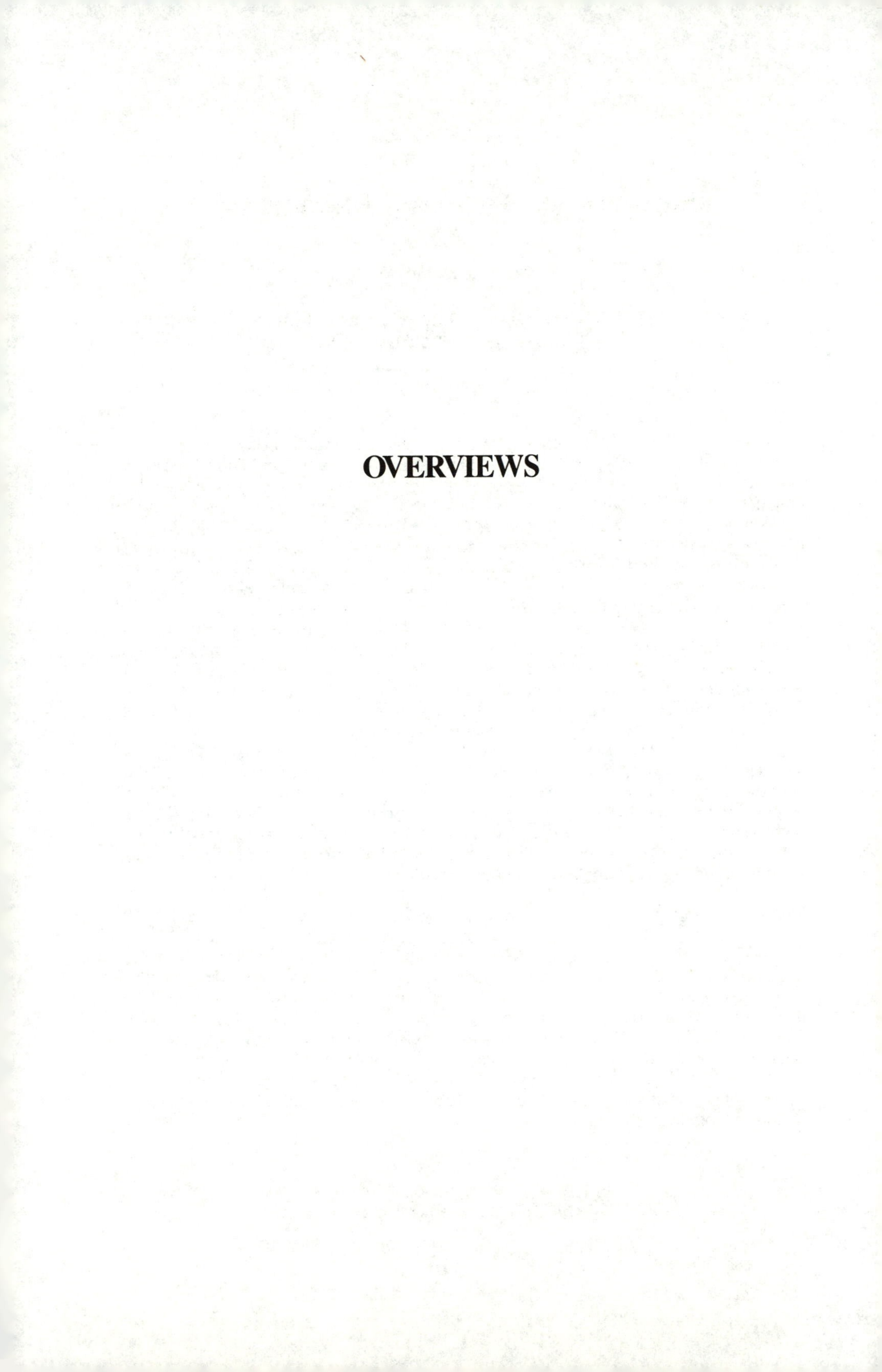

# OVERVIEWS

# Chapter 1

# Overview of Polymer Photophysics

**Charles E. Hoyle**

**Department of Polymer Science, University of Southern Mississippi, Hattiesburg, MS 39406-0076**

In order to write a complete review of polymer photophysics, one would need a series of books dedicated to each of the important aspects of this continually developing field. There are already a number of excellent collections and reviews (1-17) on polymer photophysics to which the reader is referred. The book by Guillet (1) is particularly instructive as an overview of polymer photophysics and for that matter to the whole field of polymer photochemistry and is highly recommended to anyone who wishes to gain a rapid, but thorough introduction to the area.

This introductory chapter is designed to introduce the reader to the current status of polymer photophysics. By analogy with small molecule photophysics, luminescence can be classified as fluorescence or phosphorescence depending on whether emission occurs from a singlet state or a triplet state, respectively. Polymer photophysics, both in its historical development as well as its current practice, can be divided into relatively few categories: excimer formation, luminescence anisotropy, luminescence quenching, luminescent probes and excited state energy migration. After introducing each of these basic categories, this chapter is concluded by a short analysis of the future of polymer photophysics. The fundamental and/or applied aspects of polymer photophysics will be noted where appropriate in each section.

## Excimer Formation

Excimers are excited state complexes which consist of two identical species, one of which is in the excited state prior to complexation (See Scheme I). The subject has been thoroughly reviewed for polymers in a recent article by Semerak and Frank (4). Briefly (Scheme I), an excited monomer species M* combines with an identical ground state molecule M to produce an excimer E*. Both excited species M* and E* may undergo the normal processes for deactivation of excited states, i.e., non-radiative decay, radiative decay, or product formation.

0097-6156/87/0358-0002$06.00/0

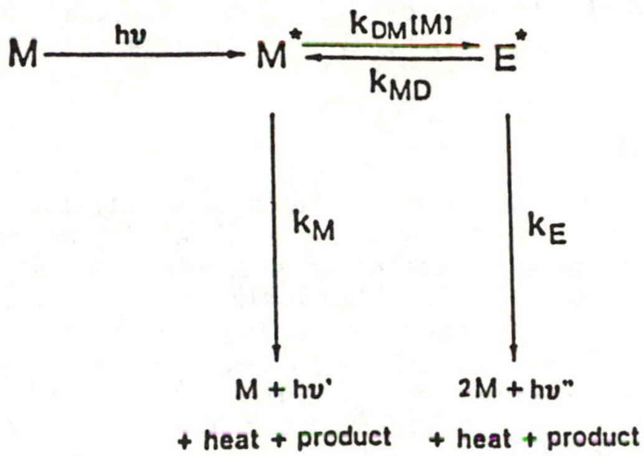

$$M \xrightarrow{h\nu} M^* \underset{k_{MD}}{\overset{k_{DM}[M]}{\rightleftharpoons}} E^*$$

$$M^* \xrightarrow{k_M} M + h\nu' + \text{heat} + \text{product}$$

$$E^* \xrightarrow{k_E} 2M + h\nu'' + \text{heat} + \text{product}$$

Scheme I

In the publications on excimer formation in polymers to date, the vast majority have concentrated on homopolymers or copolymers having pendant aromatic chromophores such as phenyl or naphthyl groups. Polymers and copolymers based on 1-vinylnaphthalene, styrene, 2-vinylnaphthalene and N-vinylcarbazole have probably received the most attention while polymers based on vinyltoluene, acenaphthalene, vinylpyrene, 2-naphthylmethacrylate, and a number of other monomers have also been studied, but to a lesser extent. Excimer formation in such polymer systems is especially favorable when the interacting species are "nearest neighbors" pendant to the polymer backbone and separated by three carbon atoms. However, excimers have also been reported for copolymer systems where the interactive chromophores are separated by a larger number of atoms.

Theories dealing with the photophysics of excimer formation and decay involving the "isolated monomer" and "energy migration" concepts have been developed in order to explain the complex fluorescence decay curves observed for polymer systems (5). Application of these fluorescence decay laws continues to be a topic of interest as will be demonstrated in chapters throughout this book.

## Luminescence Anisotropy

If a randomly distributed ensemble of anisotropic fluorescent chromophores absorbs light which is linearly polarized, the resultant fluorescence will retain, to a degree, the polarization of the exciting light source. The "depolarization" of the emission is dependent on several factors including the inherent degree of anistropy of the fluorescent chromophore, the degree of energy migration, and the rotation of the chromophore during its excited state lifetime. From steady-state and transient emission anisotropic measurements, rotational relaxation times can be deduced. Depending on the location of the anisotropic chromophore in the polymer, attached as a pendant group or incorporated as an integral part of the polymer backbone, the rotational relaxation times reflect either main chain or side group rotational properties. Since rotational relaxation times are directly related to rotational diffusion coefficients, luminescence anisotropic measurements provide information on conformational states, chain dynamics and microviscosity which might otherwise be difficult to acquire. In general, luminescence anisotropic measurements are extremely important in evaluating critical properties of both synthetic and natural polymers. Examples are presented throughout this book.

## Luminescence Quenching

Luminescence quenching involves deactivation of either an excited singlet (fluorescence quenching) or triplet state (phosphorescence quenching) by long range or short range interaction with a quencher molecule. The quenching process efficiency is determined by a variety of factors which include both orientational and interactive (electron transfer, dipole-dipole, etc.) parameters. In general, quenching can be described as a process in which two participating species (the excited state molecule and the quencher) interact either by a diffusion (collision) controlled (Stern-Volmer kinetics) or a static mechanism (Perrin kinetics) (see Ref. 1 for details). In polymer systems, it is possible for the excited state chromophore and the quencher to exist in several combinations. The following are representative of the combinations one might expect to encounter in quenching experiments involving polymers:

(a) The luminescent species is attached to the polymer and the quencher is a small molecule.

(b) The luminescent species is detached from the polymer and the quencher is bound to the polymer.

(c) Both the luminescent species and the quencher are attached to the same polymer.

(d) The luminescent species and the quencher are attached to different polymer chains.

Depending on the particular combination, quenching studies generate detailed data concerning energy migration in polymers, degree of interpenetration of polymer chains, local quencher concentration, diffusion of small molecules in polymer networks, and phase changes.

## Excited State Energy Migration

Upon absorption of a photon of light by a particular chromophore on a polymer chain, one option of the resulting excited state species is to transfer its energy to an equivalent neighboring group in the ground state. This second species may then transfer its energy to another group. The probability of the transfer process occurring (versus deactivation) is dictated by, among other factors, its proximity to an equivalent neighbor in the ground state and the orientation of the two species involved in the energy transfer step. The sequential transfer of excited state energy from one chromophore to the next can result in energy migration over a large number of equivalent groups. This energy migration phenomenon has been compared to the antenna effect in photosynthesis (1). Theories to describe singlet and triplet energy migration in polymer systems have been developed and are included in several chapters in the book.

## Luminescent Probes

One of the basic goals in polymer science is to identify the nature of the local environmental domains (viscosity, hydrophobicity, etc.) domains within a polymer solid or solution. This may be accomplished by using a variety of photophysical techniques all of which involve mixing or covalently attaching a small amount of a luminescent molecule (probe) into a polymer system. The probe molecule is designed such that one or more of its photophysical properties is directly dependent on some aspect of its environment. For example, both the fluorescence lifetime and the relative intensities of the vibrational structure of pyrene are altered when exposed to an aqueous, as opposed to a hydrophobic, medium. Polyelectrolytes such as poly(acrylic acid), under the proper conditions of pH and added electrolyte, have hydrophobic domains which are capable of solubilizing pyrene. The net result is pyrene molecules which behave photophysically as though they are in a hydrophobic environment, even though the primary medium is water. The techniques utilized to study hydrophobic domains in polymer solutions are derived, in part, from the large body of literature dealing with luminescent probes in micellar solutions (11). Other examples of the use of luminescent molecules to probe micro-environments in polymer systems are listed below.

(a) Molecules with twisted excited states can act as sensitive probes of polymer microviscosity.

(b) Highly anisotropic luminescent molecules can be used as probes to investigate orientational effects in stretched films or fibers.

(c) Bichromophoric small molecules for which the rate of excimer formation is viscosity dependent can be used to detect the local microviscosity of polymer solutions. Such probes can also be used to monitor viscosity changes which occur during polymerization.

(d) Luminescent chromophores whose excited state properties are solvent polarity or viscosity dependent can be incorporated as part of the polymer during polymerization and thereby act as a direct probe of the local environment. (An anisotropic probe bound to the polymer has already been discussed in an earlier section as one example of this phenomenon).

(e) Small charged molecules which change their luminescent properties upon complexation with polyelectrolytes in aqueous media can be used to probe the effect of polymer conformation on binding.

One can certainly imagine other uses for luminescent probe molecules (either attached or free) in addition to those listed. This area is advancing rapidly and examples are given in individual chapters in the book.

## Conclusions and Future

As the contents of this book will attest, the field of polymer photophysics continues to expand. The first decade of research in polymer photophysics was primarily, but certainly not entirely, directed toward characterization of the basic excited state properties of macromolecules. During the past ten years there has been an intensified effort to employ photophysics to solve basic questions about the nature of polymer systems. There is certainly no reason to suspect anything other than an increasing use of polymer photophysics as a fundamental tool to investigate phenomena such as blending, microviscosity, hydrophobic domains, polymer conformational structure, diffusion, and polymerization. Polymer photophysics appears to have evolved from an academic curiosity into a useful discipline capable of solving both fundamental and applied problems in the polymer field. It is hoped that the collection of articles presented in this book will furnish the reader with an appreciation for polymer photophysics and the potential it has for contributing to the future development of polymer science.

## Literature Cited


1. Guillet, J. E., Polymer Photophysics and Photochemistry, Cambridge University Press, Cambridge (1985).
2. E. V. Anufrieva and Y. Y. Gotlib, Adv. Polym. Sci., 40, 1 (1981).
3. K. P. Ghiggino, A. J. Roberts, and D. Phillips, Adv. Polym. Sci., 40 69 (1981).
4. S. N. Semerak and C. W. Frank, Adv. Polym. Sci., 54, 33 (1984).
5. Polymer Photophysics, ed., D. Phillips, Chapman and Hall, New York, NY (1985).
6. Photophysical and Photochemical Tools in Polymer Science, ed., M. Winnik, Reidel, Dordrecht, Netherlands (1987).
7. Photochemistry and Photophysics of Polymers, eds., N. S. Allen and W. Schnabel, Applied Science, Ltd., London (1984).
8. D. A. Holden and J. E. Guillet, in Developments in Polymer Photochemistry-1, ed., N. S. Allen, (1980), Applied Science, Ltd., London.
9. E. D. Owen, in Developments in Polymer Photochemistry-1, ed. N. S. Allen, (1980), Applied Science, Ltd., London.
10. A. C. Somersall and J. E. Guillet, J. Macromol. Sci., Rev. Macromol. Chem. C, 13, 135.
11. J. Kerry Thomas, The Chemistry of Excitation at Interfaces, ACS Monograph 181, American Chemical Society, Washington, DC (1984).
12. H. Morawetz, Science, 203, 405 (1979).
13. Y. Nishijima, J. Polym. Sci., Part C, 31 353 (1970).
14. Y. Nishijima, Progr. Polym. Sci. Jpn., 6, 199 (1973).
15. Y. Nishijima, J. Macromol. Sci., Phys., 8, 407 (1973).
16. Ann. N. Y. Acad. Sci., eds., H. Morawetz and I. Z. Steinberg, 366 (1981).
17. I. Soutar in Developments in Polymer Photochemistry, ed., N. S. Allen, 3 (4), Applied Science, Ltd., London (1982).


RECEIVED August 4, 1987

# Chapter 2

# Study of Complex Polymer Materials

## Fluorescence Quenching Techniques

**Mitchell A. Winnik**

**Department of Chemistry and Erindale College, University of Toronto, Toronto M5S 1A1, Canada**

Fluorescence quenching techniques (1) provide a battery of useful tools for the study of morphology in complex polymer systems. One of the most important applications of these techniques is to the study of interfaces and interphases in polymer systems.

Many polymer materials contain polymer-polymer interfaces. These include polymer blends, interpenetrating networks, core-shell polymer colloids, and polymer micelles. The properties of these materials depend, one believes, on the nature of the interface and on factors which operate within very short distances (50Å - 100Å) of the interface. These are the dimensions of polymer molecules, which means that a proper understanding of the performance of these materials requires understanding of the interface at the molecular level.

Among the tools that permit one to obtain molecular information about interfaces [e.g., x-ray and neutron scattering, solid state nmr (2)], fluorescence quenching methods (3) offer some important advantages. They are sensitive. The equipment is readily available and relatively inexpensive. There is scope and versatility to those methods. There are many sources in the literature one can turn to for ideas for new experiments to study systems composed of synthetic polymers, because of the wide-spread applications of fluorescence techniques in the biological sciences (4). This chapter provides a brief introduction to some applications of fluorescence quenching to study interfaces in polymer systems.

## STRATEGY

Mechanism. Fluorescence quenching is defined operationally as any interaction between a species $F^*$ in an electronically excited state and another species Q which leads to a decrease in the fluorescence intensity or fluorescence decay time of $F^*$. This definition encompasses all quenching mechanisms.

0097-6156/87/0358-0008$06.00/0

All fluorescence quenching processes between non-diffusing pairs of excited states and quencher molecules are characterized by rate constants $k_q(r)$ which are a function of the distance r between them. Orientation may also be important. The sensitivity of each quenching process to distance and orientation depends upon the quenching mechanism. Energy transfer by dipole coupling (5,6) (Förster energy transfer) can occur over distances up to 100Å. Electron transfer (6) can occur over 15Å to 20Å. Quenching by paramagnetic species requires orbital overlap and is a short range process (6).

A convenient way of classifying these processes is to define a parameter $R_0$ which represents the distance at which the quenching rate (for randomly oriented F/Q pairs) equals the unquenched decay rate of $F^*$. $R_0$ is a measure of the span of the quenching process. The value of $R_0$ will depend upon the details of the quenching mechanism and the particular F/Q pair under consideration. A selection of $R_0$ ranges for different quenching processes is collected in Table I.

Table I. Bimolecular Excited State Quenching Processes

| | interaction mechanism | effective distance[a,b] |
|---|---|---|
| 1. | Energy transfer by | |
| | dipole coupling | 10Å to 100Å |
| | electron exchange | 4Å to 15Å |
| | reabsorption | as far as emission reaches |
| 2. | Electron transfer | 4Å to 25Å |
| 3. | Exciplex formation | 4Å to 15Å |
| 4. | Excimer formation | ca. 4Å |
| 5. | Non-emissive self-quenching | 4Å to 15Å |
| 6. | Heavy atom effect | ca. 4Å |
| 7. | Chemical bond formation | ca. 2Å to 4Å |

[a]The minimum interaction distance is arbitrarily taken to be 4Å except where new chemical bonds are formed.

[b]Each pair of chromophores, for each interaction mechanisms, has its own characteristic distance $R_0$ at which the interaction rate equals the decay rate of the unquenched excited state. These values are estimates of the range of $R_0$ for randomly oriented non-diffusing pairs of species.

The experimenter chooses the quenching process and F/Q pair according to the distance scale he or she wishes to explore. For example, in the study of simple polymer blends, energy transfer (here $R_0$ = 22Å) was much more sensitive than exciplex formation ($R_0$ = ca. 7 to 10Å) at detecting small amounts of chain interpenetration at the interface (7,8).

Synthesis. Chemical synthesis is an essential part of any application of fluorescence quenching to the study of interfaces. Three quite different approaches have been described in the literature. First, one can label polymer-1 with F and polymer-2 with Q. In a blend of the two polymers $F^*$ and Q interact only at the interface between the polymers. Second, polymer-2 may fortuitously quench fluorescence of a dye F attached to polymer-1. For example, poly(vinylmethyl ether) [PVME] quenches anthracene fluorescence (9). This lucky observation has been used in conjunction with anthracene-labelled polystyrene [PS-Ant] to map out the phase diagram of mixtures of PS with PVME and to study spinodal decompostion kinetics (9). Finally, one can label either polymer-1 or polymer-2 with F, and choose as a quencher a small molecule which is preferentially soluble in one of the polymers. Examples of this approach are described below. This approach works best when one of the polymers is in the glassy state and the other is rubbery.

Span. The span of a quenching experiment is the distance scale sensed by the experiment. It represents the resolution of the measurement. There are in fact two quite different distance scales involved in fluorescence quenching experiments. One is determined by $R_0$. In experiments in which diffusion is unimportant, $R_0$ is the only important distance scale. If diffusion of F or Q occurs over a distance comparable to or larger than $R_0$ on a time scale of the lifetime of $F^*$, the span will be somewhat larger that $R_0$.

In some experiments, mixing or demixing occurs on a much longer time scale. Here, where one studies changes in fluorescence intensity (I) over a period of minutes or hours, the relevant distance scale of the experiment is much larger. For demixing of PS-Ant and PVMF (9), the span is larger than the dimensions of individual macromolecules and probably reflects the size of the initially formed phase domains. In sorption experiments of a small molecule Q into a thin film or small particle labelled with F, the span is the film thickness or particle radius (10).

## EXAMPLES

The examples which follow are chosen from my laboratory. I do not wish to imply that they are the most interesting or most important experiments which use fluorescence quenching applications to rather complicated materials. They represent a prototype of the kinds of experiments and the kinds of materials one would study in an industrial laboratory.

The materials we studied are non-aqueous dispersions of polymer particles. Colloidal stability of these particles in hydrocarbon solvents is conferred by a surface covering of a highly swollen polymer (the stabilizer) on a second polymer, insoluble in the medium (the core polymer), which comprises $90^+$% of the material (11). These particles are prepared by dispersion polymerization : polymerization of a monomer soluble in the medium to yield an insoluble polymer, carried out in the presence of a soluble polymer which becomes the stabilizer. In the examples discussed here, the core polymer is formed by free radical polymerization. Hydrogen abstraction from the soluble polymer present in the reaction medium leads to formation of a graft copolymer. In this way we have prepared poly(methyl methacrylate) [PMMA] particles stabilized by polyisobutylene [PIB] (12,13) and poly(vinyl acetate) [PVAc] particles stabilized by poly(2-ethylhexyl methacrylate) [PEHMA] (14,15).

PMMA Particles. PIB-stabilized PMMA particles were prepared containing naphthalene [N] groups covalently attached to the PMMA chains. This was effected quite simply by adding 1-naphthylmethyl methacrylate to the MMA polymerization step of the particle synthesis. From reactivity ratios, one knows that the N groups are randomly distributed along the PMMA chains. The particles were purified by repeated centrifugation, replacement of the supernatant serum with fresh solvent (isooctane) and redispersion. A fluorescence spectrum of the dispersion was typical of that of a 1-alkyl-naphthalene. Chemical analysis indicated a particle composition IB/MMA/N of 13/100/10.

When small amounts of anthracene [bulk concentration ca. $1 \times 10^{-4}$ M] were added to particle dispersions, energy transfer from $N^*$ to Ant could be observed (cf. Figure 1)(12). At first we thought this occurred only from N groups near the particles surface. When the concentration [Ant] was altered, the fluorescence decay time of the $N^*$ also changed (12,13). Light penetrates into these 2 $\mu$m diameter particles to a depth equal to at least its wavelength (ca. 3000Å). If the lifetime of all the N groups is decreased by Ant added outside the particle, the Ant groups must have diffused into the particle interior during sample preparation.

This curious observation is at odds with a solid core structure for the particle. One can estimate the translational diffusion coefficient of Ant in PMMA ($<10^{-16}$ cm$^2$ sec$^{-1}$) (16) and predict that it would take years for Ant to penetrate so deeply into a solid PMMA core. Here the span of the experiment is the range of Ant penetration, limited by the depth of light penetration into the particle.

The change in $I_N(t)$ as [Ant] is varied fits the Stern-Volmer expression for these particles. Here $\tau_N$ is the lifetime measured

$$\frac{1}{\tau_N} = \frac{1}{\tau_A} = \frac{1}{\tau_N^\circ} + k_q[\mathrm{Ant}]\alpha \tag{1}$$

for $N^*$ in the presence of a bulk concentration [Ant] of quencher. $\tau_A$ is the component of the Ant* decay due to energy transfer (some Ant absorbs light at 280 nm.) The term $\alpha$ is the partition coefficient of Ant inside the particle. For diffusion controlled quenching, $k_q$ can be related to the mutual diffusion coefficient D of the reactants.

In a typical experiment, $\tau_N$ = 40 ns, [Ant] = $1 \times 10^{-4}$ M and $\alpha k_q = 1 \times 10^9$ M$^{-1}$ s$^{-1}$. If $\alpha$ is approximately unity, one calculates a diffusion coefficient D = ca. $1 \times 10^{-6}$ cm$^2$ sec$^{-1}$. Since N is bound and Ant is free to diffuse, we associate D with Ant diffusion. In 40 ns, Ant diffuses on average 20Å. This distance is comparable to $R_0$, equal to 22Å for Förster energy transfer for this pair of chromophores. We measure $I_N(t)$ over several lifetimes. The span of this experiment is thus on the order of 100Å.

To explain these results, my co-workers and I proposed a "microphase" or "interpenetrating network" model for the particle structure (13). A sketch of this model is given in Figure 2. We envision that much of the PIB stabilizer is trapped inside the particle where it forms an interconnecting network of channels. These are swollen with isooctane when the particles are dispersed. Ant is presumably quite soluble in this phase and moves readily throughout these channels. If $D_{Ant}$ is on the order of $1 \times 10^{-6}$ cm$^2$ sec$^{-1}$, it would take only seconds for Ant to reach the particle center; but much longer to diffuse out of the PIB phase into the (glassy) PMMA phase. Consistent with this model, chemical analysis by thin layer chromatography indicates that most of the PIB is covalently grafted to PMMA.

This morphology has also been confirmed by X-ray scattering studies on particles containing as a probe a heavy metal derivative introduced specifically into the PIB phase (17).

In these experiments, the gross morphology of the particles was elucidated primarily from the sorption rate of Ant into the particle as determined by fluorescence quenching. The experiment was effective because $R_0$ of the energy transfer process was large enough to permit the quencher Ant to communicate with $N^*$ across the phase boundary. More detailed information is possible by comparing these results with those of a quenching process characterized by a much smaller $R_0$ value.

Oxygen Quenching in Poly(vinyl acetate) Particles. In order to study a system in which fluorescent groups could be introduced into both the stabilizer and core polymers, we turned our attention to poly(vinyl acetate) [PVAc] particles sterically stabilized by poly(2-ethylhexyl methacrylate) [PEHMA] (14,15). Phenanthrene [Phe] was chosen as the fluorescence sensor. It was introduced into the stabilizer by mixing a small amount (ca. 1%) of 9-phenanthrylmethyl methacrylate 1 with EHMA in the synthesis of PEHMA. It was introduced into the core polymer by mixing a trace (ca. 0.01%) of 1 with VAc in the presence of unlabelled PEHMA in the particle synthesis step.

Phe fluorescence is relatively insensitive to its environment, and Phe has little tendency to form excimers. As a consequence, its fluorescence decay I(t) normally has a simple exponential form. This property contributes two important features to the experiment. First, the fluorescence decay time $\tau^{\circ}$ (in the absence of quencher) is a well-defined quantity, facilitating interpretation of steady state fluorescence experiments. Second, in complex systems, non-exponential decay traces may be observed. These can be interpreted in terms of a non-uniform distribution of quenchers in the system.

The particles we prepared were relatively small, with mean diameters of ca. 300 nm. The chemical composition was 5.6 monomer mol% PEHMA and 94.4% PVAC.

Information about the particles is obtained by comparing four

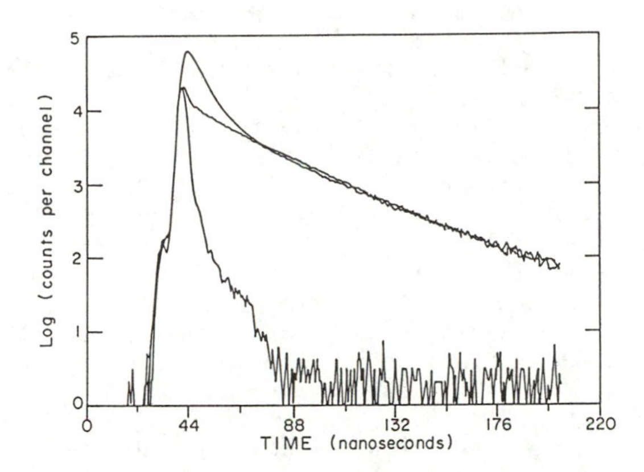


Figure 1. Fluorescence decay profile of N* (lower curve, measured at 337 nm) and Ant* (upper curve, 450 nm) for dispersions of PIB-stabilized PMMA particles, labelled with N in the PMMA phase, in isooctane containing 2 x $10^{-4}$ M Ant.

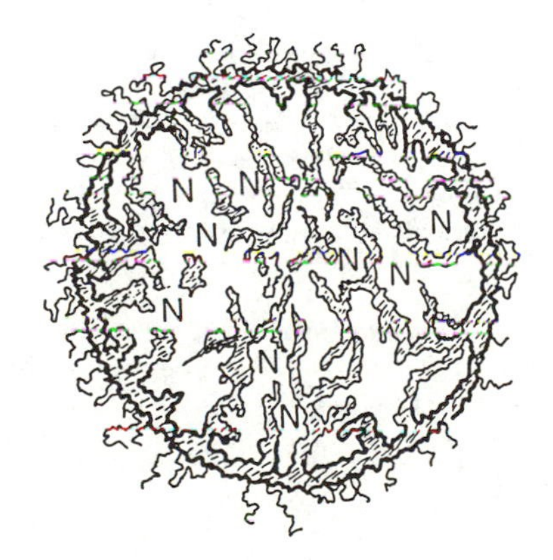


Figure 2. Our conception of the "microphase" or "interpenetrating network" model for non-aqueous dispersion morphology.

systems. First, in order to establish a benchmark, one looks at a small molecule, here 9-phenanthrylmethyl pivalate, 2, as a model for Phe-SP to establish the behavior of the polymer-bound dye. These form the basis for interpreting experiments on the two particle systems: that labelled in the stabilizer [SLP-Phe] and that labelled in the core [CLP-Phe].

Exposure of solutions of the model compound 2 and the copolymer Phe-SP to various oxygen concentrations causes both I and $\tau$ to decrease. The data in Figure 3 follow the Stern-Volmer relationship. Oxygen quenching is diffusion controlled. The decrease in slope for the copolymer is due to its smaller diffusion coefficient compared to 2. When the stabilizer-labelled particles are exposed to oxygen, the data also follow Equation 1. This result is somewhat unexpected, since I(t) here, in the presence of oxygen, deviates from an exponential form, implying a distribution of oxygen solubilities and mobilities in the stabilizer layer of the particle. The data recover their simple form when one calculates mean lifetimes $\langle\tau\rangle$ from the I(t) measurements and plots them according to Equation 1 (15). One can conclude from the particle data in Figure 3 that the entire region containing the stabilizer polymer is highly permeable to oxygen. Oxygen quenching is almost equally effective for the particle-bound stabilizer as for the free Phe-PEHMA copolymer in soluton.

One would anticipate no quenching of Phe fluorescence by oxygen in the core-labelled particle. Not only does one predict this result - the solubility and diffusion of oxygen are both substantially reduced in PVAc compared to cyclohexane - but measurements on films of PVAc containing Phe show no more than 1% quenching by oxygen (18).

In the dispersions of CLP-Phe we observe substantial oxygen quenching. The Stern-Volmer plot is curved (Figure 4, top), implying that some Phe are more accessible to quenching than others. The obvious interpretation is that some Phe are protected against quenching by virtue of being enclosed in a PVAc environment. The others are exposed to oxygen because they lie at the interface or are incorporated into a more extensive solvent-swollen interphase (15).

Quantitative data are available by fitting the data to the fractional quenching model (15). The Stern-Volmer model is modified by assuming that only a fraction $f_a$ of the fluorophores can be quenched. The remaining $(1-f_a)$ are protected:

$$\frac{I^\circ}{I^\circ - I} = \frac{1}{f_a} + \frac{1}{f_a \alpha k_q \tau^\circ [Q]} \qquad (2)$$

When our data are replotted according to Equation 2, Figure 4b, we obtain a straight line with an intercept of ca. 2. Hence 50% of the Phe groups in these particles are protected against oxygen quenching. Those that are exposed to oxygen are quenched with great effectiveness. If Phe is statistically distributed in the PVAc chains and serves as a marker for the location of those chains, we conclude that 50% of the PVAc is incorporated into the interphase.

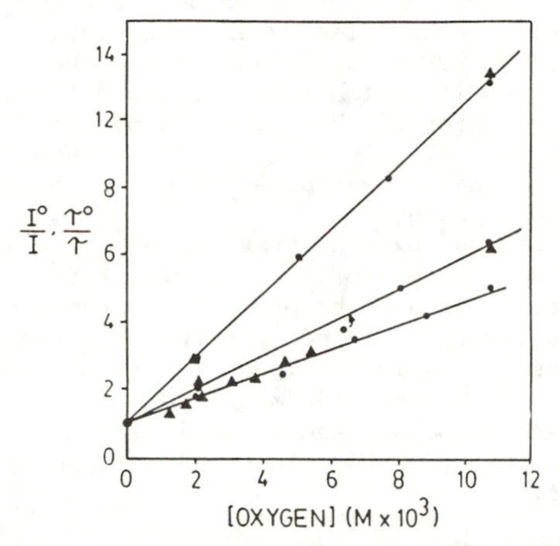


Figure 3. Plots of I°/I and $\tau°/\tau$ vs. bulk oxygen concentration for: the small molecule 2 (top line); the labelled PEHMA copolymer [Phe-SP, middle line]; the labelled particle [SLP-Phe, bottom line].

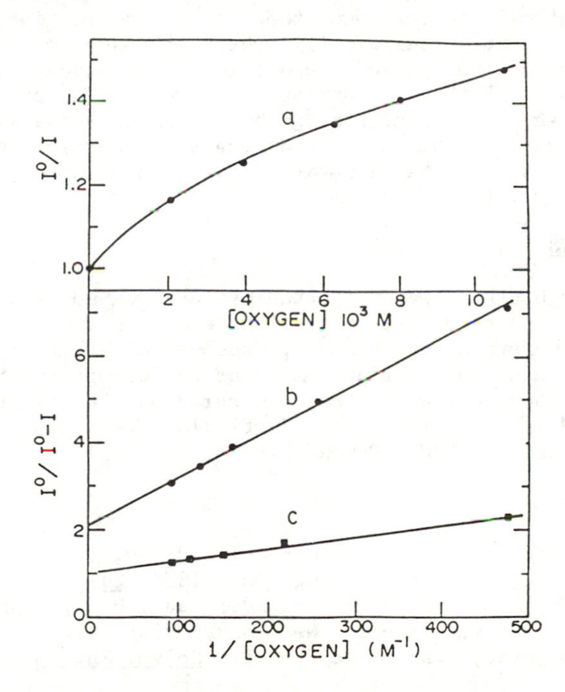


Figure 4. Fluorescence quenching by oxygen of the core-labelled particle CLP-Phe in cyclohexane. Top curve: plot of data according to Equation 1; bottom curve: plot of data according to the fractional quenching model, Equation 2.

When these experiments are repeated on dry powder samples of the particles, we observe much less quenching (18). For SLP-Phe, exposed to pure oxygen, 20% of the Phe* are quenched, compared to 80% for the dispersions in cyclohexane. these data also fit the Stern-Volmer equation. For CLP-Phe, only 5% of the Phe* are quenched in the presence of pure $O_2$ at 1 atm. These results emphasize the role of the solvent penetration in determining the properties of the interphase.

In the dry particles, oxygen uptake occurs on a measurable time scale (minutes). Here the span of the experiment is the particle radius. Such experiments hold the promise of providing sorption rates into the stabilizer phase and into the interphase. We are investigating whether it is in fact possible to obtain information in sufficient detail to measure the magnitude of the interphase in the dry particles and to obtain the solubilities and diffusion coefficients of various sorbents within the various individual phases of the material.

## SUMMARY

There are many different ways of using fluorescence quenching techniques to study the morphology of polymer materials, even those of rather complicated composition. Important considerations are (i) the use of synthesis to introduce the dye as a sensor into a specific phase of the material; (ii) the spectroscopic characteristics of the dye and the details of the quenching mechanism; and (iii) the span of the experiment. The span can be determined by the $R_0$ characterizing the quenching mechanism, the diffusion distance of the quencher during the fluorescence decay time of the sensor, or the penetration distance of a sorbent-quencher into the system.

## ACKNOWLEDGMENTS

The work described here was supported by NSERC Canada and by the Xerox Research Centre of Canada. All experiments were carried out in close collaboration with Dr. M.D. Croucher of XRCC. I am pleased to acknowledge his contribution. Most of the credit for this work belongs to my coworkers who actually carried out the experiments. They are cited in the references. Particular credit is due to Dr. Luke Egan and Dr. Onder Pekcan.

## REFERENCES

1. Fluorescence studies of polymer colloids 10.
2. (a) Sperling, L.H. Polym. Eng. Sci. 1984, 24, 1.
   (b) Stein, R.S. In Polymer Blends; Paul, D.R.; Newman, S., Eds.; Academic Press: New York, 1978; Vol. 1.
   (c) Djordjevic, M.B.; Porter, R.S. Polym. Eng. Sci. 1982, 22, 1109.
3. (a) Winnik, M.A., Ed. In Photophysical and Photochemical Tools in Polymer Science; D. Reidel: Dordrecht, Netherlands, 1987.

(b) Guillet, J.E. In Polymer Photochemistry and Photophysics; Cambridge University Press: Cambridge, UK, 1985.

4. (a) Chen, R.F.; Edelhoch, H., Eds. In Biochemical Fluorescence; Marcel Dekker: New York, 1975.
   (b) Schlessinger, J.; Elson, E.L. Methods Expt. Phys. 1982, 30 182.
   (c) Lakowicz, R. In Principles of Fluorescence Spectroscopy; Plenum Press: New York, 1983.
5. Berlman, I.B. In Energy Transfer Parameters of Aromatic Compounds; Academic Press: New York, 1973.
6. (a) Birks, J.B. In Photophysics of Aromatic Molecules; Wiley-Interscience: New York, 1971.
   (b) Inokuti, M.; Hirayama, F. J. Chem. Phys. 1965, 43, 1978.
   (c) Miller, J.R.; Beitz, J.V.; Huddleston, R.K. J. Am. Chem. Soc. 1984, 106, 5057.
7. Morawetz, H. Ann. N.Y. Acad. Sci. 1981, 366, 404.
8. Ledwith, A. Pure Appl. Chem. 1982, 54, 549.
9. Stein, R.S. Polymer 1984, 25, 956.
10. Stanton, D. M.Sc. Thesis University of Toronto, 1986.
11. (a) Barrett, K.E.J. In Dispersion Polymerisation in Organic Media; Wiley-Interscience: London, 1975.
    (b) Napper, D.H. In Polymeric Stabilization of Colloidal Dispersions; Academic Press: New York, 1983.
12. Winnik, M.A. Polym. Eng. Sci. 1984, 24, 87.
13. (a) Pekcan, O.; Winnik, M.A.; Croucher, M.D. J. Polym. Sci. Polym. Lett. 1983, 21, 1011.
    (b) Sperling, L.H. Polym. Mater. Sci. Eng. 1984, 50, 19.
14. Winnik, M.A., Egan, L.S.; Croucher, M.D. J. Polym. Sci. Polym. Chem. Ed. 1986, 24, 1895.
15 (a) Egan, L.S.; Winnik, M.A.; Croucher, M.D. Polym. Eng. Sci. 1986, 26, 15.
   (b) Winnik, M.A.; Egan, L.S., Croucher, M.D. Langmuir, submitted.
16. Berens, A.R.; Hopfenberg, H.B. J. Membrane Sci. 1982, 20, 183.
17. Winnik, M.A., Williamson, B.; Russell, T.R.; Macromolecules, 1987, in press.
18. Egan, L.S. Ph.D. Thesis, University of Toronto, 1987.


RECEIVED May 22, 1987

# Chapter 3

# Morphology in Miscible and Immiscible Polymer Blends

## Interplay of Polymer Photophysics with Polymer Physics

**Curtis W. Frank and Wang-cheol Zin**

**Department of Chemical Engineering, Stanford University, Stanford, CA 94305**

Applications of excimer fluorescence as a molecular level probe of chain configuration for isolated polymer chains and as a morphological tool for the study of phase separation are reviewed. In the first section, previous studies on the use of a strictly one-dimensional random walk model to describe energy migration in miscible blends are critically examined for blends of polystyrene (PS) with poly(vinyl methyl ether) (PVME) and poly(2-vinyl naphthalene) (P2VN) with either poly(cyclohexyl methacrylate) (PCMA) or polystyrene. The apparent energy migration efficiency is two orders of magnitude higher than expected, leading to the conclusion that energy migration in dilute blends is at best quasi-one-dimensional with a non-negligible amount of cross-loop hopping. In the second section, the PS/PVME phase diagrams for monodisperse PS of molecular weights 100,000 and 1,800,000 blended with PVME of number average molecular weight 44,600 and $M_w/M_n$=1.4 are determined. A temperature and concentration dependent binary interaction parameter is evaluated from the cloud point data and used to generate spinodal curves. Previous fluorescence results on spinodal decomposition in PS/PVME are reinterpreted using this new phase diagram to yield good agreement with the deGennes-Pincus theory. In the final section, annealing experiments for blends cast from tetrahydrofuran, chlorobenzene and toluene are used to demonstrate that toluene casting leads to the equilibrium binary blend morphology.

It has been estimated that 15 to 20 percent of all engineering plastics produced in the United States today are polymer blends or alloys, which are physical mixtures of homopolymers and/or copolymers. A serious problem exists, however, in that the bulk properties of the blend will depend on the degree of mixing achieved. Such mixing is difficult to accomplish on the molecular level and

0097-6156/87/0358-0018$06.00/0

existing experimental techniques exhibit widely varying sensitivities to the distance scale over which inhomogeneities are detectible. A major objective of research in our laboratory has been to establish excimer fluorescence as a quantitative photophysical tool for the elucidation of morphology and dynamics in polymer blends. An extensive review of the photophysics of excimer formation in the aryl vinyl polymers has been published. [1]

The application of this technique as a morphological tool requires that there be a close coupling between polymer photophysics and polymer physics. In the photophysical studies described in this paper emphasis will be placed on the development of analytical models for electronic excitation transport (EET). The areas of polymer physics that we will consider involve the configurational statistics of isolated chains and phase separation in multicomponent polymer systems. The polymer system of primary interest is the blend of polystyrene (PS) with poly(vinyl methyl ether) (PVME). Here excimer fluorescence from phenyl-phenyl interactions in PS is the main experimental observable. This blend was selected because it has been demonstrated by numerous other techniques that miscible one-phase blends may be prepared by solution casting from toluene solvent. [2,3] Moreover, the blend may be forced to phase separate by thermal means, leading to a two phase system. In addition, we will consider results for blends of poly(2-vinyl naphthalene) (P2VN) with low molecular weight poly(cyclohexyl methacrylate) (PCMA) and polystyrene (PS).

There is general agreement that delocalization of electronic excitation energy may occur in solid solutions of aryl vinyl polymers as a result of dipole-dipole interactions between aromatic rings covalently bound to the polymer chain. [4] In fact, the EET process is exceedingly complex with several modes of energy migration possible. Since the polymer chain contains a locally high concentration of aromatic groups, with separation distances between rings on adjacent repeat units of the order of five to seven Angstroms, one migration channel will be along the contour of the polymer chain. If the test chain of interest is totally isolated from identical chains, and if the chain is immobilized in a thermodynamically good matrix such that it exhibits an expanded configuration, it is possible that the energy migration could take on a one-dimensional character. As the concentration of the aryl vinyl polymer is increased, energy migration between chromophores on different chains is expected; this will approach a three-dimensional process at sufficiently high concentration.

In the first section of this paper we review a strictly one-dimensional random walk model developed by Fitzgibbon. [5] The model treats excitation migration on a finite size lattice with randomly distributed excimer forming sites acting as traps. It has been applied to explain the efficiency of trapping at excimer forming sites in three sets of apparently miscible systems. We first consider a series of PS/PVME blends in which the PS is at low concentration but of variable molecular weight. We then examine two other classes of blends in which the aryl vinyl polymer of interest is P2VN and blends are prepared with PCMA or PS as the host polymer. By contrast to the PS/PVME system, which is thermodynamically compatible over the whole concentration range and for all molecular weights of the two components, the P2VN blends are miscible only as a result of the low molecular weight of the host polymer. Our objective is to provide a critical assessment of the application of the strictly one-dimensional random walk model to the solid state blends.

In concentrated systems such as in films of the pure aryl vinyl polymer or in domains rich in the fluorescent polymer for a phase separated system it is likely that EET between rings on different chains could be as likely as between rings on the same chain. In the limit for which there is no anisotropic bias for energy migration along the chain, the EET may be considered to be three-dimensional. Two EET models have been developed to treat the problem of three-dimensional migration to pre-formed excimer sites in solid polymers. The first is a spatially periodic lattice model that restricts migration to nearest neighbors and does not permit the excitation to re-visit any earlier locations occupied during the random walk. [6] The second is a more sophisticated many-body EET theory that was originally developed by Loring, Andersen and Fayer (referred to as the LAF model) to treat isotopically doped molecular crystals with no restrictions on the chromophores sampled during the migration process. [7] Semerak has recently applied both the spatially periodic lattice model and the LAF theory to analyze excimer fluorescence in pure PS and P2VN. [8]

Although the theoretical study of EET in polymer systems has led to considerable recent advances, we choose in the second section of this paper to take a somewhat more pragmatic approach to the study of phase separation kinetics. To do so, we temporarily set aside any consideration of the details of any particular EET model and simply rely on a fundamental relationship between the ratio of the excimer to monomer emission intensities, $I_D/I_M$, and the probability M that an absorbed photon is eventually emitted by a radiative or nonradiative process of the monomer. Use of this general relation along with experimental data for the effect of concentration on excimer fluorescence in a miscible blend allows M to be determined experimentally. We will present a simple morphological model that combines the miscible blend results with an experimentally determined phase diagram to relate the absolute value of the $I_D/I_M$ ratio to the local concentration of chain segments containing aromatic groups. Our objective is to re-examine earlier data on spinodal decomposition for the PS/PVME blend using a more accurate phase diagram determined in the present study.

The preceding studies on the configuration of aryl vinyl polymer chains in dilute, miscible blends and on the kinetics of phase separation in concentrated blends were based on the implicit assumption that the initial solvent cast blend represented an equilibrium state. In the final section of this paper we explore this question with new data on the effect of the casting solvent on the fluorescence behavior of PS/PVME blends. Our objectives are to determine first whether the fluorescence observables are sensitive to differences in as-cast films and then to identify the true equilibrium state.

## Experimental

In order to improve the quantitative interpretation of the earlier studies on phase separation kinetics performed in our laboratory, [9,10] a new phase diagram was determined with narrower distribution PVME and an improved turbidimetric light scattering apparatus. The PVME was first fractionated by a precipitation procedure and the molecular weight distribution was determined on a Waters Model 244 gel permeation chromatograph having microStyragel columns. The resulting PVME had $M_w/M_n$=1.4 with $M_w$=55,000. This is to be compared with the starting material having polydispersity of about 1.9.

The cloud point curves were measured using a newly designed sample chamber that may also be used for the fluorescence measurements. This chamber has excellent thermal control and can be evacuated and back-filled with nitrogen or argon in order to minimize oxidative degradation of the sample. The sample was illuminated by a 5 mW He-Ne laser in the backface mode with the film at 45 degrees to the laser axis. The light scattered at right angles to the incident beam was passed through an attenuator and detected by an RCA 1P28 photomultiplier. Cloud point temperatures were determined by heating at a slow rate until commencement of phase separation, as evidenced by a rapid increase in the scattered light from the sample. The temperature was then decreased by several degrees and the decrease in scattering intensity monitored. This was followed by a temperature jump, smaller than the preceding temperature drop, and the process was repeated a sufficient number of times such that equilibrium was ensured. This could take many hours for the high PS concentration blends.

## Results and Discussion

Isolated Chain Statistics in Miscible Blends. Early work on the application of fluorescence spectroscopy to the study of polymer blends in this laboratory emphasized the correlation of variables expected to influence the free energy of mixing with changes in $R=I_D/I_M$. These studies included the effects of solubility parameter differences between the fluorescent guest polymer, generally poly(2-vinyl naphthalene), and the nonfluorescent host polymer matrix, [11,12] guest concentration, [12-14] molecular weight of both host and guest, [15-18] and the temperature at which the polymer blend is solvent cast [13]. In each case, an increase in R was interpreted as an increase in the local segment density, and could be rationalized on the basis of equilibrium Flory-Huggins thermodynamics.

These early studies demonstrated that excimer fluorescence is a useful addition to the battery of experimental tools available to study solid state polymer blends. However, the longer range goal of explaining the significance of the absolute value of R was not realized because there was insufficient companion information about the thermodynamics of the blends. The PS/PVME blend does not suffer from this limitation, and thus provides an excellent system for characterization of the photophysics under conditions for which miscibility or immiscibility are firmly established. In this section we examine results for PS/PVME as well as more recent work on dilute blends containing P2VN that are believed to be miscible.

The first stage of the development of excimer fluorescence as a quantitative morphological tool was to understand the photophysics of miscible blends. We began with a consideration of the temperature and molecular weight dependence of R for blends containing low concentrations of PS. As noted in the Introduction, energy migration in an isolated aryl vinyl polymer chain may be idealized as a one-dimensional random walk between adjacent pendent chromophores. [5] At each step in the walk, the excitation can be lost radiatively or nonradiatively from an isolated excited chromophore. Alternatively, the excitation can be trapped at an excimer forming site (EFS) composed of two adjacent chromophores suitably aligned in a coplanar sandwich configuration, leading to radiative or nonradiative emission from the excimer. The overall probability that an excitation absorbed anywhere in the chain will eventually be dissipated through radiative or nonradiative decay through a monomer channel is given by the quantity M and through an excimer channel by 1 - M. M may be predicted analytically, and is related to the experimental observable R through the expression

$$R = \frac{Q_D}{Q_M} \frac{1 - M}{M} \tag{1}$$

where $Q_D$ and $Q_M$ are the intrinsic quantum yields for excimer and monomer emission.

M is a complex function of composition, molecular weight and temperature. Two analytical approaches have been directed toward obtaining expressions for M. The simplest one goes back to Levinson who treated a one-dimensional random walk on an infinite lattice with randomly distributed traps. [19] Although this is clearly unrealistic for real, finite length polymer chains, it does help to set certain critical concepts. Thus, we spend some time on it before considering the results applicable to real polymers.

In the hypothetical case of an infinitely long polymer chain, the analytical expression for M is the relatively simple form

$$M = 1 - q_D - \frac{q_D^{\,2}}{\tanh(\Gamma)} \sum_{x=1}^{\infty} (1 - q_D)^x \tanh(\Gamma x) \tag{2}$$

$$\text{where} \qquad \Gamma = 0.5 \ln \left[ \frac{1 + 2E + (1 + 4E)^{1/2}}{2E} \right]$$

In this expression the first critical parameter is $q_D$, the fraction of chain dyads that are suitable excimer forming site traps. At the low PS concentration for which this model may be appropriate, $q_D$ is simply the intramolecular site fraction calculated from rotational isomeric state theory. The second parameter E is a measure of the energy transfer efficiency, being the product of the single step energy migration rate W and the lifetime of the excited monomer state $\tau$.

A more recent and exact solution to the one-dimensional, nearest neighbor transport and trapping problem has been given by Movaghar. [20] Its predictions for M are indistinguishable from the series solution. [21] In Figure 1 we present the dependence of M, calculated for $q_D = 0.027$, on the energy transfer parameter E. Simply interpreted, the model shows that facile energy migration (large E) leads to extensive sampling of the aromatic rings such that the probability of finding an EFS trap is large (M is small).

From consideration of molecular models and the geometric requirements for excimer formation, we have concluded that the trans, trans meso rotational dyad is the predominant EFS in the aryl vinyl polymers. [5,22] The identification of the EFS trap with a particular rotational dyad state was a critical factor because it opened the way to utilize the powerful rotational isomeric state theory of Flory [23] to calculate the EFS population for the isolated PS chains. This trap population is relatively small in polystyrene. For example, assuming that an atactic PS chain has 45% meso dyads, the EFS concentration at 300 K was determined to be of the order of 0.026. [22] Since the EFS dyad configuration is at higher potential energy than the preferred ground state meso configurations, an increase in temperature will lead to an increase in the EFS population as long as conformational equilibrium may be maintained. At a fixed temperature only a slight molecular weight dependence of the EFS concentration was calculated. [22]

Although the extension of the infinite lattice model to the finite lattice corresponding to real polymer chains is straightforward in principle, the analytical expression for M is algebraically complex [5] and will not be reproduced here. It is of interest, however, to consider the predictions for the dependence of the efficiency of sampling of EFS, represented by the function (1 - M)/M, on the molecular weight of the aryl vinyl polymer. This is shown in Figure 2. [24]

The one-dimensional random walk model predicts that there should be a strong dependence of the observed R on the molecular weight of the aryl vinyl polymer. At very low molecular weights, there will be many chains that contain no EFS traps and there will be very little excimer fluorescence. As the PS chain length increases, the probability of finding an EFS trap within the ensemble of chromophores that is sampled by the random walking exciton will increase and the value of R will increase. At high molecular weights, however, the chain is sufficiently long such that not all chromophores may be sampled by the hopping excitation during the excited state lifetime. This will lead to a saturation effect with the value of R leveling off at some plateau characteristic of the trap concentration within some photophysically equivalent chain segment length.

Three sets of experiments have been analyzed in terms of the one-dimensional excitation migration model. First, the random walk model predictions were compared with fluorescence spectra of dilute PS/PVME blends containing monodisperse PS with molecular weights ranging from 2200 to 390,000. Measurements of R were made at temperatures between 286 and 323 K. [22] A good fit was found to the experimental results for blends examined at temperatures of less than or equal to 300 K, but a consistent positive deviation of R above the high molecular weight plateau value was observed for the highest molecular weights.

In the second set of experiments solvent cast films containing 0.1 to 1.0 weight percent P2VN of molecular weight 70,000 and 0.0 to 2.0 weight percent 2-ethyl naphthalene (2EN) in poly(cyclohexyl

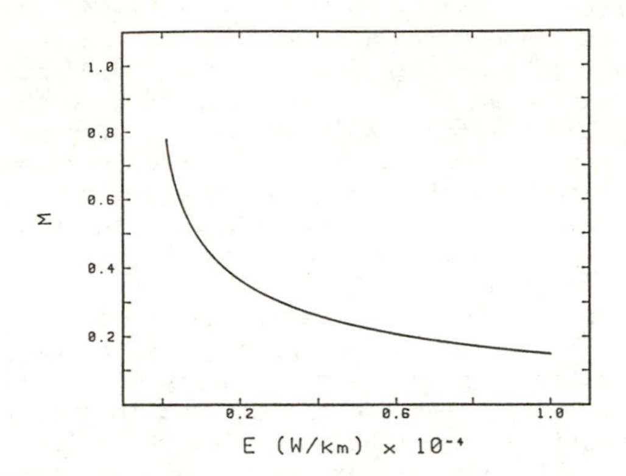


Fig. 1. Dependence of the probability M that an absorbed photon will eventually lead to radiative or nonradiative decay from an isolated aromatic ring on the energy transfer efficiency E. (Reproduced from Reference 21. Copyright 1985 American Chemical Society.)

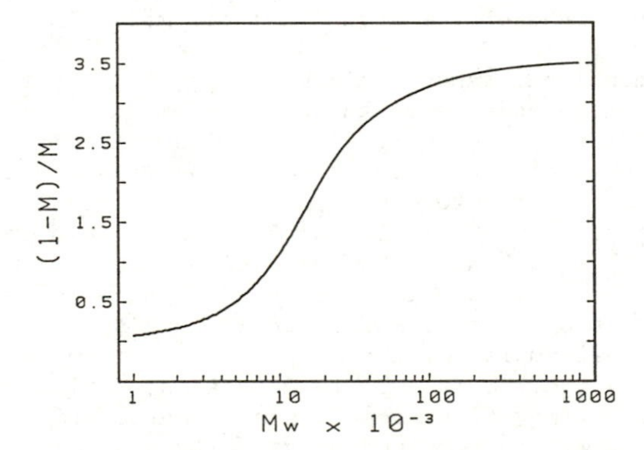


Fig. 2. Effect of molecular weight of the aryl vinyl polymer on the probability that an absorbed photon leads to excimer decay (1-M) divided by the probability of monomer decay M. This is directly proportional to the observed ratio of integrated excimer to monomer intensities. Reproduced from Reference 24.

methacrylate) (PCMA) were examined. [21] The PCMA is of low molecular weight and is believed to be miscible with P2VN, based on differential scanning calorimetry measurements. [25] At the concentrations of interest in the fluorescence experiments, EFS traps should be formed in the polymer but not among the 2EN model compounds. In addition, energy migration is expected to occur between the model and the polymer and within the polymer coil but not among the model compound rings. Photophysical analysis leads to an expression for R in terms of $Q_D/Q_M$, M and a factor describing the probability of eventual energy migration from the model to the polymer. Using the value of M determined from these experiments in the analytical expression of Equation (2), we determined that E = 5900 ±800 for the P2VN/PCMA blend. [21] This result is surprisingly large compared to the value of 51 estimated earlier for P2VN by Semerak for simple Forster transfer. [1].

The third set of experiments that have been analyzed using the strictly one-dimensional random walk model were for 0.3 weight percent P2VN(70,000) and P2VN(265,000) blended with PS(2200). [8] Here the viscosity average molecular weight is in parentheses. The P2VN(70,000)/PS(2200) blends were earlier confirmed to be miscible for P2VN concentrations at least up to 35% by weight using DSC. [17,18] The situation was somewhat ambiguous for P2VN(265,000)/PS(2200). [18] Nevertheless, at the very low concentration examined here, this blend could be considered to be marginally compatible. Application of the 1-dimensional EET model led to values of E = 2400 and 4500 for the P2VN(70,000) and P2VN(265,000), respectively. As was true for P2VN/PCMA this is much more efficient than expected from the simple Forster transfer calculation. Similar analysis of earlier PS/PVME data gave E = 600, compared to 2.7 from the Forster analysis. [8]

Each of the three experiments described above supports the general conclusion that energy migration in dilute blends does not proceed via a strictly one dimensional random walk. In the case of the PS/PVME blends this may be attributed to a decrease in the PS coil dimensions with increasing temperature as a result of adverse thermodynamic interactions with the PVME matrix upon approach of the phase separation binodal. As the local chromophore density increases due to incipient coil collapse, the excitation almost certainly will cease to follow a strictly one-dimensional random walk and frequent cross-loop hops could take place. This will lead to an increase in the dimensionality of the walk and an improved efficiency of sampling the existing EFS traps. In addition, it is likely that there will be additional EFS traps generated as the local chain segment density increases.

For the P2VN blends that are known to be incompatible at high host molecular weights and for high P2VN concentrations, it is quite possible that the P2VN chains are contracted with the blend close to phase separation on the scale of an individual chain. It may be significant to note that the P2VN(265,000)/PS(2200) blend, which is expected to be less thermodynamically compatible than P2VN(70,000)/ PS(2200), [17,18] has a higher value of E. Resolution of this problem will require development of a quasi one-dimensional model, perhaps based on a random walk on a fractal lattice. Research along these lines is being pursued by Webber. [26]

Phase Separation Kinetics. The one-dimensional random walk analysis that was earlier applied to the 5% PS/PVME and the lower concentration P2VN/PCMA and P2VN/PS blends is not expected to be valid as the concentration of the aryl vinyl polymer increases. Assuming that a three-dimensional random walk process would apply at high concentrations, we developed a simple spatially periodic lattice model to calculate the probability function M. [6] This worked quite well to fit R data for PS/PVME blends at PS concentrations greater than 60%. For pure PS and pure P2VN we have applied both the spatially periodic lattice model and the LAF model to determine the total concentration of EFS traps in the solid state. [8] The two models yielded surprisingly good agreement, in spite of their difference in sophistication. The EFS trap fraction was found to be about 0.3 for PS and 0.1 for P2VN. This total EFS population is to be compared with the intramolecular EFS concentration of 0.027, taken to be the population of trans, trans meso dyads as predicted by rotational isomeric state theory. [6] These results suggest that pairwise interactions between aromatic rings are quite probable in PS. The reduced EFS concentration for P2VN seems reasonable by comparison due to the increased ring size and consequent additional steric hindrance.

Although the success of the photophysical predictions is certainly encouraging, we prefer to emphasize the polymer physics in this section and will describe a method to examine phase separated systems without any consideration of which energy migration model is more appropriate. This more pragmatic approach requires that data for R as a function of concentration first be obtained for a miscible blend. The probability function M is then determined experimentally through use of Eq. (1). A typical plot of M as a function of the volume fraction of PS is shown in Figure 3 for a PS(100,000)/PVME blend. [9] The solid line is simply a smooth curve through the data. The significant qualitative feature is that the probability of monomer emission decreases rapidly with increasing PS concentration. This is a consequence of the increase in the number of EFS traps, which is considerable in the case of PS, and the expected increase in efficiency of excitation migration due to the increased dimensionality of the random walk. The combination of these two effects leads to considerable nonlinearity.

Once curves like that of Figure 3 have been established from fluorescence measurements on miscible blends, they may be used in multiphase photophysical models. For example, if a blend has phase separated into two phases of compositions rich in PS, $\phi_R$, and lean in PS, $\phi_L$, Figure 3 may be used to determine the corresponding M function for the rich, $M_R$, and lean, $M_L$, phases. To proceed further, we need two additional items: a morphological model that prescribes how the fluorescence from the individual phases contribute to the overall experimental observable R and a phase diagram that presents the equilibrium compositions as a function of temperature and polymer molecular weight.

The simplest possible model is to assume that the volume fractions of PS in the rich and lean phases are independent of the bulk concentration and that there is no energy migration between phases. [6] This allows one to calculate the individual contributions to the excimer and monomer fluorescence from PS chains in both the rich and lean phases. The resulting expression relates the concentration of PS in the rich phase, to the observed R.

$$\frac{I_D}{I_M} = \frac{Q_D}{Q_M} \frac{[X_R(1 - M_R) + (1 - X_R)(1 - M_L)]}{[X_R M_R + (1 - X_R)M_L]} \tag{3}$$

where $X_R = \dfrac{\phi_R (\phi_B - \phi_L)}{\phi_R(\phi_B - \phi_L) + \phi_L(\phi_R - \phi_B)}$

and $\phi_B$ is the bulk composition.

Polymer blends typically show a decrease in miscibility with increasing temperature. [27] McMaster has used a modified Flory equation of state thermodynamic model to show that the existence of a lower critical solution temperature (LCST) is caused mainly by differences in the pure component thermal expansion coefficients. [28] The phase diagram may be characterized by two types of curves. The binodal, defined as the locus of points for which the chemical potentials of each component are equal in both phases, separates the stable one-phase region from the region in which two phases coexist at equilibrium. The spinodal, which is the locus of points for which the second derivative of the Gibbs free energy of mixing equals zero, divides the two-phase region into metastable and unstable portions.

The mechanism of the phase separation depends upon the region of the phase diagram that is occupied immediately after a temperature jump above the LCST. [27]In the metastable region between the binodal and spinodal curves small fluctuations in volume fraction decay with time. Only if there is a large fluctuation leading to formation of a nucleus of critical concentration will phase separation take place; this is referred to as the nucleation and growth mechanism. In the unstable region within the spinodal there is no thermodynamic barrier against phase separation and small fluctuations in concentration may grow with time; this is the spinodal decomposition mechanism. It is this mechanism that has been of our primary interest for the PS/PVME blend.

Interpretation of the phase separation fluorescence results requires that accurate values of the equilibrium binodal compositions for a particular temperature be available. These are necessary in order to calculate the volume fraction of the rich phase in the two component system. This volume fraction is assumed to be constant during the early stages of spinodal decomposition. Two earlier attempts were made in our laboratory to measure the PS/PVME phase diagram using light scattering turbidity measurements. In the first study [9] we determined the phase diagram simply by reporting the minimum temperature at which a given blend showed visual signs of phase separation after fifteen minutes of annealing in an oil bath. In our second paper on phase separation kinetics, the initial cloud point measurements were improved by using the same sapphire substrate as for the fluorescence measurements and by taking the cloud point temperature as the point at which the first signs of visual opalescence occurred upon linear heating of the films at 5 K/min. Unfortunately, it is quite difficult to ensure that equilibrium is achieved for PS/PVME blends at high PS concentrations due to the high $T_g$ of the PS. It is quite possible, for example, that time lags generated by slow macroscopic diffusion could lead to phase separation temperatures that were higher than actual for the high molecular weight PS at high concentrations. In fact, the cloud point curves for the two PS molecular weights were superimposable above PS concentrations of 60%, [10] a result that would be unexpected on thermodynamic grounds.

For this reason we determined in this study, for the third time and with the highest accuracy of our studies to date, the cloud point curves for PS(100,000)/PVME and PS(1,800,000)/PVME blends. Using the turbidimetric technique described in the Experimental Section, we determined cloud point temperatures for each of the PS molecular weights over the whole concentration range. The results are shown in Figure 4, where the solid points represent the experimentally determined cloud point temperatures. As proof of the closer approach to equilibrium for these measurements as compared to the previous determinations, both sets of data were approximately parallel to each other and did not cross at high PS concentration.

An important extension of the experimental analysis for this phase diagram determination was the evaluation of an expression for the PS/PVME binary interaction parameter from the cloud point curves. This was done by assuming a functional form for the interaction parameter and then calculating the binodals using a free energy of mixing calculated from the Flory-Huggins configurational entropy of mixing combined with an enthalpy of mixing term based on the assumed interaction parameter. The criterion for the fit was that the same expression for the interaction parameter should yield binodal curves that passed through both sets of cloud point curves for the two PS molecular weights. Although we initially attempted to use an expression that was only temperature dependent, as was done in the previous work [9,10], we found that inclusion of a small concentration dependent term was essential to obtain good fits.

The result of the analysis was that the binary interaction parameter may be represented as

$$\chi = \frac{\Lambda V_r}{RT} \tag{4}$$

where $V_r$ is the reference volume taken to be the smaller of the two polymer repeat units and $\Lambda$ is given by

$$\Lambda = -0.3067 + 0.000833T - 0.0086C \tag{5}$$

with temperature T in K and concentration C in weight fraction. It is of interest to examine the magnitude of the temperature dependent and concentration dependent contribution to the interaction parameter. For example, at a polystyrene weight fraction of 0.5 and temperature of 400K, the concentration dependent term is only slightly greater than one percent of the temperature dependent term. Since the concentration dependence is small, it may be useful for

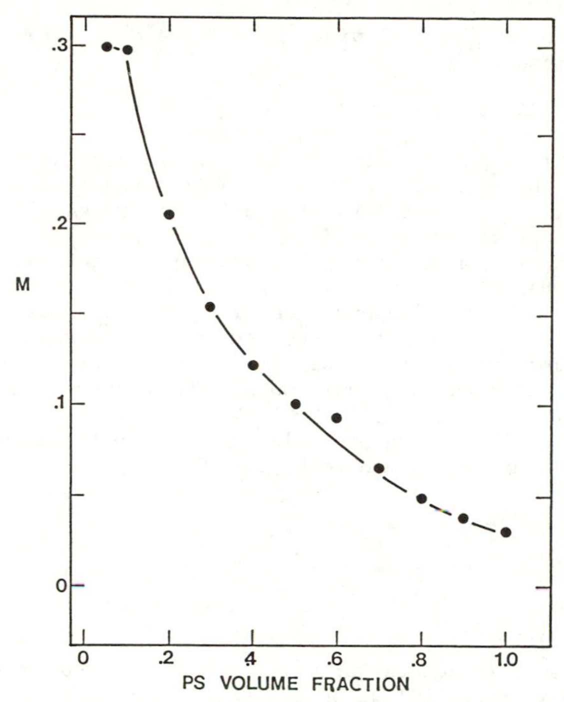


Fig. 3. Dependence of the probability of eventual monomer decay M on the concentration of polystyrene in a PS/PVME blend. (Reproduced from Reference 9. Copyright 1982 American Chemical Society.

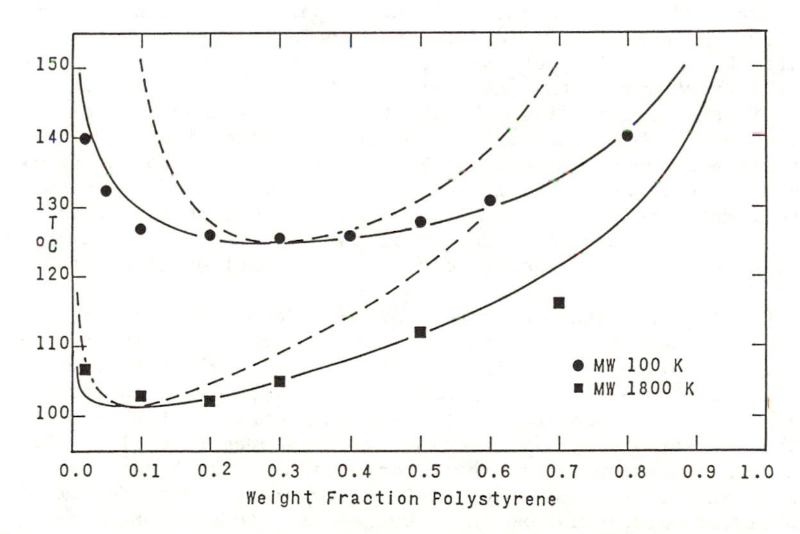


Fig. 4. Phase diagrams determined using a turbidimetric technique for PS/PVME blends having monodisperse PS molecular weights of 100,000 and 1,800,000 and PVME molecular weight of 44,600.

some purposes to use the following approximate expression for the interaction parameter.

$$\chi = 0.04165 - 15.5/T \tag{6}$$

We note from this expression that $\chi$ will become zero at a temperature of 372K. Thus, one would expect that two phase behavior would be observed for mixtures of very high molecular weight PS and PVME at temperatures greater than 372K.

Once an expression for the interaction parameter was determined, we then calculated the spinodal curves for each PS series. These curves are shown as dashed lines in Figure 4. The phase diagram thus determined has the expected increasing asymmetry as the difference in molecular weights becomes larger. The critical concentrations at which the spinodal and binodal curves coincide are about 0.3 and 0.1 for the PS(100,000)/PVME and PS(1,800,000)/PVME blends, respectively. As an approximate check on these numbers, we may apply the relationship due to Scott [29]

$$(C_A)_{cr} = \frac{(x_B)^{1/2}}{(x_A)^{1/2} + (x_B)^{1/2}} \tag{7}$$

where $x_i$ refers to the degree of polymerization of component i. From our earlier paper [10] we obtain the degrees of polymerization to be 769 for the PVME and 1650 and 29,800 for the two PS samples. These lead to critical PS concentrations of 0.41 and 0.14 for the PS(100,000)/PVME and PS(1,800,000)/PVME blends, respectively. These are reasonably close to the experimental values.

Two sets of experiments on phase separation kinetics have been performed. The first series of experiments was designed to demonstrate the feasibility of using excimer fluorescence to test Cahn's kinetic treatment of the early stages of spinodal decomposition. [30] This is a linearized theory that predicts that the volume fractions of the two phases remain constant with time while the compositions change gradually. This is in contrast to the nucleation and growth mechanism in which the composition of the growing phase remains constant but the phase itself increases in size. Analysis of the very early stages of the phase separation allowed estimation of the macroscopic diffusion coefficient, found to be of the order of $\sim 10^{-11}$ $cm^2 sec^{-1}$. [9]

Cahn's linearized theory of spinodal decomposition was extended recently by deGennes [31] and Pincus [32] through scaling techniques applied to polymer blends in the melt. A critical feature of the modified analysis is the proposal by Pincus [32] that the growth rate of the concentration fluctuation that controls the kinetics during the early stages of decomposition is proportional to the melt reptation diffusion coefficient. As a result, the kinetics of the early region of spinodal decomposition is predicted to depend strongly upon molecular weight. Our second study on phase separation kinetics of PS/PVME blends was designed to test this prediction. [10]

Since the deGennes-Pincus analysis was developed for blends in which the two components are the same molecular weight, we made a scaling modification of their theory to account for the fact that our experimental blends have different PS and PVME molecular weights. Thermally induced phase separation was followed in the same manner as previously [9] for 10% and 50% PS blends of both molecular weights. It is significant to note that excimer fluorescence is a more sensitive measure of phase separation at short times because films annealed for about one minute showed no visual signs of opalescence while R increased between 10 and 50% for the different blends.

Using the phase diagram established in the present study, we re-evaluated fluorescence kinetic data determined in the earlier phase separation work. [10] Before examining the results, it is worthwhile to look more closely at Figure 4. We note that the phase separation experiments were intended to be done under conditions such that the mechanism of phase separation was spinodal decomposition, not nucleation and growth. This may be accomplished most easily for a particular blend by selecting a system at the critical composition. At that point there would be the minimum tendency for nuclei to form during the time necessary to perform the temperature jump. As we have shown, however, the critical composition is different for the two sets of blends. An optimum PS concentration that would minimize the extent of the metastable region that must be traversed during the temperature jump would be in the region of 0.3 to 0.4 PS weight fraction. Thus, of the two sets of available data, the 0.5 concentration should be more appropriate to test the deGennes-Pincus theory.

On this basis, we redetermined the rate of growth of the fastest growing concentration fluctuation, an important parameter for spinodal decomposition, for each blend series. The growth rate for the PS(100,000)/PVME blend is predicted to be 3.9 times larger than that for the PS(1,800,000)/PVME blend; the experimentally observed factor is 2.7, which is quite reasonable agreement and is considerably better than earlier work using the previously determined phase diagram. [10] The remaining difference could be attributed to the slight polydispersity of the PVME in which the presence of short unentangled PVME chains that can diffuse rapidly during the early stages of phase separation is expected to weaken the molecular weight effect.

Nonequilibrium Effects in Solvent Casting. The third major objective of this paper is to consider the influence of the casting solvent on the morphology of the resulting blend. To do so, we examined blends cast from tetrahydrofuran and have compared the results to blends prepared from toluene and chlorobenzene. The first indication of possible kinetic limitations to the attainment of thermodynamic equilibrium was the observed dependence of the optical appearance on the rate of solvent evaporation from the THF casting solution. Rapid drying yields very cloudy films while slow drying produces films with extremely finely dispersed domains such that the films appear almost clear. We note that the former case might lead to a polymer "skin" being formed on the surface of the casting solution, which could actually impede the solvent evaporation and lead to more entrapped solvent than in the latter case. The rapid and slow cast films were kept under vacuum at room temperature for five days before measurement of the fluorescence spectra. We first consider $I_D/I_M$ for the slow-cast THF and toluene cast films which is presented as a function of PS concentration in Figure 5. Additional experiments with chlorobenzene as the casting solvent gave results identical to the toluene. THF cast films gave R values that were higher than the toluene cast blends suggesting that the local concentration of phenyl rings was higher for blends prepared from THF. Of course, this is consistent with the observed inhomogeneous nature of the film from the qualitative observations of optical clarity. Although such information is often difficult to interpret, the occurrence of cloudiness is definite evidence for phase separation; conversely, optically clear films could be miscible or immiscible. [13] The observation that casting from THF leads to phase separation is similar to that reported earlier by Gelles and which was analyzed using the two phase morphological model. [10]

A second spectral parameter that receives much less attention than R is the excimer band position. In Figure 6 we present excimer band position data relative to that for the toluene cast films after three days of annealing at 393K for 30% PS and 70% PS blends. For the moment we focus only on the excimer band position of the as-cast film and note that the THF cast blend was 700 $cm^{-1}$ larger for the 30% PS and 900 $cm^{-1}$ larger for the 70% PS blend. Although this is difficult to interpret in the absence of information on the binding energy for the excimer complex, it does suggest that the excimer forming site is less stable for blends cast from THF than for those cast from toluene.

In order to determine whether the effects of the THF could be removed by annealing, the films were placed under vacuum for 10 hours at 383 K and then sealed between two sapphire plates before annealing an additional 1 1/2 days at 383 K. This led to a very slight reduction in R for the toluene and chlorobenzene cast films but the initially clear THF cast films dropped sharply to almost equal the toluene results, as shown in Figure 5. By contrast, whereas the initially cloudy THF cast films did become clear upon annealing and the R value did decrease to about 1/3 of its initial value, it was still 30 to 100% higher than the toluene or chlorobenzene results. At the same time time the excimer band position shifted red leveling out at about 500 $cm^{-1}$ higher energy than the toluene and chlorobenzene films. Further annealing did not change either R or the excimer band position. However, when the top sapphire plate was removed exposing the film again to the ambient atmosphere and the films annealed for two more days at 383 K, the R value and the excimer peak position for all films, regardless of casting solvent and processing history, came to within 15% of common values. Similar results were obtained for a fresh set of samples cast from THF and annealed for three days under vacuum without any cover seal at 383 K.

Two qualitative conclusions may be drawn from these fluorescence results. The first point is related to whether there is residual casting solvent that might affect the morphology of the film. It is not possible to comment on the presence of residual chlorobenzene or toluene from the results of this study because of the lack of any dramatic change in R or excimer band position upon annealing. No separate gravimetric studies of solvent evaporation were conducted in this work, although such experiments have been

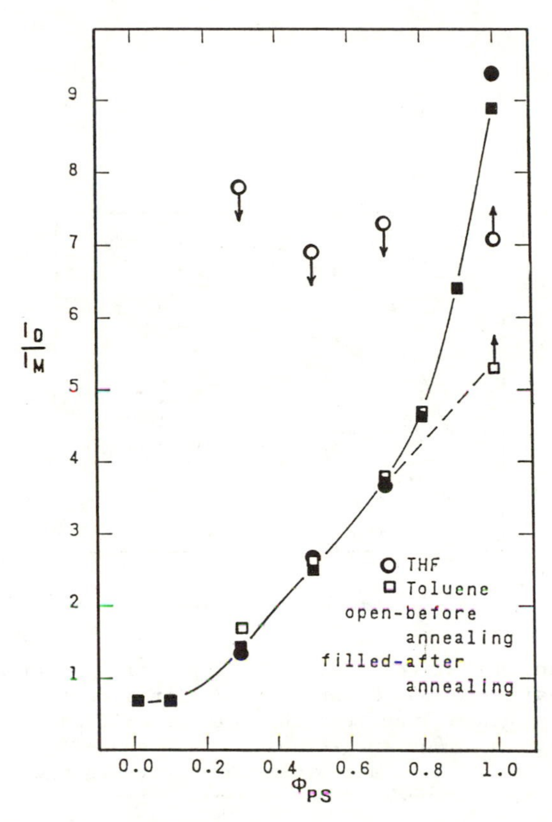


Fig. 5. Dependence of the observed excimer to monomer intensity ratio on polystyrene concentration for PS/PVME blends cast from tetrahydrofuran (circles) and toluene (squares) with the fluorescence spectra measured before (open symbols) and after (filled symbols) annealing at 383 K for 10 hours.

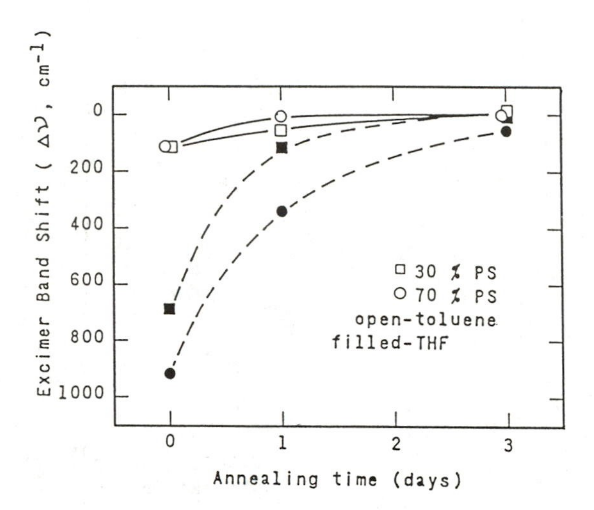


Fig. 6. Dependence of the excimer band position on time of annealing at 383 K for PS/PVME blends cast from toluene (open symbols) and tetrahydrofuran (filled symbols) for polystyrene concentration of 30% (squares) and 70% (circles). Data are plotted relative to the excimer band position of annealed toluene cast films.

performed previously for room temperature casting of blends in which the host polymer was either polystyrene or poly(methyl methacrylate). [18] Although the initially clear THF cast films did change to become almost identical to the other blends, it is not possible to separate any residual solvent effects from local equilibration consisting of short chain segmental motion as well as large scale cooperative diffusion. The situation is less ambiguous for the THF cast films that were prepared by a rapid evaporation process and that were initially cloudy. The fact that the approach of R and the excimer band position toward the values exhibited by the toluene cast films was retarded when the blend was sealed between the two sapphire plates, but which proceeded to completion when the top plate was subsequently removed, suggests that there indeed was residual THF in the ten micron thick PS/PVME films even after ten hours annealing under vacuum at 383K. Thus, even highly volatile solvents can be retained in the film, particularly if there is some source of specific interaction between the solvent and one of the polymers. We have considered this point separately. [33]

The major conclusion regards the true equilibrium state of the binary PS/PVME blend. This seems the most straightforward and probably the strongest result of this study on nonequilibrium casting effects. From the lack of any significant change in the fluorescence behavior of the toluene cast results upon annealing, and from the tendency of the THF results, whether initially for clear or cloudy films, to approach the toluene cast films, we conclude that toluene casting leads to the equilibrium morphology. Since there are no large specific interactions of the toluene with either of PS or PVME or, more importantly, the interactions that exist are not asymmetric, the presence of residual toluene does not differ appreciably from the situation in the casting solution. This is in distinct contrast to the THF case for which the interactions are quite asymmetric.

Acknowledgment

This work was supported by the Army Research Office under contract DAAG29-82-K-0019.

References

1. Semerak, S.N.; Frank, C.W.; Advances in Polymer Science **1983**, 54, 31.
2. Bank, M.; Leffingwell, J.; Thies, C.; Macromolecules **1971**, 4, 43.
3. Bank, M.; Leffingwell, J.; Thies, C.; J. Polymer Sci. A-2 **1972**, 10, 1097.
4. Klopffer, W.; Ann. N.Y. Acad. Sci. **1981**, 366, 373.
5. Fitzgibbon, P. D.; Frank, C. W.; Macromolecules **1982**, 15, 733.
6. Gelles, R. G.; Frank, C. W.; Macromolecules **1982**, 15, 747.
7. Loring, R.F.; Andersen, H.C.; Fayer, M.D.; J. Chem. Phys. **1982**, 76, 2015.
8. Semerak, S.N., Frank, C.W.; Canadian J. Chemistry, **1985**, 63, 1328.

9. Gelles, R. G.; Frank, C. W.; Macromolecules **1982**, 15, 1486.
10. Gelles, R. G.; Frank, C. W.; Macromolecules **1983**, 16, 1448.
11. Frank, C. W.; Gashgari, M. A.; Macromolecules **1979**, 12, 163.
12. Frank, C. W.; Gashgari, M. A.; Chutikamontham. P.; Haverly, V.; Studies in Physical and Theoretical Chemistry **1980**, 10, 187.
13. Gashgari, M. A.; Frank, C. W.; Macromolecules **1981**, 14, 1558.
14. Frank, C. W.; Gashgari, M. A.; Annals of New York Academy of Science **1981**, 366, 387.
15. Semerak, S. N.; Frank, C. W.; Macromolecules **1981**, 14, 443.
16. Semerak, S. N.; Frank, C. W.; Macromolecules **1984**, 17, 1148.
17. Semerak, S. N.; Frank, C. W.; Advances in Chemistry Series **1983**, 203, 757.
18. Semerak, S. N.; Frank, C. W.; Advances in Chemistry Series **1984**, 206, 77.
19. Levinson, N.J.; Soc. Indust.„Appl. Math., **1962** 10m 442,
20. Movaghar, B.; Sauer, G.W.; Wurtz, D.J.; J. Stat. Phys. **1982**, 27, 473.
21. Thomas, J.; Frank, C. W.; Macromolecules, **1985**, 18, 1034.
22. Gelles, R. G.; Frank, C. W.; Macromolecules **1982**, 15, 741.
23. Flory, P. J.; Statistical Mechanics of Chain Molecules; Interscience Publishers: New York, **1969**.
24. Thomas, J.W.; Ph.D. Dissertation, Stanford University, **1985**.
25. Thomas, J.W.; Frank, C.W.; Manuscript in preparation.
26. Webber, S.; private communication.
27. Paul, D.R.; Newman, S.; Polymer Blends, Vol. 1, 2.
28. McMaster, L. P.; Macromolecules **1973**, 6, 760.
29. Scott, R.L.; J. Chem. Phys. **1949**, 17, 279.
30. Cahn, J. W.; J. Chem. Phys. **1965**, 42, 93.
31. de Gennes, P.-G.; J. Chem. Phys. **1980**, 72, 4756.
32. Pincus, P.; J. Chem. Phys. **1981**, 75, 1996.
33. Leigh, C.; Frank, C.W.; Manuscript in preparation.

RECEIVED September 28, 1987

# Chapter 4

# Applications of Fluorescence Techniques for the Study of Polymer Solutions

**Herbert Morawetz**

**Department of Chemistry, Polytechnic University, Brooklyn, NY 11201**

Intramolecular excimer emission, the polarization of fluorescence, nonradiative energy transfer and the use of medium-sensitive fluorophores has been used to study the conformational mobility of polymers in dilute solution, the interpenetration of chain molecules, the association of polymers with each other or with small species and the cooperative transition of certain polycarboxylic acids from a compact to an expanded state.

The use of fluorescence techniques for the characterization of the behavior of chain molecules in solution offers a number of important advantages: (a) Since fluorescence can be detected at extreme dilution of the emitting species, labeling polymers with fluorophores need not result in significant modifications of their other properties. (b) Since the lifetime of the excited chromophore, $\tau$, is generally of the order of 10 ns, the kinetics of fast processes whose relaxation times are comparable to $\tau$ may be studied by fluorescence techniques. (c) The phenomenon of nonradiative energy transfer over relatively long distances may be used to characterize the spatial relation of donor and acceptor labeled chain molecules. (d) The medium sensitivity of the emission from the "dansyl" label has been used to study polymer complex formation and the transition of chain molecules from a contracted to an expanded state.

## Conformational Mobility.

In considering the rate at which a flexible chain molecule changes its shape in dilute solution, we are facing a conceptual difficulty illustrated in Figure 1. If a hindered rotation around a bond in the middle of the chain takes place with no change in the other internal angles of rotation or bond angles, then a long segment of the chain would have to swing through the viscous solvent with a prohibitive expenditure of energy. If, on the other hand, we avoid this by having two hindered rotations take place simultaneously in what has been called a "crankshaft-like motion", then two potential energy barriers have to be surmounted in the transition state and this was intuitively believed to lead to a doubling of the activation energy.[1] With such strict simultaneity of the passage over the two energy barriers hindered rotations in the backbone of long chain

0097–6156/87/0358–0037$06.00/0

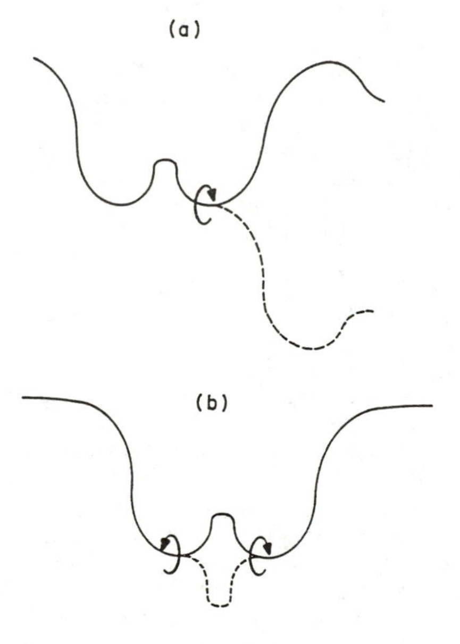


Figure 1. Schematic representation of conformational transitions in a flexible chain molecule. (a) A single hindered rotation. (b) Two correlated rotations in a "crankshaft-like motion".

molecules would be expected to be at least a hundred times slower than in analogous small molecules where only a single energy barrier has to be passed and even if the condition of simultaneity is somewhat relaxed, a substantial difference in the rate of conformational change of polymers and their analogs might have been expected.

The question whether hindered rotation around a given type of bond is slowed down when this bond occurs in the middle of a long chain may be resolved unambiguously by the use of a structure containing two aromatic residues which may be made to lie parallel to each other by a single conformational transition. If this transition occurs during the lifetime of the excited chromophore, excimer emission will be observed. Dibenzylacetamides substituted in the para positions are suitable for such an experiment:

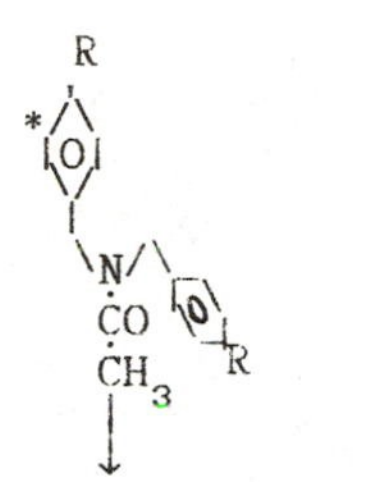


"monomer" emission

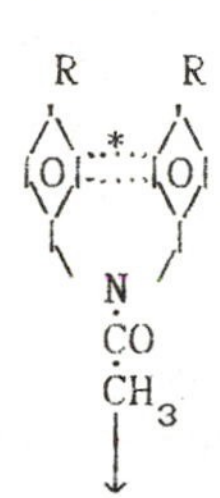


excimer emission

It was found that excimer emission was reduced only by 27% when the dibenzylacetamide was built into the center of a long polyoxyethylene chain as compared to the low molecular weight analog in which R were methyl groups.[2] This proves that hindered rotation in long polymer chains takes place at similar rates as in small molecules.

A powerful method for the study of conformational mobility of polymers utilizes polymer chains carrying at their ends pyrene residues. Although excimer emission from molecules containing two phenyl residues is observed only if these phenyls are separated by three atoms, no such restriction exists with molecules containing two pyrene moieties, since the excited pyrene has a much longer life time. Thus, if one of the pyrenes at the end of a polymer chain is excited, "monomer" emission will be reduced if the two chain ends meet each other during the excited life time to form an excimer. This principle was used to derive the dependence of the rate constant of chain cyclization on the length of the polymer chain[3,4] and the solvent power of the medium.[5]

Another method for the study of the conformational mobility of polymers utilized the dependence of the polarization of the emitted light on the rotational diffusion of the fluorophore. This technique was pioneered by Anufrieva and her collaborators[6] who used polymers with anthracene built into the chain backbone. They showed, for instance, that conformational mobility in polymethacrylates is much smaller than in polyacrylates.

Polymer Chain Entanglements in Semidilute Solutions.

Since entanglements of flexible chain molecules in solution lead to a decrease of configurational entropy because of the interdependence of the chain conformations, such interpenetration is strongly resisted in dilute solution. As the concentration of the solution increases, a point is eventually reached where the swollen polymer coils can no longer be accommodated without interpenetration. This concentration is generally designated by c* and it is said to divide dilute from "semidilute" solutions, marking characteristic changes in the concentration dependence of thermodynamic and hydrodynamic properties.[7]

While c* may be related to the dimensions of polymer chains derived from light scattering or intrinsic viscosity, estimates of the degree of chain interpenetration at concentrations beyond c* present a more difficult problem. We approached it by employing the phenomenon of nonradiative energy transfer. If a system contains two fluorophores such that the first (the "donor") has an emission spectrum which overlaps the absorption spectrum of the second (the "acceptor"), then the excitation energy of the donor can be transferred to the acceptor over relatively long distances. The efficiency of this transfer is given by [8]

$$\text{Eff.} = R_o^6/(R_o^6 + r^6) \qquad (1)$$

where r is the distance between donor and acceptor and $R_o^6$ is proportional to the "overlap integral" of donor emission and acceptor absorption spectra. If a solution of a mixture of polymers labeled with donor and acceptor, respectively, is rapidly frozen, it may be assumed that the degree of chain entanglement is not changed during the freezing process. If the frozen solvent is now removed by sublimation and the polymer pressed into a pellet, the fluorescence spectrum of the sample will reflect the proximity of donor and acceptor labels and, therefore, the extent of polymer chain interpenetration which existed in the solution from which the sample was derived.

After establishing the feasibility of the method[9], we have recently applied it to monodisperse polystyrene in good and poor solvents[10] using carbazole as the donor and anthracene as the acceptor. With these labels $R_o$ = 2.75 nm. The ratio of anthracene and carbazole emission intensities, $I_A/I_C$, in the freeze-dried sample was compared with this ratio in films cast from solutions containing the two labeled polymers, $(I_A/I_C)_f$, where it could be assumed that the two polymers were randomly mixed.

Typical results are illustrated in Figure 2. In very dilute solutions polymer chains do not interpenetrate and the freeze-dried sample consists of globules, each containing one polymer chain. Energy transfer is minimal, $I_A/I_C$ very small and independent of the solvent in which the polymer had originally been dissolved. $I_A/I_C$ begins to increase when the original solution concentration was close to $1/[\eta]$, which is frequently taken as a measure of c*. The surprising aspect of our results was the leveling off of the $I_A/I_C$ at around $3/[\eta]$, suggesting that at this point a strong resistence is encountered to the penetration of the molecular chains into the inner core of other molecular coils. It is particularly significant that this behavior is also seen with samples derived from cyclohexane solutions which are close to the theta point. In the Flory/Krigbaum theory of the second virial coefficient of flexible chain polymers[11] the polymer chains are modeled as clouds of disconnected ch in segments, so that the excluded volume vanishes when it is zero for the isolated chain segment. This has led to the general assumption[12] that polymer coils are "freely interpenetrating" in theta solvents. However, a computer simulation study by Olaj and Pelinka[13] has shown that the vanishing excluded volume of chain molecules under theta conditions is due to a compensation of the effect of an attractive potential for small chain interpenetrations, due to the attraction between the chain segments, and a repulsive potential for deep interpenetration where the unfavorable entropy, due to the interdependence of the chain conformations, is the dominant effect. Our study seems to have provided an experimental verification of this concept.

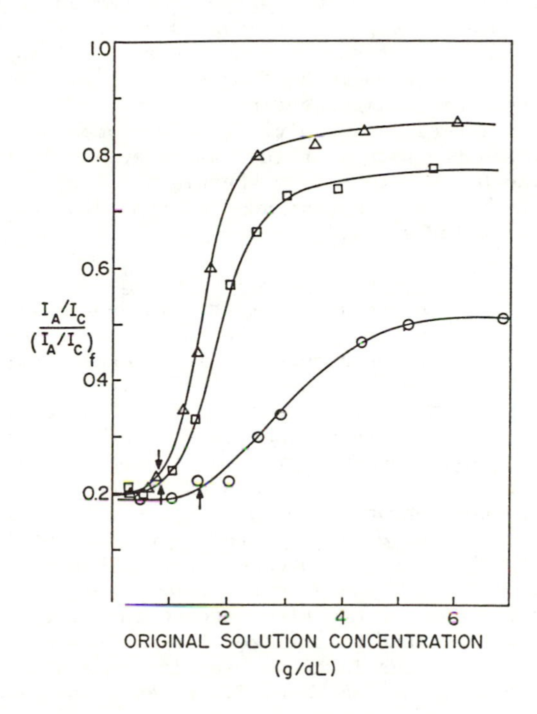


Figure 2. Ratio of anthracene and carbazole emission intensities from freeze-dried mixtures of monodisperse polystyrene (M = 410,000) carrying 0.0092 moles/kg of the labels, relative to this ratio in films cast from solutions of a mixture of these polymers. Excitation at the carbazole absorption peak at 294 nm. Original solution in benzene (Δ); dioxane (□) and cyclohexane (○). Arrows indicate 1/[η] in the various solvents.

Formation of Polymer Association Complexes.

Polycarboxylic acids have long been known to associate in aqueous solution with polymers acting as hydrogen bond acceptors such as polyoxyethylene (POE) or poly(N-vinyl pyrrolidone)(PVP). Since such association leads to reduced mobility, such processes can be monitored by the change in the polarization of fluorescence of a label attached to one of the interacting polymers[6]. In a more striking experiment, a "complex interchange" was followed by the decreasing polarization of the emitted light when the fluorophore labeled component of the complex was displaced by an unlabeled polymer[14].

We have studied polymer association in solution both by the increase in the energy transfer between donor and acceptor labels of interacting species[15] and by use of the medium-sensitive "dansyl" label. This chromophore fluoresces strongly in organic media but very weakly in water solution. Thus, when dansyl-labeled poly(acrylic acid) associates in aqueous solution with POE or PVP, water is displaced from the microenvironment of the label and the emission intensity increases sharply. This phenomenon allowed us to study the extent and the nature of such associations.[16,17] This technique can also be used to follow the kinetics of complex interchange[16] and with the use of a stopped-flow apparatus the much faster complex formation can also be studied[18].

Cooperative Transitions of Polycarboxylic Acids.

When alternating copolymers of maleic acid and alkyl vinyl ethers are ionized, the expansion of the chain, due to the electrostatic repulsion of the anionic charges, is strongly resisted by the hydrophobic bonding between the alkyl substituents if they contain at least four carbons. If the copolymer is labeled by dansyl groups, this effect can be observed by the rapid decrease of fluorescence intensity accompanying chain expansion as shown by Strauss and his collaborators[19,20]. A similar transition from a compact to an extended conformation takes place in poly(methacrylic acid) at a critical charge density along the polymer chain. In this case also the chain extension which removes the shielding of dansyl labels, attached to the chain backbone, from the water molecules leads to a decrease in the fluorescence intensity. We have shown[21] that this effect can be utilized for a study of the kinetics of the conformational change in a stopped-flow apparatus.

Recently, we have investigated by fluorimetry the association of poly(methacrylic acid) with the cationic dye Auramine O. Oster showed many years ago[22] that such association leads to a strong fluorescence of the dye which emits only very weakly in the free state. Anufrieva et al.[23] found that the strong fluorescence is characteristic of the binding of the dye to the contracted form of the polymeric acid and suggested that this be used as a tool for the study of the transition from the contracted to the expanded state.

In studying the association of Auramine O with poly(methacrylic acid) kinetically in a stopped-flow apparatus[24] we found an unexpected effect. With the polymer in large excess, the approach of the fluorescence intensity to its equilibrium value was independent of the polymer concentration. This suggested that the association of the dye to the polymer does not by itself lead to the increase in emission intensity. This seems to be the result of a contraction of the polymer chains for which the adsorbed dye acts as a nucleus, so that the process is monomolecular.

Acknowledgment.
We are grateful for the support of this research by the National Science Foundation by Grant DMR 85-00712.

References

(1) Allegra,G. J. Chem. Phys. 1974, 61, 4910.
(2) Liao,T.P.; Morawetz,H. Macromolecules 1980, 13, 1228.
(3) Winnik,M.A.; Redpath,T.; Richards, D.H. Macromolecules 1980, 13, 328.
(4) Cuniberti,C.; Perico,A. Eur. Polym. J. 1977, 13, 369.
(5) Winnik,M.A.; Guillet,J.A.; Li,X.-B. J. Polym. Sci., Polym. Symp. 1985, 73, 113.
(6) Anufrieva,E.V.; Gotlib,Yu.Ya. Adv. Polym. Sci. 1981, 40, 1.
(7) Daoud,M.; Cotton,J.P.; Farnoux,B.; Jannink,G.; Sarma,G.; Benoit,H.; Duplessis,J.P.;Picot,C.;De Gennes,P.-G Macromolecules 1975, 8, 804.
(8) Förster,T. Discuss. Faraday Soc. 1959, 27, 7.
(9) Jachowicz,J.; Morawetz,H. Macromolecules 1982, 15, 828.
(10) Chang,L.P.; Morawetz,H. Macromolecules 1987, 20, 428.
(11) Flory,J.P.; Krigbaum,W.R. J. Chem. Phys. 1950, 18, 1086.
(12) Yamakawa,H. "Modern Theories of Polymer Solutions", Harper and Row, San Francisco, 1971, P.169.
(13) Olaj,O.F.; Pelinka,K.H. Makromol.Chem. 1976, 177, 3413.
(14) Anufrieva,E.V.; Pautov,V.O.; Papisov,I.M.; Kabanov,V.A. Dokl.Akad.Nauk USSR 1977, 232,1096.
(15) Nagata,I.; Morawetz,H. Macromolecules 1981, 14, 87.
(16) Chen,H.-L.; Morawetz,H. Eur. Polym. J. 1983, 19, 923.
(17) Bednář,B.; Li,Z.; Huang,Y.; Chang,L.-C.P.; Morawetz,H. Macromolecules 1985, 18, 1829.
(18) Bednář,B.; Morawetz,H.; Shafer, J.A. Macromolecules 1984, 17, 1634.
(19) Strauss,U.P.; Vesnaver,G. J. Phys. Chem. 1975, 79, 1558.
(20) Strauss,U.P.; Schlesinger,M.S. J. Phys. Chem. 1978, 82, 1627.
(21) Bednář,B.; Morawetz,H.; Shafer,J.A. Macromolecules 1985, 18, 1940.
(22) Oster, G. J. Polym. Sci. 1955, 16, 235.
(23) Anufrieva,E.V.; Birshtein,T.M.; Nekrasova,T.N.; Ptitsyn,O.B.; Sheveleva,T.V. J. Polym. Sci. Pt.C. 1986, 16, 3519.
(24) Wang,Y.; Morawetz,H. Macromolecules, 1986, 19, 1925.

RECEIVED September 28, 1987

# POLYMER DYNAMICS AND COMPLEXATION

# Chapter 5

# Spectroscopic Investigation of Local Dynamics in Polybutadienes

**L. Monnerie, J. L. Viovy, R. Dejean de la Batie, and F. Lauprêtre**

**Laboratoire de Physicochimie Structurale et Macromoleculaire, associe au Centre National de la Recherche Scientifique, ESPCI, 10 rue Vauquelin, 75231 Paris, Cedex 05, France**

The fluorescence anisotropy decay technique and $^{13}C$ spin-lattice magnetic relaxation have been used to investigate the local dynamics of bulk polybutadienes at temperatures at least 60K higher than the glass-rubber transition temperature. The orientation autocorrelation function required to account for the experimental data agrees with the Hall-Helfand expression proposed for the local dynamics of polymer chains. In addition, the elementary motions, observed via the considered spectroscopic techniques, have a temperature dependence of their correlation times which is close to the prediction of the William, Landel, Ferry equation, proving that they are involved in the glass-rubber transition phenomenon of the polymer.

The local dynamics of polymers in solution have been extensively studied during the last decade. On the other hand, for polymer melts many questions are still unanswered, such as, for example, the nature of the orientation autocorrelation function (OACF) which is involved, and the relationship of the segmental motions occuring at high frequency in the melt with the elementary processes responsible for the glass-rubber transition.

Spectroscopic techniques such as fluorescence anisotropy decay (FAD), and $^{13}C$ spin-lattice magnetic relaxation ($T_1$ NMR) are well suited to investigation of the local dynamics of polymer melts.

In this paper we present results recently obtained in our laboratory on bulk polybutadienes.

## MATERIALS AND METHODS

Polybutadiene Firestone "Diene 45 NF" with the following microstructure, % cis = 37, % trans=59, % 1-2 = 12, and Mn = $1.7.10^5$, Mw = $4.1.10^5$ was used as a matrix for FAD experiments. An anthracene labeled polybutadiene with the same microstructure was synthetized by anionic polymerization as

0097-6156/87/0358-0046$06.00/0

previously reported for polystyrene ( 1 ). Monofunctional "living" chains of molecular weight $10^5$ were prepared and deactivated by 9,10-bis (bromomethyl)anthracene. The resultant chains of molecular weight $2.10^5$ contain a dimethylene anthracene fluorescent group in their middle, as shown in Figure 1. Because the fluorescence transition moment (represented by a double arrow in Figure 1) lies along the chain axis, fluorescence anisotropy will be insensitive to the rotation of the label around the 9,10 axis of the anthracene moiety and reflect only the motions of the backbone. The labeled polybutadiene (1 % by weight) and Diene 45 NF (99 %) were mixed in solution, then the solvent was removed by evaporation. The optical density of the films was less than 0.1 to avoid energy transfer and reabsorption. The polymer films were placed in a cell specially designed for elastomers ( 2 ). A sequence of molding and stoving operations in an argon atmosphere was carried out to avoid bubbles, to ensure perfect adhesion at the interfaces and to relax stresses.

Polybutadiene Bayer Uran the microstructure of which is %cis = 98, %trans = 1, % 1-2 = 1 and Mn=323000, Mw=1096000 was used for $^{13}C$ NMR experiments.

FAD measurements were performed on the cyclosynchrotron LURE-ACO at Orsay (France). The apparatus is described elsewhere ( 3 ). The continuous spectrum from the synchrotron allowed for the matching of the last absorption peak of the dye (401 nm). This excitation wavelength was selected by a double holographic grating with 2-nm slits, and the most intense emission peak (435 nm) was selected by a single holographic grating with 2-nm slits. This procedure greatly improves the rejection of spurious fluorescence, which is one of the major problems of fluorescence studies in bulk polymers. The negligible level of spurious fluorescence was checked by using blank samples of the unlabeled matrix. The transmission of the emission monochromator was calibrated in both polarization directions with a 0.5% precision.

The polarized emission spectra and the apparatus response (recorded at emission wavelength) were sampled with a 0.12 ns channel width by the single-photon counting technique. Thanks to the stability of the pulses, the short-time limit of the experimental window is about 0.1 ns. The upper limit, imposed by the repetition rate of the pulses and the lifetime of the dye, is 62 ns.

Spin-lattice magnetic relaxation times $T_1$ on $^{13}C$ nuclei were measured at 25.15 MHz and 62.5 MHz using Jeol PS 100 and Bruker WP 250 spectrometers respectively. DMSO-d6 was used as an external lock. $T_1$ was obtained from the 180°-t-90° sequence with an accuracy of 7 %.

## FLUORESCENCE ANISOTROPY DECAY

Under the action of a suitable electromagnetic field, polarized along the P direction as shown in Figure 2, the absorption of light is proportional to the scalar product of the incident electric field and the transition moment. In the same way, the emission of light is proportional to the scalar product of the direction of the analyzer and the transition moment. Thus, excitation of an isotropic population of fluorescent species by polarized light generally creates a temporary anisotropic population of excited molecules. Molecular motions progressively destroy this anisotropy and affect

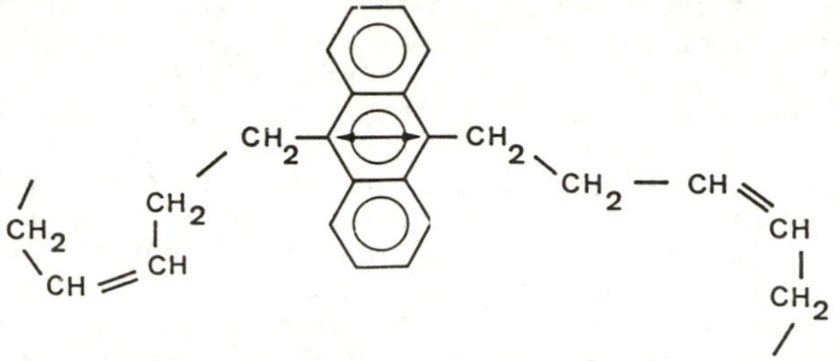

Figure 1. Polybutadiene with anthracene in the middle of the chain. The transition moment of anthracene is represented by a double arrow.

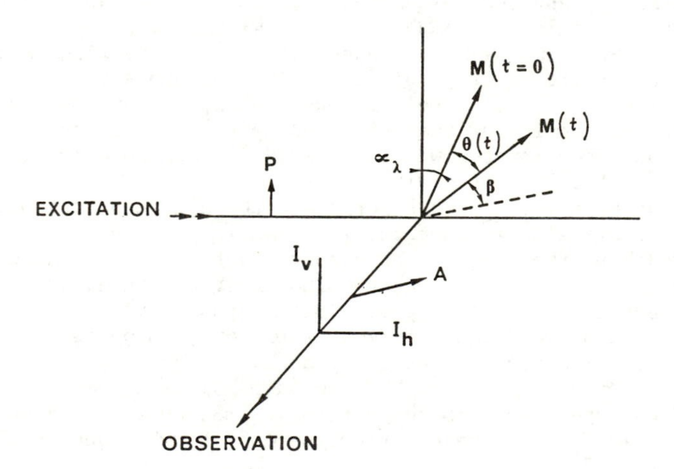


Figure 2. Polarized absorption and fluorescence emission : P, polarizer; A, analyzer.

the polarization of the reemitted fluorescence light. The quantity of interest, the fluorescence anisotropy, r, is defined as :

$$r = (I_v - I_h)/(I_v + 2\ I_h) \qquad (1)$$

where $I_v$ and $I_h$ correspond to fluorescence intensities for analyzer direction parallel and perpendicular respectively to the vertical polarization of the incident beam. In this expression $(I_v + 2\ I_h)$ represents the total fluorescence intensity. The fluorescence anisotropy emitted at time t will progressively decrease as a function of time and finally reach a zero value. A complete analysis of the phenomenon ( 4 ) shows that the evolution of r(t) as a function of time is directly proportional to the second moment of the OACF of the transition moment :

$$r(t) = M_2(t) = \langle 3 \cos^2 \Theta(t) - 1 \rangle /2 \qquad (2)$$

where $\Theta(t)$ is the angle through which the vector under consideration (the transition moment in FAD) rotates during time t, the brackets mean an ensemble average.

The time interval t during which the evolution of $M_2(t)$ can be recorded is directly related to the fluorescence lifetime ($10^{-10}$ to $10^{-8}$ s); experimentally $M_2(t)$ can be obtained until approximately $10^{-7}$ s (100 ns).

It should be pointed out that FAD is virtually unique in its ability to obtain the OACF.

A detailed study of the FAD of anthracene labeled polybutadiene in a matrix of Diene 45 NF has been performed in the temperature range 240K-353K ( 2 ).

At 335.7K, the FAD curve shown in Figure 3 is particularly suitable for comparison with various molecular models of local chain dynamics. Indeed, at this temperature the decay is not too fast, in such a way that an accurate comparison with the models can be made over the full time window available, but nevertheless the FAD curve goes close to zero. It appears that the isotropic model (single exponential for the OACF) does not fit the data and that OACF expressions specifically proposed for polymer dynamics (for a review of them, see Ref. 3 and 5 ) are required. The specific behavior of polymer chains is due to the chain connectivity requirement and the description of local dynamics requires consideration of the following features:

- "elementary motions" with a characteristic time, $\tau_1$, which diffuse along the chain sequence and lead to a non-exponential short time term in the OACF.
- a damping or a truncation of this diffusion, which yields in the OACF an exponential loss, with a characteristic time $\tau_2$ ($\tau_2 > \tau_1$)

For bulk polybutadiene, the best fit, shown in Figure 3, is reached in using the OACF expression proposed by HALL and HELFAND ( 6 ) :

$$M_2(t) = \exp(-t/\tau_2)\ \exp(-t/\tau_1) I_o(t/\tau_1) \qquad (3)$$

where $I_o$ represents the modified Bessel function of order 0.

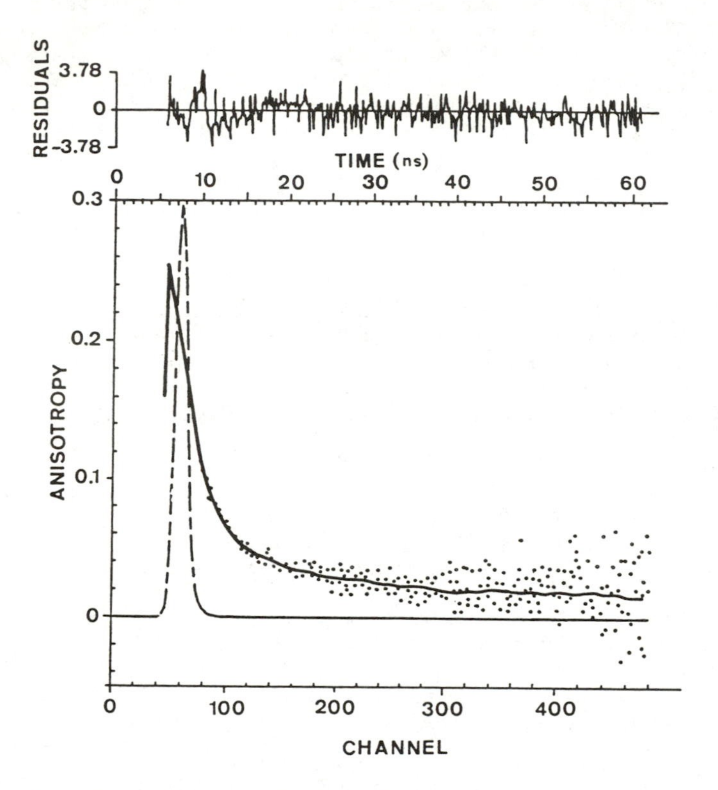


Figure 3. Comparison of the best fit obtained from the Hall-Helfand expression reconvoluted by the measured instrumental function (excitation pulse) (continuous line) with the experimental anisotropy (dots) of labeled polybutadiene at 335.7K. The excitation pulse is plotted as a dash-dot line (arbitrary scaled). The upper graph represents the weighted residuals.

It is worth noting that this expression also gives the best fit for FAD curves of labeled polystyrene in solution ( 5 ).

In bulk polybutadiene, the ratio $\tau_2/\tau_1$ does not vary significantly with temperature and remains rather high (=30). This implies that the processes responsible for the damping in polybutadiene are slow compared to the diffusive ones.

The temperature dependence of $\tau_1$ has been studied in the range 240K- 353K. It is interesting to compare this dependence with the prediction of the well-known William, Landel, Ferry time-temperature superposition equation ( 7 ) which can be written as :

$$\log \tau_T = Cte + C_1^g C_2^g/(T-T_\infty) \qquad (4)$$

$$T_\infty = Tg - C_2^g$$

where Tg is the glass-rubber transition temperature (172K for Diene 45 NF) and $C_1^g$, $C_2^g$ are phenomenological parameters taken from low frequency viscoelastic measurements (For Diene 45 NF, $C_1^g = 11.2$ and $C_2^g = 60.5K$ ( 7 )). In Figure 4, the correlation time $\tau_1$ is plotted versus $1/(T-T_\infty)$ and the dash-line corresponds to the W.L.F. equation. The experimental data satisfactory fit a linear dependence with a slope fairly close to the predicted one. This means that the local reorientation processes observed by FAD are involved in the glass-rubber transition phenomenon.

It should be pointed out that a comparison of the FAD curves of labeled polybutadiene to the ones obtained for free 9,10 (dialkyl) anthracene probes in a Diene 45 NF matrix led to the conclusion ( 8 ) that the local motions observed by the FAD experiments performed on labeled polybutadiene involve about 6 butadiene units.

## CARBON 13 N.M.R.

The local dynamics of polymer chains can be studied by $^{13}C$ NMR through measurements of the spin-lattice magnetic relaxation time, $T_1$. The spin of a given $^{13}C$ relaxes by dipolar relaxation mechanism with the bonded $^1H$ so that the corresponding $T_1$ is related to the reorientation motion of the involved CH internuclear vector through the following expression ( 9 ) :

$$\frac{1}{T_1} = \frac{h^2}{4\pi^2} \frac{\gamma_C^2 \gamma_H^2}{10} \frac{1}{r_{CH}^6} [J(\omega_H - \omega_C) + 3 J(\omega_C) + 6 J(\omega_H + \omega_C)] \qquad (5)$$

$$J(\omega) = \int_o^\infty M_2(t)\, e^{-i\omega t}\, dt$$

where h is the Planck constant, $\gamma_C$ and $\gamma_H$ are the gyromagnetic ratios of $^{13}C$ and $^1H$ respectively, $\omega_C$ and $\omega_H$ are the Larmor resonance frequencies of $^{13}C$ and $^1H$, $J(\omega)$ is the spectral density at the frequency $\omega$ of the OACF of the CH motion, and $r_{CH}$ is the length of the CH bond.

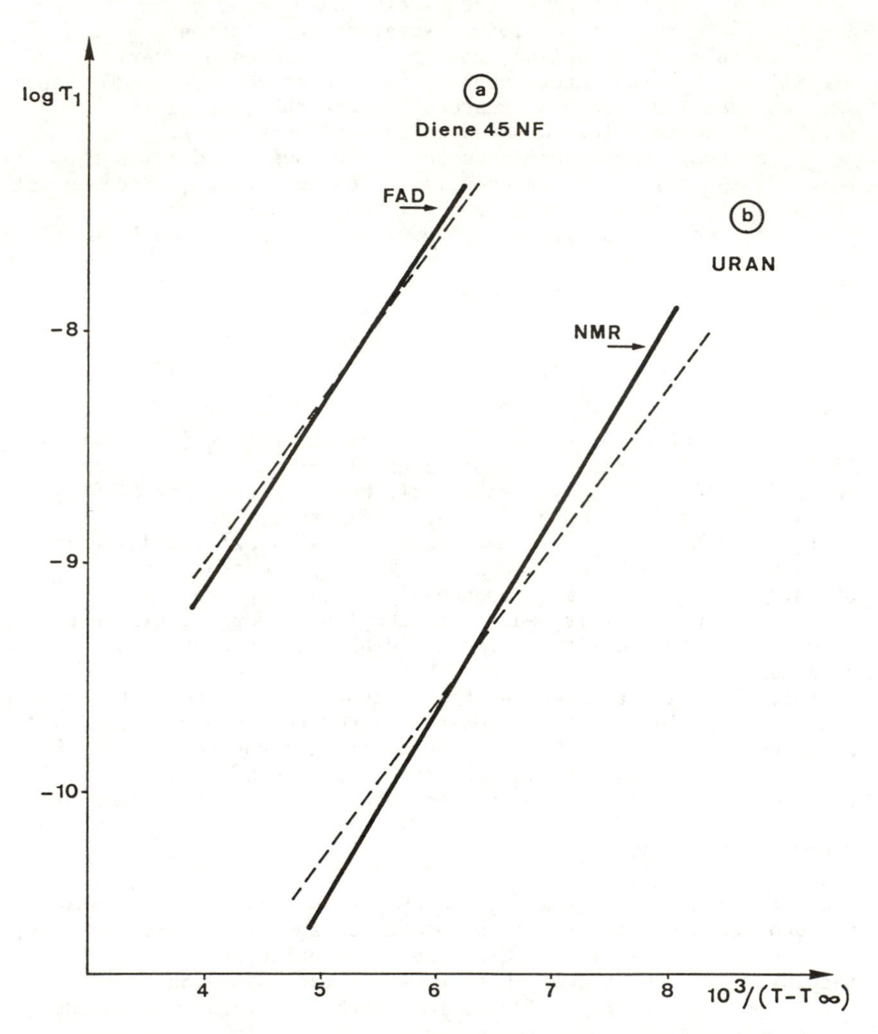


Figure 4. Logarithmic plot of the correlation time $\tau_1$ vs $(T-T_\infty)^{-1}$.
a/ $\tau_1$ determined from FAD for labeled polybutadiene in a Diene 45 NF matrix.
b/ $\tau_1$ determined from $^{13}C$ $T_1$ measurements for polybutadiene Uran. Dashed lines correspond to Equation 4 using in each case the coefficients appropriate to the considered matrix.

The main interest of $^{13}C$ NMR spectroscopy is that it allows one to investigate the local motions performed by the various groups of the polymer chain. Thus, it yields more detailed information than other techniques. On the other hand, to derive the correlation times of the motions from the $T_1$ measurements, it is necessary to use an "a priori" expression of the OACF, $M_2(t)$.

In the case of polymer melts at temperatures higher than (Tg+60K), the segmental mobility is high enough to get narrow resonance lines in the NMR spectrum and thus to perform the $T_1$ measurements on the same spectrometers which are commonly used for high resolution NMR in liquids.

Polybutadiene Uran has been studied at 25.15 MHz and 62.5 MHz in the temperature range 224 K - 358 K.

The $nT_1$ values obtained for CH and $CH_2$ groups of Uran, where n = 1 for CH and n = 2 for $CH_2$, are shown as a function of $10^3/T$ in Figures 5 and 6. On the same figures are plotted the best fits obtained by using the Hall-Helfand OACF expression. It clearly appears that the experimental values of $T_1$ at the minimum are much higher than those expected from this OACF expression. It should be noticed that the comparison would be even worse with the single exponential OACF corresponding to an isotropic motional model. Secondly in the whole temperature range, the ratio $T_1(CH)/T_1(CH_2)$ is different from the expected value of 2, its value lies around 1.5 at both observation frequencies.

These two discrepancies can be accounted for by considering, in addition to the damped diffusion of elementary motions along the chain sequence (which leads to the Hall-Helfand expression for the OACF), either a Brownian motion of the CH bonds inside a cone of half angle $\Theta$, or a specific anisotropic motion of the cis 1,4 sequences of the polybutadiene chain ( 9 ). In both cases, the resulting OACF can be written as :

$$M_2(t) = (1 - a) \exp(-t/\tau_2) \exp(-t/\tau_1) I_o(t/\tau_1) + a \exp(-t/\tau_o) \exp(-t/\tau_1) I_o(t/\tau_1) \qquad (6)$$

where the parameter a describes the contribution of the anisotropic motion with correlation time $\tau_o$. The best fits (shown on Figures 5 and 6) are obtained with $\tau_1/\tau_o \geqslant 150$, $\tau_2/\tau_1 \geqslant 600$ and a(CH) = 0.27, a($CH_2$) = 0.46. These values indicate that the additional anisotropic motion (characterized by $\tau_o$) is very fast as compared to the elementary segmental motions (described by $\tau_1$ ). In the explored temperature range, $\tau_o$ should be shorter than $10^{-11}$ s at 224K and $10^{-13}$ s at 358K.

Recent $^{13}C$ $T_1$ measurements performed on various bulk polymers with low Tg, such as polyisoprene, polyisobutylene, polyvinylmethylether and polypropyleneoxide, have shown that the experimental value of $T_1$ at the minimum is always much higher than the prediction from Hall-Helfand expression. Thus, this discrepancy cannot be assigned to a specific motional anisotropy of the cis 1,4 sequences of the polybutadiene chain. It seems more likely to assign it to a fast anisotropic motion of the CH bonds inside a cone. In the case of polybutadiene, the corresponding half angle of the cones should be 26° and 36° for CH and $CH_2$

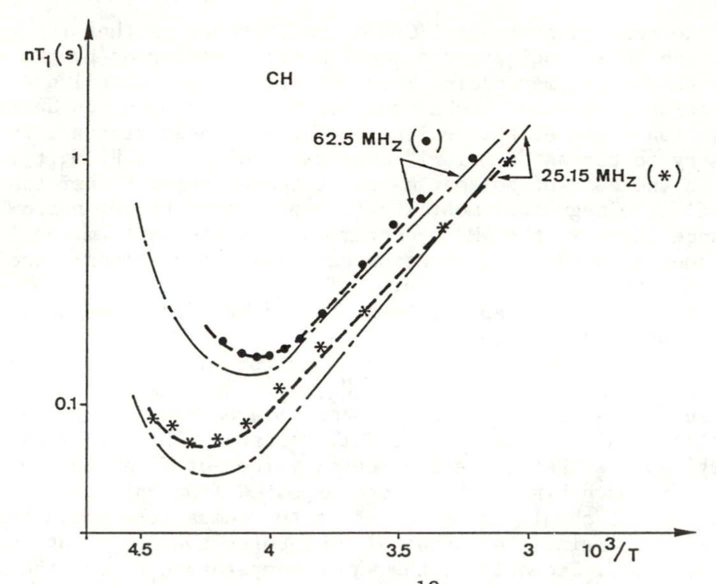


Figure 5. Temperature dependence of $^{13}C$ $T_1$ at 25.15 MHz and 62.5 MHz for CH of Uran polybutadiene. Dots are experimental data. The dash-dot lines correspond to the Hall-Helfand fit (Equation 3). The dashed lines are the best fit obtained by using Equation 6.

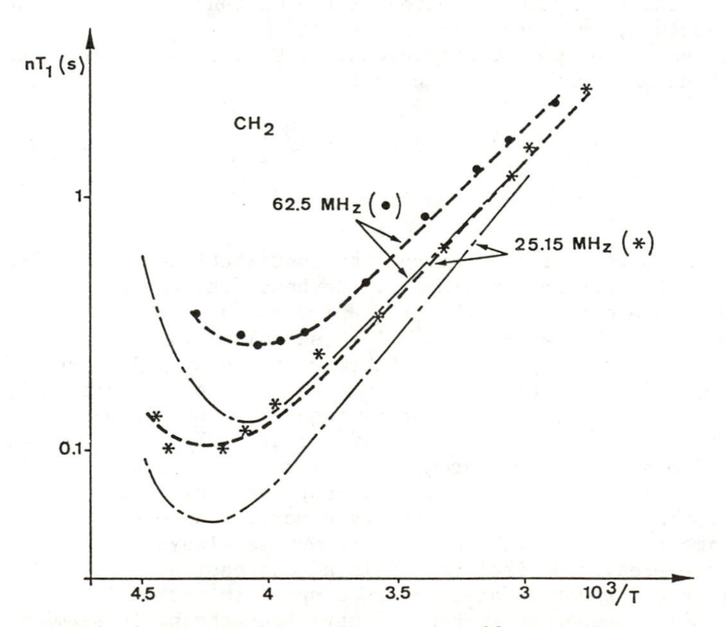


Figure 6. Temperature dependence of $^{13}C$ $T_1$ at 25.15 MHz and 62.5 MHz for $CH_2$ of Uran polybutadiene (same representation as in Figure 5).

respectively. This result is quite satisfactory, the constraints due to the size and the rigidity of the double bond decreasing the amplitude of the libration motion of the CH bonds relative to that of the $CH_2$ groups which are linked to the chain backbone through single bonds only.

Another interesting feature deals with the high value obtained for the ratio $\tau_2/\tau_1$. This means that in high cis content polybutadiene, the damping of the diffusion of the elementary motions along the chain sequence is very weak.

The correlation time $\tau_1$ derived from $T_1$ values of Uran is plotted vs $(T-T_\infty)^{-1}$ on Figure 4. A satisfactory agreement is found with the prediction of Equation 4, using the appropriate coefficients ($T_\infty$ = 101K, $C_1^g$ = 11.35, $C_2^g$ = 59.6K ( 7 )). Thus, the elementary motions of the polybutadiene chain which are investigated by $^{13}C$ NMR are involved in the glass-rubber transition phenomenon.

## COMPARISON OF THE CORRELATION TIMES

The spectroscopic techniques which have been used to investigate the segmental motions of bulk polybutadiene in a temperature range far enough from the glass-rubber transition show that these motions are involved in the glass-rubber transition phenomenon. Thus, it is interesting to go further in comparing the absolute values of the obtained correlation times. At a given $(T-T_\infty)$ interval, for example 192.3K corresponding to a value of 5.2 for $10^3/(T-T_\infty)$, the correlation time of the anthracene label in the middle of the chain is $6.5\times10^{-9}$ s. At the same temperature the value of $\tau_1$ corresponding to the elementary motions responsible for $^{13}C$ relaxation is $4.4\times10^{-11}$ s. Thus, there is about 2 decades of difference in the correlation times.

It is worthnoting that such a difference in correlation time values does not originate from the difference of microstructure between Diene 45 NF and Uran. Indeed, preliminary measurements have shown that for the same $(T-T_\infty)$ interval, similar $T_1$ values are found for both polymers.

As the $^{13}C$ $T_1$ measurements do not imply any labeling, the corresponding correlation time has to be considered as the actual characteristic time of the elementary motions of the polybutadiene chain. The longer correlation time observed with FAD for anthracene labeled polybutadiene might originate either from an inertial effect of the anthracene group or, more likely, from the larger volume which is required to rotate significantly the anthracene transition moment lying along the local chain axis, compared to the volume which is involved in the conformational changes leading to $^{13}C$ spin-lattice relaxation. Such a statement is consistent with the result ( 8 ) that about 6 monomer units are involved in the motions observed by FAD for labeled polybutadiene.

## CONCLUSION

In this paper we have investigated the segmental motions of bulk polybutadiene, in a temperature range higher than (Tg + 60K), by using fluorescence anisotropy decay, and $^{13}C$ spin-lattice magnetic relaxation time, $T_1$.

The FAD technique has shown that the orientation autocorrelation function of local motions in a polymer chain does not correspond to the single exponential associated with isotropic rotation. On the other hand, the Hall-Helfand expression which accounts for the chain connectivity requirement leads to a very satisfactory agreement. Thus the segmental dynamics of polymer melts should be described by elementary motions undergoing a damped diffusion along the chain sequence, in a similar way to that is found in polymer solutions. This indicates that the same types of conformational changes are involved in both cases, the slowing down of these jumps in polymer melts, compared to the case of polymer solutions, arises from the intermolecular constraints.

In addition to these elementary motions, the $^{13}C$ $T_1$ measurements have shown that a much faster anisotropic motion occurs, corresponding to a libration motion of the CH bonds inside a cone. Recent $^{13}C$ $T_1$ measurements ( 10 ) performed on bulk polybutadiene containing both cis 1,4 and trans 1,4 sequences have shown a lower mobility of the trans sequences relative to the cis units. This result illustrates the main interest of $^{13}C$ studies, allowing one to reach more detailed information on the dynamics of each group in the polymer chain and thus yielding a way of setting up a relationship between the chemical structure and the dynamic behavior of bulk polymers.

Another important result deals with the temperature dependence of the correlation times of the elementary motions, which agrees fairly well with the prediction of the William, Landel, Ferry equation, using the phenomenological coefficients obtained from low frequency viscoelastic measurements. This means that the elementary motions which are observed by FAD and $^{13}C$ $T_1$ are involved in the glass-rubber transition phenomenon.

In this paper, the considered spectroscopic techniques have been applied to bulk polybutadiene. Similar studies on other polymer chains are under progress.

## LITERATURE CITED


1. Valeur, B.; Monnerie, L. J. Polym. Sci. , Polym. Phys. Ed. , 1976, 14 , 11.
2. Viovy, J.L.; Monnerie, L.; Merola, F. Macromolecules , 1985, 18 , 1130.
3. Brochon, J.C. In Protein Dynamics and Energy Transduction ; Shin'ishi Ishiwata,Ed.; Taniguchi Foundation: Japan, 1980.
4. Monnerie, L. In Static and Dynamic Properties of the Polymeric Solid State ; Pethrick,R.A.; Richards, R.W., Ed.; NASI Series: Reidel Publ., 1982.
5. Viovy, J.L.; Monnerie, L.; Brochon, J.C. Macromolecules , 1983, 16 , 1845.
6. Hall, C.K.; Helfand, E. J.Chem.Phys. , 1982, 77 , 3275.
7. Ferry,J.D. Viscoelastic Properties of Polymers , 3rd ed.; Wiley: New-York, 1980.
8. Viovy, J.L.; Frank, C.W.; Monnerie, L. Macromolecules , 1985, 18 , 2606.
9. Gronski,W. Makromol. Chem. , 1977, 178 , 2949.
10. Dejean de la Batie, R. Dr.Sc. Thesis, Université Pierre et Marie Curie, Paris, 1986.


RECEIVED March 13, 1987

# Chapter 6

# Excluded Volume Effects on Polymer Cyclization

**Mitchell A. Winnik**

**Department of Chemistry and Erindale College, University of Toronto, Toronto M5S 1A1, Canada**

Recent theory suggests that cyclization phenomena (1) are much more sensitive to excluded volume effects than other properties of polymer chains. Intramolecular fluorescence quenching processes in molecules containing appropriate end groups permit one to study both the dynamics and thermodynamics of end-to-end cyclization. As a consequence, the sensitivity of polymer cyclization to excluded volume can be examined.

Excluded volume effects (2) in polymers are defined as those effects which come about through the steric interaction of monomer units which are remotely positioned along the chain contour. Each individual interaction has only a small effect, but, because there can be many such interactions in a long polymer, excluded volume effects become very large. One consequence of excluded volume is to expand the polymer coil dimensions over that predicted from simple random walk models. The "unperturbed" values of the root-mean-squared end-to-end distance $R_F°$ and radius of gyration $R_G°$ for ideal (random walk) chains expand (to $R_F$ and $R_G$) in the presence of excluded volume.

In solution, excluded volume effects on a polymer can be suppressed by decreasing the quality of the solvent for the polymer. Shrinkage of the mean dimensions occur. At various individual combinations of solvent and temperature, $R_G^2$, as measured from light or neutron scattering, exactly equals $(R_G°)^2$. In these "theta-solvents," other properties of polymers are expected to follow ideal behaviour.

Recent theory suggests that cyclization equilibria are doubly sensitive to excluded volume effects (3). Not only does chain expansion increase the mean chain end separation, but a second factor due to pair correlations also acts to decrease the probability of the chain ends being in proximity: The two chain

0097-6156/87/0358-0057$06.00/0

ends cannot, obviously, occupy the same space. In addition, the adjacent segments on the two chain ends also interfere with each other. Their positions are correlated with those of the chain ends. These effects sum up in such a way as to make it very difficult for the chain ends even to get near one another.

In poor solvents, unfavorable solvent-polymer interactions lead to net attraction between the chain ends and their adjacent segments. This tends to overcome excluded volume repulsion. At the theta-point, these two factors should be exactly in balance. The chain should recover ideal behaviour.

Ideal behaviour for chain conformation is represented by a Gaussian distribution W(r) of end-to-end distances. Excluded volume effects give a much different end distance distribution function, Figure 1, with the biggest differences operating on chains which happen to have their chain ends in proximity (4).

Cyclization probability depends upon the radial distribution function

$$W(0) = \lim_{r \to 0} 4\pi r^2 W(r) = 4\pi r^2 [3/2\pi R_F^2]^{3/2} \tag{1}$$

For ideal chains W(0) should decrease as chain length $N^{-3/2}$. For very long chains experiencing full excluded volume, the exponent is predicted to take the value -1.92. Measurements of cyclization equilibria should allow these predictions to be tested. The theory of cyclization dynamics is less advanced. The prediction of the classic treatment of Wilemski and Fixman (5) suggests that the diffusion-controlled cyclization rate constant $k_1^D \sim N^{-1.5}$ in the absence of excluded volume. The following sections of this chapter examine experimental results about excluded volume effects on experimental values of the cyclization equilibrium constant $K_{cy}$ and the rate constant for end-to-end cyclization $k_1$ (6).

<u>The Polymer</u>. The focus here is on the cyclization behaviour of polystyrene in dilute solution (ca. 2 x $10^{-6}$ M) as studied through measurements of intramolecular excimer formation in <u>1</u> (7) and exciplex formation in <u>2</u> (8).

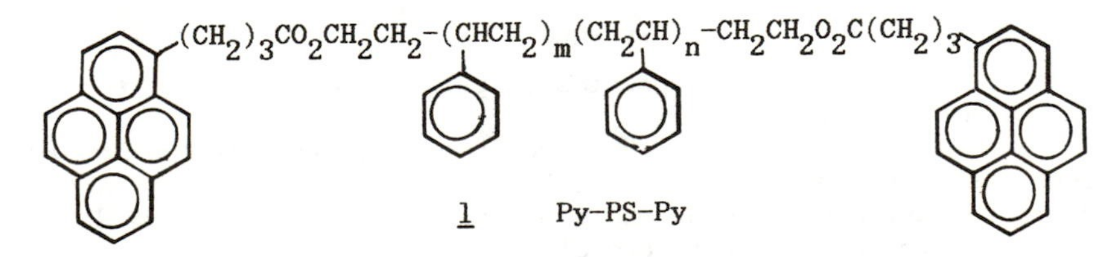


<u>1</u> Py-PS-Py

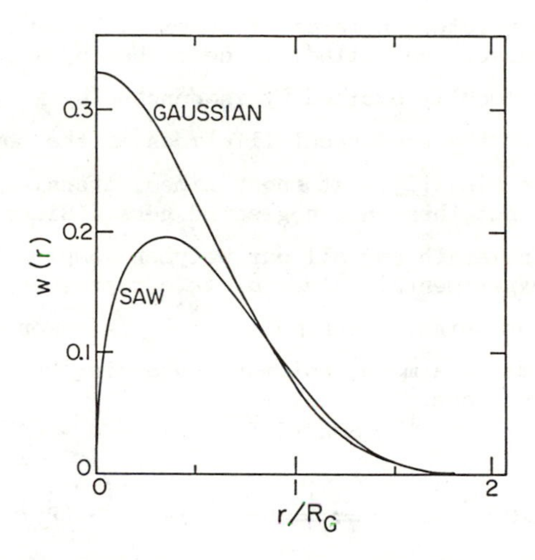


Figure 1. A plot of the end-distance distribution function W(0) vs. the ratio of the end separation divided by $R_G$ for a Gaussian chain and for self-avoiding walk [SAW, full excluded volume], following Ref. 4.

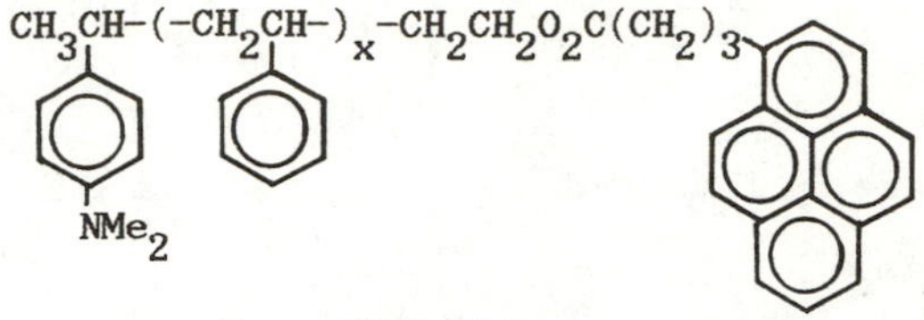

$CH_3CH\text{-}(\text{-}CH_2CH\text{-})_x\text{-}CH_2CH_2O_2C(CH_2)_3\text{-}$

NMe$_2$

2 DMT-PS-Py

Samples to be discussed range in molecular weight from 3000 to 30,000 and have quite narrow molecular weight distributions ($\bar{M}_w/M_n \leq 1.15$).

Data are interpreted in terms of Scheme I below (8,9). Cyclization (diffusion controlled) is described by $k_1$; ring opening to regenerate the locally excited Py is denoted by $k_{-1}$. The terms $k_M$ and $k_E$ represent the reciprocal lifetimes of the monomer and excimer [1] or exciplex [2]. At short times, transient effects will contribute to $k_1$, but these are neglected here. Since $k_1$ is very sensitive to chain length and all our polymer samples have a finite polydispersity, experimental values of $\langle k_1 \rangle$ represent an average over this molecular weight distribution. $k_M$ is taken to be the reciprocal lifetime of a model polymer containing Py at one end, but no quencher at the other.

SCHEME I

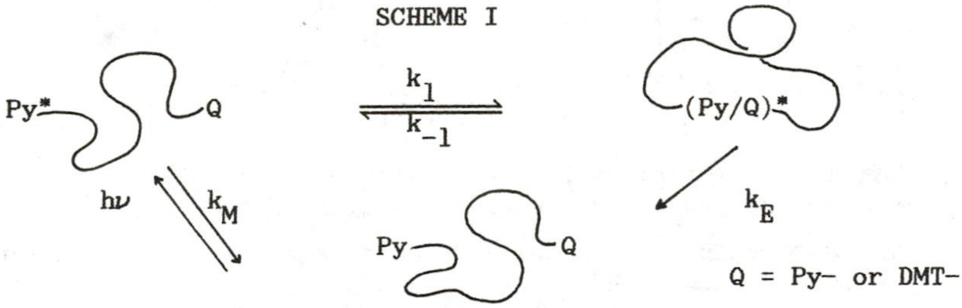

Py* Q $\underset{k_{-1}}{\overset{k_1}{\rightleftharpoons}}$ (Py/Q)*

$h\nu$ $k_M$ $k_E$

Py Q

Q = Py- or DMT-

Rate constants are obtained from fluorescence decay analyses of the monomer decay profile $I_M(t)$ and the excimer/exciplex profile $I_E(t)$. These are fit to sums and differences of two exponential terms, respectively, and analyzed in a manner previously described in detail (7,9,10). Steady state measurements of $I_E$ and $I_M$ are related to $\langle k_1 \rangle$

$$\frac{I_E}{I_M} = \frac{k_{fE}}{k_{fM}} \frac{\langle k_1 \rangle}{(k_{-1} + k_E)} \tag{1}$$

where $(k_{fE}/k_{fM})$ is the ratio of the radiative rate constant of the excimer (or exciplex) to that of the monomer. This expression is particularly useful under conditions that one knows or can assume that $k_{-1}$ and $k_E$ remain invariant.

Exciplex Formation in Cyclopentane. This section begins on a cautionary note. Fluorescence decay measurements on the two lowest molecular weight samples of DMT-PS-Py [M = 3000, 5000] give data that differ in one key respect from the predictions in Scheme I. The long component from the exciplex decay always has a longer lifetime than that from the monomer decay (10). $\langle k_1 \rangle$ values depend on which decay time is used in the calculation, with $\langle k_1 \rangle_M$ larger than $\langle k_1 \rangle_E$. The former falls on the line defined by a plot of log $\langle k_1 \rangle$ vs log N for other samples, and is arbitrarily chosen as the proper value for the discussion which follows. The general trends raised in the discussion are not affected by this difficulty. Nevertheless, these differences point to problems in Scheme I for short chains, which we hope eventually to be able to sort out. At short times $k_1$ ought to show a time dependence. We have some indication that this transient contribution is particularly important for short chains.

Log-log plots of $\langle k_1 \rangle$, $k_{-1}$ and $k_E$ determined for samples of 2 of M = 3000 to 28,000 in cyclopentane at 35° and at 50° are shown in Figure 2. Straight lines can be fitted through each set of data points. The rate constants $k_{-1}$ and $k_E$ are approximately constant with increasing chain length for M > 5000. At the lower temperature $k_{-1}$ is known with relatively poor precision, since it is determined by a subtraction step and is nearly ten-fold smaller than $k_E$. As the temperature is raised, $k_{-1}$ increases markedly whereas $k_E$ does not change.

At 22° the slope of the log $\langle k_1 \rangle$ vs log N plot is -1.48, within experimental error of the value predicted from theory (5) for cyclization dynamics of polymer chains in the absence of excluded volume.

Upper and lower theta-temperatures. Many polymers precipitate from solution when heated. This "lower critical solution temperature" [LCST] lies above the normal (1 atm) boiling temperature for the solvent. The phenomenon is observed for samples in sealed tubes. For polystyrene in cyclopentane (bp 49°) the LCST occurs at ca. 150°C (11). The exact temperature varies with molecular weight. Many scientists believe that the LCST in the limit of very high molecular weight corresponds to a second theta temperature. Because of experimental difficulties, there is not much evidence of this point.

The lower theta temperature corresponds to the minimum solution temperature extrapolated to infinite chain length. Polymer precipitation at low temperatures comes about because of a poor heat of mixing between polymer and solvent. In a sealed tube at high temperatures, solvent volume expands much more than that of the polymer. Entropic factors make mixing more difficult when there is a large free volume mismatch between solute and solvent. One believes that the polymer dimensions contract as the LCST is approached. Phase separation occurs when it is exceeded.

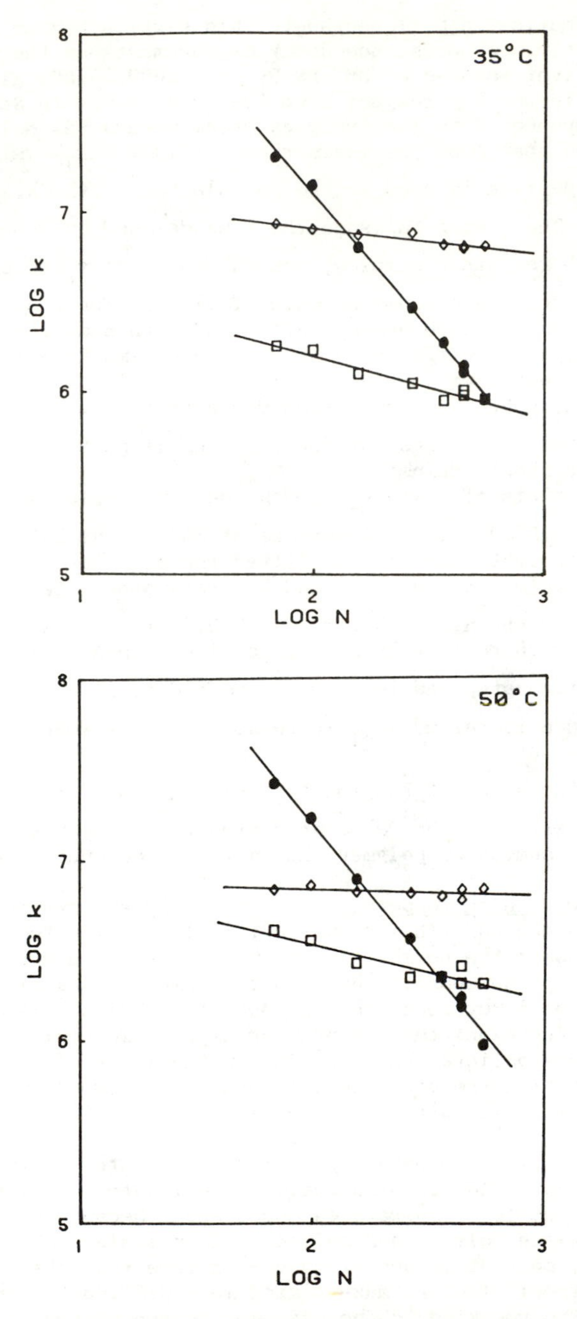


Figure 2. Log-log plots of $\langle k_1 \rangle$, $k_{-1}$ and $k_E$ vs. chain length for samples of 2 in cyclopentane at 35° and at 50°.

For polystyrene in cyclopentane, one predicts that there should be two theta-temperatures, at 22° and 150° respectively. In the intermediate temperature range, cyclopentane should be a good solvent for PS.

These temperature effects on solvent quality of cyclopentane for polystyrene should have interesting consequences on the cyclization behaviour of the polymer. We carried out experiments on DMT-PS-Py over a wide temperature range. These are the same samples shown in Figure 2. Since the polymer concentrations in these samples were so low, they could be studied at temperatures where higher concentrations would precipitate from solution. At each temperature, plots of log $\langle k_1 \rangle$ vs log N were linear.

The slopes of these plots ($-\gamma$) change with temperature. At the lowest and highest temperatures, $\langle k_1 \rangle$ varies as $N^{-1.4}$. In the intermediate range of temperatures, $\langle k_1 \rangle$ is most sensitive to chain length, varying as $N^{-1.6}$ at 63°. This behaviour is shown in Figure 3.

The most significant feature of this plot is the maximum in the $\gamma$ vs temperature plot. The increase in $\gamma$ is anticipated, since there is good evidence that above 22° cyclopentane becomes a reasonably good solvent for PS. In a very good solvent such as toluene, $\langle k_1 \rangle$ varies as $N^{-1.8}$. Here the maximum absolute value found is 1.6. The plot in Figure 3 indicates that above 75°, cyclopentane becomes a progressively poorer solvent for PS. The excluded volume effects that lead to an increase in $\gamma$ in the expression $k_1 \sim N^{-\gamma}$ become suppressed as the solutions are heated toward the upper theta-temperature.

<u>Excimer Formation in Mixed Solvents</u>. The term "cosolvency" is used to describe the principle whereby mixtures of certain pairs of poor solvents for a polymer can act as a good solvent for that polymer (12). The preferred explanation is that the heat of mixing of the two solvents in unfavorable. The presence of the polymer leads to a solution in which the number of contacts between solvent 1 and solvent 2 molecules can be minimized. This in turn leads to more polymer-solvent contacts and fewer polymer-polymer contacts than would occur for polymer solutions in either individual solvent. We are interested in examining the influence of cosolvency on PS cyclization.

Acetone is a very poor solvent for PS. High molecular weight samples of PS will not dissolve in acetone at room temperature. Intrinsic viscosity measurements on samples of PS indicate that mixtures of acetone with cyclohexane lead to expanded coil dimensions (13). Two theta compositions are found, one containing 5 mol % and one containing 83 mol % acetone. In the intermediate range of concentrations, the mixture behaves as a good solvent for PS.

A fluorescence spectrum of a sample of Py-PS-Py of $M_n$ = 4500 is shown in Figure 4. There is less excimer emission in the mixed solvent than in either pure solvent (14). In Figure 5 values are presented for $\langle k_1 \rangle$ (here corrected for solvent viscosity $\eta_0$ effects) for this sample and one of $M_n$ = 25,000. Cyclization rates decrease in mixed solvents. A curious feature of these data is that acetone as a poor solvent seems to have a much larger accelerating effect on the higher molecular weight polymer. We attribute this (14) difference in behaviour to the larger relative importance of end group solvation by acetone in the sample of M = 4500.

The most important observation is that $k_{-1}$ increases in the mixed solvents: good solvents both retard ring-closure dynamics and promote ring-opening. Large effects should be found for the cyclization equilibrium constant $\langle K_{cy} \rangle = \langle k_1 \rangle / k_{-1}$ .

In Figure 6 values of $K_{cy}$ are plotted as a function of solvent composition. These go through a minimum in a solvent composed of 50 mol % acetone. The decrease in $K_{cy}$ over that in pure cyclopentane is a factor of 2.0.

One anticipates that cyclization probability will decrease as coil dimensions increase. Older theory, which neglects pair correlation effects, predicts that

$$W(0) \sim R_G^{-3/2}$$

and

$$K_{cy} \sim [\eta]^{-1}$$

since the intrinsic viscosity $[\eta]$ varies as $R_G^{3/2}$. We can use the data of Maillols et al. (13) on intrinsic viscosity measurements in acetone-cyclohexane measurements to estimate the change in $[\eta]$ between theta- and good solvent limits for PS of $\bar{M}_n$ = 4500. These data predict a decrease in $K_{cy}$ of not more than 30%. We observe a much larger change in $K_{cy}$. Some other factor is responsible. We attribute this to the pair correlations effects cited by recent theory (3,4) and illustrated in Figure 1. These experiments provide one of the few experimental verifications of the validity of this prediction of the theory.

## SUMMARY

Cyclization is much more sensitive to excluded volume effects than other properties of polymer chains. These effects operate both to retard ring-closure dynamics and to accelerate the ring-opening step of cyclized chains.

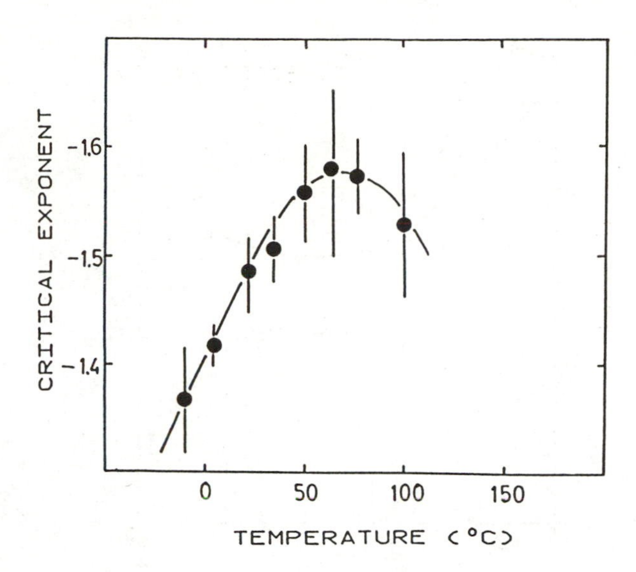


Figure 3. The cyclization exponent from the expression $\langle k_1 \rangle = aN^{-\gamma}$ as a function of temperature for samples of 2 in cyclopentane.

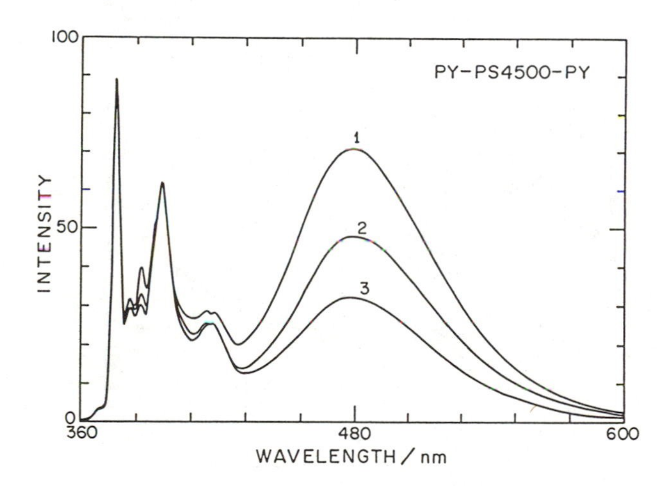


Figure 4. Fluorescence spectra of Py-PS-Py ($M_n$ = 4500) in cyclopentane [1], acetone [2], and a 1:1 mixture [3] of these solvents.

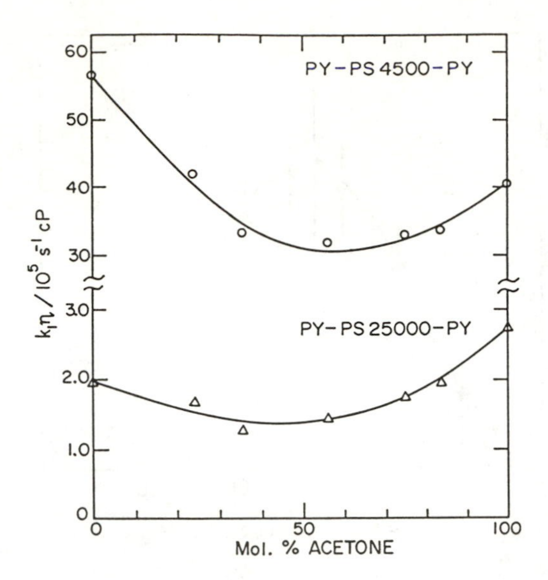


Figure 5. Viscosity corrected $\langle k_1 \rangle$ values for Py-PS-Py samples of $M_n$ = 4500 and 25,000 as a function of solvent compostion in cyclopentane-acetone mixtures.

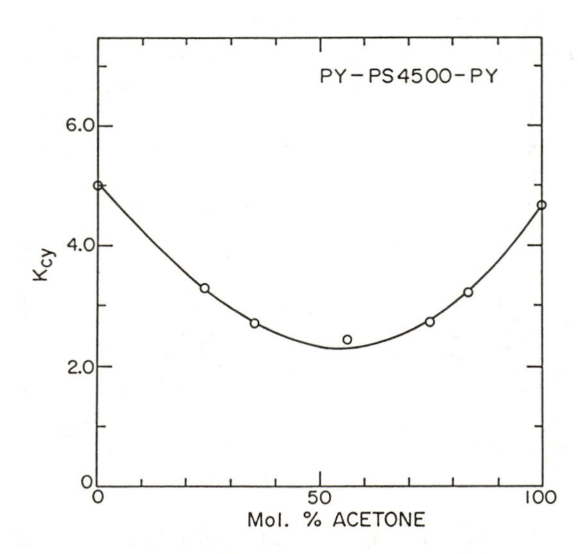


Figure 6. The cyclization equilibrium constant $K_{cy}$ vs. solvent composition for Py-PS-Py (M = 4500) in cyclopentane-acetone mixtures.

Acknowledgment. We are grateful for the support of NSERC Canada and the donors of the Petroleum Research Fund, administered by the American Chemical Society. The real credit for the results described here belongs to Dr. A. Sinclair and Dr. J.M.G. Martinho, who carried out the experiments described here. They are, of course, cited in the references.

REFERENCES

1. Cyclization Dynamics of Polymers 23.
2. Flory, P.J. In Statistical Mechanics of Chain Molecules; Wiley-Interscience: New York, 1969.
3. deGennes, P.G. In Scaling Concepts in Polymer Physics; Cornell University Press: Ithaca, New York, 1979.
4. Oono, Y.; Freed, K. J. Phys. A Math. Gen. 1982, 15, 1931.
5. (a) Wilemski, G.; Fixman, M. J. Chem. Phys. 1964, 60, 866, 878.
   (b) Doi, M. Chem. Phys. 1975, 9, 455; Sunagawa, S.; Doi, M. Polym. J. 1975, 7, 604; 1976, 8, 239.
6. Cuniberti, C.; Perico, A. Prog. Polym. Sci., 1984, 10, 271.
7. (a) Winnik, M.A. Accounts Chem. Res. 1985, 18, 73.
   (b) Winnik, M.A. In Cyclic Polymers; Semlyen, A.J., Ed.; Applied Science Publishers: London, U.K., 1986.
8. (a) Beinert, G.; Winnik, M.A. Can. J. Chem. 1986, 64, 1743.
   (b) Winnik, M.A.; Sinclair, A.M.; Beinert, G. Macromolecules 1985, 18, 1517.
   (c) Sinclair, A.M.; Winnik, M.A.; Beinert, G. J. Am. Chem. Soc. 1985, 107, 5798.
9. Winnik, M.A.; Redpath, A.E.C.; Paton, K.; Danhelka, J. Polymer 1984, 25, 91.
10. Sinclair, A.M. Ph.D. Thesis, University of Toronto, 1986.
11. Saeki, S.; Kuwahara, N; Konno, S.; Kaneko, M. Macromolecules 1973, 6, 246, 589.
12. (a) Palit, S.R.; Colombo, G.; Mark, H. J. Polym. Sci. 1951, 6, 295.
    (b) Ishikawa, K.; Kawai, T. J. Chem. Soc. Jpn. Ind. Chem. Sec. 1952, 55, 173.
13. Maillols, H.; Bardet, L; Grombo, S. Eur. Polym. J. 1978, 14, 1015.
14. Martinho, J.M.G.; Winnik, M.A. Macromolecules 1986, 19, 2281.

RECEIVED May 22, 1987

## Chapter 7

# Time-Resolved Optical Spectroscopy as a Probe of Local Polymer Motions

**Dean A. Waldow[1], Patrick D. Hyde[1], M. D. Ediger[1], Toshiaki Kitano[2], and Koichi Ito[2]**

[1]Department of Chemistry, University of Wisconsin, Madison, WI 53706
[2]Department of Materials Science, Toyohashi University of Technology, Toyohashi 440, Japan

A picosecond holographic grating technique has been used to observe the local segmental dynamics of anthracene-labeled polyisoprene in dilute hexane, cyclohexane, and 2-pentanone solutions. The transition dipole of the anthracene label lies along the chain backbone, assuring that only backbone motions are detected. The experimental observable, the orientation autocorrelation function of a backbone bond, is compared with theoretical predictions. The observed correlation functions are consistent with models proposed by Hall and Helfand and by Bendler and Yaris. The effects of temperature and solvent viscosity upon the local dynamics of polyisoprene are investigated. The activation energy for the observed local segmental motions is ~ 7 kJ/mole.

The relationship between the structure of a polymer chain and it dynamics has long been a focus for work in polymer science. It is on the local level that the dynamics of a polymer chain are most directly linked to the monomer structure. The techniques of time-resolved optical spectroscopy provide a uniquely detailed picture of local segmental motions. This is accomplished through the direct observation of the time dependence of the orientation autocorrelation function of a bond in the polymer chain. Optical techniques include fluorescence anisotropy decay experiments(1-6) and transient absorption measurements(7). A common feature of these methods is the use of polymer chains with chromophore labels attached. The transition dipole of the attached chromophore defines the vector whose reorientation is observed in the experiment. A common labeling scheme is to bond the chromophore into the polymer chain such that the transition dipole is rigidly affixed either parallel(1-7) or perpendicular(8,9) to the chain backbone.

0097-6156/87/0358-0068$06.00/0

Time-resolved optical experiments rely on a short pulse of polarized light from a laser, synchrotron, or flash lamp to photoselect chromophores which have their transition dipoles oriented in the same direction as the polarization of the exciting light. This non-random orientational distribution of excited state transition dipoles will randomize in time due to motions of the polymer chains to which the chromophores are attached. The precise manner in which the oriented distribution randomizes depends upon the detailed character of the molecular motions taking place and is described by the orientation autocorrelation function. This randomization of the orientational distribution can be observed either through time-resolved polarized fluorescence (as in fluorescence anisotropy decay experiments) or through time-resolved polarized absorption.

The transient absorption method utilized in the experiments reported here is the transient holographic grating technique(7,10). In the transient grating experiment, a pair of polarized excitation pulses is used to create the anisotropic distribution of excited state transition dipoles. The motions of the polymer backbone are monitored by a probe pulse which enters the sample at some chosen time interval after the excitation pulses and probes the orientational distribution of the transition dipoles at that time. By changing the time delay between the excitation and probe pulses, the orientation autocorrelation function of a transition dipole rigidly associated with a backbone bond can be determined. In the present context, the major advantage of the transient grating measurement in relation to typical fluorescence measurements is the fast time resolution (~ 50 psec in these experiments). In transient absorption techniques the time resolution is limited by laser pulse widths and not by the speed of electronic detectors. Fast time resolution is necessary for the experiments reported here because of the sub-nanosecond time scales for local motions in very flexible polymers such as polyisoprene.

In this paper, we report measurements of the orientation autocorrelation function of a backbone bond in dilute solutions of anthracene-labeled polyisoprene. The anthracene chromophore was covalently bonded into the chain such that the transition dipole for the lowest electronic excited state lies along the chain backbone. This assures that only backbone motions are detected. Our experimental measurements of the orientation autocorrelation function on sub-nanosecond time scales are consistent with the theoretical models for backbone motions proposed by Hall and Helfand(11) and by Bendler and Yaris(12). The correlation functions observed in three different solvents at various temperatures have the same shape within experimental error. This implies that the fundamental character of the local segmental dynamics is the same in the different environments investigated. Analysis of the temperature dependence of the correlation function yields an activation energy of ~ 7 kJ/mole for local segmental motions.

## Experimental Section

Experimental Apparatus. Figure 1 shows a block diagram of the apparatus used in these experiments. A Q-switched, mode-locked

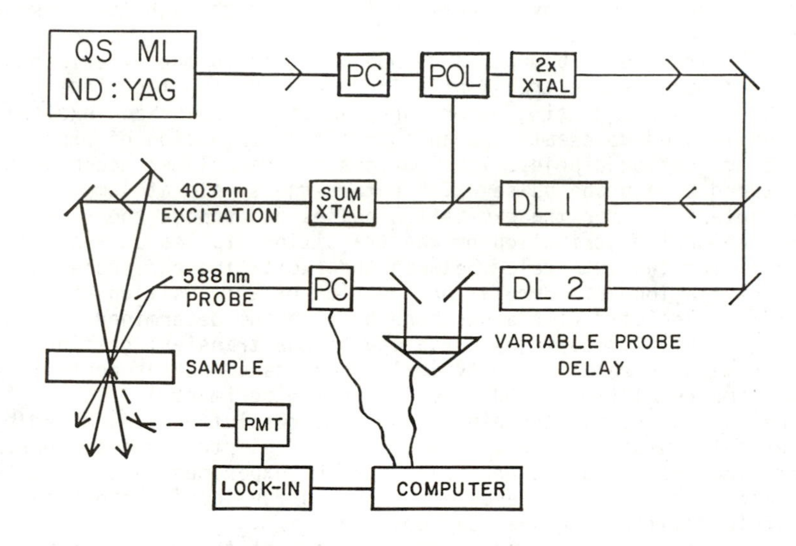


Figure 1. Q-switched, mode-locked Nd:YAG laser with two synchronously pumped dye lasers: PC = Pockels' cell; POL = polarizer with escape window; DL1, DL2 = cavity dumped dye lasers; PMT = photomultiplier tube. (Reproduced from Ref. 7. Copyright 1986 American Chemical Society.)

Nd:YAG laser synchronously pumps two dye lasers. One of the dye lasers (DL1) is operated at a wavelength of 649 nm. A cavity-dumped single pulse from this dye laser is frequency summed with a single pulse of the YAG fundamental (1064 nm) to produce the excitation wavelength of 403 nm. This pulse is beamsplit to form the grating. A second dye laser (DL2) provides the probe pulse (588 nm) which is sent down an optical delayline before entering the sample at the Bragg angle relative to the grating formed by the excitation beams. Varying the position of a retroreflector on the optical delayline varies the timing between the excitation and probe pulses and thus provides the time base for the experiment. The diffracted probe beam is detected by a photomultiplier and a lock-in amplifier. A microcomputer controls the data acquisition by varying the probe polarization and monitoring the diffracted signal strength and the delayline position. The sample temperature during any given measurement was constant to ± 0.1°C.

Experimental Technique. The picosecond transient grating technique has been extensively utilized to study electronic excitation transfer(13,14), rotational reorientation of small molecules(10), and acoustic phenomena(15). On longer time scales, the same concept is utilized in forced Rayleigh scattering experiments to measure translational diffusion(16). In a transient grating experiment, optical interference between two crossed laser pulses creates a spatially periodic intensity pattern in an absorbing sample. This results in a spatial grating of excited states which then diffracts a third (probe) beam brought into the sample at some later time. The two observable experimental quantities are the intensity of the diffracted signal for the probe beam polarized parallel ($T_{\parallel}(t)$) and perpendicular ($T_{\perp}(t)$) to the polarization of the excitation beams.

Figure 2 shows $T_{\parallel}(t)$ and $T_{\perp}(t)$ for anthracene-labeled polyisoprene in dilute hexane solution. The sharp rising edge in the data indicates the time when the excitation pulse enters the sample. At times very soon after this, there is a large difference between the diffracted signal for the two different probe polarizations. $T_{\parallel}(t)$ is larger than $T_{\perp}(t)$ since a larger number of excited state transition dipoles are oriented parallel to the excitation pulse polarization. As molecular motions of the polymer occur, the excited state transition dipoles randomize their orientations, and the difference between $T_{\parallel}(t)$ and $T_{\perp}(t)$ goes to zero. Both curves continue to decay due to the excited state lifetime (~ 8 nsec). Quantitatively, the shapes of $T_{\parallel}(t)$ and $T_{\perp}(t)$ are given in terms of the orientation autocorrelation function CF(t) and the excited state decay function K(t) as follows:

$$T_{\parallel}(t) = \{K(t)(1+2r(t))\}^2 \quad (1)$$

$$T_{\perp}(t) = \{K(t)(1-r(t))\}^2 \quad (2)$$

In these equations, r(t) is the time-dependent anisotropy function, which is proportional to CF(t) as

$$r(t) = r(0)CF(t) \quad (3)$$

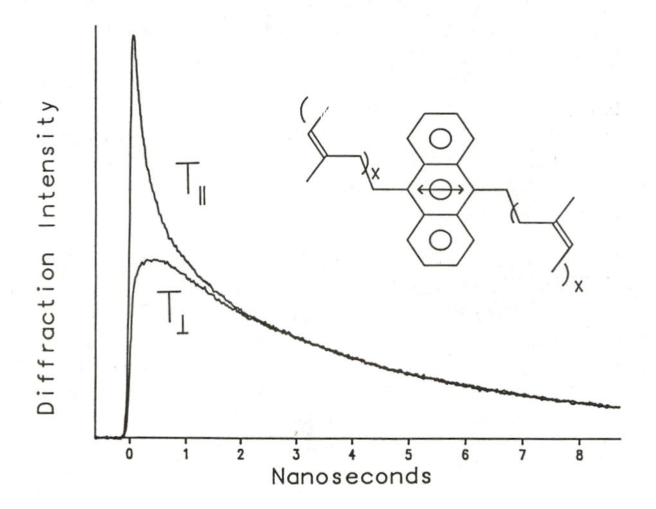


Figure 2. Transient grating decays for 9,10-bis(methylene)-anthracene labeled polyisoprene in dilute hexane solution. $T_{\parallel}$ and $T_{\perp}$ are the diffraction efficiencies of the grating for the probe beam polarized parallel and perpendicular to the excitation beams (see Equations 1 and 2). The two curves are initially different because the excitation beams create an anisotropic orientational distribution of excited state transition dipoles. As backbone motions occur, the transition dipoles randomize and the two curves coalesce. Both curves eventually decay due to the excited state lifetime. The structure of the anthracene-labeled polyisoprene is also displayed, with the position of the transition dipole indicated by a double arrow. (Reproduced from Ref. 7. Copyright 1986 American Chemical Society.)

where r(0) is the fundamental anisotropy for the transition being observed. r(t) can be obtained from the transient grating signals by the following manipulation:

$$r(t) = \frac{\sqrt{T_{\parallel}(t)} - \sqrt{T_{\perp}(t)}}{\sqrt{T_{\parallel}(t)} + 2\sqrt{T_{\perp}(t)}} \qquad (4)$$

The experimental anisotropy contains information about molecular motion, but is independent of the excited state lifetime. Equation 4 indicates that the orientation autocorrelation function can be obtained directly and unambiguously (within the multiplicative constant r(0)) from the results of a transient grating experiment.

By setting the time delay between the excitation and probe beams to a given value and then alternating the polarization of the probe beam, the experimental anisotropy r(t) for time t can be obtained using Equation 4. In this mode of data acquisition, the experimental anisotropy is obtained at a relatively small number of time points but with quite good precision at each point. The results of such an experiment on a dilute solution of labeled polyisoprene in hexane are shown in Figure 3. The smooth curve running through the data points is the best fit theoretical curve using the Hall-Helfand model. In general, the fitting of the anisotropy involved three adjustable parameters: r(0), and the two parameters in each correlation function. The finite lengths of the excitation and probe pulses were accounted for in the fitting procedure. Further details on data acquisition, error analysis, and curve fitting are presented in Ref. 7.

Materials. A polyisoprene prepolymer was prepared by anionic polymerization in benzene at room temperature. The prepolymer was coupled with 9,10-bis(bromomethyl) anthracene to form a labeled polymer of twice the molecular weight of the prepolymer. Fractionation of the resulting products allowed polyisoprene chains containing exactly one anthracene chromophore in the chain center to be isolated. The labeled polymer was $M_n$ = 10,800 and $M_w/M_n$ = 1.16. The structure of the labeled chain is shown as an inset in Figure 2 (see below). The chain microstructure was determined to be 39% cis-1,4, 36% trans-1,4, and 25% vinyl-3,4. All solvents were used as received and showed negligible absorption in the wavelength region of interest. The hexane and cyclohexane were spectrophotometric grade (99+%) while the 2-pentanone was 97%. Solvent viscosities were obtained from Ref. 17.

Label Effect. In order to assess at least partially the effect of the label on the chain dynamics, we also performed measurements on dilute solutions of 9,10-dimethyl anthracene. The reorientation time for the free dye in cyclohexane was ~10 psec, 50 times faster than the time scale for motion of the labeled chain in cyclohexane. Hence we conclude that the observed correlation functions are not dominated by the hydrodynamic interaction of the chromophore itself with the solvent, but can be attributed to the polymer chain motions.

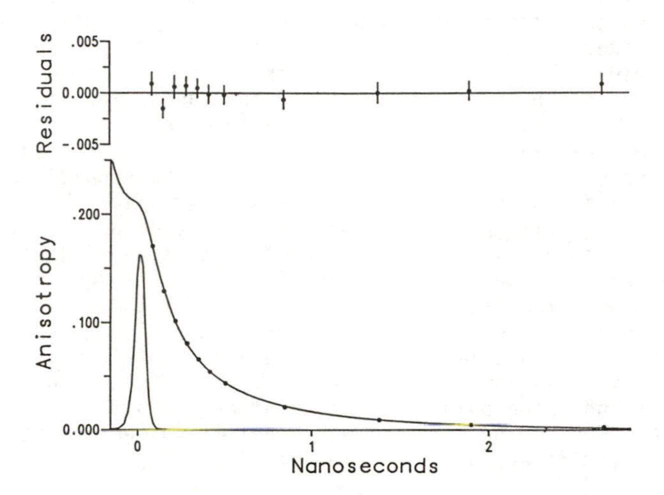


Figure 3. Time-dependent anisotropy for anthracene-labeled polyisoprene in dilute hexane solution. The experimental anisotropy was obtained by setting the delay between the excitation and probe pulses to a given position and then varying the polarization of the probe beam. In the bottom portion of the figure, the smooth curve through the data is the best fit to the Hall-Helfand model($\tau_1$=236 ps, $\tau_2$=909 ps, and r(0)=0.250). Unweighted residuals for the best fit to this model are shown along with the experimental error bars in the top portion of the figure. Note that the residuals are shown on an expanded scale (10x). The instrument response function is indicated at the left.

## Results and Discussion

Theoretical Models for Local Segmental Motions. The second order orientation autocorrelation function measured in this experiment is defined by:

$$CF(t) = \langle P_2(\hat{\mu}(t) \cdot \hat{\mu}(0)) \rangle \quad (5)$$

In this expression, $P_2$ is the second Legendre polynomial and $\hat{\mu}(t)$ is a unit vector with the same orientation as the transition dipole at time t. The brackets indicate an ensemble average over all transition dipoles in the sample. The correlation function has a value of one at very short times when the orientation of $\mu(t)$ has not changed from its initial orientation. At long times, the correlation function decays to zero because all memory of the initial orientation is lost. At intermediate times, the shape of the correlation function provides detailed information about the types of motions taking place. Table I shows the three theoretical models for the correlation function which we have compared with our experimental results.

Table I. Correlation Functions

| | |
|---|---|
| $\exp(-t/\tau_1)\exp(-t/\tau_2)I_0(t/\tau_1)$ | Hall and Helfand(11) |
| $0.5\sqrt{\pi/t}\ \{1/\sqrt{\tau_2}-1/\sqrt{\tau_1}\}^{-1}\ \{\text{erfc}\sqrt{t/\tau_1}-\text{erfc}\sqrt{t/\tau_2}\}$ | Bendler and Yaris(12) |
| $\exp(-t/\tau_1)\exp(-t/\tau_2)\{I_0(t/\tau_1)+I_1(t/\tau_1)\}$ | Viovy, et al.(4) |

In a previous publication(7), we compared the experimental anisotropies for dilute solutions of labeled polyisoprene in hexane and cyclohexane to several theoretical models. These results are shown in Table II. The major conclusions of the previous study are: 1) The theoretical models proposed by Hall and Helfand, and by Bendler and Yaris provide good fits to the experimentally measured correlation function for both hexane and cyclohexane. The model suggested by Viovy, et al. does not fit as well as the other two models. 2) Within experimental error, the shape of the correlation function is the same in the two solvents (i.e, the ratio of $\tau_2/\tau_1$ is constant). 3) The time scale of the correlation function decay scales roughly with the solvent viscosity.

The results of an experiment on a dilute solution of labeled polyisoprene in hexane are shown in Figure 3. The smooth curve running through the points is the best fit to the anisotropy data using the Hall-Helfand model. The data points are indicated by closed circles which are drawn larger than the error bars for clarity. The error bars can be seen in the residuals plots at the top of the figure. (The vertical scale for the residuals plots is expanded by a factor of 10.) Figure 4 shows the anisotropy for a dilute cyclohexane solution with the best fit using the Bendler-Yaris model.

Temperature Dependence of Local Segmental Motions. In this paper, our major focus is on the temperature and viscosity dependence of

Table II. Best fit parameters

| | r(0) | $\tau_1$(ps) | $\tau_2$(ps) | $\tau_2/\tau_1$ | T(°C) | $\eta$(cp) |
|---|---|---|---|---|---|---|
| Hexane | | | | | | |
| Hall and Helfand | 0.254±0.016 | 246±30 | 890±70 | 3.6±0.5 | 21.5 | 0.32 |
| Bendler and Yaris | 0.258±0.019 | 82±10 | 1160±90 | 14.2±2.0 | | |
| Viovy, et al. | 0.287±0.016 | 81±9 | 720±60 | 8.9±1.2 | | |
| Cyclohexane | | | | | | |
| Hall and Helfand | 0.244±0.016 | 610±70 | 2210±150 | 3.6±0.5 | 21.5 | 0.94 |
| Bendler and Yaris | 0.244±0.016 | 205±25 | 2870±200 | 14.0±2.0 | | |
| Viovy, et al. | 0.260±0.020 | 230±30 | 1790±130 | 7.8±1.2 | | |
| 2-Pentanone | | | | | | |
| Hall and Helfand | 0.264±0.016 | 284±40 | 1000±90 | 3.6±0.6 | 35.1 | 0.42 |
| | 0.254±0.020 | 350±50 | 1320±100 | 3.8±0.6 | 22.8 | 0.49 |
| | 0.250±0.016 | 750±120 | 2600±180 | 3.5±0.6 | -8.6 | 0.72 |
| | 0.255±0.016 | 1340±150 | 4850±600 | 3.6±0.7 | -26.5 | 0.95 |

the correlation function in 2-pentanone. For the sake of clarity, we have chosen to fit this data to only one model. The model of Hall and Helfand was chosen because it provided a good fit to data from all three solvents. In addition, the model is easily understood in physical terms(11). This model considers a two state system involving both correlated and isolated conformational transitions. Correlated transitions allow small internal sections of chains to change their conformational states without requiring reorientation of the chain ends. Isolated conformational transitions involve some sort of translation or reorientation of larger sections of the polymer chain. In this model, $\tau_1$ and $\tau_2$ are time constants for correlated and isolated transitions, respectively (see Table I). Figure 5 shows time dependent anisotropies for dilute solutions of labeled polyisoprene in 2-pentanone at various temperatures. Also shown are the best fits to the anisotropies using the Hall-Helfand model. The best fit parameters are given in Table II. Within experimental error, the shape of the correlation function in 2-pentanone at all temperatures studied is the same as the shape of the correlation functions observed in hexane and cyclohexane.

We can analyze the temperature dependence of the local segmental dynamics in terms of the Kramers picture of diffusion over a barrier(18,19). Presumably, more than one fundamental process contributes to the local motion observed in these experiments. In this case, we can write an expression for the rate $w_i$ of the $i^{th}$ process which contributes to the reorientation of the chromophore's transition dipole, in terms of the viscosity $\eta$ and the activation energy $E_i$:

$$w_i \propto \frac{1}{\eta} \exp(-E_i/kT) \tag{6}$$

The fact that the observed correlation functions do not change shape as a function of temperature implies that the relative contributions of these various fundamental processes do not change significantly with temperature. Hence we can replace Equation 6 with

$$w \propto \frac{1}{\eta} \exp(-E/kT) \tag{7}$$

where w is inversely proportional to the time scale of the correlation function decay and E represents an average activation energy for the local segmental motions being observed. If we let $\tau$ represent the time required for the correlation function to reach 1/e of its original value, then a plot of $\ln(\tau/\eta)$ versus 1/T will yield E. Figure 6 displays this plot, with the slope giving E = 7.4 = 0.7 kJ/mole. (Of course, other procedures for picking a characteristic $\tau$ would yield the same activation energy.) To the best of our knowledge, other measurements of the activation energy for the local segmental dynamics of polyisoprene have not been reported. Our estimate of the activation energy is in reasonable accord with the 8.3 kJ/mole torsional barrier for propene(20).

The above treatment implicitly assumes that the local segmental dynamics are not affected by changes in global chain

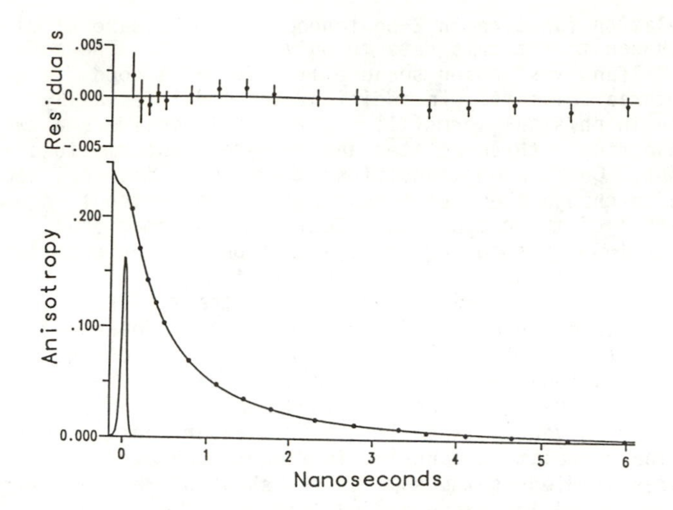


Figure 4. Time-dependent anisotropy for anthracene-labeled polyisoprene in dilute cyclohexane solution. The smooth curve through the data is the best fit to the Bendler-Yaris model($\tau_1$=210 ps, $\tau_2$=2750 ps, and r(0)=0.243).

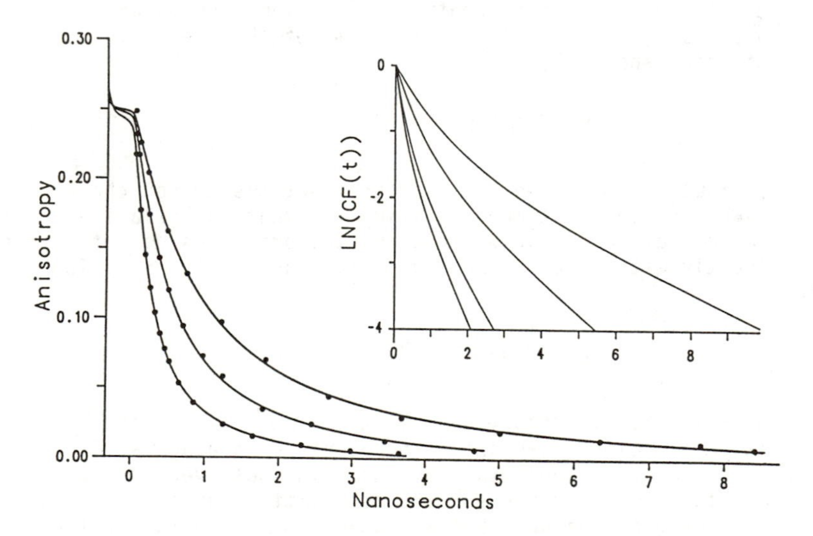


Figure 5. Time-dependent anisotropies for labeled polyisoprene chains in dilute 2-pentanone solutions. The smooth curves through the data points are the best fits to the Hall-Helfand model for 22.8 °C, -8.6 °C, and -26.5 °C (bottom to top). The data at 35.1 °C is omitted for clarity. Semilog plots of the best fit correlation functions are shown in the inset. Note that all the correlation functions are quite non-exponential.

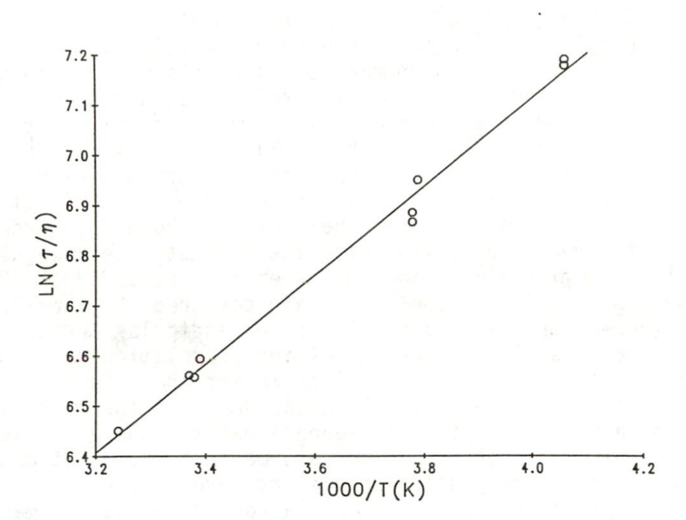


Figure 6. Arrhenius plot for dilute solutions of labeled polyisoprene in 2-pentanone. The activation energy calculated from the slope of the best fit line is 7.4 kJ/mole. On the vertical scale, $\tau$ represents the 1/e point of the best fit correlation functions using the Hall-Helfand model. The data points represent results of independent experiments. The units for $\tau$ and $\eta$ are ps and centipoise, respectively.

dimensions that may result from temperature changes. This assumption is expected to be valid for the chains used in this study because of their low molecular weight. Experiments currently in progress are examining the molecular weight dependence of the local segmental dynamics of polyisoprene.

## Comparison to Previous Work

Hatada, et al. have reported C-13 NMR experiments on a dilute solution of polyisoprene in methylene chloride(21). It is difficult to make a rigorous comparison with these measurements for two reasons: 1) The NMR measurements were made at a single frequency and temperature. Hence no information about the shape of the correlation function is available. 2) The NMR experiment senses reorientation of a C-H vector which is perpendicular to the chain backbone. Our optical experiments measure the correlation function for a vector parallel to the chain backbone. In order to make at least a rough comparison with the NMR data, we make use of the Brownian dynamics simulations of Weber and Helfand(22). These simulations for a polyethylene-like chain compared the correlation functions for vectors both parallel and perpendicular to the chain backbone. Their results indicate that the correlation function for the perpendicular bond decays ~ 4 times faster than the correlation function for the parallel bond. If we assume that the correlation function for a C-H vector in polyisoprene has the shape obtained in the computer simulation, and further assume that the relationship between parallel and perpendicular vectors revealed by the simulations is valid for polyisoprene, a comparison can be made between the optical and NMR experiments.

Following the above procedure, we used the NMR data of Hatada, et al. to estimate the correlation function for a bond in the chain backbone. The resulting estimate was ~ 2.5 times faster than results obtained in the optical experiments. The difference represents either the crudeness of the method used for comparison, or the perturbation of the chain dynamics caused by the presence of the label in the optical experiments. In the future, we will perform measurements with both techniques on systems where the two sets of results can be more easily correlated.

Viovy, Monnerie, and Brochon have performed fluorescence anisotropy decay measurements on the nanosecond time scale on dilute solutions of anthracene-labeled polystyrene(4). In contrast to our results on labeled polyisoprene, Viovy, et al. reported that their Generalized Diffusion and Loss model (see Table I) fit their results better than the Hall-Helfand or Bendler-Yaris models. This conclusion is similar to that recently reached by Sasaki, Yamamoto, and Nishijima(3) after performing fluorescence measurements on anthracene-labeled poly(methyl methacrylate). These differences in the observed correlation function shapes could be taken either to reflect the non-universal character of local motions, or to indicate a significant difference between chains of moderate flexibility and high flexibility. Further investigations will shed light on this point.

In spite of the difference in the shape of the observed correlation functions, it is interesting to compare the timescale of the decays in the polystyrene and polyisoprene systems. At the same viscosity, our results for labeled polyisoprene indicate motions occuring ~ 9.5 times faster than for the labeled polystyrene investigated by Viovy, et al. This is in very good agreement with results from C-13 NMR(21,23) which under comparable conditions give a ratio of $T_1$(polyisoprene)/$T_1$(polystyrene) of ~ 10. This comparison indicates that the hydrodynamic interaction of the label with the solvent does not dominate the observed dynamics.

## Concluding Remarks

In this paper, we have shown the utility of time-resolved optical techniques for the investigation of local segmental motions in polymer chains on a sub-nanosecond time scale. Detailed information about chain motions is contained in the time dependence of the orientation autocorrelation function of a backbone bond. The constant shape of the correlation function in various solvents at different temperatures implies that the same mechanisms are involved in local motions under all conditions investigated. In terms of the Hall-Helfand model, the ratio of correlated to uncorrelated transitions is constant. Analysis of the temperature dependence of the labeled polyisoprene yields an activation energy of 7.4 kJ/mole for local segmental motions.

In experiments currently in progress, the techniques used in this paper are being applied to the observation of local polymer dynamics in concentrated solutions and in the bulk. Measurements can be made on time scales as long as several triplet lifetimes (~ 100 msec) because the transient grating technique utilizes absorption and not fluorescence. This long time window will allow the investigation of local motions in the bulk as a function of temperature from the rubbery state to the glass transition.

## Acknowledgments

This work was supported in part by the Graduate School of the University of Wisconsin, the Shell Foundation (Faculty Career Initiation Grant), Research Corporation, Dow Chemical USA, and the National Science Foundation (DMR-8513271). In addition, acknowledgment is made to the donors of the Petroleum Research Fund, administered by the American Chemical Society, for partial support of this research. We thank Mr. H. Tominaga and Mr. N. Ota at Toyohashi University of Technology for their help in the preparation and characterization of the polymer.

## Literature Cited

1. Ricka, J.; Amsler, K.; Binkert, Th. Biopolymers 1983, 22, 1301.
2. Phillips, D. Polymer Photophysics: Luminescence, Energy Migration, and Molecular Motion in Synthetic Polymers; Chapman and Hall: London, 1985.
3. Sasaki, T.; Yamamoto, M.; Nishijima, Y. Makromol. Chem., Rapid Commun. 1986, 7, 345.

4. Viovy, J.L.; Monnerie, L.; Brochon, J.C. Macromolecules 1983, 16, 1845.
5. Viovy, J.L.; Frank, C.W.; Monnerie, L. Macromolecules 1985, 18, 2606; Viovy, J.L.; Monnerie, L.; Merola, F. Macromolecules 1985, 18, 1130.
6. Viovy, J.L. Monnerie, L. Polymer 1986, 27, 181.
7. Hyde, P.D.; Waldow, D.A.; Ediger, M.D.; Kitano, T.; Ito, K. Macromolecules 1986, 19, 2533.
8. Biddle, D. Nordstrom, T. Arkiv Kemi 1970, 32, 359.
9. Valeur, B. Monnerie, L. J. Polym. Sci. 1976, 14, 11.
10. Moog, R.S.; Ediger, M.D.; Boxer, S.G.; Fayer, M.D. J. Phys. Chem. 1982, 86, 4694.
11. Hall, C.K. Helfand, E. J. Chem. Phys. 1982, 77, 3275.
12. Bendler, J.T. Yaris, R. Macromolecules 1978, 11, 650.
13. Ediger, M.D.; Domingue, R.P.; Fayer, M.D. J. Chem. Phys. 1984, 80, 1246.
14. Fayer, M.D. Ann. Rev. Phys. Chem. 1982, 33, 63.
15. Nelson, K.A.; Miller, R.J.D.; Lutz, D.R.; Fayer, M.D. J. Appl. Phys. 1982, 53, 1144.
16. Nemoto,N.; Landry, M.R.; Noh, I.; Yu, H. Polym. Comm. 1984, 15, 141.
17. International Critical Tables; McGraw-Hill: New York, 1930, VII, 211.
18. Kramers, H.A. Physica 1940, 7, 284.
19. Helfand, E. J. Chem. Phys. 1971, 54, 4651.
20. Flory, P.J. Statistical Mechanics of Chain Molecules; Wiley Interscience: New York, 1969; p.52.
21. Hatada, K.; Kitayama, T.; Terawaki, Y.; Tanaka,Y.; Sato,H. Polym. Bull. 1980, 2, 791.
22. Weber, T.A. Helfand, E. J. Phys. Chem. 1983, 87, 2881.
23. Inoue, Y. Konno, T. Polymer J. 1976, 8, 457.

RECEIVED April 27, 1987

# Chapter 8

# Phosphorescence Decay and Dynamics in Polymer Solids

**Kazuyuki Horie**

**Institute of Interdisciplinary Research, Faculty of Engineering, University of Tokyo, 4-6-1, Komaba, Meguro-ku, Tokyo 153, Japan**

Non-exponential phosphorescence decay is frequently observed for various aromatic chromophores molecularly dispersed in polymer matrices. Various possible mechanisms for non-exponential decay are reviewed, and a dynamic quenching mechanism by polymer matrices including the effect of a time-dependent transient term in the rate coefficient is discussed in some detail. The biphotonic triplet-triplet annihilation mechanism is also introduced for the non-exponential decay under high-intensity and/or repeated laser irradiation.

The influence of molecular structure and motion of polymer matrices on the photophysical and photochemical processes of molecularly dispersed chromophores is a topic of increasing interest. There are several reasons for this activity. Polymer matrices have been considered as convenient media for spectroscopic investigations of excited triplet states over wide temperature ranges, and conversely, the use of photophysically detectable probes allows the investigation of the structure of polymer matrices and of photophysical transitions connected with changes in the mobility of certain structural units. Another reason for such studies is connected with the practical interests concerning the reactivity of low-molecular-weight compounds embedded in polymer matrices in photomemory and photosensitive polymer systems, and of the reactivity of additives admixed to polymers as stabilizers against photodegradation and thermal degradation.

The main difference between solid-state reactions and those in solution is that of freedom of molecular motion (1-3) due to restriction of mobility of reactants in solids. Another important feature is the heterogeneous progress of reactions (3,4) frequently observed in solid states due to the microscopically heterogeneous states of aggregation or free volume distribution of the reaction media. In the case of poly(methyl methacrylate) (PMMA), which is an organic glass and is usually regarded as an inert matrix for photophysical and photochemical processes, a marked deviation from

0097-6156/87/0358-0083$06.00/0

the exponential decay of benzophenone phosphorescence has been observed (5,6). Arrhenius plots of phosphorescence intensity (7), lifetime and depolarization (5,6,8,9) of various chromophores in polymer matrices showed breaks at temperatures corresponding to the glass transition ($T_g$) and subglass transitions ($T_{\alpha'}$, $T_\beta$, $T_\gamma$) of the matrix polymers.

In the present paper, we discuss mainly the non-exponential decay of phosphorescence and its origin in polymer matrices. The effect of multi-photon processes on the decay curves is also discussed.

Deviations of Phosphorescence Decay from Exponentiality in Polymer Solids

The general photophysical behavior of excited triplet states is in various textbooks (10,11). The main processes for a chromophore, A, are the following:

| | | | |
|---|---|---|---|
| $^3A^* \longrightarrow A + h$ | | phosphorescence ($k_{PT}$) | (1) |
| $^3A^* \longrightarrow A$ | | nonradiative deactivation ($k_{IT}$) | (2) |
| $^3A^* + B \longrightarrow A + {}^3B^*$ | | triplet energy transfer | (3) |
| $^3A^* + A \longrightarrow A + {}^3A^*$ | | triplet energy migration | (4) |
| $^3A^* + {}^3A^* \longrightarrow {}^1A^* + A$ | | triplet-triplet annihilation | (5) |

Triplet-triplet energy transfer was first clearly demonstrated by Terenin and Ermolaev (12,13) who showed the phosphorescence of naphthalene (acceptor, B) resultant upon excitation of benzophenone (donor, A) followed by triplet energy transfer from A to B in rigid solution at 77 K. Triplet energy transfer requires molecular orbital overlap between the donor and acceptor, and the transfer efficiency depends on the energy gap between the energy levels of the excited triplet states, $E_T$, of the donor and acceptor. Such energy transfer due to the electron exchange interaction was theorized by Dexter (14), after whom the mechanism is named.

Stern-Volmer Model. When the chromophores are sufficiently mobile as in fluid solution, the bimolecular quenching process of A by a quenching molecule, B, including the triplet energy transfer and some collisional quenching, will result in the single exponential

$$^3A^* + B \longrightarrow A + B \qquad \text{triplet quenching } (k_q) \qquad (6)$$

phosphorescence decay of the donor chromophore, A. Thus, the phosphorescence decay profile, I(t), is expressed by eq (7)

$$I(t) = C\exp(-t/\tau) = C\exp[-t(1/\tau_o + k_q[B])] \qquad (7)$$

where C is a constant, and $\tau$ is a single lifetime, which is related to the phosphorescence lifetime in the absence of quencher, $\tau_o = 1/(k_{PT} + k_{IT})$, the quenching rate constant, $k_q$, and the quencher concentration, [B]. The relative phosphorescence yield, $I/I_o$, defined as the ratio of the yields in the presence of acceptor to that in its absence, has a so-called Stern-Volmer concentration dependence (15).

$$I/I_o = 1/(1 + k_q \tau_o [B]) \qquad (8)$$

Perrin Model. At the other extreme, Perrin (16) considered the case where the donor and acceptor molecules are immobile, and energy transfer occurs instantaneously when the two molecules lie within a critical transfer distance, $R_o$, and does not occur at all at large intermolecular separations.

The decay function for donor phosphorescence is given by

$$\begin{aligned} I(t) &= 1 \qquad (t=0) \\ I(t) &= \exp\left[-(t/\tau_o) - (C_B/C_o)\right] \qquad (t>0) \end{aligned} \qquad (9)$$

where $C_B$ is the acceptor concentration, $C_o$ is a parameter called the critical transfer concentration (which is defined as the reciprocal of the spherical volume of the radius $R_o$). The relative yield of donor phosphorescence is given by eq (10) for this case.

$$I/I_o = \exp(-C_B/C_o) \qquad (10)$$

Inokuti-Hirayama Model. The Perrin model is too simplified, although it is convenient for practical use. The static triplet-triplet energy transfer between immobile chromophores dispersed in solids can be well described by the Inokuti-Hirayama theory (17).

Based on the Dexter mechanism, with a distance-dependent rate coefficient for triplet energy transfer, a non-exponential decay function for donor phosphorescence in a rigid solution was derived as

$$I(t) = \exp[-(t/\tau_o) - \gamma^{-3}(C_B/C_o)G(e^{\gamma}t/\tau_o)] \qquad (11)$$

where $\gamma$ is related to Dexter's quantities by

$$\gamma = 2R_o/L \qquad (12)$$

$$e^{\gamma}/\tau_o = 2\pi k^2/h \int F_A(E)\varepsilon_{\beta}(E)dE \qquad (13)$$

$$G(z) = (\ln z)^3 + 1.732(\ln z)^2 + 5.934(\ln z) + 5.445 \qquad (14)$$

Mataga et al. (18) studied energy transfer from the excited triplet of benzophenone to naphthalene by laser flash photolysis at 77K and showed that the non-exponential decay curves of the benzophenone triplet obey the Inokuti-Hirayama equation (eq (11)). Inokuti and Hirayama (17) themselves compared the data on triplet-triplet transfer between certain aromatic molecules obtained by Terenin and Ermolaev with eq (11) and reported that a good fit was found with an appropriate choice of the parameters $C_o$ and $\gamma$.

Thus, it is thought that the non-exponential phosphorescence decay of donor in the presence of an acceptor in an immobilized system is well explained by the Inokuti-Hirayama model based on the static triplet-triplet energy transfer mechanism. This model expects a single-exponential decay for the phosphorescence of a chromophore in the absence of acceptor molecules. However, the phosphorescence decays of organic molecules molecularly dispersed in polymer matrices are known to be non-exponential in some cases even in the absence of other additives. Consequently, other reasons should be considered for such deviations from exponentiality.

The deviation from a single exponential curve for the phosphorescence decay of some chromophores dissolved in plastics was first noted by Oster et al.(19) Nonexponential decay curves were obtained for naphthalene and triphenylene phosphorescence in poly(methyl methacrylate) (PMMA) at room temperature (20), while exponential decays at room temperature were observed in PMMA for anthracene triplet (21), pyrene (22), and coronene phosphorescence (20). Graves et al. (23) analyzed the temperature dependence of phosphorescence parameters for a number of aromatic hydrocarbons in PMMA from 77 to 400K and suggested the existence of intermolecular thermally assisted energy transfer from the chromophore to the host plastic in the higher temperature region.

El-Sayed et al. (20) interpreted the non-exponential decay profiles of naphthalene and triphenylene in PMMA at room temperature in terms of a triplet-triplet annihilation mechanism. The decrease of the extent of non-exponential behavior with decreasing excitation intensity and the observation of delayed fluorescence were the bases of the interpretation. They suggested that the non-exponential decays were observed for molecules having first-order lifetimes of the order of several seconds. However, coronene with a triplet lifetime of 8.5s gave an exponential decay. They also suggested rather high concentrations of the excited triplet state of the chromophore based on the diffusion-controlled triplet-triplet annihilation mechanism. Later, Jassim et al. (24) proposed another type of triplet-triplet annihilation mechanism for the non-exponential phosphorescence decay consisting of energy transfer from the chromophore to the matrix polymer, triplet energy migration through the matrix polymer, and triplet-triplet annihilation between the chromophore triplet and the polymer triplet.

Horie and Mita (5) measured the phosphorescence decay of benzophenone in PMMA over a wide temperature range (80 to 433K). Non-exponential decays were observed for temperatures between $T_\beta$ (onset of ester side group rotation of the matrix polymer) and $T_g$ (glass transition temperature), and the decay profile was independent of the intensity of the excitation laser pulse over a 100 times change of the intensity(25). Thus, the non-exponential decay was attributed to a single photon process comprised of intermolecular dynamic quenching of the benzophenone triplet by ester groups in the side chain of PMMA. Detailed discussion of the dynamic quenching mechanism will be given in the next section.

It is noteworthy that the triplet decay curves at room temperature of various chromophores in PMMA observed in the literature can be divided approximately into two groups according to the $E_T$ of the chromophores as is shown in Table I. Non-single-exponential phosphorescence decay has been observed for chromophores with an $E_T$ smaller than that of PMMA (297 to 301 kJ) (23,24), while single-exponential phosphorescence or T-T absorption decay is obtained for chromophores with lower triplet energy levels. These results also support the occurrence of dynamic quenching for excited-triplet-state chromophores by matrix PMMA due to an endothermic triplet energy transfer mechanism.

Recently, Richert and Baessler(37) regarded the non-exponential decay of benzophenone as a dispersive triplet transport phenomenon to trap sites, and approximated it by a stretched exponential fit ($\ln I(t) + t/\tau_o = -C(t/t_o)^{\alpha}$) with a time dependent dispersion parameter $\alpha$.

## Phosphorescence Decay of Benzophenone in Polymer Solids

Typical decay curves of benzophenone (BP) phosphorescence (analyzed at 450 nm) at various temperatures in PMMA excited by a 10-ns nitrogen laser pulse at 337 nm are shown in Fig. 1 (6). The phosphorescence intensity, I(t), decreases as a single exponential below the temperature corresponding to the ester side-group rotation ($T_{\beta}$ = -30°C for PMMA). Deviations from a single-exponential decay are observed for $T>T_{\beta}$, which increase with increasing temperature, but the deviation becomes less marked above the glass transition temperature, $T_g$, of the matrix polymer and disappears at 150°C. A similar temperature dependence of the decay profile was also observed for benzophenone phosphorescence in other acrylic polymers (28) and in polystyrene (PS) and polycarbonate (PC) (29).

Transient spectra showed that the whole course of the decay curve is due to benzophenone triplet (6), and the irradiation intensity independence shown in Fig. 2 excluded the occurrence of T-T annihilation under the present experimental conditions (25). Since the deviation is not observed at temperatures below $T_{\beta}$ of each acrylic polymer or $T_{\gamma}$ of polystyrene and polycarbonate, this quenching is not suggested to be of a static character like the Inokuti-Hirayama type mentioned in the previous section. Instead, it is suggested to be of a dynamic character due to collision between the functional groups. The occurrence of quenching was ascertained by comparison with the model quenching rate constant of benzophenone triplet by methyl acetate in acetonitrile solution ($3.9\times10^{3}\ M^{-1}s^{-1}$ at 30°C) (28). Quenching of benzophenone triplet by phenyl or phenylene groups in polystyrene or polycarbonate has also been studied ($1.2\times10^{6}\ M^{-1}s^{-1}$ for polystyrene in benzene at 30°C) (30). These orders of magnitude for the quenching rate constants in nonviscous solution are reasonable for an uphill-type endothermic triplet energy transfer mechanism.

As the dynamic quenching for a phosphorophore gives single exponential decay in nonviscous solution (Stern-Volmer model), it is necessary to consider why the dynamic quenching in polymer solids especially below $T_g$ results in a non-exponential decay profile.

## Kinetics for Non-exponential Decay Due to Dynamic Quenching (6,29)

The decay process of triplet benzophenone, $^3BP$, is given by the following:

Table I. Type of Triplet Decay Curves and Triplet Energies, $E_T$, for Various Chromophores Dispersed in PMMA at Room Temperature

| chromophore | type of triplet decay | $E_T$, (27) kJ/mol | ref |
|---|---|---|---|
| fluorene | non-single-exponential | 284 | 20 |
| benzophenone | non-single-exponential | 283 | 5,6 |
| triphenylene | non-single-exponential | 278 | 20,24 |
| phenanthrene | single-exponential | 260 | 24 |
| naphthalene | non-single-exponential | 255 | 20 |
| | single-exponential | | 24 |
| coronene | single-exponential | 228 | 20 |
| benzil | single-exponential | 227 | 26 |
| pyrene | single-exponential | 202 | 22 |
| anthracene | single-exponential | 179 | 21 |

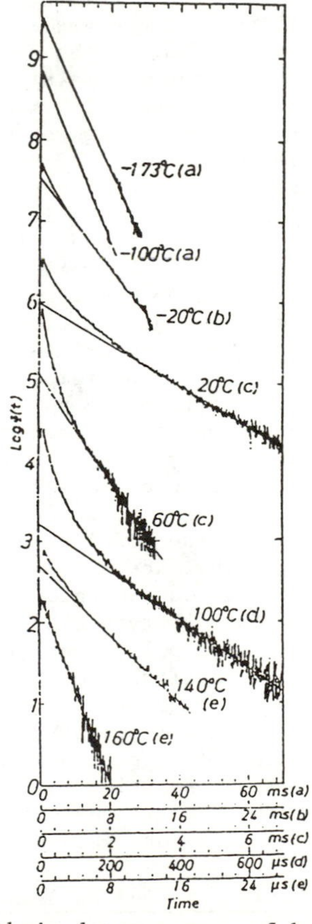


Figure 1. Semilogarithmic decay curves of benzophenone phosphorescence in PMMA excited by 10-ns nitrogen laser pulse at 337 nm. Temperature and symbols for time scales are given beside the curves. (Reproduced from Reference 6. Copyright 1984 American Chemical Society.

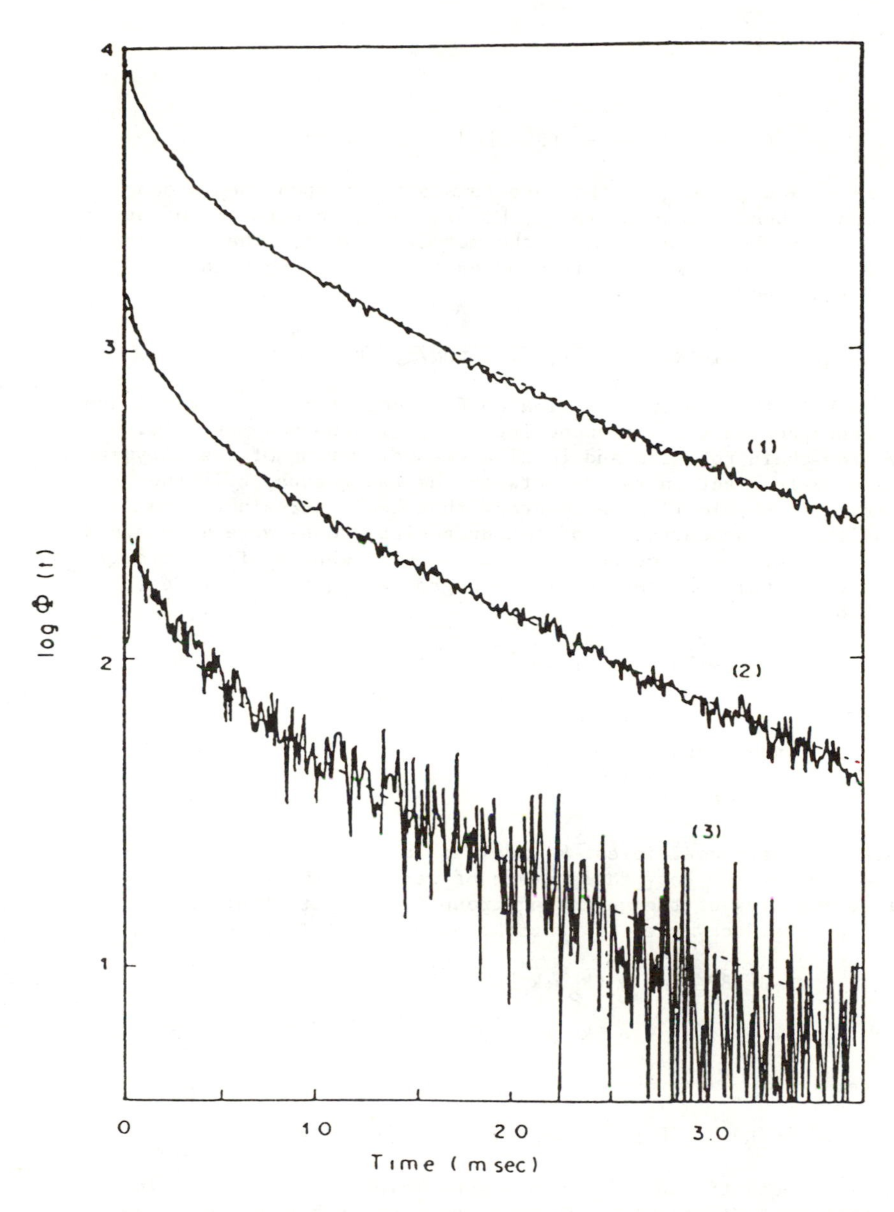


Figure 2. Phosphorescence decay profiles of BP in PMMA at 20°C. Excitation intensity: (1) $4.0 \times 10^{15}$, (2) $6.4 \times 10^{13}$, (3) $2.8 \times 10^{12}$ photon/pulse. (Reproduced with permission from Reference 25. Copyright 1984, Pergamon.)

$$^3BP^* \xrightarrow{k_o} BP \qquad (15)$$

$$^3BP^* + [Q] \xrightarrow{k_q} BP + [Q] \qquad (16)$$

where $k_o = k_{PT} + k_{IP}$ is the rate constant for spontaneous deactivation of benzophenone triplet, [Q] is the concentration of ester, phenyl, or phenylene group in the matrix polymer. The bimolecular rate coefficient, $k_q$, is given by eq (17) including both the diffusion and chemical steps (30):

$$k_q = \frac{4\pi RDN}{1 + 4\pi RDN/k}\left[ 1 + \frac{R}{(1 + 4\pi RDN/k)(\pi DT)^{\frac{1}{2}}} \right] \qquad (17)$$

where D is the sum of diffusion coefficients for the carbonyl groups in benzophenone and for quenching groups in the polymer, limited by side-chain rotation and local segmental motion of the polymer chain, R is reaction radius between the two groups, k is the intrinsic (chemical) rate constant that would pertain if the equilibrium concentration of the quenching groups were maintained, and N is Avogadro's number divided by $10^3$. When $k_q$ is controlled by the diffusion process of the two groups, eq (17) is reduced to eq (18):

$$k_q = 4\pi RDN(1 + R/(\pi Dt)^{\frac{1}{2}} = A + B/t^{\frac{1}{2}} \qquad (18)$$

with

$$A = 4\pi RDN$$

$$B = 4R^2(\pi D)^{\frac{1}{2}}N$$

Thus, the rate coefficient $k_q$ includes a time-dependent term that is important at the very early stage of reaction where the steady-state diffusive flux of the quenching group is not yet attained.

As the decay rate of benzophenone triplet is given by eq (19)

$$-d[^3BP^*]/dt = (k_o + k_q[Q])[^3BP^*]$$
$$= (k_o + A[Q] + B[Q]t^{-\frac{1}{2}})]^3BP^*] \qquad (19)$$

we get

$$[^3BP^*] = [^3BP^*]_o \exp[-(k_o + A[Q])t - 2B[Q]t^{\frac{1}{2}}] \qquad (20)$$

for the concentration of benzophenone triplet, $[^3BP^*]$, at time t, where $[^3BP^*]_o$ is the initial concentration of benzophenone triplet. The phosphorescence intensity, I(t), is proportional to $k_{PT}[^3BP^*]$, so we get finally

$$\ln I(t) = -(k_o + A[Q])t - 2B[Q]t^{\frac{1}{2}} + \ln I_o$$
$$= -(t/\tau) - C(t/\tau)^{\frac{1}{2}} + \ln I_o \qquad (21)$$

where

$$1/\tau = k_o + A[Q] = k_o + 4\ RDN[Q] \qquad (22)$$

$$B = C\tau^{-\frac{1}{2}}/(2[Q]) = 4R^2(\pi D)^{\frac{1}{2}}N \qquad (23)$$

The curve fitting for the phosphorescence decay curves with eq (21) gives the values of reciprocal lifetime $1/\tau = k_o + A[Q]$ and a parameter B, which are shown in Fig. 3 for the cases of PMMA and polystyrene. The breaks reflect the changes in the molecular motions at $T_g$, $T_{\alpha'}$ (PMMA) or $T_\beta$ (PS, PC) corresponding to the local mode relaxation of the main chain, and $T_\beta$ (PMMA) or $T_\gamma$ (PS, PC) corresponding to the onset of rotation of the side-chain ester or phenyl group or the main-chain phenylene group. Arrhenius plot of $1/\tau$ for benzophenone in poly(methyl acrylate) (PMA) (6) showed another break at 40°C (above $T_g$ of PMA) which corresponds to an activation-controlled reaction. The diffusion coefficient, D, for reacting carbonyl groups calculated from the values of $1/\tau$ and B at each temperature also showed breaks at each transition temperature, as exemplified in Fig. 4 for the cases of PMMA, polystyrene and polycarbonate. It should be noted that D in Fig. 4 is defined as the translational diffusion coefficient for the reacting functional groups but not for the molecule. The diffusion process at temperatures below $T_g$ would be caused by rotation of the benzophenone molecule and segmental motion within a few monomer units of matrix polymers. Nevertheless, the values of D in these polymers at 100°C are compared to the value of $D = 5.6 \times 10^{-13}$ $cm^2/s$ (32) for mass diffusion of ethylbenzene in polystyrene at 30°C. The reaction radius, R, amounts to 3-5 Å. The transition temperatures of the matrix polymers monitored by phosphorescence decay of benzophenone including the case in poly(vinyl alcohol) are summarized in Table II. The $\alpha$ transition for PMMA and other acrylic polymers and $\beta$ transition for polystyrene, polycarbonate, and poly(vinyl alcohol) are attributed to a local mode relaxation of the main chain. The phosphorescence probe technique is effective for the detection of this sub-glass transition as well as other rotation mode transitions ($T_\beta$ or $T_\gamma$) of polymer matrices.

## Hydrogen Abstraction of Benzophenone Triplet in Poly(vinyl alcohol)

The phosphorescence of BP (0.1%) in poly(vinyl alcohol) (PVA) film (250 micron thickness) excited by a 10-ns nitrogen laser pulse at 337 nm decays exponentially for $T < T_\gamma$ (-100°C) or $T > T_g$ (85°C), but deviates from single exponential for $T_\gamma < T < T_g$. The deviation was attributed to the diffusion-controlled hydrogen abstraction reaction between benzophenone triplet and the PVA matrix (33). The occurrence of non-exponential decays in poly(vinyl alcohol) where there is no possibility of triplet energy migration is an additional proof of the absence of the T-T annihilation mechanism in the phosphorescence decay of benzophenone in polymer matrices.

The diffusion-controlled rate coefficient for hydrogen abstraction, $k_a$, of benzophenone triplet for PVA and the phosphorescence intensity, I(t), were given by eqs (24) and (25), in a similar manner to the cases of physical quenching of the benzophenone phosphorescence by matrix polymers.

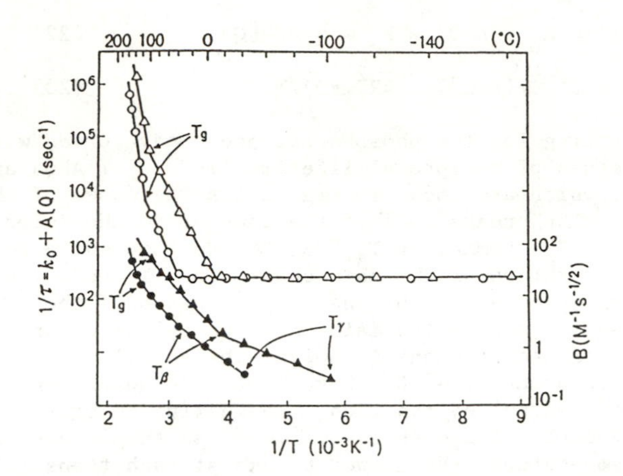


Figure 3. Temperature dependence of reciprocal lifetime, $1/\tau$, (O, Δ) and contribution of non-exponential term, B, (●,▲) for benzophenone phosphorescence in PMMA (O,●) and in polystyrene (Δ,▲).

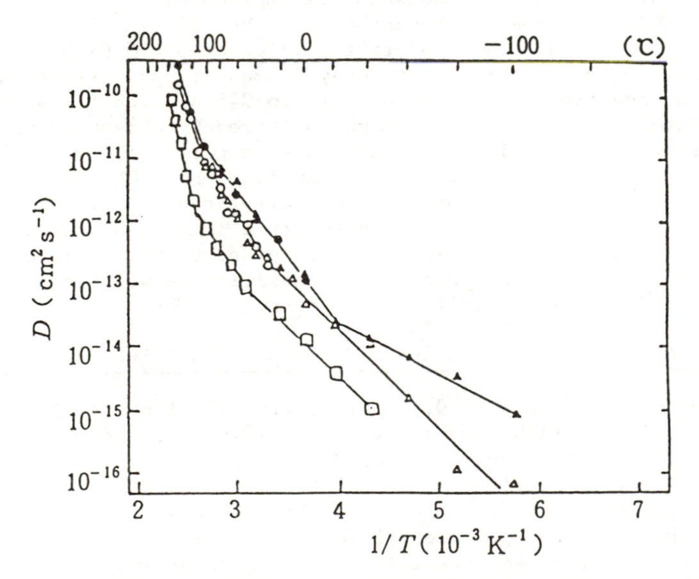


Figure 4. Arrhenius plots of diffusion coefficient, D, of interacting groups for dynamic quenching of benzophenone triplet by phenyl, phenylene, and ester groups in polystyrene (●,▲), polycarbonate (O,Δ), and PMMA (□), respectively.

$$k_a = 4\pi RDN(1 + R/(\pi Dt)^{\frac{1}{2}}) = A + B/t^{\frac{1}{2}} \quad (24)$$

$$\ln I(t) = -(k_o + A[PVA])t - 2B[PVA]t^{\frac{1}{2}} + \ln I_o \quad (25)$$

The curve fitting for the experimental phosphorescence decay with eq (25) gives the values of $1/\tau = k_o + A[PVA]$ and B. The breaks at $T_g = 85°C$ and $T_\beta = 30°C$ and the appearance of B at $T_\gamma = -100°C$ clearly reflect the change in molecular motions of the matrix poly(vinyl alcohol).

In order to know the net quantum yield for the benzophenone disappearance, $\Phi$(-BP), by hydrogen abstraction in poly(vinyl alcohol), the change in UV spectra of benzophenone at 256 nm in the poly(vinyl alcohol) film was followed during continuous irradiation of 365 nm UV light. The $\Phi$(-BP) given in Table III is very small for $T < T_g$, suggesting the predominant occurrence of backward reaction ($k^g_{back}$) of benzophenone ketyl radical in the cage for the temperature range below $T_g$. The T-T absorption spectra and lifetime of the ketyl radical were observed for $T > 120°C$ (34). The supposed reaction scheme for the photochemistry of benzophenone in poly(vinyl alcohol) including the backward cage reaction is summarized in Fig. 5.

## Phosphorescence Decay of Benzophenone under Multi-photon Conditions

Salmassi and Schnabel (35) measured the decays of phosphorescence and triplet absorption of benzophenone in PMMA and polystyrene under the high intensity irradiation of 347 nm frequency-doubled ruby laser single pulse. The initial triplet concentration amounted up to $6x10^{-4}$ mol/l in comparison with the value of less than $6x10^{-6}$ mol/l for the case (6,25) in the preceding sections with the nitrogen laser pulse. In PMMA a single triplet decay mode following first-order kinetics was observed at T < 150K and at T > 410K. Two distinct modes of triplet decay were observed in the intermediate temperature range: a fast first-order process, the lifetime (ca. 4μs at 295K) being independent of the initial triplet concentration, $[T]_o$, and a slow second-order process, the first half-lifetime being proportional to $[T]_o^{-1}$. Similarly, two distinct modes of triplet decay were also obtained with polystyrene matrices for 180K < T < 350K, and it was concluded that triplet-triplet (T-T) annihilation is an important deactivation route in these experiments at temperature ranges between $T_g$ and $T_\gamma$, where the rotation of α-methyl groups (PMMA) or phenyl groups (polystyrene) commences.

The decay rate of triplet concentration in the presence of a T-T annihilation mechanism is given by

$$-d[T]/dt = \Sigma k_1[T] + k_{TT}[T]^2 \quad (26)$$

where $\Sigma k_1$ denotes the sum of all first-order or pseudo-first-order rate constants of triplet decay, and T-T annihilation becomes dominating when $\Sigma k_1 << k_{TT}[T]$. As to the origin of two distinct modes of triplet decay, Salmassi and Schnabel (35) attributed the "fast" decaying triplets to those which were formed in close proximity. The "slowly" decaying triplets were thought to be formed at loci separated by relatively large distances and to need diffusive processes for their interactions. By using the value of $k_{TT}$ =

Table II. The transition temperature of matrix polymers monitored by benzophenone phosphorescence and quenching reaction radius, R

| | $T_\beta$(C) | $T_\alpha$,(°C) | $T_g$(°C) | R(A) |
|---|---|---|---|---|
| Poly(metyl methacrylate) | -40 | 40 | 110 | 5.0 |
| Poly(isopropyl methacrylate) | -70 | 20 | 80 | 5.3 |
| Poly(methyl acrylate) | -70 | | 10 | (3.0) |
| | $T_\gamma$(°C) | $T_\beta$(°C) | $T_g$(°C) | R(A) |
| Polystyrene | -100 | -20 | 100 | 5.8 |
| Polycarbonate | -100 | 20,100 | 150 | 5.2 |
| Poly(vinyl alcohol) | -100 | 30 | 85 | 3.5 |

Table III. Quantum yield, Φ(-BP), for benzophenone disappearance during 365 nm irradiation and the fraction of ketyl radical formation, $f_a$, from transient measurements (33)

| Temperature (°C) | $I_0$ (einstein/cm$^2$·s) | Φ(-BP) | $f_a$ |
|---|---|---|---|
| 0 | $2.6\times10^{-8}$ | 0.018 | 0.85 |
| 30 | $1.9\times10^{-8}$ | 0.042 | 0.95 |
| | $1.4\times10^{-8}$ | 0.056 | |
| | $8.1\times10^{-9}$ | 0.049 | |
| | $8.0\times10^{-9}$ | 0.052 | |
| 60 | $2.6\times10^{-8}$ | 0.075 | 0.99 |
| 100 | $1.9\times10^{-8}$ | 0.32 | 1.0 |
| 140 | $1.9\times10^{-8}$ | 0.49 | 1.0 |

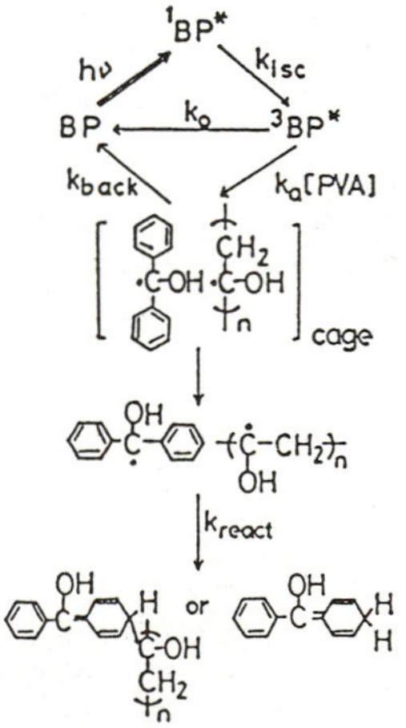

Figure 5. Reaction scheme of photo-excited benzophenone in poly(vinyl alcohol).

$5x10^7$ $M^{-1}s^{-1}$ at 295K in PMMA and the Smoluchowski equation for the diffusion controlled T-T annihilation rate constant, $k_{TT}$, the value of D = $1x10^{-8}$ $cm^2/s$ was obtained for the diffusion coefficient of benzophenone in PMMA. This value is of the same order of magnitude as for D of oxygen in PMMA, but is very large compared to D = $5.6x10^{-13}$ $cm^2/s$ (32) of ethylbenzene in polystyrene at 100°C. Consequently, triplet energy migration through the matrix polymer resulting in a T-T annihilation between the chromophore triplet and polymer triplet (24) might be a necessary process for the occurrence of T-T annihilation in polymer matrices below $T_g$.

Salmassi and Schnabel (35) also noticed the hydrogen abstraction reaction by benzophenone triplet from PMMA and polystyrene, which became more evident at elevated temperatures. The quantum yield for ketyl radical formation for benzophenone in PMMA at 430K was estimated to be Φ(BPH) = 0.06, which means that only about 12% of the benzophenone triplets were converted to ketyl radicals. There is a possibility that hydrogen abstraction by benzophenone triplet from PMMA proceeds mainly through a two-photon process (6). The fraction of ketyl radical formation was about 20% in polystyrene matrix.

The non-exponential decay of benzophenone phosphorescence in PMMA at room temperature was also observed by Fraser et al. (36), under the condition of repeated irradiation of a nitrogen laser pulse, and they proposed a triplet-triplet annihilation mechanism in which the polymer matrix itself participated as an energy acceptor from benzophenone triplet and a medium for the triplet energy migration.

## Conclusion

Various possible mechanisms for the non-exponential phosphorescence decays of aromatic chromophores in rigid glasses and polymer solids have been reviewed. In the presence of effective triplet energy acceptor, static Dexter type energy transfer to acceptor molecule results in the Inokuti-Hirayama type non-exponential triplet decay (eq (11)). The non-exponential decay of phosphorophores in the absence of other additives in polymer solids can be attributed to a dynamic quenching by the polymer matrices with the diffusion controlled rate coefficient including a time-dependent transient term (eq (21)). Under high-intensity and/or repeated laser pulse irradiation, non-exponential phosphorescence decay (due to a biphotonic triplet-triplet annihilation mechanism which probably involves triplet energy migration through the polymer matrix) was also observed.

## Literature Cited


1. Guillet, J. E. Pure Appl. Chem. 1977, 49, 249.
2. Williams, J. L. R.; Daly, R. C. Prog. Polym. Sci. 1977, 5, 61.
3. Horie, K.; Mita, I. Kobunshi 1985, 34, 443.
4. Smets, G. Adv. Polym. Sci. 1983, 50, 17.
5. Horie, K.; Mita, I. Chem. Phys. Lett. 1982, 93, 61.
6. Horie, K.; Morishita, K.; Mita, I. Macromolecules 1984, 17, 1746.
7. Somersall, A. C.; Dan, E.; Guillet, J. E. Macromolecules 1974, 7, 233.

8. Rutherford, H.; Soutar, I. J. Polym. Sci., Polym. Phys. Ed. 1977, 15, 2213.
9. Rutherford, H.; Soutar, I. J. Polym. Sci., Polym. Phys. Ed. 1980, 18, 1021.
10. Birks, J. B. Photophysics of Aromatic Molecules; Wiley-Interscience, New York, 1970.
11. Turro, N. J. Modern Molecular Photochemistry; Benjamin, Menlo Park, 1978.
12. Terenin, A. N.; Ermolaev, V. L. Dokl. Akad. Nauk. SSSR 1952, 85, 547.
13. Terenin, A. N.; Ermolaev, V. L. Trans. Faraday Soc. 1956, 52, 1042.
14. Dexter, D. L. J. Chem. Phys. 1953, 21, 836.
15. Stern, O.; Volmer, M. Physik. Z. 1919, 20, 83.
16. Perrin, F. Compt. Rend. 1924, 178, 1978.
17. Inokuti, M.; Hirayama, F. J. Chem. Phys. 1965, 43, 1978.
18. Kobashi, H.; Morita, T.; Mataga, N. Chem. Phys. Lett. 1973, 20, 376.
19. Oster, G.; Geacintov, N.; Khan, A. U. Nature 1962, 196, 1089.
20. El-Sayed, F. E.; MacCallum, J. R.; Pomery, P. J.; Shepherd, T. J. Chem. Soc. Faraday Trans. II 1979, 75, 79.
21. Melhuish, W. H.; Hardwick, R. Trans. Faraday Soc. 1962, 58, 1908.
22. Jones, P. F.; Siegel, S. J. Chem. Phys. 1969, 50, 1134.
23. Graves, W. E.; Hofeldt, R. H.; McGlynn, S. P. J. Chem. Phys. 1972, 56, 1309.
24. Jassim, A. N.; MacCallum, J.R.; Moran, K. T. Eur. Polym. J. 1983, 19, 909.
25. Horie, K.; Mita, I.Eur. Polym. J. 1984, 20, 1037.
26. Horie, K.; Tsukamoto, M.; Mita, I. Prepr. 1st SPSJ IPC 1984, 96.
27. Murov, S. L. Handbook of Photochemistry; Marcel Dekker, New York, 1973, p. 27.
28. Horie, K.; Morishita, K.; Mita, I. Kobunshi Ronbunshi 1983, 1983, 40, 217.
29. Horie, K.; Tsukamoto, M.; Morishita, K.; Mita, I. Polym. J. 1985, 17, 517.
30. Mita, I.; Takagi, T.; Horie, K.; Shindo, Y. Macromolecules 1984,17, 2256.
31. Noyes, R. M. Prog. React. Kinet. 1961, 1, 129.
32. Vrentas, J. S.; Duda, J. L. J. Polym. Sci., Polym. Phys. Ed. 1977, 15, 417.
33. Horie, K.; Ando, H.; Morishita, K.; Mita, I. J. Photograph. Soc. Jpn. 1984, 47, 345.
34. Horie, K.; Ando, H.;Mita, I. Polym. Prepr. Jpn. 1984, 33, 1471.
35. Salmassi, A.; Schnabel, W. Polym.Photochem. 1984, 5, 215.
36. Fraser, I. M.; MacCallum, J. R.; Moran, K.T. Eur. Polym. J. 1984,20, 425.
37. Richert, R.; Baessler, H. J. Chem. Phys. 1986, 84, 3567.

RECEIVED May 22, 1987

Chapter 9

# Fluorescence Probes for the Study of Solvation and Diffusion of Reagents in Network Polymers

K. J. Shea, G. J. Stoddard, and D. Y. Sasaki

Department of Chemistry, University of California, Irvine, CA 92717

A dansyl monomer, 1, prepared by the condensation of p-vinyl benzyl amine with dansyl chloride has been used as a fluorescent marker to probe the microenvironment of styrene-divinylbenzene networks. The probe readily reveals the degree of solvation of polymer chains which is found to correlate with the degree of crosslinking. Subtle yet important differences in polymer morphology are also uncovered by this method. The fluorescence emission intensity of polymer bound probe 1 is found to be quenched upon treatment with strong electrophiles ($Ph_3C^+BF_4^-$). Monitoring the dimunition in fluorescence emission intensity permits study of the rate of diffusion of electrophilic reagents through styrene-divinylbenzene networks.

Macroporous styrene-divinyl benzene (S-DVB) copolymers are widely used as supports for chemical reactions (1). The surface area, pore volume, and pore size of these materials can be manipulated by a judicious choice of reaction conditions (2). It is recognized that reaction cosolvent and the ratio of monomer to cosolvent are important variables and considerable speculation has been offered regarding the relationship between polymerization conditions and polymer morphology (3). On the basis of these studies a model has emerged to account for macroporosity in these materials (4). The continuous or gel phase is found to consist of aggregated microspheres. The macropores are defined by voids created by these aggregated microspheres.

The gel or continuous phase of these materials is produced by phase separation that occurs during polymerization. Properties of the gel phase are influenced by (a) the degree of crosslinking, (b) cosolvent and (c) the monomer-cosolvent ratio. The last two factors will affect the phase separation and thus the dimensions of the continuous phase as well as the degree of solvation of the polymer chains during phase separation.

The objective of the present study is to develop a diagnostic that will enable us to evaluate solvation and chemical transport in

0097-6156/87/0358-0097$06.00/0

the continuous or gel phase of highly crosslinked macroporous materials. In previous studies (5) we had noted that polymerization conditions, cosolvent, and crosslinking monomer can exercise a dramatic effect on the chemical reactions of these supports. We undertook, therefore, a systematic study of these variables using a series of macromolecules that contain an increasing degree of crosslinking.

Fluorescence spectroscopy was utilized as the diagnostic to evaluate these phenomena. It will be shown that a single probe molecule, a derivative of dimethylaminonaphthalenesulfonamide, (dansyl, 1) can function both as an environmental probe to evaluate the degree of solvation of polymer chains in the gel phase and also serve as a sensitive indicator for the diffusion of ionic reagents through the crosslinked gel network.

## Results and Discussion

Preparation of Materials. Dansyl probe 1 was prepared by condensation of p-vinyl benzyl amine with dansyl chloride (Equation 1)

$CO_2H$ $CH_2NH_2$ $N(CH_3)_2$ $SO_2Cl$ $N(CH_3)_2$ $SO_2NHCH_2$ 1 (1)

Network polymers were prepared both by bulk and suspension free radical polymerization techniques employing AIBN as initiator. The composition and method of preparation of these materials is summarized in Table I. Non-porous "glass bead" polymers were prepared by suspension polymerization from neat mixtures of styrene and technical grade DVB with nominal crosslinking ratios of 5, 20, and 50 mole %. Macroporous materials were also prepared by suspension polymerization using toluene as diluent ($f_m$ = 0.5) (2a) with crosslinking ratios of 5, 20, and 50 mole %. Macroporous polymers prepared by bulk polymerization were also synthesized; the degree of crosslinking for these materials was 50 mole %. One type of bulk polymerization was formulated identically with material prepared by suspension techniques (toluene diluent and DVB as the crosslinking agent). The second type was prepared using acetonitrile as diluent. The polymers provide a spectrum of materials for analysis with varying degrees of crosslinking and a range of polymer morphologies. The particle size (125-150 μ) of the solid materials were kept uniform by sizing. The ratio of probe 1 to monomer in all of the above materials was kept uniform ($10^{-4}$).

Table I

Summary of Polymerization Conditions

| Designation | Monomers | % Crosslinking Monomer | Diluent $(f_m)^a$ | Polymerization Conditions |
|---|---|---|---|---|
| DVB-5-S-N | ST-DVB | 5 | None | suspension[b] |
| DVB-20-S-N | ST-DVB | 20 | None | suspension |
| DVB-50-S-N | ST-DVB | 50 | None | suspension |
| DVB-5-S-T | ST-DVB | 5 | toluene (0.5) | suspension |
| DVB-20-S-T | ST-DVB | 20 | toluene (0.5) | suspension |
| DVB-50-S-T | ST-DVB | 50 | toluene (0.5) | suspension |
| DVB-50-B-T | ST-DVB | 50 | toluene (0.5) | bulk[c] |
| DVB-50-B-A | ST-DVB | 50 | aceto-nitrile (0.5) | bulk |

a. $f_m$ is the volume fraction of diluent to monomer + diluent used during polymerization

b. Typical polymerizations were carried out in a morton flask containing a mixture of water (200 mL), diluent (20 mL), monomers (20 g), methocel (90 mg) as dispersant, and AIBN (200 mg) as initiator. The mixture was stirred rapidly for 8 h at 70°C. The polymer beads obtained were washed (refluxing acetone), dried (high vacuum), and sized with sieves (100-120 mesh).

c. Polymerizations were carried out in 35 mL capacity, medium-walled glass tubes containing monomers (8 g), diluent (8 mL), and AIBN (80 mg) as initiator. The mixture was freeze-thaw degassed three times, sealed, and heated to 80°C for 14 h. The temperature was increased to 125°C for an additional 12 h. The polymer obtained was crushed, washed (refluxing acetone, 12h), and sized with sieves (100-120 mesh).

Solvation Studies. The fluoresence emission maximum of polymer bound probe provides a measure of the ability of solvent molecules to solvate polymer chains in the network. The solvatochromic shift of probe 1 has been linearly related to a wide variety of organic solvents using the empirical relationship shown in Equation 2 (6).

$$\lambda^{f1}_{calc}(nm) = 53.45\ \pi^* + 20.48\ \alpha + 9.932\ \beta + 457.1 \qquad (2)$$

The terms $\pi$, $\alpha$, and $\beta$ are empirical solvent parameters developed by Taft and Kamlet (7). The fluoresence emission maximum of probe 1 in pure organic solvent defines the pure solvent reference line in Figure 1. When probe 1 is covalently attached to a polymer backbone which in turn is immersed in solvent, the deviation in fluorescence emission wavelength from the pure solvent correlation line reveals how the polymer perturbs the microenvironment of the probe. This microenvironment can vary from pure solvent-like to one dominated by the polymer backbone. The fluorescence emission of the probe in dry polymer is also indicated in the figure. A summary of the solvatochromatic data is discussed below.

Solvation of "Glass Beads". The fluoresence emission of solvent-equilibrated "glass-beads" displays a predictable trend along the lines of crosslink density (Figure 1). The 50% glass beads (DVB-50-S-N) parallel the "dry" polymer correlation line indicating little probe solvation. However, as the crosslinking is decreased solvation increases. Thus the probe 1 readily reveals that these glassy, gel-like, materials are solvated to a degree which is controlled in part by the number of crosslinks. A departure from this trend is noted in poor swelling solvents (EtOH) where all polymers exhibit solvatochromic shifts that approximate the dry state (no solvation), a situation that indicates the networks remain collapsed and the polymer backbone dominates the probe microenvironment. It should be noted, however, that all polymers exhibit a blue shift upon changing from $CH_2Cl_2$ to EtOH. This finding indictes that all polymer chains, even DVB-50-S-N, are solvated to some degree in good swelling solvents (i.e., $CH_2Cl_2$).

Macroporous Polymers Prepared by Suspension Polymerization. Unlike the non-porous "glass beads", all macroporous polymers prepared by suspension polymerization with toluene as cosolvent exhibit fluoresence emission that parallels the solvent correlation line (Figure 2). This finding indicates a substantial degree of probe solvation. Indeed, even the 50% crosslinked material (DVB-50-S-T), exhibits solvation behavior similar to the lightly crosslinked 5% glass beads. An important difference in this series occurs in the poor swelling solvents (EtOH). In this solvent the 5 and 20% crosslinked materials (DVB-5-S-T, DVB-20-S-T) exhibit a fluoresence emission that parallels the "dry" polymer, however, the emission of 50% crosslinked material (DVB-50-S-T) remains close to the pure solvent correlation line. This observation reveals a truly permanent micropore structure that remains intact, regardless of solvent. Thus even the highly polar solvent EtOH can penetrate and solvate the probe in 50% macroporous materials (DVB-50-S-T).

Macroporous Polymers Prepared by Bulk Polymerization. With toluene as diluent (DVB-50-B-T) the highly crosslinked networks closely parallel the pure solvent correlation line indicating the probe is highly solvated even in poor polymer solvents (Figure 3). This result indicates a gel phase with a high degree of permanent micropore structure. There is essentially no difference between this material and that prepared by suspension polymerization (DVB-50-S-T).

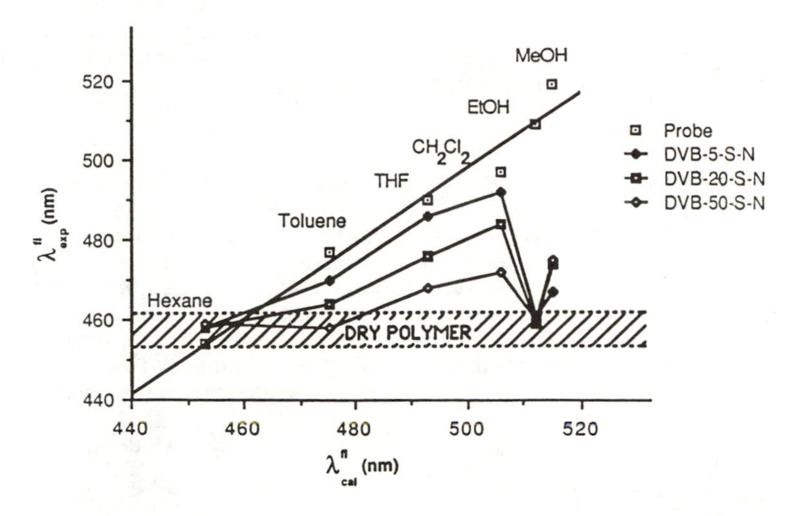


Figure 1. Fluorescence emission of solvent equilibrated "glass beads".

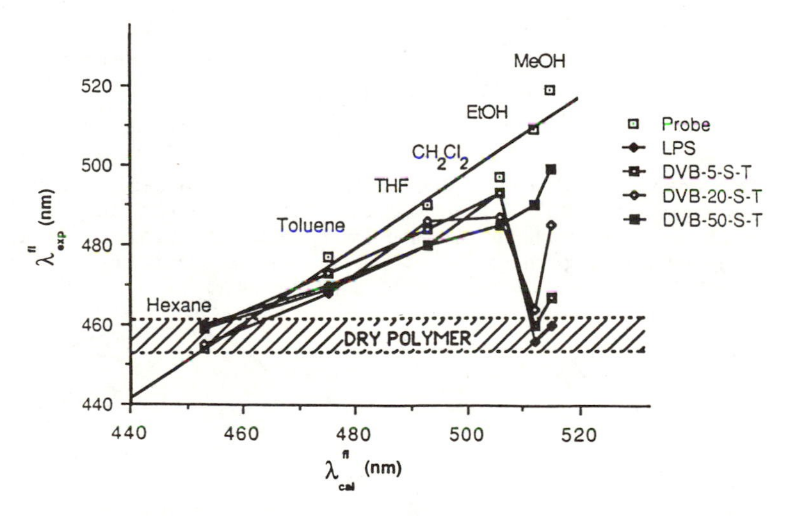


Figure 2. Fluorescence emission of solvent equilibrated macroporous styrene-divinylbenzene copolymers prepared by suspension polymerization.

Quite interestingly, when the bulk polymerization is run with $CH_3CN$ as diluent, the resulting polymer does not parallel the solvent correlation line, rather it parallels very closely the "dry" polymer region, indicating very little solvation in all solvents. This implies these materials are comprised of a gel phase that is substantially less pervious to all solvents.

Diffusion Studies

We have observed that the fluoresence emission intensity of the dansyl probe is diminished in acidic solvents. Indeed the fluoresence of 1 can be completely suppressed in the presence of strong acids ($CF_3CO_2H$) and by treatment with a variety of electrophiles. On the basis of NMR investigations, the mechanism of the fluoresence quenching is found to involve reaction of the proton (electrophile) with the dimethylamino group of the dansyl probe (Equation 3). The protonated or alkalated ammonium ion 2 does not fluoresce when excited at 350-360 mm, the absorption maximum of 1.

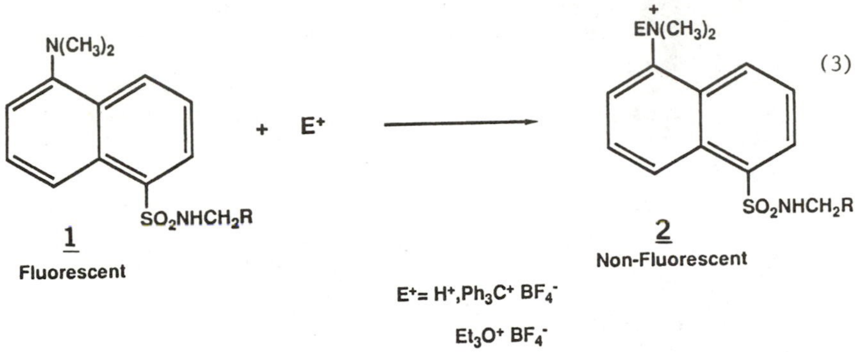

$N(CH_3)_2$ ... $SO_2NHCH_2R$ + $E^+$ ⟶ $\overset{+}{E}N(CH_3)_2$ ... $SO_2NHCH_2R$ (3)

1 Fluorescent    2 Non-Fluorescent

$E^+ = H^+, Ph_3C^+\ BF_4^-$

$Et_3O^+\ BF_4^-$

The response of the fluoresence probe to added electrophile offers the potential for utilizing this reaction to study the migration of electrophiles in macromolecules containing the dansyl probe by evaluating the rate at which the electrophile diminishes the net fluoresence intensity (8), Equation 4.

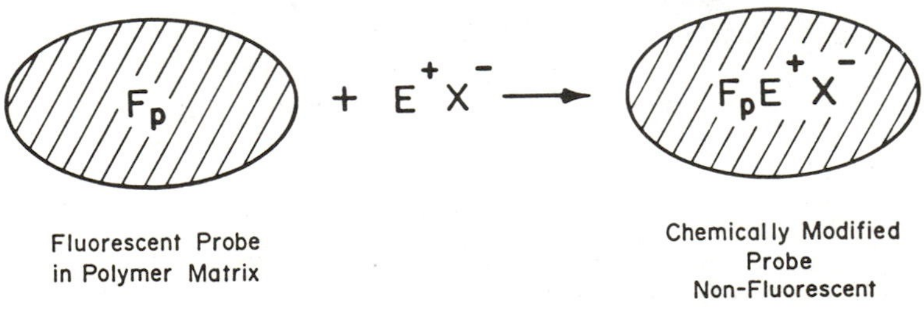

$$F_p + E^+X^- \longrightarrow F_pE^+X^- \quad (4)$$

Fluorescent Probe in Polymer Matrix    Chemically Modified Probe Non-Fluorescent

Thus the dansyl probe can serve two functions, first as a solvatochromic diagnostic to evaluate solvation of polymer chains and second as a probe to monitor the diffusion of electrophiles in polymer networks.

The technique used to study the rate at which electrophilic reagents diffuse into the network domain is straightforward. A stirred solution or suspension of the probe-labeled polymer in $CH_2Cl_2$:hexane (13:4) is continuously irradiated in a fluorescence spectrometer cell. The fluorescence emission intensity is adjusted to 100%. At $t_o$, a solution of the electrophile is added via syringe and the fluoresence emission intensity is continuously recorded. A trace of the dimunition of fluorescence intensity will reflect the overall rate for the diffusion of reagent into the polymer domain and reaction with the fluorescence probe. In control experiments, the fluorescence emission of 1 was found to be quenched "instantaneously" when electrophile was added to homogeneous solutions of 1, thus the principal contribution to the intensity-time curve will be to reveal the influence of the polymer on impeding transport of electrophile through the network. To facilitate a comparative study of materials, the ratio of polymer bound fluorescence probe to added electrophile was held constant at 10:1. This ratio was chosen to yield convenient fluorescence quenching times while revealing the differences in penetrability of related polymeric materials. Triphenylmethyltetrafluoroborate ($Ph_3C^+BF_4^-$) was used as the electrophile in these reactions. The results are summarized in Figures 4-6.

Glass Beads (DVB-5, 20, 50-S-N). A rather straightforward trend is noted for this series of materials - the rate of fluorescence quenching is inversely proportional to the nominal crosslinking density (Figure 4). The fluorescence emission of 5% crosslinked glass beads (DVB-5-S-N) is quenched fastest ($t_{1/2}$ = 15 sec) and completely while the 50% material (DVB-50-S-N) has lost less than 50% of its intensity after 9 min.

Macroporous Suspension Polymers (DVB-5, 20, 50-S-T). Copolymerization in the presence of diluents results in formation of macroporous materials. As can be seen in Figure 5, macroporosity exerts a dramatic effect upon the quenching rate. Not surprisingly, LPS is quenched "instantly" upon addition of electrophile, what is most interesting however, is that for all practical purposes, there is no difference in quenching rate between DVB-5-S-T and DVB-20-S-T macroporous materials. Quite remarkably, all domains are readily accessible since the fluorescence emission is completely quenched in these materials. The leveling off of fluorescence emission intensity for the 50% crosslinked material (DVB-50-S-T) reveals kinetically inaccessible domains under the conditions of the experiment.

Bulk Macroporous Polymers (DVB-50-B-T, DVB-50-B-A). Comparison of DVB-50-B-T with DVB-50-S-T reveals there is little difference between polymers prepared in bulk and by suspension techniques. Both materials, prepared with toluene as cosolvent, reveal fluorescence quenching traces that are almost superimposable. There is, however, a very dramatic difference between material prepared by bulk polymerization using different cosolvents. Polymer DVB-50-B-T reveals a large fraction of all sites are quenched within 15 sec (Figure 6), a small tail on this curve indicates a component (<5%) of inaccessible sites. In contrast, the material prepared with

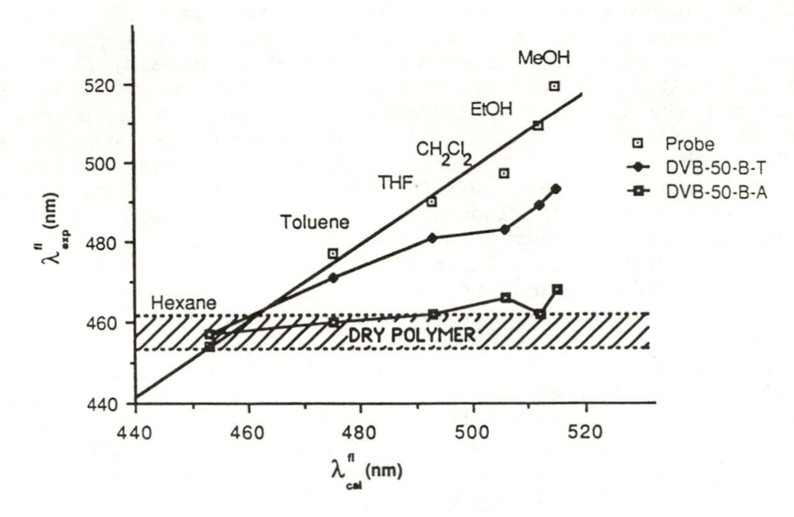


Figure 3. Fluorescence emission of solvent equilibrated macroporous sytrene-divinylbenzene copolymers prepared by bulk polymerization.

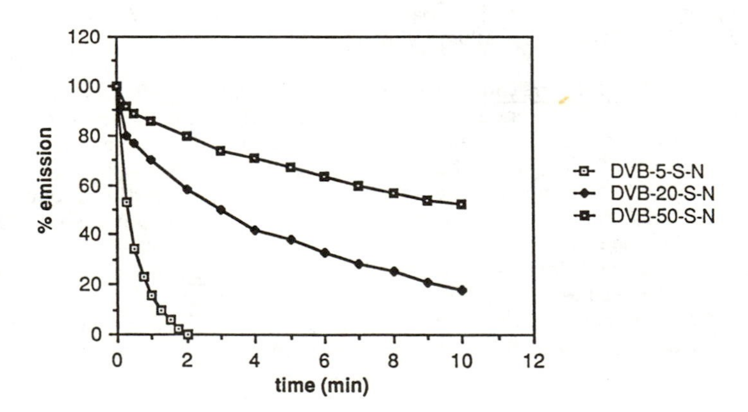


Figure 4. Quenching results for styrene-divinylbenzene "glass beads".

$CH_3CN$ as diluent (DVB-50-B-A) shows a fast burst of quenching (~40% within 60 sec) then relatively little quenching thereafter. This finding is consistent with the view that in $CH_3CN$, a macroporous material is produced, but this material has a relatively impenetrable gel phase in contrast to the material prepared with toluene cosolvent.

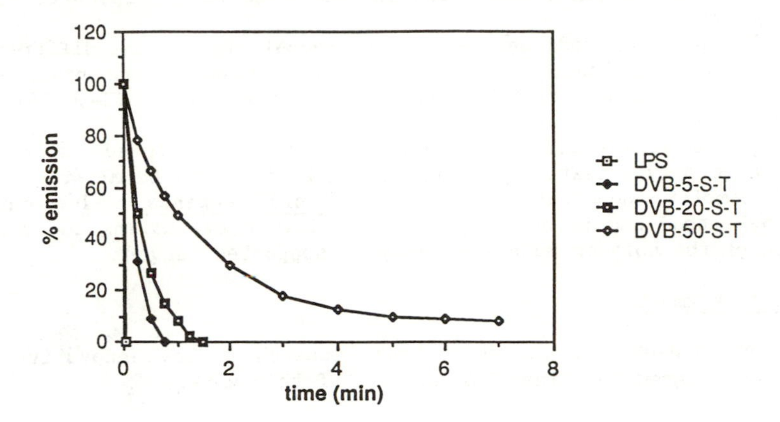


Figure 5. Quenching results for macroporous styrene-divinylbenzene copolymers prepared by suspension polymerization.

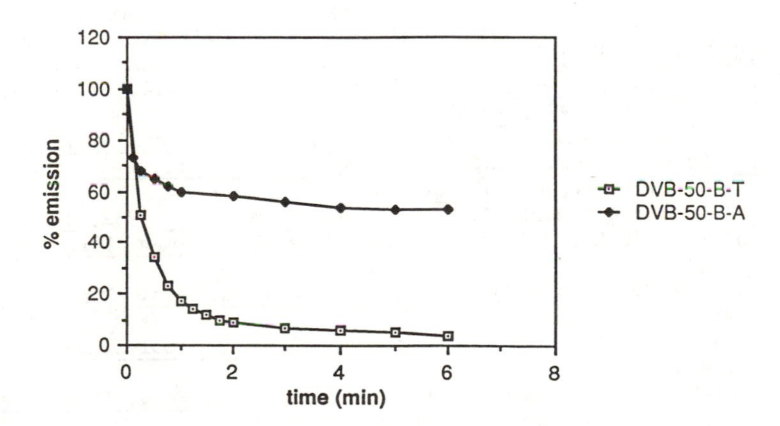


Figure 6. Quenching results for macroporous styrene-divinylbenzene copolymers prepared by bulk polymerization.

## Conclusions

The fluorescence probe 1 offers a valuable handle for a detailed analysis of the solvation and penetrability of highly crosslinked network polymers.

Parallels can be drawn between the solvatochromic shift data and the rates of diffusion. This permits a link to be established between polymer chain solvation and the rate of transport of ionic reagents.

The probe method can graphically reveal significant differences in the microenvironment between highly crosslinked materials that contain the same nominal crosslink density, i.e. DVB-50-B-T and DVB-50-B-A.

It is also noted that even in highly crosslinked materials, suitable polymerization conditions can be found to permit the rapid and complete penetrability of even ionic reagents. The dansyl probe method should be of general utility in evaluating the influence of the polymer matrix on polymer supported reagents.

## Acknowledgments

We are grateful to the U.S. Army Research Office Innovative Research Program for financial support of this work.

## Literature Cited

1. (a) Albright, R.L. Reactive Polymers 1986, 4, 155; (b) Sherrington, D.C., in Hodge, P., and Sherrington, D.C. (Eds.), "Polymer-Supported Reactions in Organic Synthesis," John Wiley and Sons Ltd., New York, 1980, Ch. 1.
2. (a) Millar, J.R.; Smith, D.G.; Marr, W.E.; Kressman, T.R.E. J. Chem. Soc. 1963, 218; (b) Lloyd, W.G.; Alfrey, T. J. Pol. 1962, 62, 301; (c) Seidl, J.; Malinsky, J.; Pusek, K.; Heitz, W. Adv. Polymer Sci. 1967, 5, 113; (d) Kun, K.A.; Kunin, R. Polymer Lett. 1984, 2, 587.
3. (a) Sederel, W.L.; DeJong, G.J. J. Appl. Polym. Sci. 1973, 17, 2835; (b) Kun, K.A.; Kunin, R. J. Pol. Sci., A-1 1968, 6, 2689; (c) Dusek, K. Polymer Lett. 1965, 209.
4. (a) Guyot, A.; Bartholin, M. Prog. Polym. Sci. 1982, 8, 277; (b) Howard, G.J.; Midgley, C.A. J. Appl. Polymer Sci. 1981, 26, 3845.
5. Shea, K.J., Dougherty, K.D., J. Am. Chem. Soc. 1986, 108, 1091.
6. Shea, K.J.; Okahata, Y.; Dougherty, T.K. Macromolecules 1984, 17, 296.
7. Kamlet, M. J.; Abboud, J. L. M.; Taft, R. W. Prog. Phys. Org. Chem. 1981, 13, 485.
8. For a related approach to the chemical modification of a fluorescent probe, see Pan, S.; Morawetz, H. Macromolecules 1980, 13, 1157-1162.

RECEIVED September 3, 1987

# Chapter 10

# Light-Induced Conformational Changes of Polymers in Solution and Gel Phase

**Masahiro Irie**

**Institute of Scientific and Industrial Research, Osaka University, Ibaraki, Osaka 567, Japan**

Attempts have been made to control the polymer chain conformation reversibly by photoirradiation. The aim is attained by incorporating photochromic chromophores into the pendant groups or main chain. It was found from the results in solution that the intramolecular electrostatic force of repulsion between photogenerated pendant cations is a more effective driving force for the conformational change than trans-cis geometrical isomerization of unsaturated linkages in the polymer backbone. The large conformational change at the molecular level is amplified into the shape change of polymer gels at the visible macro level. Polyacrylamide gels having photoionizable triphenylmethane leucocyanide groups dilated 2.2 times in each dimension by ultraviolet irradiation. Electric field effect on the gel was also examined. By applying alternating electric field(0.5 Hz), the rod-shaped gel showed oscillating motion.

A polymer chain conformation is well known to depend on the environment, such as solvent or temperature. In good solvents, polymers have an extended conformation, while they shrink in poor solvents at low temperature. Polyelectrolytes change their conformation with changes in pH and salt concentration(1). Our interest is to control the polymer chain conformation by "photochemistry" rather than by changing the environment(2,3). It is obviously a tedious method to change the environment to control the chain conformation. The photochemical method is much superior in the response time, reversibility and easy procedure.

Among numerous photochemical reactions, photochromic reactions are useful for this purpose. The photochromic reaction is defined as a reversible change in a chemical species between two forms having different absorption spectra,

$$A \underset{h\nu',\Delta}{\overset{h\nu}{\rightleftharpoons}} B$$

0097–6156/87/0358–0107$06.00/0

Besides the absorption spectral change, the isomerizations are always accompanied by certain physical property changes, such as dipole moment and/or geometrical structural changes. The property changes may be utilized as a driving force to induce the conformational changes by incorporating the chromophores into the polymers.

The aim to control the polymer chain conformation by photo-irradiation was attained by using following photochromic reactions; 1) trans-cis geometrical isomerization of unsaturated linkages in the polymer backbone, 2) reversible generation of strong dipoles in the polymer pendant groups, and 3) photoionization of the pendant groups. Representative examples of each system are polyamide with backbone azobenzene residues (4-8), poly(methyl methacrylate) with pendant spirobenzopyran groups (9-11), and poly(N,N-dimethylacrylamide) with pendant triphenylmethane leucohydroxide groups(12). The first part describes some details of these examples.

It seems possible to amplify the photostimulated conformational changes in solution at the molecular level into shape changes of polymer gels or solids at the visible macro level. The first proposal to use the structural changes at the molecular level for direct conversion of photon energy into mechanical work has been made by Merian (13) in 1966. Since then, many materials, most of which contained azobenzene chromophores, have been reported to show photostimulated deformation(14). Till now, however, the reported deformations were limited to less than 10%. In addition, Matějka et. al. have pointed out that in many cases photo-heating effect instead of photochemical reaction plays a dominant role in the deformation(15,16).

In due consideration of these results, we have decided to employ electrostatic forces to achieve a large reversible deformation of gels. The electrostatic force is expected to be a more effective driving force for the conformational changes of polymer chains than trans-cis geometrical isomerization of unsaturated linkages. The second part describes the photostimulated dilation of polymer gels.

During the course of experiments to reveal an electric field effect on the behavior on the photogenerated mobile ions, we found a peculiar phenomenon, reversible bending motion of the rod-shaped gels. The result will also be briefly described.

## Photostimulated Conformational Changes in Solution - Molecular level

Figure 1 illustrates the proposals to induce the conformational changes of polymer chains by using photochromic reactions. The mechanism (1) utilizes trans-cis geometrical isomerization as a tool to enforce the conformational changes. When the trans-cis photoisomerizable chromophores are incorporated into the polymer backbone, the isomerization from trans to cis form kinks the extended polymer chains, resulting in a compact conformation. The compact conformation returns to the initial extended conformation by the thermal or photochemical isomerization of the chromophores from the cis to trans form. Polyamides with azobenzene groups in the polymer backbone are among the earliest in which trans-cis isomerizable chromophores are used to regulate the conformation of polymer chains(4,5).

The azobenzene chromophore is known to change the geometry as follows(17):

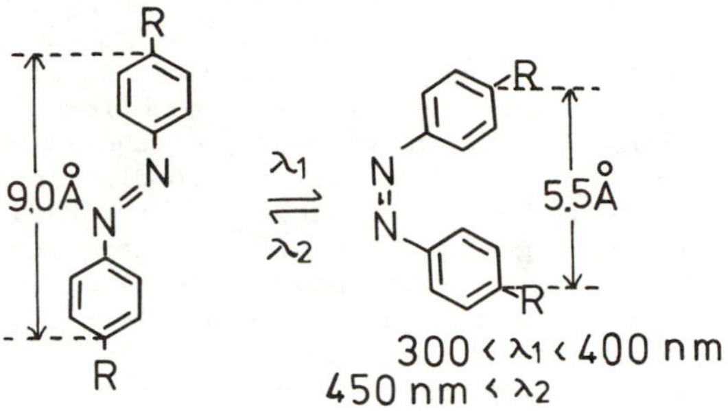

9.0Å $\underset{\lambda_2}{\overset{\lambda_1}{\rightleftharpoons}}$ 5.5Å

$300 < \lambda_1 < 400$ nm
450 nm $< \lambda_2$

A typical example of the polymer having azobenzene residues is (4,5):

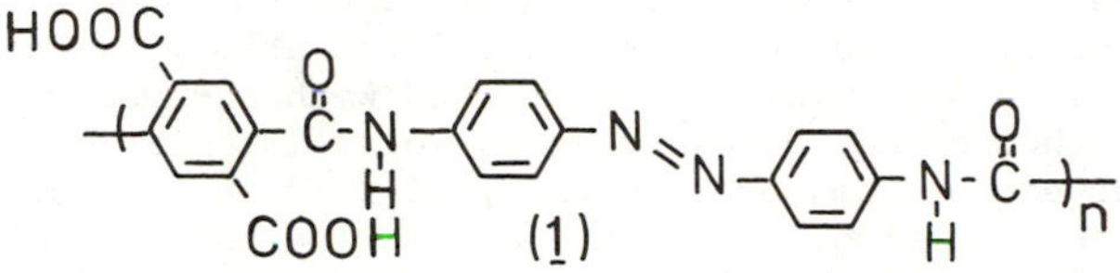

(1)

The intrinsic viscosity,[η],of polyamide(1) in N,N-dimeythylacetamide was found to decrease from 1.22 to 0.50 dl/g on ultraviolet irradiation ($410 > \lambda_1 > 350$ nm) and to return to the initial value in 30 h in the dark at 20°C. The slow recovery of the viscosity in the dark was accelerated by visible light irradiation($\lambda_2 > 470$ nm). On alternate irradiation of ultraviolet and visible light, the viscosity reversibly changed by as much as 60%.

Mechanism(2) employs an electrostatic force of repulsion among photogenerated charges as a driving force for a conformational change. Triphenylmethane leucoderivatives are the most convenient chromophores to produce positive charges in the pendant groups of polymers, because the quantum yield of the photodissociation is reasonably high ( $\Phi > 0.2$) and the photogenerated positive charges have a rather long lifetime( $\tau \sim$ min). Upon ultraviolet irradiation, the chromophore dissociates into an ion pair with production of an intensely colored triphenylmethyl cation. The cation recombines thermally with the counter ion(18):

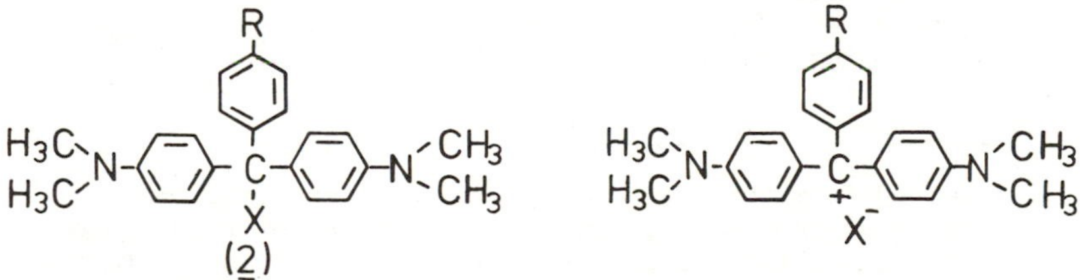

(2)

Triphenylmethane leucohydroxide residues were introduced into the pendant groups by copolymerizing the vinyl derivative (2, X= OH, R= $CH{=}CH_2$)with N,N-dimethylacrylamide(12). In the dark before irradiation, a methanol solution containing the copolymer exhibited a pale green color. Upon ultraviolet irradiation ( $\lambda > 270$ nm), the solution became deep green, which color disappeared slowly in the dark with a halflife time of 3.3 min (Figure 2A). The appearance of a deep green color means that the pendant triphenylmethyl leucohydroxide residues dissociate into triphenylmethyl cations and hydroxide ions. The photogenerated positive charges recombine with the dissociated hydroxide ions to reproduce the colorless leuco form. Concurrently with the coloration, the reduced viscosity of the solution, $\eta_{sp}/c$, showed a remarkable increase from 0.55 to 1.6 dl/g as depicted in Figure 2B. After removal of the light, $\eta_{sp}/c$ returned to the initial value with a half-life time of 3.1 min. The viscosity change indicates that the polymer chain expands upon ultraviolet irradiation and shrinks in the dark. The good correlation between the viscosity change and the absorption intensity at 620 nm implies that expansion of the polymer conformation is induced by the electrostaic repulsive forces among the pendant triphenylmethyl cations.

The concentration dependence of $\eta sp/c$ confirmed the above expansion mechanism. In the dark before photoirradiation, the dependence was linear; the reduced viscosity decreased with decreasing the concentration of the polymer. During ultraviolet irradiation, this dependence showed an anomalous behavior; $\eta_{sp}/c$ steeply increased at low polymer concentration. The viscosity during photoirradiation was 4 times larger than the viscosity in the dark at 0.04g/dl. At low polymer concentration, screening of the electrostatic potential by the counter ions becomes weak and consequently the increase of the repulsive forces of the positive charges along the polymer chain expands the dimension of the chain. The photo-effect due to the electrostatic forces is much larger than the effect observed for polyamides having azobenzene residues in the backbone. It is worthwhile to note that the photostimulated increase of $\eta sp/c$ was strongly suppressed by the presence of salt($10^{-3}$ M LiBr).

The ratio of the specific viscosity during photoirradiation to that in the dark, $\eta_p/\eta_d$, increased with increasing content of triphenylmethane leucohydroxide residues in the pendant groups and reached a maximum of 3.3 at 0.18 mole fraction. Above the content,the ratio decreased, the decrease being due to the low solubility of the residues in methanol.

The concept to adopt the electrostatic repulsive force as a driving force for a photostimulated expansion of the polymer chain is useful and widely applicable to other polymer systems. Polystyrene and polyacrylamide having pendant leucohydroxide and leucocyanide groups were found to change their solution viscosity in methylene chloride and in water, respectively.

## Photostimulated Dilation of Polymer Gels - Macro Level

It is inferred from the above results on the conformational changes in solution that the electrostatic force of repulsion between photogenerated charges, mechanism(2), is a more effective driving force for conformational changes than trans-cis geometrical isomerization of unsaturated linkages, mechanism(1). In attempting to amplify the large conformational changes due to the electrostatic repulsive forces in solution at the molecular level to the visible macro level, we introduced the mechanism(2) into the gel phase.

Acrylamide gels(3) containing a small amount of triphenyl-

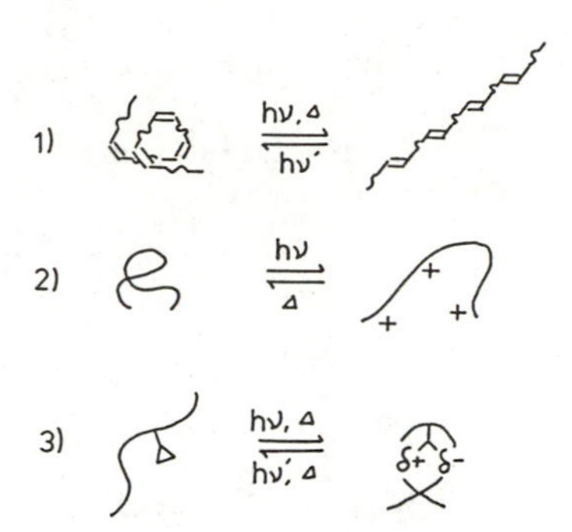


Figure 1. Schematic illustration of photostimulated conformational chages of polymer chains.

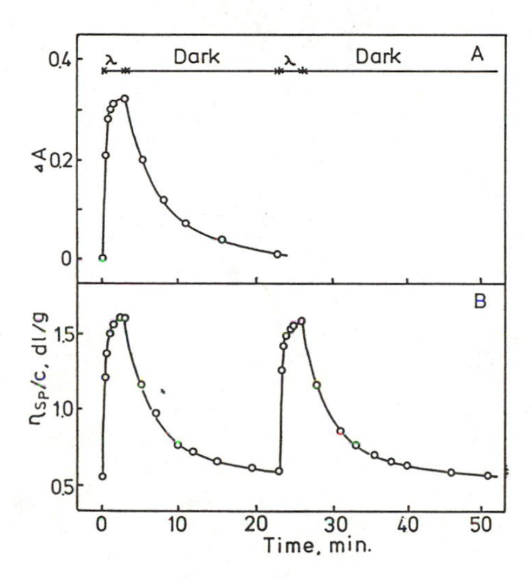


Figure 2. Photostimulated (A) conformational change and (B) viscosity change of poly(N,N-dimethylacrylamide) having pendant triphenylmethane leucohydroxide groups (9.1 mol%) in methanol at 30°C. Polymer concentration was 0.06 g/dl. $\lambda$ >270 nm. (Reproduced with permission from Ref.12. Copyright 1985, Huethig & Wepf Verlag.)

methane leucohydroxide(2, X= OH,R= CH=$CH_2$) or leucocyanide(2, X= CN, R= CH=$CH_2$)residues were prepared by free radical copolymerization of acrylamide and di(N,N-dimethylaniline)-4-vinylphenylmethane leucohydroxide(2, X= OH,R= CH=$CH_2$) or leucocyanide(2, X= CN,R= CH=$CH_2$) in DMSO in the presence of N,N-methylenebisacrylamide(19,20).

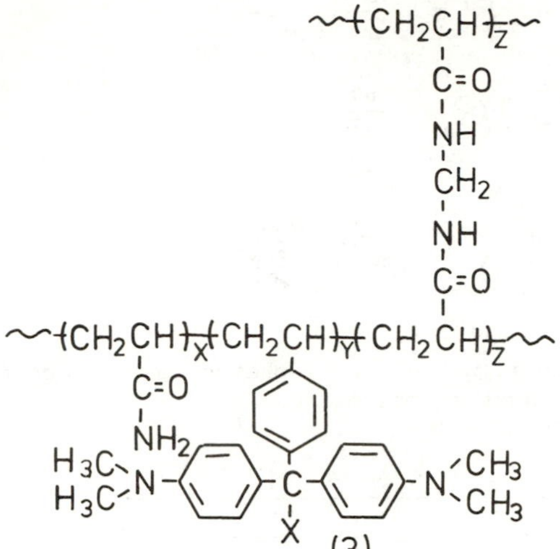

X: OH, CN

(3)

The gels were swollen to the equilibrium condition on standing in water overnight. Then the weight or dimension change of the gels induced by ultraviolet light was measured.

A disk-shaped gel(10 mm in diameter and 2 mm in thickness) having 3.7 mole% triphenylmethane leucohydroxide residues showed photostimulated reversible dilation in water. Upon ultraviolet irradiation( $\lambda$>270 nm), the gel swelled and the weight increased by as much as 3 times within 1 h. The dilated gel deswelled in the dark to the initial weight in 20 h. The cycles of dilation and contraction of the gel could be repeated several times. The gel having the leucohydroxide residues swelled even in the dark when the aqueous solution became acidic. At pH 3.8, the gel has a strong green color and an 11 fold weight increase was observed compared with the weight at pH 8.0. The weight increase in the dark is due to chemical ionization of weakly basic leucohydroxide residues.

In order to make gels insensitive to the pH changes, the hydroxide residues were replaced by the cyanide groups. The weight of the leucocyanide gel remained constant in the range of pH 4 - 9. Figure 3 shows the photoresponsive behavior of the gel having leucocyanide residues(1.9 mole%) in water. Upon ultraviolet light irradiation, the gel weight increased as much as 18 times. In the dark, the dilated gel contracted again slowly to the initial weight. The low degree of swelling of the leucocyanide gel in the dark contributes to the larger expansion ratio.

Figure 4 shows the size and shape changes of the gel before and after photoirradiation. When the whole gel was irradiated with ultraviolet light, both the gel length and diameter expanded by as much as two times(Figure 4B). When the irradiation beam was localized to one side of the rod shaped gel, the gel showed bending motion (Figure 4D).

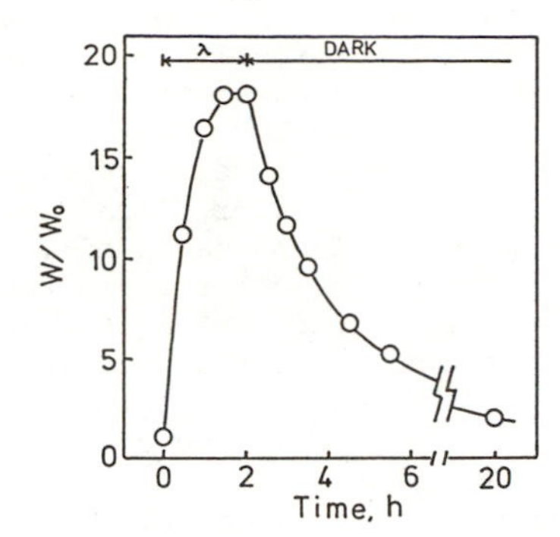


Figure 3. Photostimulated dilation and contraction of polyacrylamide gel having triphenylmethane leucocyanide groups (1.9 mole%) with light of wavelength longer than 270 nm at 25°C. Initial pH of the external water phase was 6.5. $W_0$ is the weight of the gel before photoirradiation. (Reproduced from Ref. 20. Copyright 1986 American Chemical Society.)

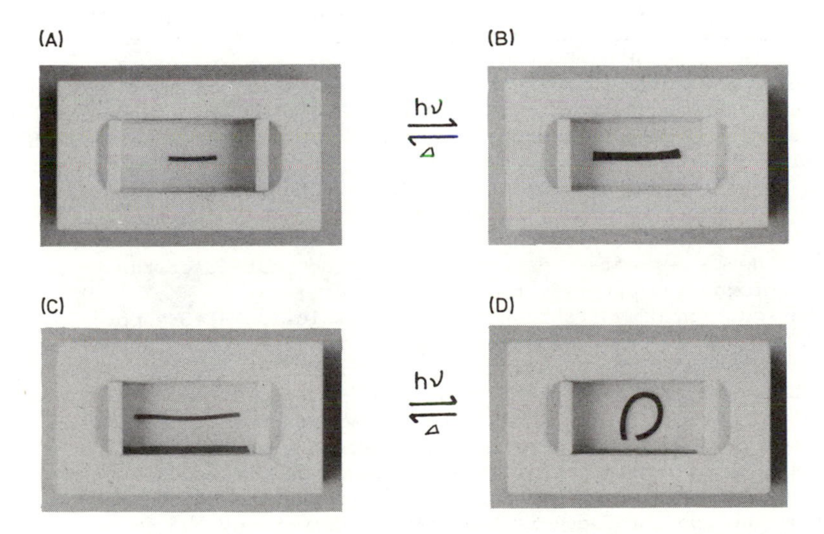


Figure 4. Photostimulated size and shape changes of polyacrylamide gels having triphenylmethane leucocyanide groups (3.1 mole %):(A) before photoirradiation, (B) the whole gel being irradiated with ultraviolet light, (C)before photoirradiation, (D) upper side of the rod shaped gel being irradiated.

Figure 5 A and B show the rate of coloration at 660 nm and the gel dimension expansion rate under continuous light irradiation. The triphenylmethyl cation is well known to have very strong absorption at 622 nm(18). The absorbance band at the wavelength corresponds to the amount of positive ions generated in the gel. Upon ultraviolet irradiation, the color of the gel changed quickly from pale green to deep green in less than 3 min. and then remained almost constant. In the dark, the color again returned to the initial pale green in several hours. The size of the gel,on the other hand,increased slowly and reached a saturated value in around 2 h. The photostimulated dilation was 2.2 times. The large difference in the photoresponse time between the coloration and the size change implies that the rate controlling step of the gel dilation is not the rate of ionization of the leucocyanide residues but the diffusion of the gel network into water. According to Tanaka et.al. (21) the characteristic time of gel expansion is proportional to $a^2/D$, where a and D are the final radius of gel sphere in equilibrium and the diffusion coefficient. D is defined as $D = E/f$, where E is longitudinal bulk modulus of the network in the gel fluid. The relation suggests that the slow rate can be improved by decreasing the gel size. For example, the response time is expected to become shorter than 1 sec, when the size is decreased smaller than 0.1 mm.

The amount of gel dilation strongly depended on the content of the leucocyanide residues in the gel network. The dilation ratio measured in weight, $W/W_0$, where W and $W_0$ are the gel weight after and before photoirradiation,increased with increasing leucocyanide content and reached the maximum at around 2 mole%. Above 2 mole%, the photo-induced dilation was suppressed. The bell shape dependence suggests that the leucocyanide residues have two competitive functions in the photo-induce dilation process. One essential function of the chromophore is to produce positively charged groups attached to the gel network. The formation of charges, fixed cations and free anions, generates osmotic pressure differentials between the gel inside and the outer solution, which is considered to be the main driving force for gel expansion(see below). The leucocyanide residues, on the other hand, have a tendency to contract the gel network because of their hydrophobic nature. At higher leucocyanide content, the latter effect is considered to dominate over the expansion force and suppress the dilation.

The addition of salts to the external solution also suppresses the gel expansion. The gels were immersed in the salt solution and swollen to the equilibrium condition on standing in solution overnight, then the gels were photoirradiated. Both NaCl and KBr reduced the photostimulated swelling ratio to a similar extent. No photostimulated dilation was observed in the presence of $10^{-2}$M NaCl or KBr for both gels having leucohydroxide or leucocyanide residues. The salt effect gives strong evidence that the expansion of the gel is caused by the ions produced in the polymer network by photoirradiation.

Mechanism of Gel Dilation by Photoirradiation The swelling of gels with fixed charges in the network can be quantitatively understood by the osmotic pressure differentials obtained from Donnan equilibrium(22,23). The equilibrium value of the volume fraction of the gel network, $\Phi$ , can be obtained from the pressure balance equation.

$$\Pi(\Phi)_{conf} + \Pi(\Phi)_{cont} + \Pi(\Phi)_{ion} + \Pi(\Phi)_{coul} = 0$$

$\Pi(\Phi)_{conf}$ represents the swelling pressure due to the conformational entropy of the network, i.e. mixing entropy and rubber elasticity term. $\Pi(\Phi)_{cont}$ arises from the interaction among polymer segments and solvent molecules. $\Pi(\Phi)_{ion}$ is the osmotic pressure differentials resulting from the difference in the ion concentration in the gel and the external solution. The term $\Pi(\Phi)_{coul}$ takes into account the contribution from the electrostatic forces of the fixed charges, which would enhance the gel expansion.

In the photoresponsive gel system, the Donnan term, $\Pi(\Phi)_{ion}$ plays the most important role. $\Pi_{conf}$ and $\Pi_{cont}$ are considered not to be affected by photoirradiation. The term $\Pi_{coul}$ is expected to contribute to the expansion, when the charged groups are in close proximity to one another in the network and the electrostatic force of repulsion is not effectively screened by the counter ions. This term becomes important in an electric field effect, which will be discussed in the next section. Here, attention is restricted to the $\Pi_{ion}$ term. An ion distribution in the gel and in the external solution is given in Table I. LCN and $L^+$ represent leucocyanide residues in gel network and the cations, respectively. $C_0$ is the initial concentration of the residues in the gel network. $I_0$ , $\Phi$ and t are the absorbed light intensity, ionic dissociation quantum yield and the irradiation time, respectively. z and y represent the concentration of $CN^-$ and $H^+$ ions in the gel phase, respectively. z and x are the respective concentration in the external solution. In the present example, KBr was used as the salt and the concentration of $K^+$ in the external solution and in the gel were m and n, respectively.

Accordingly to Donnan theory, distribution of mobile ions between the gel and the solution is equilibrium for each ion as follows,

$$\lambda = \frac{[H^+]_g}{[H^+]_s} = \frac{[OH^-]_s}{[OH^-]_g} = \frac{[K^+]_g}{[K^+]_s} = \frac{[Br^-]_s}{[Br^-]_g}$$

,where g and s refer to the gel and the external solution, respectively. $\lambda$ is the distribution constant. Here, adopting the approximation to use concentrations rather than activities and substituting the values from Table I, $\lambda$ can be derived as follows,

$$\lambda = 1 \Big/ \sqrt{1 + \frac{I_0 \Phi t}{n + y}}$$

The difference in concentration of all mobile ions, E, between the gel and the external solution can be expressed by using $\lambda$ ,

$$E = \frac{1 - \lambda}{1 + \lambda} I_0 \Phi t$$

E is directly proportional to the osmotic pressure differentials as

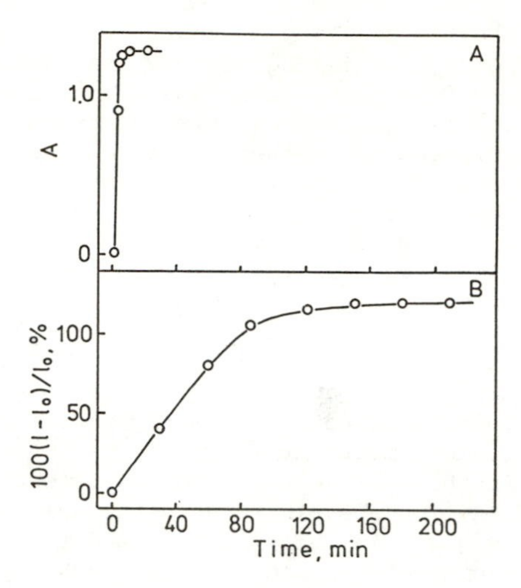


Figure 5. Photostimulated (A) color change and (B) dimension change of polyacrylamide gel having triphenylmethane leucocyanide groups (1.9 mole%) in water. (Reproduced from Ref. 20. Copyright 1986 American Chemical Society.)

Table I. Ion Distribution between the Gel Inside and External Solution

| Gel Inside | | External Solution | |
|---|---|---|---|
| LCN | $C_0 - I_0 t$ | | |
| $L^+$ | $I_0 \Phi t$ | | |
| $CN^-$ | z | $CN^-$ | z' |
| $H^+$ | y | $H^+$ | x |
| $OH^-$ | $K_w/y$ | $OH^-$ | $K_w/x$ |
| $K^+$ | n | $K^+$ | m |
| $Br^-$ | $I_0 t + y + n - K_w/y - z$ | $Br^-$ | $x + m - K_w/y - z'$ |

E is directly proportional to the osmotic pressure differentials as stated in the van't Hoff equation.

$$\Pi_{ion} = RTE$$

$$= \frac{1 - \lambda}{1 + \lambda} RTI_0\Phi t$$

The degree of swelling of the gel with fixed charged groups is mainly controlled by $\Pi_{ion}$ and can be expressed as follows,

$$\text{Degree of swelling} \propto \frac{1 - \lambda}{1 + \lambda} RTI_0\Phi t$$

The equation indicates that photogeneration of ions, the increase of $I_0\Phi t$, causes the decrease of $\lambda$, resulting in the increase of the degree of swelling. The suppression of the gel expansion by the addition of salt can be explained in the following way. The increase of the salt concentration increases in n and m, giving rise to the increase of $\lambda$, resulting in the decrease of swelling. The deswelling behavior in the presence of salt is due to the increase of the Donnan distribution constant of ions.

The photostimulated dilation of the gel is well understood by the osmotic pressure mechanism. Although the electrostatic repulsive forces between the positive ions play an important role in the conformational changes in solution, they are considered to make a minor contribution to the gel dilation.

## Reversible Bending of Rod-Shaped Gels in an Electric Field - Macro Level

Polyelectrolyte gels have been reported to shrink in an electric field in the dark conditions. Polyacrylamide gels having acrylic acid groups collapses by applying the field of 0.72 V/cm in an acetone-water mixture(24). The contraction behavior was accounted for by analyzing the minimization conditions of the total free energy, free energy associated with the deformation of the gel and the energy with the work done by the negatively charged acrylic acid groups in the electric field. The deswelling of gels in an electric field was also observed for a water swollen poly(2-acrylamide-2-methyl-1-propane-sulfonic acid)(25). In these experiments, the gels are in contact with electrodes. These conditions make it difficult to distinguish the electric field effect from the electrochemical reactions on the electrodes, which will change pH of the gel inside and the solution.

We have tried to measure pure electric field effect on the photoresponsive acrylamide gels by using a small water pool(teflon, 36 x 19 x 15 mm) with two parallel platinum electrodes(26). The rod-shaped gel was placed parallel to the electrodes to avoid the contact of the gels with the electrodes. The acrylamide gels containing triphenylmethane leucocyanide groups(25 mm in length and 2 mm in section diameter) were prepared in capillary tubes by the same method as described in the previous section.

Figure 6 shows photostimulated bending motion of an acrylamide

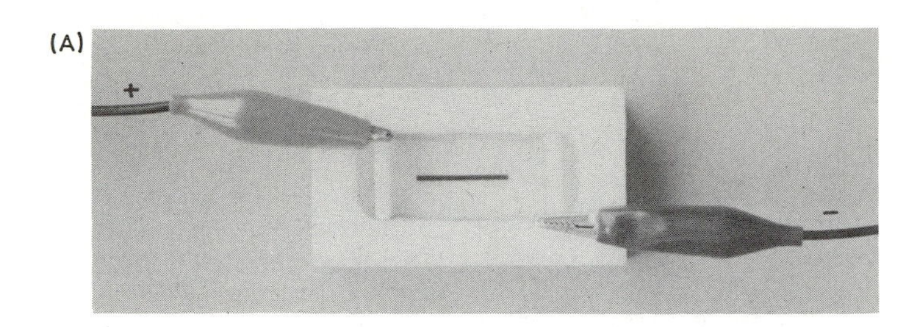


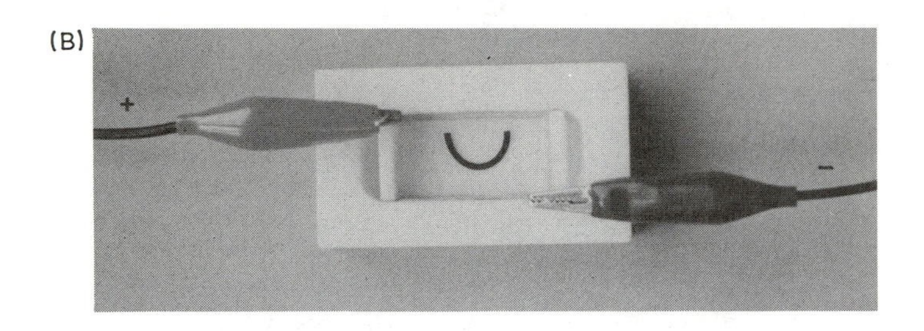


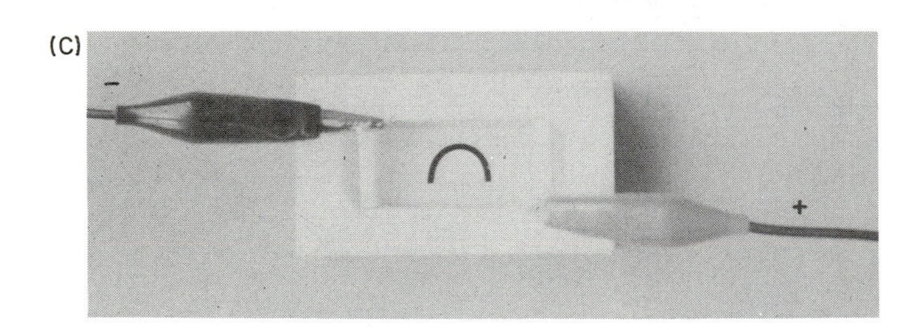


Figure 6. Photostimulated bending motion of a rod shaped acrylamide gel (25 mm in length and 2 mm in section diameter) having 3.1 mole% triphenylmethane leucocyanide groups in an electric field (10 V/cm) in water: (A) before photoirradiation, (B)under ultraviolet irradiation, (C) under ultraviolet irradiation, polarity of the electric field being opposite to that in (B).

gel having 3.1 mole% triphenylmethane leucocyanide groups in water in an electric field. In the dark, the gel shape did not change in an electric field. 10 V/cm(Figure 6A). Upon ultraviolet irradiation ( $\lambda$ >270 nm),the gel quickly bent in 1 min. The gel end moved to the direction of a positive electrode(Figure 6B). During the bending motion, the center of gravity of the gel remained at the initial position. Translational movement of the entire gel to the negative electrode was not observed. By changing the polarity of the electric field, the gel again becomes straight and then bends to the another direction(Figure 6C). The response time of the gel shape change was around 2 min. After switching off the light, the gel slowly returned to the initial straight shape in the electric field. The result suggests that photodissociation of the leucocyanide groups in the gel is indispensable to the gel bending motion.

In order to determine quantitatively the response time of the motion, one end of the rod-shaped gel was fixed to the wall and the moving distance of the other free end, l, from the initial position was measured as a function of irradiation time. Figure 7 shows the photostimulated bending motion of the gel(26 mm in length and 2 mm in section diameter) in an electric field(10 V/cm). The free end moved toward the positive electrode with a initial speed of 0.40 mm/sec. By changing the polarity of the electric field, the end moved to another direction. Upon switching off the electric field, the bent gel returned to the initial position with a speed of 0.075 mm/sec.

The bending rate depends on the applied field. Upon increasing the field, the response time increases in proportion to the applied field. Although the bending rate became very slow, the bending motion was observed in a very weak field of 1.25 V/cm. In this case, effective voltage applied to the gel was only 0.25 V.

In the above experiments, deionized water was used as the external solution. As described in the previous section, the addition of salts to the solution decreased the photostimulated dilation of the gels. If the bending motion in the electric field was due to the osmotic pressure mechanism, the addition of salt would also suppress the motion. This is not the case. On the contrary, the response time of the bending motion was accelerated by the addition of salts to the external solution. The bending rate in the solution containing 2 x $10^{-3}$mole/l NaCl was 1.5 mm/sec, which is 4 time faster than the rate in the absence of NaCl. The result indicates that the bending motion in the electric field is not due to the osmotic pressure mechanism.

When a solution containing salts is used, it is difficult to examine pure electric field effect without being disturbed by the electrochemical reactions on the electrodes. The reactions on the electrodes produce pH gradient in the solution. Although the leucocyanide gels are rather insensitive to the pH change, the correlation between the bending motion and the pH change was examined by adding a pH indicator, phenol red, into the external solution. The bending motion was found to be much faster than the color change on the electrode. The result suggests that the rapid bending motion is independent of the pH change. This was further confirmed by applying alternating electric field, as shown in Figure 8. The rod shaped gel, one end of which is fixed on the wall, vibrates in response to the alternating electric field of 0.5 Hz. under ultraviolet irradiation.

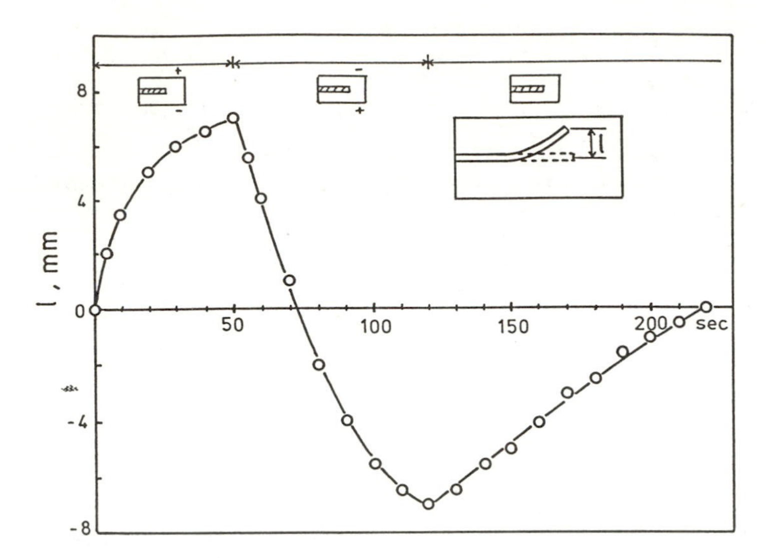


Figure 7. Photostimulated bending motion of a rod shaped acrylamide gel (26 mm in length and 2 mm in section diameter) having 3.1 mole% triphenylmethane leucocyanide groups in an electric field (10 V/cm) in water. The electric field was removed after 120 sec. (Reproduced from Ref. 26. Copyright 1986 American Chemical Society.)

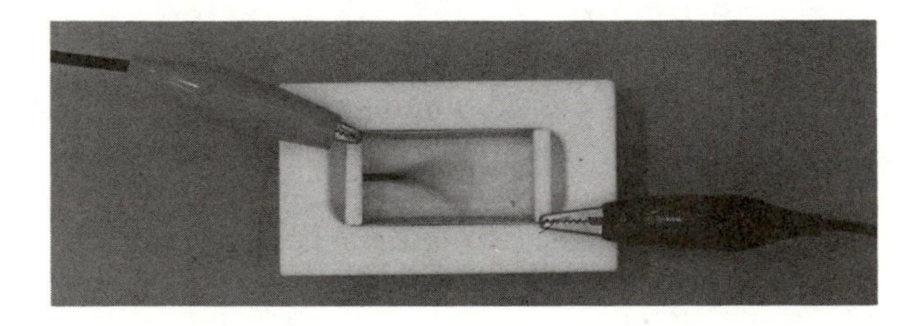

Figure 8. Photostimulated vibratinal motion of a rod shaped acrylamide gel having 3.1 mole% triphenylmethane leucocyanide groups in an alternating electric field (0.5 Hz, ± 8V/cm) in water.

In this case, the pH of the solution remained in the neutral value (around 6.5).

The bending behavior of the gel suggests inhomogeneous expansion of the gel in the electric field. The negative electrode side of the gel expands larger than the other side. Since the electric field is applied perpendicular to the gel axis, mobile negative ions, $CN^-$, are attracted to the positive electrode side in the gel. Consequently, excess positive charges are left on the other side. Internal repulsive force between the positive charges, which are fixed in the gel network, is considered to cause the expansion of the negative electrode side of the gel.

## Other Properties of Photoresponsive Polymers

Because of limitation of this chapter, only a few properties of photoresponsive polymers are described. Table II includes several other properties so far reported(27-50). All of these physical and chemical properties are found to be controlled reversibly by photo-irradiation. It is now generally accepted that photochromic reactions are useful as a tool to photo-control the properties of synthetic polymers. The photoresponsive polymers have potential applications for many photoactive devices, such as sensors, switches, memories, photo-mechanical transducers and so on.

Table II. Photocontrol of Physical and Chemical Properties of Polymer Solutions and Solids

| Solution | Solid |
|---|---|
| Viscosity (3-12,27,28) | Membrane Potential (35) |
| pH-value (5,29) | Wettability (36) |
| Solubility (30-32) | Shape (19,20,37-45) |
| Metal Ion Capture (7,34) | Sol-Gel Transition (46,47) |
| | Tg (48) |
| | Compatibility of Polymer Blends (49) |
| | Absorptive Ability (50) |

## Acknowledgments

Acknowledgment is made to the Donor of the Petroleum Research Fund, administered by the American Chemical Society, for partial support of this activity.

## Literature Cited

1. Morawetz, H. Macromolecules in Solution; 2nd ed., Wiley: New York, 1975; p 344.
2. Irie, M. In Molecular Models of Photoresponsiveness; Montagnoli, G.;Erlanger, B. F., Ed.; Plenum: New York, 1983;p 291.
3. Irie, M. In Photophysical and photochemical Tools in Polymer Science; Winnik, M.A., Ed.,; Reidel: Dordrecht, 1986; p 269
4. Irie, M.; Hayashi, K. J. Macromol. Sci. Chem. 1979, A13, 511.
5. Irie, M.; Hirano, K.; Hashimoto, S.; Hayashi, K. Macromolecules 1981, 14, 262.
6. Blair, H.S.; Pogue, H.I.; Riordan, J.E. Polymer, 1980, 21, 1195.
7. Kumar, G.S.; DePra, P.; Neckers, D.C. Macromolecules, 1984, 17, 2463..
8. Zimmermann, E.K.; Stille, J.K. Macromolecules, 1985, 18, 321.
9. Irie, M.; Menju, A.; Hayashi, K. Macromolecules, 1979, 12, 1176.
10. Menju, A.; Hayashi, K.; Irie, M. Macromolecules, 1980, 14, 755.
11. Irie, M.; Hayashi, K.; Menju, A. Polymer Photochem. 1981, 1, 233.
12. Irie, M.; Hosoda, M. Makromol. Chem. Rapid Commun. 1985, 6, 533.
13. Merian, E. Text. Res. J. 1966, 36, 612.

14. Smets, G. Adv. Polym. Sci. 1983, 50, 18.
15. Matejka, L.; Dusek, K.; Ilavsky, M. Polymer Bull. 1979, 1, 659.
16. Matejka, L.; Ilavsky, M.; Dusek, K.; Wichterle, O. Polymer 1981, 22, 1511.
17. de Lange, J.J.; Robertson J.M.; Woodward, J. Proc. R. Soc. 1939, A171, 398.: Mampson, G.C.; Robertson, J.M. J. Chem. Soc. 1942, 409.
18. Herz, M. L. J. Am. Chem. Soc. 1975, 97, 6777.
19. Irie, M.; Kungwatchakun, D. Makromol. Chem. Rapid Commun. 1984, 5, 829.
20. Irie, M.; Kungwatchakun, D. Macromolecules 1986, 19, 2476.
21. Tanaka, T.; Fillmore, D.J. J. Chem. Phys. 1979, 70, 1214.
22. Glignon, J.; Scallan, A. M. J. Appl. Polym. Sci. 1980, 25, 2829.
23. Ricka, J.; Tanaka, T. Macromolecules 1984, 17, 2916.
24. Tanaka, T.; Nishio, J.; Sun, S-T.; Ueno-Nishio, S. Science 1982, 218, 467.
25. Osada, Y.; Hasebe, M. Chem. Lett. 1985, 1285.
26. Irie, M. Macromolecules 1986, 19, 2890.
27. Irie, M.; Schnabel, W. Makromol. Chem. Rapid Commun. 1984, 5, 413.
28. Irie, M.; Schnabel, W. Macromolecules 1986, 19, 2846.
29. Irie, M. J. Am. Chem. Soc. 1983, 105, 2078.
30. Irie, M.; Tanaka, H. Macromolecules 1983, 16, 210.
31. Irie, M.; Iwanaga, T.; Taniguchi, Y. Macromolecules 1985. 18, 2418.
32. Irie, M.; Schnabel, W. Macromolecules 1985, 18, 394.
33. Kumar, G. S.; DePra, P.; Neckers, D. C. Macromolecules, 1984, 17, 1912.
34. Shinkai, S.; Kinda, H.; Ishihara, M.; Manabe, O. J. Poly. Sci. Chem. 1983, 21, 3525.
35. Irie, M.; Menju, A.; Hayashi, K. Nippon Kagaku Kaishi 1984, 227.
36. Ishihara, K.; Hamada, N.; Kato, S.; Shinohara, I. J. Polym. Sci. Chem. 1983, 21, 1551.
37. Agolini, F.; Gay, F.P. Macromolecules 1970, 3, 349.
38. Van der Veen, G.; Prins, W. Nature, Phys. Sci. 1971, 230, 70.
39. Smets, G.; Evans, G. Pure Appl. Chem. Suppl. Macromol. Chem. 1973, 8, 357.
40. Eisenbach, C. D. Polymer 1980, 21, 1175.
41. Osada, Y.; Katsumura, E.; Inoue, K. Makromol. Chem. Rapid Commun. 1981, 2, 411.
42. Blair, H.S.; Pogue, H.I. Polymer, 1982, 23, 779.
43. Ishihara, K.; Hamada, N.; Kato, S.; Shinohara, I. J. Polym. Sci. Chem. 1984, 22, 121.
44. Aviram, A. Macromolecules 1978, 1, 1275.
45. Kohjiya, M.; Hashimoto, T.; Yamashita, S.; Irie, M. Chem. Lett. 1985, 1479.
46. Irie, M.; Iga, R. Makromol. Chem. Rapid Commun. 1985, 6, 403.
47. Irie, M.; Iga, R. Macromolecules 1986, 19, 2480.
48. Irie, M.; Mohri, M.; Hayashi, K. Polym. Prep. Jpn. 1985, 34, 716.
49. Irie, M.; Iga, R. Makromol. Chem. Rapid Commun. 1986, 7, 751.
50. Okamoto, Y; Sakamoto, H; Hatada, K; Irie, M. Chem. Lett. 1986, 983.

RECEIVED March 13, 1987

Chapter 11

# Luminescence Studies of Molecular Motion in Poly(*n*-butyl acrylate)

**J. Toynbee and I. Soutar**

**Chemistry Department, Heriot-Watt University, Riccarton, Currie, Edinburgh, EH14 4AS, Scotland**

Relaxation processes in solid poly(n-butyl acrylate) have been studied using both luminescent labelling and probe techniques. A wide range of effective test frequencies have been accessed through the employment of both singlet and triplet excited states as reporter molecules. Segmental relaxation of the polymer in the region of the glass transition has been studied by emission anisotropy measurements on samples labelled with acenaphthylene and 1-vinylnaphthalene respectively. In addition probe interrogation of the matrix microviscosity has revealed an involvement of the polymer matrix in the photophysical behaviour of the triplet state of naphthyl chromophores and information germane to speculation over the existence of triplet excimers of naphthalene is discussed.

The application of luminescence techniques to the study of macromolecular behaviour has enjoyed an enormous growth rate in the last decade. The attraction of such methods lies in the degrees of both specificity and selectivity afforded to the investigator. Consequently the polymer may be doped or labelled at sufficiently low concentration levels of luminophore as to induce minimal perturbation of the system. Polarized photoselection techniques offer particular attraction in the study of relaxation phenomena both in solution and solid states. In principle, astute labelling can allow elucidation of the molecular mechanisms responsible for the macroscopic relaxations exhibited by the macromolecular system. In addition, luminescent probes can address the microviscosity of their environment.

The use of both singlet and triplet states as molecular reporters permits access to an extremely wide range of effective test frequencies. Fluorescence depolarization techniques have been employed extensively to study the relaxation behaviour of macromolecules in solution (*cf*. references (1 - 3) and references therein). Such anisotropy measurements have also been used, to a much lesser extent, to study high frequency relaxation processes in the polymeric

0097-6156/87/0358-0123$06.00/0

solid state. (3 - 6) Phosphorescence depolarization, on the other hand, offers interesting possibilities to study the macromolecular solid state at much longer time scales upon which greater resolution of the various mechanisms which constitute the overall viscoelastic or dielectric relaxation spectra might be anticipated. Whilst the potential of phosphorescence polarization measurements has been recognized for many years very few successful applications of the technique have been reported (9 - 11) for empirical reasons.

In the current paper we extend previous work on poly(methyl acrylate) (9 - 10) and poly(methyl methacrylate) (9 - 11) in which the phosphorescence depolarization technique was shown to provide data which were consistent with reported dielectric and mechanical relaxation experiments, to the study of the molecular behaviour of poly(n-butyl methacrylate). This polymer, whilst of technological application, has received much lesser attention using conventional dynamic relaxation techniques than has been devoted to PMA and PMMA. In addition, fluorescence depolarization measurements have been employed in an attempt to provide complementary information regarding the higher frequency behaviour of the polymer.

In addition, luminescence intensity and lifetime data obtained not only from the labelled polymer but also from emission of dispersed naphthalene, acenaphthene and 1,1-dinaphthyl-1,3-propane (DNP) are briefly discussed. These measurements have provided information not only of relevance to the relaxation behaviour of the polymer matrix but also to the photophysics which occur therein and intramolecularly within the DNP.

## Experimental

Materials. Acenaphthene (Ac), acenaphthylene (ACE) and naphthalene (Np) (Aldrich) were purified by multiple recrystallization from ethanol followed by multiple sublimation under high vacuum.

1-Vinylnaphthalene (1-VN) was prepared by dehydration of methyl-1-naphthylcarbinol (Koch-Light) with 10% potassium hydrogen sulphate and purified by fractional distillation under reduced pressure.

1,1-dinaphthyl-1,3-propane (DNP) was prepared according to the method of Chandross and Dempster (12).

*n*-Butyl acrylate was freed of inhibitor, vacuum degassed, prepolymerized and purified by high vacuum fractional distillation immediately prior to use.

Poly(*n*-butyl acrylate) (PBA) was prepared by free radical polymerization (AIBN) under high vacuum to less than 10% conversion at 60%C. Labelled PBA was obtained by copolymerization with *ca.* 0.2 mole % 1-VN or ACE to produce polymers P/VN and P/ACE respectively. The polymers were purified by multiple reprecipitation from benzene into cold methanol or 40/60 petroleum ether. Molar masses of the resultant polymers estimated by gpc exceeded $10^6$ and comprized mononodal distributions.

Sample Preparation. Polymer samples for luminescence investigation were produced as thin polymer films by solvent casting from $CH_2Cl_2$

in quartz cells. Residual solvent was removed under high vacuum at $100^{o}C$ for 5-10 days prior to sealing.

Instrumentation. Steady state luminescence measurements were made using a Perkin-Elmer MPF-3L spectrofluorimeter. Samples were contained in an optical dewar and temperature variation achieved using a flow of pre-heated or cooled dry nitrogen gas. Phosphorescence measurements were made using a rotating-can phosphoroscope.

Fluorescence lifetimes and time-resolved spectroscopic measurements were obtained using an Edinburgh Instruments 199 spectrometer using the time-correlated single photon counting technique using excitation from a coaxial flash lamp with hydrogen as the discharge medium. Temperature control was achieved using an Oxford Instruments Cryostat. The 199 spectrometer was also modified to accomplish phosphorescence decay analysis using a microsecond flashlamp. The experimental arrangement has been described (13).

## Results and Discussion.

Phosphorescence Polarization Measurements. The degree of polarization was determined as a function of temperature for the systems P/ACE, P/VN and P-Np (dispersed naphthalene probe). The excitation wavelength was 310 nm and that of analysis, 490 nm. The onset of depolarization was observed at *ca.* 210, 195 and 180K for P/ACE, P/VN and P-Np respectively. The onset temperature for P/ACE is in good agreement with that of the conventional value for $T_g$, consistent with the low frequency equivalence of the time-scale imposed by the phosphorescent label at this temperature in the PBA matrix.

The polarization data were transformed into apparent relaxation times, $\tau_r$, *via* the Perrin equation

$$(P^{-1} - 1/3) = (P_o^{-1} - 1/3)(1 + 3\tau_e/\tau_r) \qquad (1)$$

where $\tau_e$ represents the lifetime of the electronically excited state and $P_o$ is the intrinsic polarization, estimated in this instance as that attained at low temperatures (77K).

The resultant data are plotted in Arrhenius form in Figure 1 and the resultant values for the activation energies of the relaxation processes are listed in Table I.

Table I. Temperature Dependence of Relaxation Times

| System | $E_a$/kJ $mol^{-1}$ |
|---|---|
| P/ACE | 98 ± 5 |
| P/VN | *ca.* 100<br>44 ± 5 |
| P-Np | 100 ± 5<br>33 ± 5 |

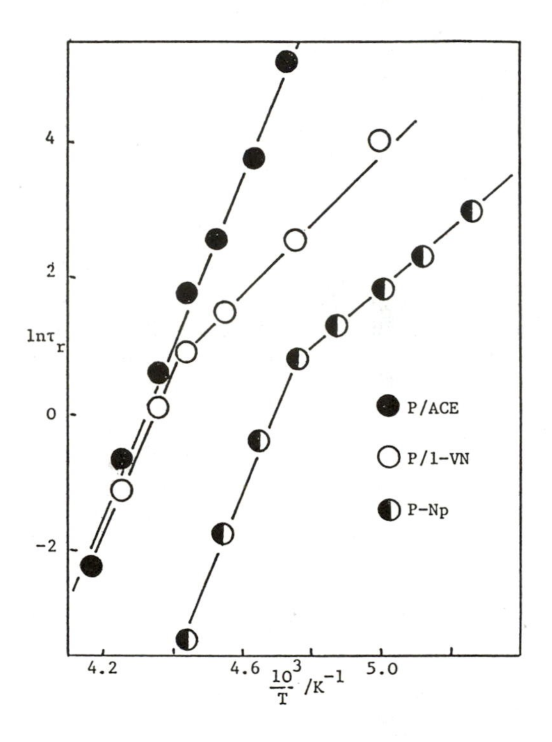


Figure 1. Temperature dependence of relaxation times derived from phosphorescence depolarization data.

The mode of attachment of the labels to the polymer chain is depicted below.

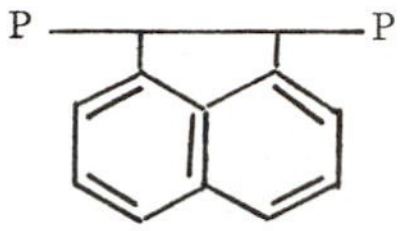

P/ACE 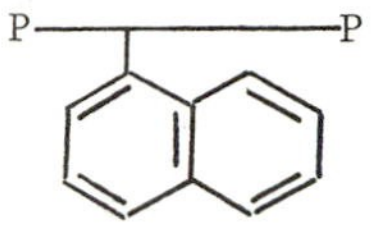 P/1-VN

The acenaphthylene label, being incapable of motion independent of the polymer chain, is expected to monitor segmental relaxation of the macromolecule. The relaxation data adhere well to an Arrhenius relationship over the relatively restricted frequency-temperature range sampled. This behaviour is similar to that observed for the PMA relaxation as determined by phosphorescence depolarization(9,10) but contrasts with that of PMMA in which backbone motion of lesser activation energy was sensible at temperatures inferior to $T_g$ (11). It is apparent that the conjunct roles of ester group and backbone substituents are important in determination of the extent to which backbone segments experience mobility as the glass transition is approached. In this respect the influence of tacticity in 1,1-disubstituted alkyl methacrylate polymers requires investigation.

The greater degree of freedom enjoyed by the 1-naphthyl label in P/VN is most likely ascribable to motion independent of the polymer chain about the bond of attachment. Since the vectors for absorption and emission are located within different planes within the molecular framework for triplet emission, such motions constitute a mechanism of enhanced depolarization for the P/VN system. In the case of PMMA it has been shown (11) that independent motion of a 1-VN label occurs in the vincinity of the relaxation of the polymer but is characterized by an apparent activation energy inferior to that sensed by an ester label or as afforded by dielectric and dynamic relaxation data for the β-process. Consequently whilst the onset of motion might be consistent with an increased degree of free volume released by the β-mechanism, the activation energy for naphthyl group reorientation within the lifetime of the excited states should not be equated with that of the β-process itself in PBA.

Motion of the dispersed naththalene triplet probe in P-Np is evident at 180K. By analogy with previous observations of the PMMA system(11) such mobility might be concomitant with the onset of the β-relaxation. It is apparent from dynamic mechanical data at 1Hz that resolution of the α and β relaxations is difficult(14). It may be that the use of molecular probes and labels such as 1-VN (as described above) offers a more sensitive sensory exploration of the matrix than is afforded by macroscopic techniques. It is also possible that the degrees of freedom for motion imbued in the matrix do not result from the motion of ester groups as generally inferred in PMA and PMMA for the β-relaxation but that conformational changes within the *n*-butyl substituent itself, constitute sufficient release of free volume within the matrix to allow rotational mobility of both the 1-VN label and the dispersed naphthalene probe.

It is apparent that our concepts of the relaxational processes in well studied systems such as the poly(n-alkyl acrylates) and poly(n-alkyl methacrylates) warrant further study and that luminescence probe and labelling techniques offer unique opportunities for the study of the less understood systems which constitute the majority of polymer solids.

Phosphorescence Intensity Measurements. The intensity of phosphorescence emission from the system P/ACE, P/VN and P/Np ($\lambda_e$ = 310 nm; $\lambda_a$ = 490 nm) was monitored as a function of temperature.

Consider the photophysical kinetic scheme:

$$^1M + h\nu_a \rightarrow {}^1M^* \qquad I_a \qquad (2.1)$$

$$^1M^* \rightarrow {}^1M + h\nu_M \qquad k_{FM}[^1M^*] \qquad (2.2)$$

$$^1M^* \rightarrow {}^1M \qquad k_{IM}[^1M^*] \qquad (2.3)$$

$$^1M^* \rightarrow {}^3M^* \qquad k_{TM}[^1M^*] \qquad (2.4)$$

$$^3M^* \rightarrow {}^1M + h\nu_T \qquad k_{PT}[^3M^*] \qquad (2.5)$$

$$^3M^* \rightarrow {}^1M \qquad k_{IT}[^3M^*] \qquad (2.6)$$

in which the rate constants $k_{IM}$ and $k_{IT}$ refer to all radiationless deactivation steps of the excited singlet and triplet states respectively and can be expanded to accommodate bimolecular quenching; $k_{FT}$ and $k_{PT}$ are the radiative rate coefficients associated with fluorescence and phosphorescence respectively.

The quantum yield of phosphorescence, $\phi_p$, may be written as

$$\phi_p = \phi_{TM}[k_{PT}/(k_{PT} + k_{IT})] \qquad (3)$$

where $\phi_{TM}$ is the quantum yield of intersystem crossing.

The intensity of phosphorescence, $I_p$, is proportional to the quantum yield of phosphorescence. Separation of the non-radiative rate coefficient $k_{IT}$ into temperature-independent and thermally activated components furnishes the expression (15,16).

$$1/I_p - 1/I_o = B \exp(-E/RT) \qquad (4)$$

where $I_o$ is the phosphorescence intensity at absolute zero and E is the "activation energy".

Somersall et al(16) have shown that plots of $\ln(I_p)$ as a function of $T^{-1}$ are useful in exhibition of "transition temperatures" in solid polymers. The limitations of such representations in evaluation of "activation energies" has been discussed(10,11). In the current work, discontinuities in such plots occurred at two "transition temperatures" for each of the systems P/ACE and P/VN and P - Np as shown in Figure 2. In the case of the labelled systems, P/ACE, and P/VN, the higher temperature discontinuity occurs at ca. 215K in good agreement with that of ca. 218K furnished by dielectric and mechanical relaxation studies at 1Hz(17) for the α-transition (glass transition) of the polymer. The onset temperature for this transition as sensed by the naphthalene probe in P - Np is 200K.

It is likely that "molecular probing" of the microenvironment accurately reflects the glass-rubber transition of the polymer in this instance as a consequence of the influence of polymer mobility upon radiative and radiationless deactivation pathways of the excited triplet state. Indeed, the correlation between "true" transition temperatures and phosphorescence intensity changes is stronger the greater the concentration of bimolecularly quenching ground state species (such as oxygen)(10,11)

Plots of $\ln(I_p - I_o)$, where $I_o$ is approximated in this study as the intensity of phosphorescence at 77K, provide apparent energies of activation of the photophysical deactivation of the triplet naphthalene species, which are in broad agreement for all of the systems studied.

A lower temperature discontinuity in the $\ln I_p$ vs. $T^{-1}$ plots is observed for all of the systems studied is apparent at ca. 140K. In view of the data furnished by polarization data (vide supra) and of the problems associated in dynamic mechanical relaxation with separation of the α- and β- transitions of the polymer(14) it is suggested that this "transition" is photophysical in origin as discussed below.

A caveat must be sounded as to the use of intensity (or lifetime) data accrued through the use of triplet probes in sensing polymeric transitions in this manner. The probe/label data are complementary but vicarious in the nature of the information relayed to the observer.

Apparent energies of activation through the use of equation (4) of ca. 20 and 14 kJ mol$^{-1}$ respectively were recorded for the upper and lower transition temperature transitions apparent from $I_p$ data.

Phosphorescence Lifetime Data The lifetime, τ, of the triplet state is related to the rate coefficients of radiative ($k_{PT}$) and nonradiative ($k_{IT}$) decay by the relation

$$\tau^{-1} = k_{PT} + k_{IT} \quad (5)$$

Separation of the radiationless decay-rate coefficient into temperature dependent and independent components yields the expression

$$\tau^{-1} - \tau_o^{-1} = C(\exp - E/RT) \quad (6)$$

where $\tau_o^{-1} = k_{PT} + k'_{IT}$, $k'_{IT}$ being the temperature independent component of $k_{IT}$.

Activation energy data were afforded by equation (6) translated into Arrhenius form. Plots of $\ln\tau_p$ as a function of $T^{-1}$ yielded apparent transition temperatures in agreement with those afforded by intensity data (as reported above). Activation energies in the high and low temperature regimes were not in general accord with those registered using intensity data being ca. 50 and 3kJ mol$^{-1}$ respectively.

These data, taken at face value, might be interpreted as being indicative of a temperature dependence of $\phi_{TM}$ by comparison of equations (3) and (5). However, the photophysics of the polymer systems are much more complex than hitherto supposed(10,11) as discussed below. In particular, the decays of phosphorescence intensity are not exponential, and the lifetime data used in equation (6) for the estimation of energies of activation are those obtained from the "tails"of the decays. Similar methods were adopted for the estimation of $\tau_e$ values in the evaluation of relaxation data *via* equation (1).

It is apparent that the use of phosphorescence data in the study of polymeric transitions is fraught with complications and the use of steady-state intensity data and low time-resolution decay data, in particular, are of limited application to the study of relaxation phenomena.

*Nature of the Lower Temperature Transition*. The complex non-exponential phosphorescence decays, apparent under the higher time resolution afforded by use of the modified 199 spectrometer are a consequence of interjection by the polymer matrix in the photophysics experienced by the chromophore. Non-exponential decays of triplet naphthalene (and other chromophore) emissions have been observed in PMMA. Horie et al.(18-20) ascribe such effects to dynamic, inter-molecular quenching of the excited state by the polymer whereas MacCallum et al(21-23) invoke an energy migrative process within the polymer following quenching of the triplet state of the naphthalene.

In the PBA matrix, exponential decays were not observed at any temperature in excess of 140K. In addition, delayed fluorescence was observable from the naphthyl chromophores at higher photon fluxes. Considering the low concentrations of chromophore and the fact that for the labelled systems, diffusive motion of the naphthyl species is prohibited, these data are supportive of the MacCallum mechanism.

These data shall be published more fully elsewhere.

*Fluorescence Measurements*. Depolarization becomes apparent at *ca*. 270K. The initial steady increase of $P^{-1}$ with temperature at temperatures in excess of 270K is followed by a dramatic enhancement at temperatures superior to *ca*. 380K. Figure 3 displays the data for the P/ACE system following transformation into relaxation times *via* equation (1). The curves for P/VN, P/ACE and P - Np may be divided into two pseudo-linear regimes yielding activation energies of *ca*. 60 and 25 kJ $mol^{-1}$ in the higher and lower temperature regions respectively.

It is apparent that;

(i) the dispersed naphthalene probe senses the same relaxation and experiences a similar temperature dependence for decay of its photo-selected orientation to those sensed by the labelled systems. This is somewhat surprizing considering the small molar volume of the probe. Jarry and Monnerie(4) have indicated that the dynamic behaviour of a particular fluorescent probe depends upon its size relative to that of the kinetic segments of the polymer involved in the transition. Similar correspondence to probe size have been reported using paramagnetic probe interrogations of polymer matrices(24).

(ii) invoking the onset of segmental motion, consistent with the occurrence of the $\alpha$ relaxation, in explanation of the depolarizations observed at 270K, this temperature is elevated some 50K compared to that obtained for the glass transition measured at low frequencies. This magnitude of time-temperature shift would be as generally expected for the shift of some eight decades of frequency to the fluorescence time base. Again, the lack of evidence for an independently resolved $\beta$ process on these time scales is to be expected.

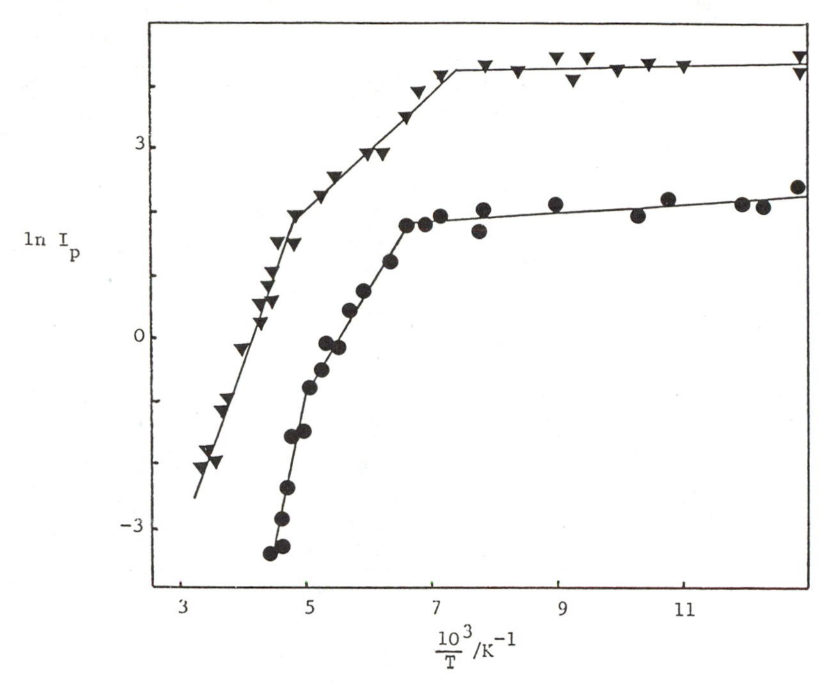


Figure 2. Temperature dependence of phosphorescence intensity in the systems P/ACE (▼) and P-Np (●)

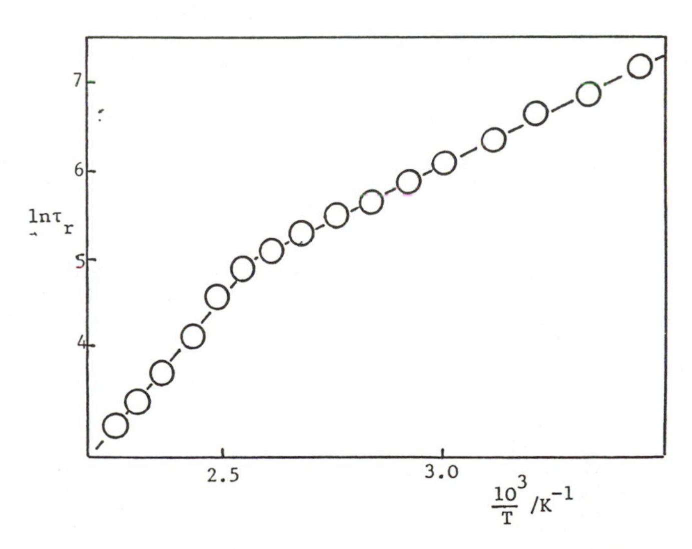


Figure 3. Temperature dependence of relaxation times derived from fluorescence depolarization data for the P/ACE system.

Consequently, using this interpretation, the higher temperature transition evident at ca. 380K might be ascribed to the occurrence of a liquid-liquid transition as the kinetic segment length involved in the relaxations of the melt alters. Alternatively the curvature of the plot may merely show that the relaxation data are poorly described, even over this limited frequency-temperature range, by an Arrhenius functional representation.
(iii) the "activation energies" obtained from the Arrhenius plots are low relative to low frequency phosphorescence probing. Whilst comparable dielectric and dynamic relaxation data are not available for PBA over a broad frequency range, the shift to lower apparent activation energy as frequency increases is generally anticipated for $\alpha$ relaxations.

Aplication of Intramolecular Excimer Formation to the Study of PBA Relaxation. DNP was dispersed in PBA and its photophysical behaviour compared to the P-Ac and P-Np systems.

Fluorescence studies of excimer formation in the probe give information regarding the microviscosity of the matrix. Unusually the probe seems to reflect the low frequency glass transition and caution must be applied in interpreting photophysical data obtained with such probes in terms of polymer relaxation mechanisms.

More interestingly, triplet state studies have produced date which expand our experience of intramolecular triplet state interactions and are relevant to the issue regarding the existence of triplet excimer species in naphthalene compounds(25-30). The data may be summarized as follows:
(a) At low temperatures, <180K, the naphthyl species exhibit similar phosphorescence spectra characteristic of an isolated naphthalene chromophore.
(b) At temperatures in excess of 180K the spectrum of the DNP is displaced to lower energies relative to those of naphthalene or acenaphthene (whose spectral maxima remain constant over the entire T range).
(c) If this behaviour is interpreted in terms of the formation of an excited state complex whose formation is dependent upon mobility for conformational change of the DNP allowed by the matrix, changes in the excitation spectrum of the DNP (in the high T region) reveal that the complex can not be considered to be truly excimeric in nature.
(d) Phosphorescence decays of the DNP are complex; due to energy transfer to the polymer matrix. The decays are wavelength dependent (being longer lived the lower the energy of analysis) consistent with excited state complex formation. However no "rising component" is discernible (in contrast to the decay profiles reported for fluid solution by Takemura et al(25,26) consistent with a ground state association.
(e) The sample of PBA/ACE (copolymerized acenaphthylene) when exposed (adventitiously) to trace amounts of oxygen exhibited anomalous decay behaviour, of the type reported as due to triplet excimers, at temperatures in excess of the $\beta$-relaxation temperature of the matrix (as shown by phosphorescence data). A "rising component" is clearly evident at longer wavelengths of analysis. These decays are similar to those observable in fluid media when oxygen concentrations are such as to allow observation of the anomalous emissions.

The results of the studies (a) - (c) described above are not in agreement with the existence of triplet state excimers of naphthalene. It is concluded that previous reports in the literature(25-29) are erroneous and that it is likely that some other type of complex such as an "oxyplex" is responsible for the observed behaviour and would be consistent with the fact that Pratte et al(30) did not observe triplet naphthalene excimers in bichromophoric systems under highly degassed fluid solutions.

Acknowledgments

The authors take pleasure in thanking the SERC for an equipment grant; SERC and ICI(UK) PLC for a CASE studentship (to J.T.) and Dr D.G. Rance (ICI Petrochemicals and Plastics Division) for many fruitful discussions.

Literature Cited

1. Soutar, I. In Developments in Polymer Photochemistry; Allen, N.S., Ed.: Applied Science: London, 1982; Vol. 3, Chapter 4.
2. Monnerie, L. In Polymer Photophysics; Phillips, D., Ed.: Chapman and Hall: London, 1985; Chapter 6.
3. Monnerie, L.; Viovy, J.-L. In Photochemical Tools in Polymer Science; Winnik, M.A., Ed.; D. Riedel: Dordrecht, 1986; NATO ASI Ser. C; Vol. 182, p 193.
4. Jarry, J. P.; Monnerie, L. Macromolecules 1979, 12, 927.
5. Viovy, J.-L.; Monnerie, L; Brochon, J. C. Macromolecules 1983, 16, 1945.
6. Viovy, J.-L., Monnerie, L; Merola, F. Macromolecules 1985, 18, 1130
7. Oster, G; Nishijima, Y; Forsch - Hochpolym. Forsch. 1964, 3, 313.
8. Nishijima, Y. J. Polym. Sci. 1970, C31, 353.
9. Rutherford, H.; Soutar I. J. Polym. Sci. Polym. Lett. Ed. 1978, 16, 131.
10. Rutherford, H.; Soutar I. J. Polym. Sci. Polym. Phys. Ed. 1977, 15, 2213.
11. Rutherford, H.; Soutar I. J. Polym. Sci. Polym. Phys. Ed. 1980, 18, 1021.
12. Chandross E. A.; Dempster C. J. J. Am. Chem. Soc. 1970, 92, 3586.
13. Soutar I. In Photophysics of Synthetic Polymers; Phillips, D.; Roberts, A. J., Eds.; Science Reviews: Northwood, 1982; p 82.
14. Kolarik, J. J. Polym. Sci. Polym. Phys. Ed. 1983, 24, 2445
15. Boustead, I. Eur. Polym. J. 1970, 6, 731.
16. Somersall, A. C.; Dan, E.; Guillet J. E. Macromolecules 1974, 7, 233.
17. McCrum N. G.; Read, B. E.; Williams G. Anelastic and Dielectric Effects in Polymer Solids; J. Wiley: London, 1967.
18. Horie, K.; Mita, I. Chem. Phys. Lett. 1982, 93, 61.
19. Horie, K.; Morishita, K.; Mita, I. Macromolecules 1984, 17, 1746.
20. Horie, K.; Mita, I. Euro. Polym. J. 1984, 20, 1037.
21. Jassim, A. N.; MacCallum, J. R.; Shepherd, T.M. J. Chem. Soc. Faraday Trans II 1979, 75, 79.

22. Jassim, N. N.; MacCallum, J. R. and Moran, K. T. Eur. Polym. J. 1983, 19, 909.
23. Jassim, A. N.; MacCallum, J. R.; Moran, K. T. Euro. Polym. J. 1984, 20, 425.
24. Kumler, P. L.; Boyer, R. F. Macromolecules 1976, 9, 903.
25. Takemura, T.; Aikawa, M.; Baba, H.; Shindo, Y. J. Am. Chem. Soc. 1976, 98, 2206.
26. Takemura, T.; Aikawa, M.; Baba, H. J. Lumin. 1976, 12/13, 819.
27. Subudhi, P. C.; Lim, E. C. Chem. Phys. Lett. 1976, 44. 479.
28. Webster, D.; Baugher, J. T.; Lim, B. T.; Lim, E. C. Chem. Phys. Lett. 1981, 77, 294.
29. Lim, B. T.; Lim, E. C. J. Chem. Phys. 1983, 78, 5262.
30. Pratte, J. F.; Webber, S. E.; De Schryver, F. C. Macromolecules 1985, 18, 1284.

RECEIVED April 27, 1987

Chapter 12

# Behavior of a Twisted Intramolecular Charge-Transfer Compound Bonded to Poly(methyl methacrylate)

**Ryuichi Hayashi [1], Shigeo Tazuke [1], and Curtis W. Frank [2]**

[1]Research Laboratory of Resources Utilization, Tokyo Institute of Technology, 4259, Nagatsuta, Midori-ku, Yokohama, Japan
[2]Department of Chemical Engineering, Stanford University, Stanford, CA 94305

A twisted intramolecular charge transfer (TICT) compound, a 4-(N,N-dimethylamino)benzoate group,was bonded to poly(methyl methacrylate)(PMMA) (the content of the TICT compound $<0.01\%$ of MMA unit), and the effects of polymer environment on its fluorescence were investigated in ethyl acetate. The contribution of the TICT state was much less for the polymer than for the monomer model compound (ethyl N,N-diamethyl-amino-benzoate), indicating restricted side chain rotation and/or reduced solvation in polymer even under extremely dilute condition. With increasing polymer concentration, the emission showed a blue shift as well as a decreased intensity up to polymer concentration of 70%. Above this concentration, the activation energy associated with twisting a dimethyl-amino group($E_1$) jumped suddenly. The twisting motion required free volume of $\sim$ 6 $cm^3$/mol, which was in good agreement with the activation volume obtained from pressure effect on the TICT phenomenon. The TICT phenomenon is shown to be sensitive to both polarity and microviscosity of surroundings so that the phenomenon is expected to be a good candidate for studying polymers by fluorescence techniques.

Twisted intramolecular charge transfer (TICT) compounds represented by N,N-dimethylaminobenzonitrile and N,N-dimethylaminobenzoate exhibit characteristic dual fluorescence spectra in polar fluid media [1]. An abnormal fluorescence band appearing at the longer wavelength region (a* band) of the normal fluorescence band (b* band) has been attributed to fluorescence from the intramolecular charge transfer state which has a twisted conformation of the dimethylamino group from the planar structure in the b* state [2]. Therefore, appearance of the two fluorescence bands is the essential condition to confirm the occurence of the TICT phenomenon. There are a number of compounds such as 9-(N,N-dimethylanilino)anthracene [3], (p-dimethylamino)-benzylidenemalononitrile [4] and so forth

0097-6156/87/0358-0135$06.00/0

exhibiting TICT-like emission but without clear separation of the a* and b* bands. The position, the shape and the intensity of these TICT-like emissions are sensitive to molecular mobility, solvation, and polarity of the microenvironment around the chromophore so that they can be candidates for studying polymers by fluorescence techniques [5]. Dual fluorescence of the TICT compounds would provide much more detailed and fertile information [6].

Basic study of the TICT phenomenon in polymers is meaningful in two ways. First, it is useful to understand both static and dynamic behaviors of the a* and b* bands in polymer as polymer photophysics. Secondly, it is useful to evaluate the TICT chromophores as fluorescence probes.

As a fluorescence probe, a TICT compound possesses several merits. It can probe both microviscosity and micropolarity and furthermore, the mode of molecular motion necessary to the TICT phenomenon is well defined and limited to the rotation around a particular bond. This is a difference from intramolecular exciplex or excimer formation in which multiple modes of bond rotation and bending are compounded [7]. Fluorescence polarization studies also can not be free from ambiguity of the mode of rotation [8].

To determine the usefulness and limits of the TICT chromophore as a fluorescence probe, we have to study the basic photophysics of the chromophore in polymers. The objective of this study is to examine the rotational motion of the amino group in polymer bonded 4-(N,N-dimethylamino)benzoate (DMAB) chromophore in polymer solutions. The effects of polymer concentration, temperature and pressure on the TICT phenomenon of DMAB in the PMMA side chain were compared with those of a small molecular model compound, ethyl N,N-dimethylaminobenzoate, in solution. Our concern is to have an inside look into how polymeric environment influences the TICT phenomenon.

## Brief Description of TICT Phenomenon

A series of conjugated compounds containing both electron withdrawing and electron releasing groups in a molecule sometime exhibit dual fluorescence in polar media. The explanation put forward by Grabowski [2] is now generally accepted. According to his TICT state hypothesis, the donor part (mostly dialkylamino group) in an initially planar electron deficient aromatic molecule rotates around the amino-phenyl bond to induce excited state charge separation within the molecule. The kinetic model is expressed in Figure 1.

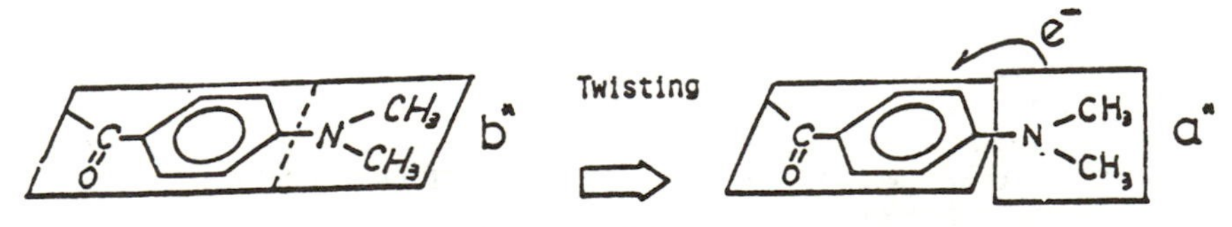


Scheme of Figure 1

The non-CT b* state produced directly by photoexcitation relaxes to the TICT a* state by crossing over a rotational barrier ($E_1$). On the basis of photostationary state approximation, the following kinetic expressions are obtained [9].

$$\Phi_b = \frac{k_{bf}(k_2+k_a)}{k_b(k_2+k_a)k_1K_a} \qquad (1)$$

$$\Phi_b = \frac{k_{af}k_1}{k_b(k_2+k_a)+k_1k_a} \qquad (2)$$

$$R = \Phi_a/\Phi_b = \frac{k_{af}k_1}{k_{bf}(k_2+k_a)} \qquad (3)$$

Being an equilibrium state, the temperature dependence of R is similar that of an excimer [10], exciplex [10], and specific cases of photoinduced electron transfer reactions [11]. The Arrhenius-type plots could exhibit both positive and negative slopes at lower and higher temperature regions, respectively. If $k_2$ is negligibly small in comparison with other deactivation processes from the a* state, that corresponds to the lower temperature region, eq. (3) is simplified as eq. (4). In the higher temperature region, the excited state equilibrium between a* and b* state is established, so that eq. (3) is rewritten as eq. (5).

$$R = \frac{k_{af}k_1}{k_{bf}k_a} \qquad (4)$$

$$R = \frac{k_{af}k_1}{k_{bf}k_2} \qquad (5)$$

## Polymer Effects on the TICT Phenomenon in Very Dilute Solution

Absorption and emission spectra of radical copolymer of methyl methacrylate with N,N-dimethylaminobenzoyloxyethylmethacrylate ( $< 0.01\%$, $Mn = 3 \times 10^5$)(I) and ethyl N,N-dimethylaminobenzoate (II) are compared in Figure 2. Except for a small red shift ($\sim 3$ nm) in I, the absorption spectra are almost identical. Although the presence of a ground state dimer or solute-solvent complex is known for N,N-dimethylaminobenzonitrile, the best studied TICT compound, there seems to be no such complications for the present studies. This is a favorable condition to study the TICT chromophore in polymers since polymers tend to enhance complicated molecular association even in extremely dilute solution [12] which overshadows the aimed photophysical phenomenon.

Contrary to absorption spectra, fluorescence spectra of I and II are largely different. While both I and II exhibit dual fluorescence at 350 nm and 450 nm, the intensity ratio of a* band to b* band (R) of II is much higher than that of I. Assignment of the a* and b*

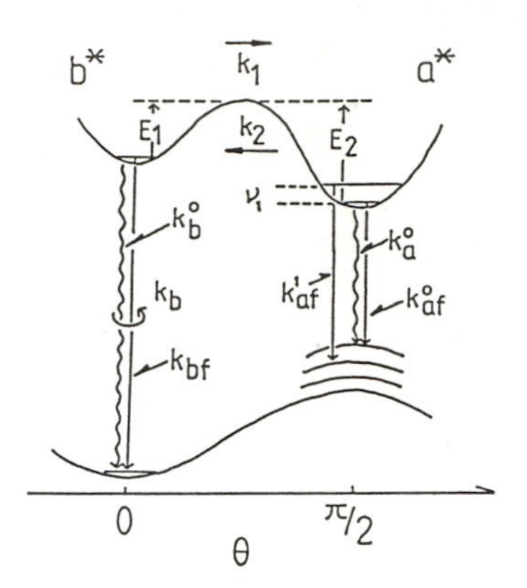


Fig. 1 Grabowski's kinetic model
The initially formed excited state (b* state) is converted to the charge separated state (a* state) by crossing over an energy barrier ($E_1$). Definitions of rate constants are as follows:

$k_1$ : forward rate constant of b*→a*.

$k_2$ : backward rate constant of b*←a*.

$k_{af}0, k_{af}1$ : radiative transition from a* state, but the radiation transition from the vibrationally lowest a* state is symmetry forbidden. Then $k_{af}1 >> k_{af}0$. The activation energy $h\nu_1$ is usually smaller than $E_1$. $k_{af} = k_{af}0 + k_{af}1 \exp(-h\nu_1/kT)$

$k_a = k_{af} + k_a^0$

$k_{bf}$, $k_b^0$ : radiative and non radiative transition from b* state, respectively.

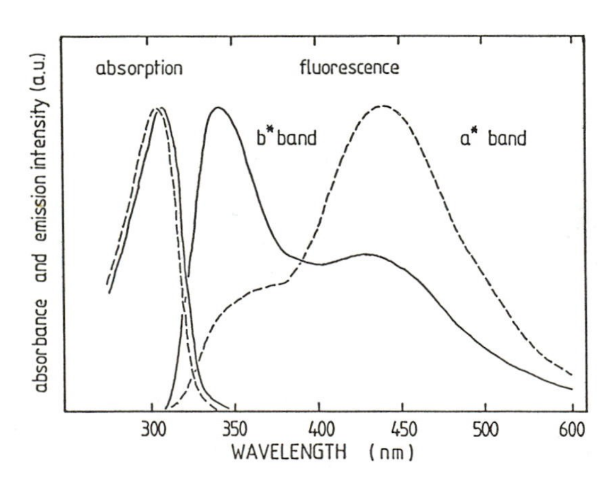


Fig. 2 Absorption and emission spectra in ethyl acetate at 25°C. [chromophore] = 1 x $10^{-4}$ M, real line: I ; dotted line: II.

bands of II has already been made [13]. The low value of R for I may be the reflection of hindered side chain motion or suppressed solvation in polymer. The shorter wavelength of the a* band in I than in II is also explainable as due to restricted solvation in I. If the a* state emits fluorescence with incomplete solvation owing to the surrounding polymer chain, the observed blue shift of the a* band in I is a reasonable consequence.

To gain further insight of the "polymer effect", temperature dependence of R is shown in Figure 3. The Arrhenius plots have a negative slope for I indicating that the condition of eq.(4) is satisfied whereas the full expression of eq.(3) is necessary to describe the temperature dependence of II. Since $k_{bf}$ is considered to be temperature independent and a small temperature dependence of $k_a$ corresponding to vibrational excitation in the a* state ($<10^2$ $cm^{-1}$) [21] would be negligible relative to $E_1$, the overall activation energies obtained from Figure 3 represent approximately $E_1$ and $E_1$-$E_2$. The back reaction ($k_2$) is thus negligible in comparison with $k_1$ for I whereas the excited state equilibrium is established above -30°C for II.

It is not well understood how torsional motion of a polymer side group is affected by the rest of the polymer chain in highly dilute solutions. In the case of intramolecular exciplex formation of 1-(1-pyrenyl)-3-(4-N,N-dimethylaminophenyl) propane bonded to polystyrene [14], the rate of exciplex formation in dilute polymer solution is much slower than that of a reference small molecule system.

Time resolved fluorescence spectroscopy provides direct information on the relation between the a* and b* states as shown in Figure 4. The spectra depicted in an arbitrary amplitude do not represent the actual time scale of rise-and-decay profile. However, it is clear that the a* band of I remains after dissipation of the b* band supporting a unidirectional process b* - a*. Both the a* and b* bands of II decay simultaneously with 3.8 nanoseconds of lifetime manifesting the equilibrium between two states.

## Polymer Concentration Effects on TICT Emission

The activation energies ($E_1$ + $h\nu_1$ $\doteq$ $E_1$) range between 10 kJ/mol in dilute solutions of I, as shown in Figure 5. With increasing the polymer concentration, the emission maximum of the a* band shows a high energy shift as much as 40 nm (Figure 6), indicating that the potential surface of the a* state is altered with concentration.

The value of $E_1$ is nearly constant up to 70% of the polymer concentration, then begins to increase with concentration. The break point in Figure 5 appearing at very concentrated regions means that the TICT probe is not sensitive to macroscopic viscosity but rather to microviscosity or local viscosity around the chromophore. Similar phenomenon has been reported for polystryene in benzene solution using fluorescence depolarization technique by Nishijima [15]. Based on the rotational relaxation time of naphthacene in solution, they indicated that the local viscosity was independent of polymer concentration below 15% and increased gradually with increasing the concentration from 15% to 60%. The results indicate a decrease in the extent of probe rotation with increasing the concentration from 15%

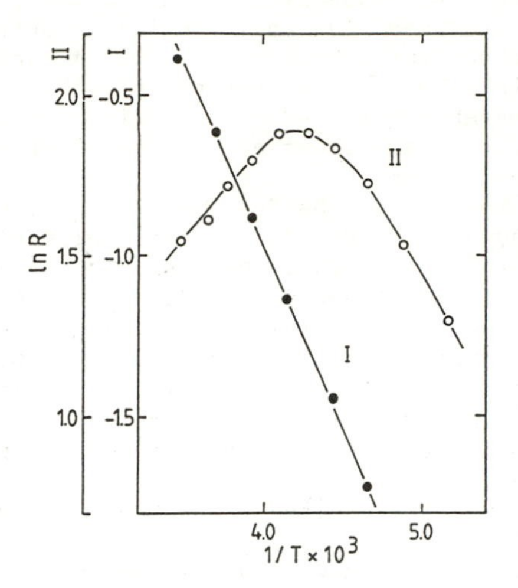


Fig. 3 Arrhenius plots of the TICT phenomenon ●: I, O: II.

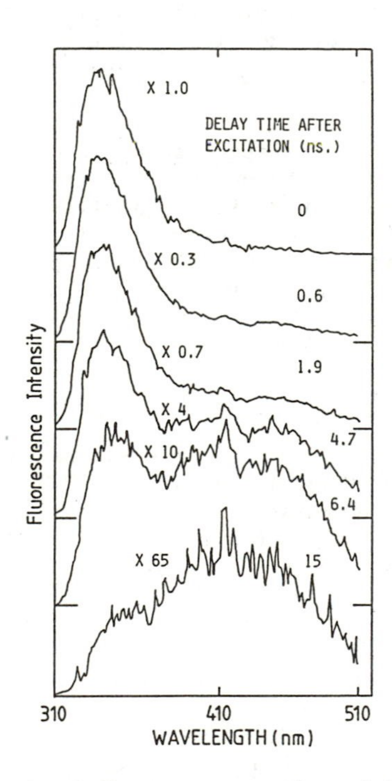


Fig. 4 Time resolved fluorescence of I with arbitrary amplitude.

to 60%. Although the probe size as well as the kind of polymer are different from the present system, their results correspond qualitatively to the results in Figure 6 rather than to those in Figure 5. The gradual blue shift of the a* band would consequently be attributed to decreased degree of solvent reorientation. An alternative explanation is also possible assuming reduced solvation with increasing polymer chain density around DMBA group which substitutes solvent molecules. The value of $\Phi_a/\Phi_b$ shows similar concentration dependence which can be explained as due to reduced solvation of the a* state giving rise to a decreased stationary concentration ratio of [a*]/[b*]. At this polymer concentration region, the solution contains enough free volume for the twisting motion, so that the value of $E_1$ is still kept constant.

Above the polymer concentration of 70%, $E_1$ starts to increase due to increased friction of rotation or reduced available free volume. The value of $E_1$ increases by a factor of three over the concentration range from 70% to 100% as shown in Figure 5, whereas the local viscosity measured by Nishijima increased ten times above 60%. As a result of extensive overlap of polymer chains in this concentration region, the rotational motion of the chromophores is highly restricted and subject to a high energy barrier to be crossed over to the a* state.

It is a difficult problem to fit the results in highly concentrated polymer solutions to the existing theories of polymer solutions. However, deGennes' mesh model and scaling concepts [16] might provide a qualitative explanation to these phenomena. According to the model, the molecular structure of a semi-dilute solution looks like a network with a certain average mesh size $\xi$ (correlation length). Inside this mesh, a small sphere with a diameter $D < \xi$ moves easily with a friction coefficient relating to the viscosity of the pure solvent. The motion of a sphere with a diameter $D < \xi$ is subject to the friction closer to the viscosity of the entangled solution. From small angle neutron scattering experiments on semi-dilute polystyrene [17], $\xi$ was determined to be in the order of $10^1$ - $10^2$ Å in the concentration range of $10^{-1}$ to $10^{-2}$ g $cm^{-3}$. This mesh size seems large enough to allow free rotation of of the DMA group, so that the friction relevant to $E_1$ is constant in this concentration region. With further increase in concentration, although it is outside the scope of the scaling treatment, it would be reasonable to propose that $\xi$ continues to decrease, eventually reaching the level of the probe size (= 10 Å). Then the hindrance imposed on the motion is expected to increase causing the dramatic rise in $E_1$.

## Interpretation by Free Volume Theory

In the highly concentrated region above 70%, the fluorescence spectra no longer depend on temperature below a characteristic temperature $T_b$ which is about 40 K above the glass transition temperature, $T_g$. The motion of the DMA group is strongly suppressed below this particular temperature. The concentration dependence of $T_b$ runs parallel with that of $T_g$ as shown in Figure 7. The free volume fraction ($V_f$) is correlated with $T_g$ according to the WLF equation (6) [18], where 0.025 represents the free volume

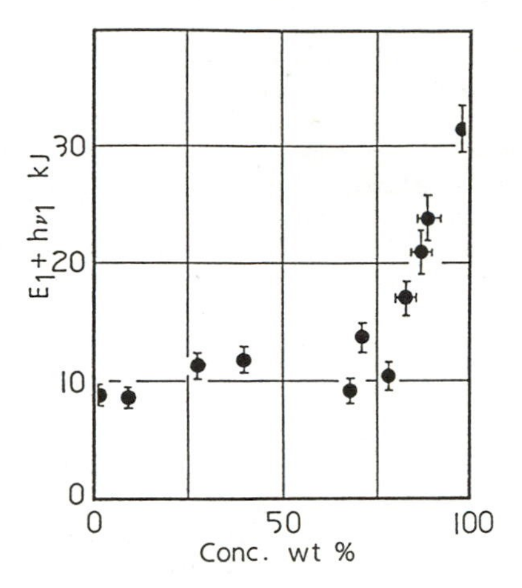


Fig. 5 Activation energy ($E_1 + h\nu_1$) of I as a function of polymer concentration.

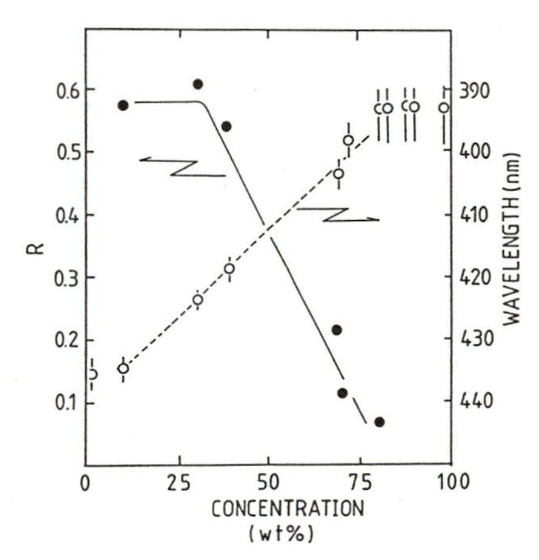


Fig. 6 Dependence of the energy and intensity of the TICT emission on polymer concentration. Sample: I at 25°C.

$$V_f = 0.025 + 4.8 \times 10^{-4}(T - T_g) \quad (6)$$

fraction at $T_g$. The solvent contribution to the free volume of a polymer solution of concentration C is given by eq. (7) [19], where

$$V_f = C[0.025 + 4.8 \times 10^{-4}(T - T_g)] + (1 - C)[0.025 + \alpha_s(T - T_g')] \quad (7)$$

$\alpha_s$ and $T_g'$ are the thermal expansion coefficient and the glass transtion temperature of the solvent. Using eq. (7), the concentration dependence of $T_g$ can be described by a line for $V_f = 0.025$ in Figure 7. Similarly, the values of $T_b$ fall in the region between two lines calculated for $V_f = 0.05$ and 0.07. This means that the rotational motion of the DMA group requires a free volume of about $5.4 \pm 0.8$ $cm^3$/mol at room temperature calculated by multiplying the volume for one mole of PMMA monomer unit by the free volume fraction.

## Pressure Effects on TICT Emission

Importance of solvent viscosity or free volume in the TICT phenomenon was discussed in the previous section. There are a number of ways to control viscosity of the medium. The easiest way is to change solvent, however, this brings about the complicated problem of influencing miscellaneous solvent effects. Use of mixed solvents also causes the ambiguity of selective solvation. The best method presumably is the study of pressure effects [20]. By applying hydrostatic pressure to a solution, the solvent reduces its free volume without much affecting other solvent properties.

Since the kinetic expressions for I and II are different, we cannot discuss the difference between I and II in terms of activation free volume for the formulation of the a* state. The pressure dependence of the rate constant $k_1$ in the case of I and the equilibrium constant $K(= k_1/k_2)$ in the case of II is expressed as eq. (8) and eq. (9) [21]. The rate constants except $k_1$ and $k_2$ in equations (4) and (5) are assumed to be independent of pressure. Applying eq. (8) and eq. (9), we obtained pressure effects on R depicted in Figure 8.

$$d(\ln k_1)/dP = -\Delta V^{\ddagger}/RT \quad (8)$$

$$d(\ln K)/dP = -\Delta V/RT \quad (9)$$

We can calculate the activation volume, $\Delta V^{\ddagger}$, for the formation of TICT state as 5.5 $cm^3$/mol for I, and volume difference, $\Delta V$, between the a* and b* states of II as 2.9 $cm^3$/mol.

$\Delta V$ is much more positive than the comparable values reported for ionization process in the ground state(-10 -- -30 $cm^3$/mol) [22]. In the excited state, the value becomes more positive. For example, the volume of ionization of nitrophenols [23] is more positive by 7-18 $cm^3$/mol than the value in the ground state. This has been reported to be due to a larger dipole moment of unionized nitrophenols in the excited state. The unionized excited state is already strongly solvated so that the loss of free volume brought about by

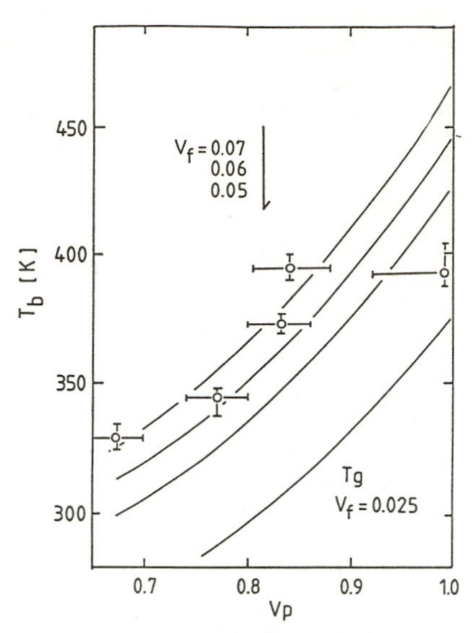


Fig. 7 Specific temperature ($T_b$) as a function of polymer fraction in solution. $T_b$ is the temperature at which linear Arrhenius plot of ln R breaks. The lines are based on eq. (7).

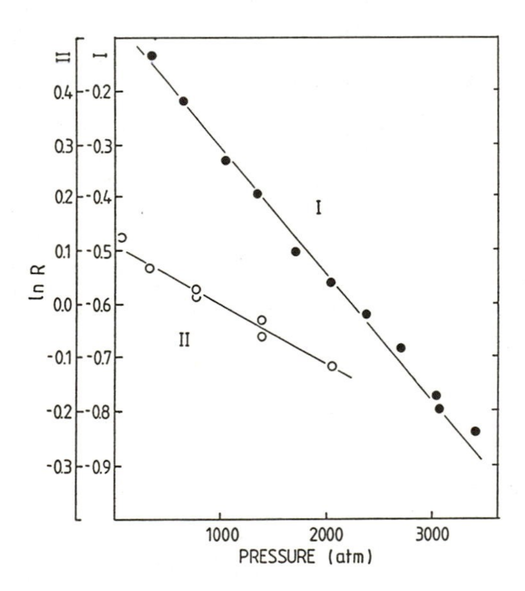


Fig. 8 Pressure dependence of the TICT emission intensity in ethyl acetate at 25°C.

●: I $\Delta V^{\ddagger}$ 5.5 $cm^3$/mol

O: II $\Delta V$ 2.9 $cm^3$/mol

ionization is less than the case of ground state ionization. More akin to the present system is the intramolecular ionization of sulphanilic acid. The ΔV of ionization was obtained as + 2 $cm^3$/mol [24] which agreed very well with + 2.9 $cm^3$/mol determined for II.

The value of activation volume for I is in excellent agreement with the free volume estimated in the previous section. The free volume is consequently a dominant factor deciding the ease of forming the TICT state. Both concentration and pressure effects can be thus interpreted by a single factor of free volume.

Knowing the pressure-viscosity relation, viscosity dependence of the R times η value for I is shown in Figure 9. The increase of Rη with increasing viscosity suggests that the torsional motion is not governed by simple Stokes - Einstein type diffusion. An empirical expression, $R\eta = A = B\eta^x$ [25] was used to explain the results in Figure 9. x is given by $x = (W_\eta - W_2/W_\eta$ where $W_2$ and $W_\eta$ are the required energies for diffusion into solvent voids and for viscous Stokes - Einstein flow, respectively [26]. The best fit was obtained when x was 0.83, corresponding to $W_2/W = 0.16$. The energy to diffuse into the solvent voids considered as the origin of free volume is only one sixth of the energy needed for macroscopic viscous flow described by Stokes. From the temperature dependence of viscosity of ethyl acetate, $W_\eta$ is estimated as 7.5 KJ/mol providing 1.2 KJ/mol of $W_2$. In the case of the excimer formation of pyrene, $W_2$ has been reported as 2.9, 3.9, and 6.8 KJ/mol for three solvents, n-nonane, cyclohexane, and i-propanol, respectively [25]. Considerably smaller $W_2$ than that of pyrene excimer formation, and the large deviation from Stokes - Einstein flow shown by the small value of x is probably relevant to the small rotor size. This chromophore will therefore be suitable to detect a small free volume change in highly viscous media.

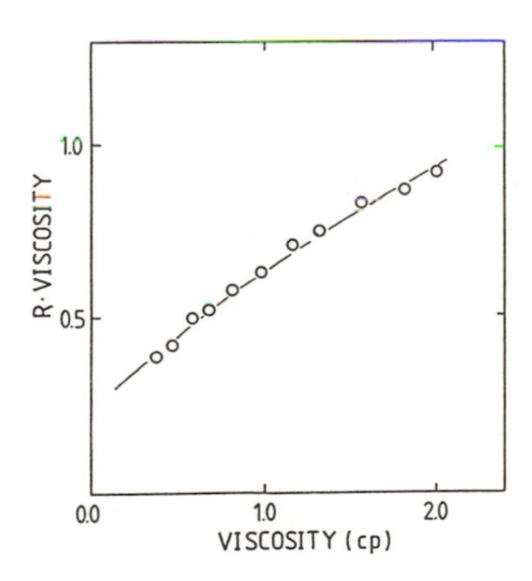


Fig. 9 Viscosity dependence of TICT emission intensity. Sample: I at 25°C. The line was drawn by computer fitting, $R\cdot\eta = 0.23 + 0.4 \times \eta^{0.83}$.

Literature Cited

1. Lippert, E.; Luder, W.; Boos, H. In Advances in Molecular Spectroscopy; Mangini, A., Ed.; Pergamon Press: Oxford, 1962 p. 443.

2. Rotkiewicz, K.; Grellman, K. H.; Grabowski, Z. R. Chem. Phys. Lett. 1973, 19, 315.

   Grabowski, Z. R.; Rotkiewicz, K.; Siemiarczuk, A.; Cowley, D. J.; Baumann, W. Nouv. J. Chim. 1979, 3, 443.

3. Okada, T.; Fujita, T.; Mataga, N. Z. Phys. Chem. N. F. 1976, 101, 57-66.

4. Loutfy, R. O.; Arnold, B. A. J. Phys. Chem. 1982, 86, 4205.

5. Loutfy, R. O. Macromolecules 1981, 14, 270.
   Loutfy, R. O. J. Polym. Sci. Polym. Phys. Ed. 1982, 20, 825.

6. Al-Hassan, K. A.; Rettig, W. Chem. Phys. Lett. 1986, 126, 273.

7. De Schryver, F. C.; Demeyer, K.; Toppet, S. Macromolecules 1983, 16, 89.

   Souter, I.; Phillips, D.; Roberts, A. J.; Rumbles, G. J. Polym. Sci.: Polym. Phys. Ed. 1982, 20, 1759.

8. Nishijima, Y. J. Macromol. Sci., Phys. 1973, 8, 407.

9. Grabowski, Z. R.; Rotkiewicz, K.,; Rubaszewska, W.; Kirlor-Kaminska, E. Acta Phys. Pol. 1978, A54, 767.
   Rettig, W. J. Lumin. 1980, 26, 21.

10. Birks, J. B. Photophysics of Aromatic Molecules; Wiley-Interscience: London, 1970; p 301.

11. Kim, H. B.; Kitamura, N.; Kawanishi, Y.; Tazuke, S. to be submitted.

12. Tazuke, S. Makromol. Chem. Suppl. 1985, 14, 145.

13. Cowley, D. J; Healy, P. J. Proc. Royal Irish Academy 1977, 397.
    Rettig, W.; Wermuth, G.; Lippert, E. Ber. Bunsenges. Phys. Chem. 1979, 83, 692.

14. Tazuke, S.; Iwasaki, R. Japan-US Polymer Symposium Preprints Oct., 1985, p 199, Kyoto, Japan.

15. Nishijima, Y.; Mito, Y. Repts. Progr. Polym. Phys. Jpn. 1967, 10, 139.

16. de Gennes, P. G.; Scaling Concepts in Polymer Physics; Cornell Univ. Press: London, 1979.

17. Dauod, M.; Cotton, J. P.; Farnoux, B.; Jamink, G.; Benoit, de Gennes, P. G. Macromolecules 1975, 8 804.

18. Williams, M. L.; Landel, R. F.; J. D. Ferry, J. D. J. Am Chem. Soc. 1955, 77, 3701.

19. Kelly, F. N.; Buech, F. J. Polym. Sci. 1961, L, 549.

20. Fitzgibbon, P. D.; Frank, C. W. Macromolecules 1981, 14, 1650.

21. Offen, H. W. In Organic Molecular Photophysics; Birks, J. B. Ed.; Wiley-Interscience: London, 1973, Vol. 1, p 103.

22. Isaacs, N. S. In Liquid Phase High Pressure Chemistry; Wiley-Interscience: 1981; p 162.

23. Hamann, S. D. Austr. J. Chem. 1975, 28, 701.

24. Asano, T.; LeNoble, W. J. Chem. Rev. 1978, 78, 407.

25. Alwatter, A. H.; Lumb, M. D.; Birks, J. B. In Organic Molecular Photophysics, Birks, J. B. Ed., Wiley-Interscience: London, 1973; Vol. 1, p 403.

26. Gierer, A.; Wirtz, K. Z. Naturf. 1953, 8a, 532.

RECEIVED May 22, 1987

# Chapter 13

# Electronic Spectroscopy of Squaraine

## Its Relationship with the Stabilization Mechanism of Squaraine Particles in Polymer Solutions

**Kock-Yee Law**

**Xerox Corporation, Webster Research Center, Webster, NY 14580**

The fluorescence emission of bis(4-dimethylaminophenyl)squaraine, **1**, in $CH_2Cl_2$ containing varying concentration of poly(vinyl formal) (PVF) has been studied. Three emission bands ($\alpha$, $\beta$ and $\gamma$ in the order of decreasing energy) are observed in $CH_2Cl_2$ solution and are found to be the emission from the excited state of **1**, from the excited state of a solute-solvent complex and from a relaxed twisted excited state of the solute-solvent complex, respectively. Model compound studies show that squaraine forms strong solute-solvent complexes with alcoholic solvent molecules. Analogous complexation process between **1** and the OH groups in PVF is also shown to occur. A model for the stabilization of particles of **1** in polymer solution is put forward where we propose that the stabilization mechanism is a steric effect achieved by adsorption of PVF macromolecules onto particles of **1** via the formation of the PVF:**1** complex.

Bis(4-dimethylaminophenyl)squaraine, **1** (1), and many of its derivatives are known to possess useful photoconductive and semi-conductive properties. Although this class of compounds exhibits sharp optical absorption in the red in solution ($\lambda_{max}$ ~620-670 nm, $\epsilon_{max}$~$3x10^5$ $cm^{-1}M^{-1}$), its absorption in the solid state is intense and panchromatic from the

0097-6156/87/0358-0148$06.00/0

visible to the near-IR (400-1000 nm). These optical characteristics have made them very attractive for xerographic photoreceptor (2) and organic solar cell applications (3). In these applications, squaraine based photoconductive devices are generally fabricated by casting various squaraine-polymer dispersions onto appropriate substrates. The stability of squaraine particles in polymer solution is thus of technological significance because it not only affects the film quality and the dispersity of particles in the squaraine layer, but also has profound influence on the photoconductivity of the resulting device. Recently, the stability of particles of 1 in various polymer solutions has been studied (Law, K.Y. J. Imaging Sci., in press). Flocculation of particles of 1 was generally observed except when poly (vinyl formal) and poly(vinyl butyral) were used. The superior dispersing effect of these poly(vinyl acetal) polymers on 1 is against general solubility principle because ethereal solvents are poor solvents for 1.

Here we report preliminary results on the multiple fluorescence emission of 1 and 2. From structure-property relationships, solvent effect and temperature effect studies, we are able to show that the multiple emission is from the emission of free squaraine in solution, the emission of the solute-solvent complex and the emission of a twisted relaxed excited state. Further solvent effect study using 2 as a model shows that squaraine forms strong solute-solvent complexes with alcoholic solvent molecules. Analogous complexation process is also detected between 1 and the hydroxy groups on the macromolecular chains of poly(vinyl formal). The important role of this complexation process on the stabilization mechanism of particles of 1 in polymer solution is discussed.

## Experimental

Squaraines 1 and 2 were synthesized from squaric acid and N,N-dialkylaniline derivatives using the procedure described by Sprenger and Ziegenbein (4). Poly(vinyl formal), PVF, formal content 82%, acetate content 12%, hydroxy content 6%, was purchased from Scientific Polymer Products Inc. Solvents were spectro grade from Fisher and were routinely stored over 3Å molecular sieves before use. Fluorescence spectra were taken on a Perkin-Elmer MPF 44A spectrofluorimeter which was equipped with a differential corrected spectra unit (DCSU-2).

## Results And Discussions

Figure 1 shows the fluorescence excitation and emission spectra of **1** in $CH_2Cl_2$. The excitation spectrum was found to be independent of the monitoring wavelengths and was identical to the absorption spectrum. Three emission bands at $\lambda_F$ 646, 660 and ~702 nm are observed in the emission spectrum and they are designated as the $\alpha$-, the $\beta$- and the $\gamma$-band respectively. Controlled experiments showed that the multiple emission is from **1** rather than from any impurities or any aggregational states of **1**.

Probable explanations for the multiple emission of **1** are (a) vibronic fine structure of the emission band; (b) emission from a relaxed excited state or (c) emission from an exciplex (with solvent) or an excited state of some kind of solute-solvent complex. In order to differentiate these possibilities, the effects of solvent and temperature on the multiple emission bands were studied. Squaraine **2** was chosen as a model compound for our investigation because of its high solubility in various organic solvents. Results in Figure 2 show that $\lambda_{max}$ of **2** undergoes a bathochromic shift as solvent parameter $\pi^*$ increases and this is accompanied with a change in the composition of the emission spectra (see inset in Figure 2). For example, in diethyl ether (Figure 3a), the emission spectrum of **2** is dominated by the $\alpha$-band with $\lambda_F$ at 641 nm. The excitation spectrum is independent of the monitoring wavelength and is identical to the absorption spectrum. Because of the good overlap and mirror image relationship between the absorption and the emission spectrum, the spectral result in Figure 3a suggests that the $\alpha$-band is an emission from the Franck-Condon excited state of **2**. An emission shoulder at $\lambda_F$ 660 nm and a small band at $\lambda_F$ 698 nm are also seen in Figure 3a and these two emission bands are assigned to the $\beta$- and the $\gamma$-band according to their Stokes shifts. If one assumes that the multiple emission is due to fine vibrational structures, the $\alpha$-band and the $\beta$-band would be from the (0,0) and (0,1) transitions respectively. From the splitting between the $\alpha$- and the $\beta$-band in Figure

1 R = $CH_3$
2 R = n-$C_4H_9$

$-(CH(CH_2)CH)_x-(CH_2-CH(OAc))_y-(CH_2CH(OH))_z-$ (formal ring: O–$CH_2$–O)

PVF x=82% y=12% z=6%

Squaraines 1 and 2

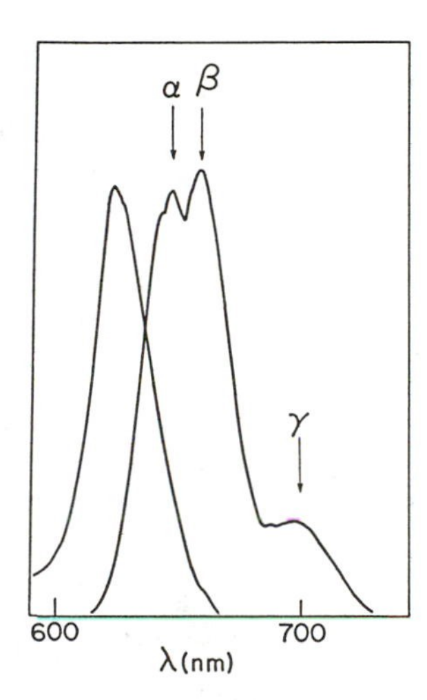


Figure 1. Corrected fluorescence excitation and emission spectra of **1** in methylene chloride ([1]~$3\times10^{-7}$ M).

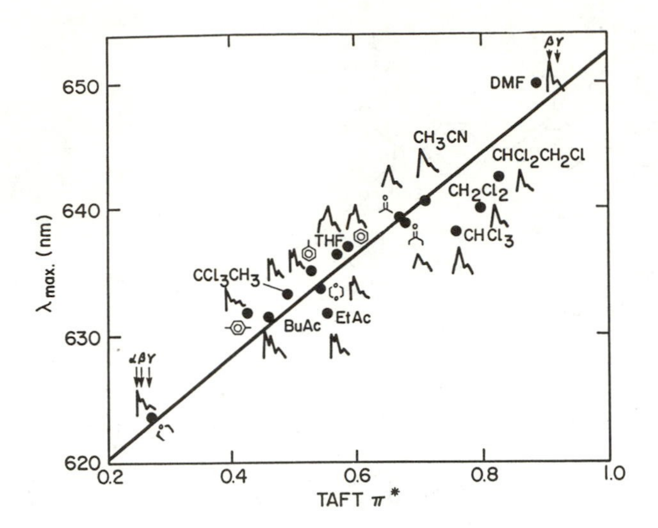


Figure 2. Plot of $\lambda_{max}$ of 2 as a function of solvent parameter $\pi^*$.

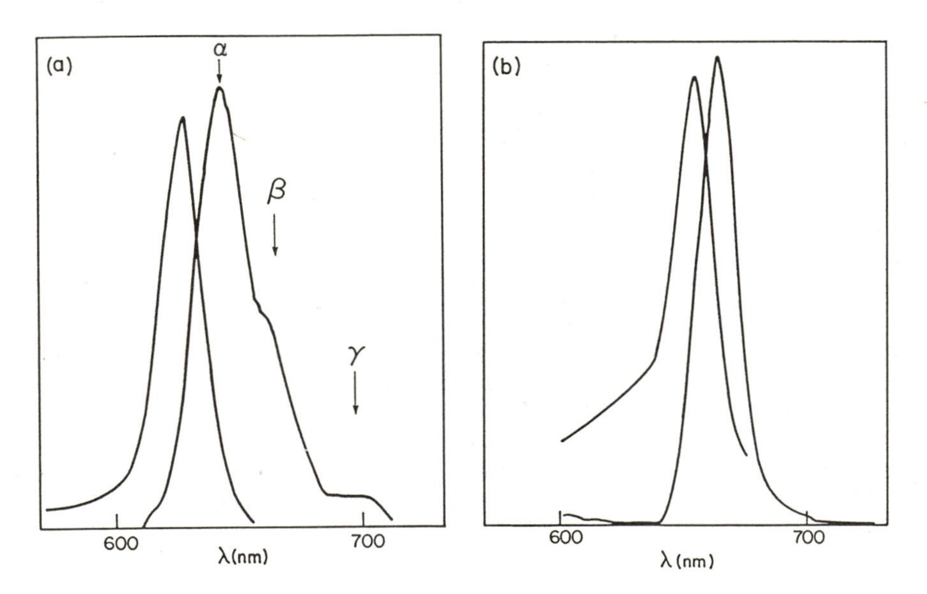


Figure 3. Corrected fluorescence excitation and emission spectra of 2 in diethyl ether (a) at 298°K (b) at 77°K (conc. ~$5 \times 10^{-7}$ M).

3a, the $\gamma$-band would be the (0,3) transition. The (0,2) transition which is expected to be ~675 nm in Figure 3a, is missing. Careful inspection of all the emission spectra taken in our laboratory fails to identify any emission band or shoulder corresponding to the (0,2) transition. Solvents are known to affect the relative intensity of vibrational bands in aromatic hydrocarbons. Notable examples are the effect of solvent on the absorption fine structure of benzene (Ham effect) (5) and the effect of solvent on the relative intensity of the fine structure in the fluorescence emission of pyrene (6-9). The variation in intensity of the $\alpha$-, the $\beta$- and the $\gamma$-band as the structure of squaraine and the solvent vary may well be vibrational in nature, but the persistent absence of the (0,2) transition in over 50 emission spectra makes such an assignment tentative, at best.

Major evidence against the vibronic structure assignment comes from the specific solvent effects observed in alcoholic solvents. The $\lambda_{max}$ of 2 is found to depart from the $\lambda_{max}$ versus $\pi^*$ relationship in Figure 2 in alcoholic solvents. Instead, $\lambda_{max}$ shifts to shorter wavelengths as the steric hindrance around the hydroxy group increases, e.g. $\lambda_{max}$ is 642.6±0.2 nm in primary alcohols, ~639.4 nm in secondary alcohols and ~634.1 nm in tertiary amyl alcohol. In conjunction with the hypsochromic shift, an accompanying increase in intensity for the $\alpha$-band is observed (Figure 4). The correlation between the absorption and the emission spectra of 2 in alcohols is identical to that seen in Figure 2. The significance of the results in alcohols lies in the steric dependence (rather than $\pi^*$ dependence) of both $\lambda_{max}$ and emission composition. Very similar short range interactions between the excited state of p-dimethylaminobenzonitrile (DMABN) and alkyl amines was reported by Wang (10) recently where he assigned the long wavelength emission, which is steric sensitive, to the exciplex emission between DMABN and alkyl amines. The similar steric dependence observed in the electronic spectra in this work suggests that the interactions between 2 and alcoholic solvent molecules is also short range. We propose that 2 forms solute-solvent complexes in alcoholic solvents. Since structural effect studies (Law, K.Y., to be published) show that the multiple emissions of squaraines are not sensitive to steric factors around the N,N-

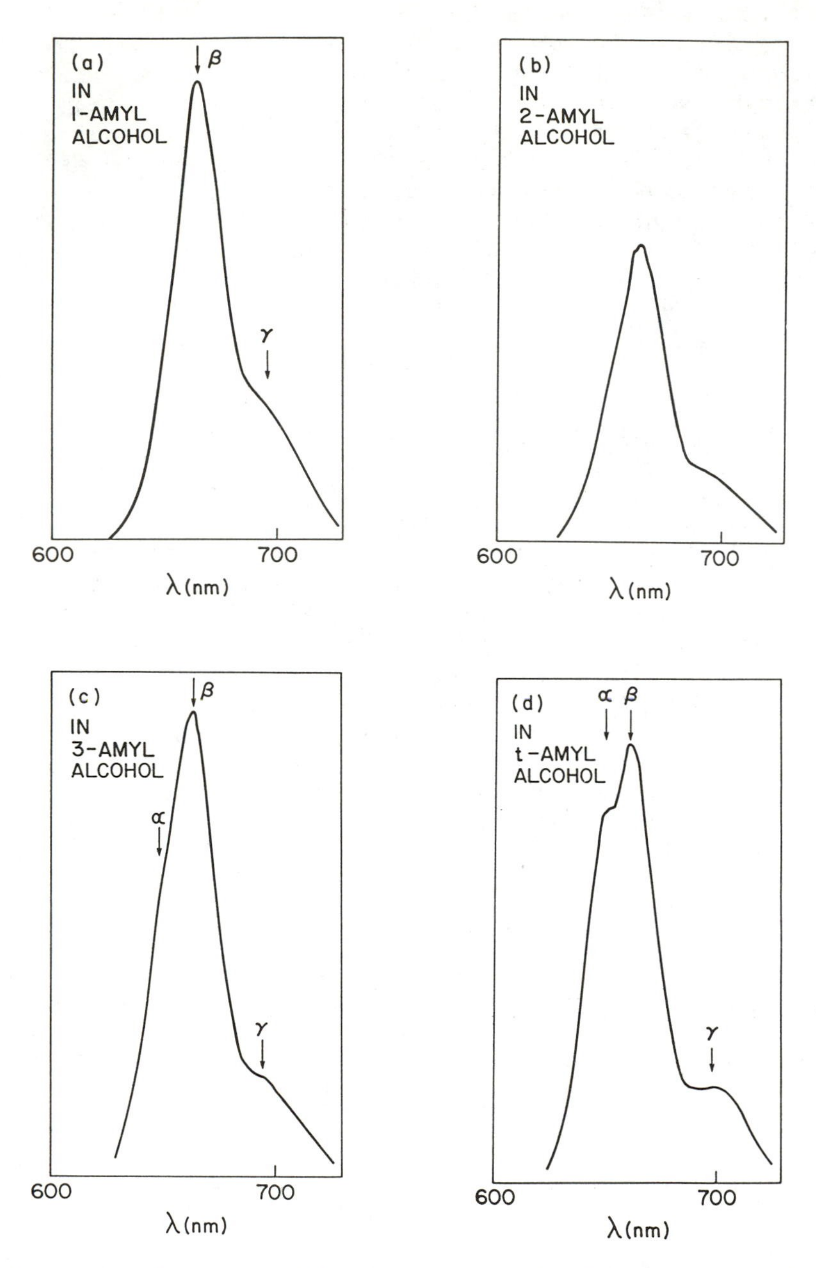


Figure 4. Corrected fluorescence emission spectra of 2 in amyl alcohols ([1]~3x10$^{-7}$M).

dialkylamino group, our results suggest that the sites of complexation in alcohols are the OH group of the alcohol and the four-membered ring of the squaraine (Scheme I).

Scheme I

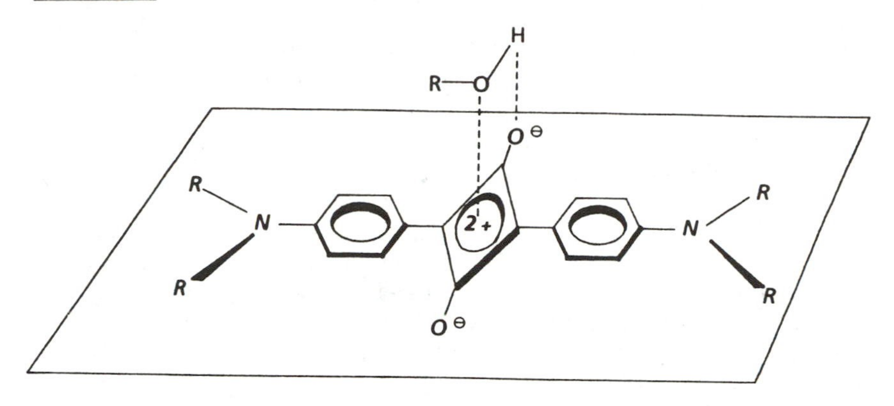


As the steric hindrance around the OH group decreases, solute-solvent complexation increases, resulting in the bathochromic shift of $\lambda_{max}$ and an increase in $\beta$-emission intensity. The general solvent effect on the $\lambda_{max}$ and the emission composition of 2 (Figure 2) suggests that the complexation process is very general and complexation becomes very pronounced in solvents of $\pi^* > 0.65$. Accordingly, $\alpha$-band is the Franck-Condon emission of the excited state of the solute and $\beta$-band is the Franck-Condon emission of the excited state of the solute-solvent complex.

Additional experimental evidence in favor of the solute-solvent complex model comes from the low temperature electronic spectra of 2 in ether. Squaraine 2 exhibits an absorption at $\lambda_{max}$ 654 nm and a single emission at $\lambda_F$ 664 nm at $77^oK$ in ethereal matrix (Figure 3b). The observation of a bathochromic shift in the emission spectrum is certainly against general expectation from vibrational fine structures where a hypsochromic shift of $\lambda_F$ should be obtained (11). The anomalously large shift lead us to conclude that the absorption is from the solute-solvent complex and that its formation is probably a temperature stabilization effect. Since the excitation spectrum at $77^oK$ is identical to the absorption spectrum and shows good overlap and mirror image relationship with the emission spectrum, the single emission at $\lambda_{max}$ 664 nm can be assigned to the Franck-Condon emission of the solute-solvent complex. Since solvent effect study of 2 shows that $\lambda_F$ is relatively insensitive towards increasing solvent polarity (e.g. $\lambda_F(\beta)$ ~660 nm in ether, $\lambda_F(\beta)$=665 nm in acetonitrile), by comparison of the spectral data in Figures 3a and b, we can conclude that the $\beta$-emission at room temperature is an emission from the solute-solvent complex. The $\alpha$-emission is an emission from the excited state of the solute. This deduction is identical with that obtained from solvent effect study.

Another significant finding in the low temperature emission spectrum is the absence of the $\gamma$-emission at ~700 nm. Since recent results on the effects of structural changes on the total fluorescence quantum yield and the multiple fluorescence emission of a number of squaraines (Law, K.Y. to be published) suggest that the $\gamma$-emission is an emission from a relaxed, twisted excited state which is formed by rotation of the C-C bond between the phenyl ring and the four membered ring of squaraine. The absence of the $\gamma$-emission in Figure 3b is complementary with our previous results, simply indicating that the C-C bond rotation is prohibited at $77^oK$.

If the above assignment of the $\alpha$- and the $\beta$-emission was correct, addition of a complexing solvent in an ethereal solution of 2 should increase the concentration of the solute-solvent complex. As a result, both $\lambda_{max}$ and $\lambda_F$ should shift to the red and the $\beta$-emission should increase. Such an experiment is, however, complicated by the fact that most complexing solvents are more polar than ether. Addition of a complexing solvent in ether not only induces complexation, but also increases the dielectric constant ($\epsilon$) of the medium, which is also known to affect the complexation process. In order to circumvent this dilemma, we performed our experiments in a ternary system, namely ether ($\epsilon$=4.43), chloroform ($\epsilon$=4.7) and n-hexane ($\epsilon$=1.9). The addition of n-hexane in the mixture is to keep the $\epsilon$ constant as the concentration of chloroform varies. Results in Figures 5a and b show that $\lambda_{max}$ and $\lambda_F$ shift to the red as [$CHCl_3$] increases at [$CHCl_3$] $\leq$1.12 M. Simultaneously, an isosbestic point at ~625 nm and an isoemissive point at ~662 nm are observed. These spectral results give positive evidence that the $\beta$-emission is indeed from the excited state of the

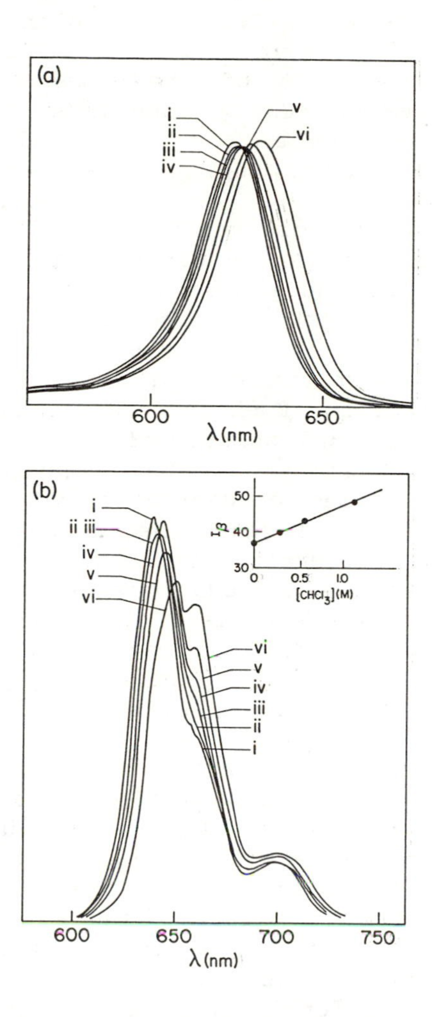


Figure 5. Effect of chloroform concentration on the (a) absorption ([2]~$10^{-5}$ M) (b) corrected fluorescence spectra of 2 ([2]~$3x10^{-7}$ M) in ether ([$CHCl_3$]=i 0 M, ii. 0.28 M, iii. 0.56 M, iv. 1.12 M, v. 2.24 M, and vi. 4.48 M).

solute-solvent complex and the α-emission is from the excited state of 2 itself. The stoichiometry of the complex is found to be 1:1 at $[CHCl_3] \leq 1.12$ M as indicated in the plot of the intensity of the β-emission ($I_\beta$) versus $[CHCl_3]$ (see inset of Figure 5b). Complexation constant between 2 and chloroform cannot be determined, however, due to the close overlapping of the α and β emission.

Further bathochromic shifts on the $\lambda_{max}$ and $\lambda_F$ and further increase in $I_\beta$ are observed at $[CHCl_3] \geq 2.24$ M. Absorption curves and emission curves no longer pass through their isosbestic or isoemissive points. This observation can either be attributed to the preferred solvation of the solute-solvent complex by chloroform as the concentration of chloroform increases or to the formation of a 1:n solute-solvent complex. The occurrence of these two events is presumably due to the highly localized concentration of chloroform in the solvation shell of squaraine, a consequence of the solute-solvent complexation process.

As noted in the introductory section, poly(vinyl acetal) polymers are excellent dispersing polymers for 1 in organic solvents. The major functionalities in these polymers are the ethereal linkages which are found to have no dispersing effect on 1 in controlled experiments. Other functional groups in the polymers are the acetate group and the OH group in which the latter is shown to be capable of forming strong complexes with squaraines (Figure 4). Using PVF as a model poly(vinyl acetal) polymer, the interaction between the OH groups in PVF and 1 in $CH_2Cl_2$ is studied. Results in Figure 1 and Figures 6a-d show that the intensity of the β-emission increases as the concentration of PVF increases, indicating that the OH groups in PVF are forming complexes with 1 in $CH_2Cl_2$. In a typical squaraine-polymer dispersion, the effective concentration of 1 is $\sim 10^{-2}$ M. Due to the low solubility of 1 in $CH_2Cl_2$ ($< 10^{-5}$ M), majority of 1 exists as particulates. This implies that the major interaction between 1 and PVF occurs on the surfaces of various squaraine particles in solution. As a result of this complexation process, PVF macromolecules adsorb onto the surfaces of particles of 1 and the OH groups in PVF are actually anchoring groups for the adsorption process. Schematic representation of the adsorption of polymer chains of

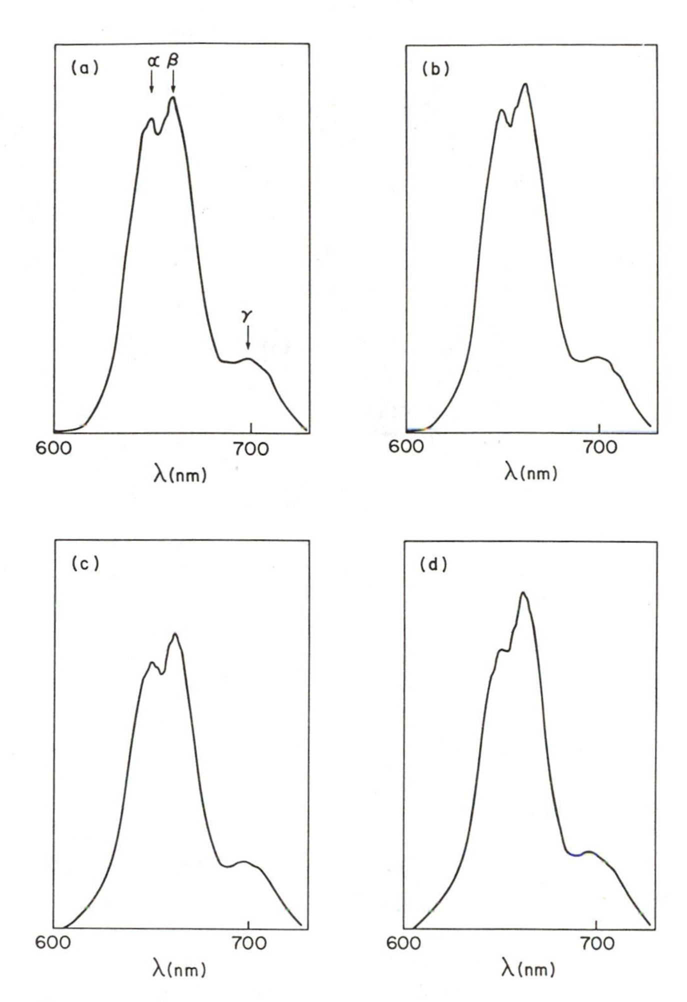


Figure 6. Corrected fluorescence spectra of 1 in $CH_2Cl_2$ ([1] ~$3 \times 10^{-7}$ M) at varying PVF concentration (a. 25 mg per mL; b. 50 mg per mL; c. 100 mg per mL and d. 200 mg per mL).

PVF onto particles of 1 in $CH_2Cl_2$ is depicted in Scheme II.

Scheme II

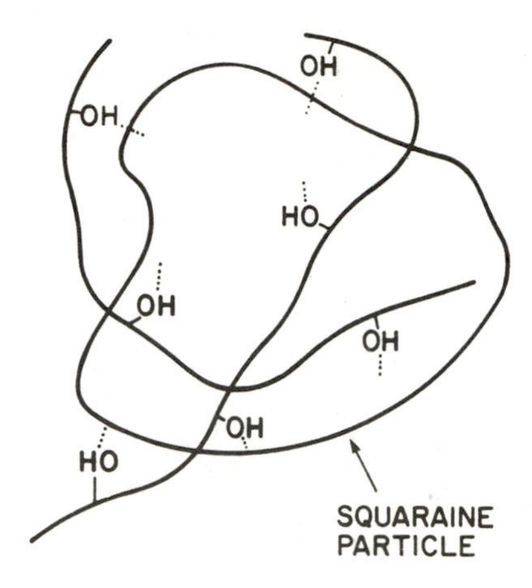


The adsorbed polymer sterically stabilizes particles of 1 in polymer solution, resulting in a stable squaraine-polymer dispersion (12,13). Analogous steric stabilization of pigment particles via specific interactions in organic solvents are known. For example, using the specific interaction between carboxylic acid groups and the polar sites on the surfaces of iron powder, Verwey and deBoer (14) were able to stabilize iron powder by oleic acid in carbon tetrachloride, benzene and n-hexane. Similarly, by utilization of the interaction between the $\pi$-orbitals on the surface of carbon black pigment and aromatic molecules, van der Waarden (15,16) was able to prepare stable dispersions of carbon black pigment in aliphatic hydrocarbon solvents using aromatic compounds with long alkyl chains as dispersing agents.

By extending the same principle, other good dispersing polymers such as styrene-allyl alcohol copolymer, vinyl chloride-vinyl acetate-vinyl alcohol terpolymer, have been identified. The stabilities of the squaraine-polymer dispersions prepared from these polymers are far superior than analogous polymers without hydroxy groups. Again, the anchoring effect of OH groups on the adsorption process, which subsequently stabilizes squaraine particles in solution, is demonstrated. Further experiments also show that other strong complexing functional groups, e.g., cyano group, can also be used as an anchoring group for the polymer in the adsorption process and stable dispersions can be prepared accordingly. For example, better dispersion stability is obtained from styrene-acrylonitrile copolymer in $CH_2Cl_2$ as compared to polystyrene itself.

In summary, the stabilization mechanism of squaraine particles in organic solvents has been understood as a steric stabilization process by spectroscopic technique. The knowledge gained in this work has enabled us to formulate other stable squaraine-polymer dispersions using polymers other than poly(vinylacetals).

## Literature Cited


1. For nomenclature, see West, R. Oxocarbon, Academic Press, New York, 1980, Chapter 10.
2. Wingard, R.E. IEEE Industry Applications, 1982, 1251.
3. Loutfy, R.O.; Hsiao, C.K.; Kazmaier, P.M. Photogr. Sci. Eng., 1983, 27,5.
4. Spenger, H.E.; Ziegenbein, W. Angew. Chem. Int. Ed. Engl., 1966, 5, 894.
5. Ham, J.S. J. Chem. Phys., 1953, 21, 756.
6. Nakajima, A. Bull. Chem. Soc. Jpn., 1971, 44, 3272.
7. Nakajima, A. J. Mol. Spectroscopy, 1976, 61, 467.
8. Hara, K.; Ware, W.R. Chem. Phys., 1980, 51, 61.
9. Dong, D.C.; Winnik, M.A. Can. J. Chem., 1984, 62, 2560.
10. Wang, Y. Chem. Phys. Lett., 1985, 116, 286.
11. Becker, R.S. Theory and Interpretation of Fluorescence and Phosphorescence, John Wiley & Sons, 1969, p. 42.
12. Parfitt, G.D. Dispersion of Powders in Liquid, 2nd Edition, John Wiley and Sons, 1973.
13. Napper, D.H. Polymeric Stabilization of Colloidal Dispersion, Academic Press, New York, 1983.
14. Verwey, E.J.; de Boer, J.H. Rec. Trav. Chim., 1938, 57, 383.
15. van der Waarden, M. J. Colloid Sci., 1950, 5, 317.
16. van der Waarden, M. J. Colloid Sci., 1951, 6, 443.


RECEIVED April 27, 1987

# Chapter 14

# Specific Interactions of (+)-Catechin and (−)-Epicatechin with Polymers that Contain the *L*-Prolyl Residue

**Wolfgang R. Bergmann[1] and Wayne L. Mattice[2]**

[1]**Department of Polymer Science, University of Akron, Akron, OH 44325**
[2]**Department of Chemistry, Louisiana State University, Baton Rouge, LA 70803**

Quantum yields for fluorescence, Q, by (+)-catechin and (-)-epicatechin are sensitive to the nature of the solvent. These compounds and their oligomers form complexes with poly(vinylpyrrolidone). Complex formation is accompanied by a dramatic collapse of the dimensions of the poly(vinylpyrrolidone) chain, and the value of Q changes in the manner expected for transfer of the fluorophore from an aqueous to a nonpolar medium. Much different behavior is seen when poly(L-proline) is substituted for poly(vinylpyrrolidone). Emission by the fluorophores exhibits discrimination between local right- and left-handed helices in the poly(L-proline) chain.

Proteins, nucleic acids, and polysaccharides are the most widely known of the polymers that occur in nature. Representatives of the plant kingdom produce other types of polymers with properties that are less well understood. A notable example is provided by polymers of (+)-catechin and (-)-epicatechin, which are found in a variety of plants. The structures of the monomers are depicted in Figure 1. Polymers are usually formed via carbon-carbon bonds from C(4) of one monomer to C(8) of its neighbor (1,2). Other linkage patterns, such as C(4) to C(6), are also known (3-5). The polymers isolated from some plants are branched (6). Weight-average molecular weights as large as 13300 have been measured for peracetylated samples (7). Much larger polymers are produced in nature because the samples isolated have been polydisperse.

Polymers of (+)-catechin and (-)-epicatechin form comlexes with many macromolecules. The formation of complexes with proteins is the basis for their use by plants as a defense mechanism (8). Most of the previous studies of these complexes have been performed at concentrations that produce precipitation (9-11). The sensitivity with which fluorescence can be detected permits study of complex formation at concentrations so low that all species remain in solution (12,13). Experimental results reported here show that this technique can also reveal subtle aspects of the complexation

0097-6156/87/0358-0162$06.00/0

process. In some complexes the fluorescnece responds to a simple change in the medium surrounding the fluorophore, but in other cases there is a more complicated response that is intimately connected with the conformation of the macromolecule.

## EXPERIMENTAL DETAILS

Poly(vinylpyrrolidone), poly(L-proline), (+)-catechin, (-)-epicatechin, and trifluoroethanol were purchased from Sigma Chemical Co. Two of the four poly(L-proline) samples studied contained an unidentified fluorescent material that was removed with activated charcoal. A fluorescent compound in the trifluoroethanol was removed by distillation. Dr. Richard W. Hemingway kindly supplied highly purified samples of synthetic (-)-epicatechin-(4B-6)-(+)-catechin, (-)-epicatechin-(4B-8)-(+)-catechin, and (-)-epicatechin-(4B-8)-(+)-catechin decaacetate. Dioxane and 1-propanol were obtained from Aldrich Chemical Co.

Fluorescence measurements were performed with an SLM 8000 fluorometer. Quantum yields were obtained with quinine bisulfate as the standard (14,15). Excitation was at 272-280 nm, and the range of the integration used in the determination of I/I(0) was 285-420 nm. Circular dichroism measurements were performed with a Jasco J-500C spectropolarimeter.

## QUANTUM YIELDS FOR FLUORESCENCE

Monomers and dimers exhibit maximal emission at 314-324 nm upon excitation at 272-280 nm. Quantum yields for fluorescence, denoted by Q, are collected in Table I. Both monomers exhibit increasing fluorescence as the solvent changes from water to trifluoroethanol to dioxane. There is no significant difference in the emission of the two monomers in a common solvent. Dimers show less intense emission that do monomers. There is a further reduction in emission upon acetylation of all phenolic and aliphatic hydroxyl groups.

## COMPLEX FORMATION by POLY(VINYLPYRROLIDONE)

Integrated fluorescence intensity in the absence and presence of an added polymer are denoted by I(0) and I, respectively. The values of I/I(0) for (+)-catechin and (-)-epicatechin in water are enhanced significantly in the presence of poly(vinylpyrrolidone), as is shown in Table II. There is a more dramatic increase in emission for a mixed dimer than for either monomer. Previous studies (10,11) of the interactions with proteins have found that the stability of the complexes rises rapidly as one proceeds from (+)-catechin or (-)-epicatechin through a series of oligomers of increasing size. The behavior of I/I(0) in water shows that (+)-catechin and (-)-epicatechin form complexes with poly(vinylpyrrolidone) in dilute solution. It suggests that the interactions becomes stronger when an oligomer is substituted for either monomer. Previous studies based on the development of turbidity (9) or the inhibition of the precipitation of a radioactively labelled tracer protein (10) have shown that poly(vinylpyrrolidone) forms complexes with proanthocyanidins at much higher concentrations.

Table I. Quantum Yields for Fluorescence (Excitation at 272-280 nm)

| Compound | Solvent | Q |
|---|---|---|
| (+)-Catechin | Water | 0.11 |
| | Trifluoroethanol | 0.19 |
| | Dioxane | 0.30 |
| (-)-Epicatechin | Water | 0.10 |
| | Dioxane | 0.29 |
| (-)-Epicatechin-(4B-6)-(+)-catechin | Dioxane | 0.05 |
| (-)-Epicatechin-(4B-8)-(+)-catechin | Dioxane | 0.07 |
| (-)-Epicatechin-(4B-8)-(+)-catechin (Decaacetate) | Dioxane | 0.03 |

Table II. I/I(0) in Water Containing Poly(vinylpyrrolidone) at a Molar Monomer Concentration of 0.009

| Fluorophore | Molarity x $10^{-5}$ | I/I(0) |
|---|---|---|
| (+)-Catechin | 4.7 | 1.31 |
| (-)-Epicatechin | 4.7 | 1.24 |
| (-)-Epicatechin-(4B-6)-(+)-catechin | 0.9 | 2.45 |

The sample of poly(vinylpyrrolidone) used in these experiments has an intrinsic viscosity of 1.72 dl/g in water at 30° C. In the presence of 0.0014 M (+)-catechin, the intrinsic viscosity falls to 0.13 dl/g. Since it is the (+)-catechin that induces the collapse of the poly(vinylpyrrolidone) chain, it is assumed that this molecule finds itself in an environment that is not completely aqueous, but instead has a high local concentration of poly(vinylpyrrolidone) segments. The solvent dependence of the quantum yield for fluorescence, reported in Table I, shows that Q increases as (+)-catechin is transferred from water to dioxane. Collapse of the poly(vinylpyrrolidone) chain about (+)-catechin therefore produces an increase in Q, which is responsible for I/I(0) > 1 in Table II.

## COMPLEX FORMATION WITH POLY(L-PROLINE)

Precipitation studies at comparatively high concentrations have shown that proanthocyanidins form complexes with poly(L-proline) as well as with poly(vinylpyrrolidone) (9,10). Figure 2 depicts the repeat units of poly(vinylpyrrolidone) and poly(L-proline). The L-prolyl unit contains one fewer methylene and has the remaining atoms rearranged so that the amide and five-membered ring become parts of the backbone of the chain. The results reported in the preceeding section might lead to the expectation that addition of poly(L-proline) to an aqueous solution of one of the fluorophores would produce I/I(0) > 1. This expectation is not realized. Poly(L-proline) with M = 60000 does not produce I/I(0) > 1. Instead it quenches the fluorescence. For example, the values of I/I(0) for $3.6 \times 10^{-5}$ M (+)-catechin in water are 0.75 and 0.41, respectively, at L-prolyl concentrations of 0.026 and 0.10 M. The Q in Table I do not permit rationalization of the quenching ability of poly(L-proline) as a "solvent" effect. It therefore is pertinent to inquire whether there might be a more specific interaction of (+)-catechin with poly(L-proline).

The test fof specificity exploits the ability of the poly(L-proline) chain to exist in either of two different conformations. Some thirty years ago these conformations were characterized by the analysis of the x-ray scattering patterns of poly(L-proline) in the solid state. The conformation known as Form I is a right-handed helix with 3.33 residues per turn, a translation along the helix axis of 1.9 A/residue, and a cis configuration at all amide bonds (16). The conformation of Form II in the solid state is a left-handed helix with 3.00 residues per turn, a translation along the helix axis of 3.12 A/residue, and a trans configuration at the amide bonds (17,18). Dilute solutions in solvents such as water, acetic acid, and trifluoroethanol support a local structure with a strong similarity to Form II (19), but there is sufficient flexibility so that a poly(L-proline) chain of high molecular weight behaves in these solvents as a random coil with a characteristic ratio of 14-20 (20).

In a suitably chosen mixed solvent system, such as acetic acid and 1-propanol (21), a sharp reversible transition between Form I and Form II can be achieved with a small change in solvent composition. Measurements of optical activity provide a convenient way to follow the transition in dilute solution. There are large changes in optical activity because Forms I and II are helices of opposite handedness. The left panel in Figure 3 depicts the reversible transition that is detected by circular dichroism measurements in mixtures of trifluoroethanol and 1-propanol. Form II is the only conformation present in trifluorethanol. A solution of poly(L-proline) in 35 : 65 trifluoroethanol : 1-propanol exhibits the same circular dichroism pattern as does a solution where the solvent is pure trifluoroethanol. However, further addition of 1-propanol produces a dramatic change in the circular dichroism. At 20 : 80 trifluoroethanol : 1-propanol the circular dichroism pattern is that characteristic of Form I. Data in Figure 3 do not extend beyond 10 : 90 trifluoroethanol : 1-propanol because of the low solubility of poly(L-proline) in 1-propanol.

The right panel in Figure 3 depicts I/I(0) for (+)-catechin in mixtures of trifluoroethanol and 1-propanol that contain poly(L-proline) at a concentration of 0.90 mg/ml. In pure trifluoroethanol I/I(0) is slightly less than one. At the opposite extreme of solvent composition, I/I(0) is greater than one and increases over a period of several days. The greatest sensitivity of I/I(0) to solvent composition occurs near 25 : 75 trifluoroethanol : 1-propanol, which is where the Form I - Form II interconversion is detected by circular dichroism. Measurements of the fluorescence of (+)-catechin in the mixed solvents, but in the absence of poly(L-proline), show no unusual dependence of emission on solvent.

We conclude that (+)-catechin can discriminate between the two Forms of poly(L-proline). The emission from (+)-catechin becomes very much stronger if the local structure of the poly(L-proline) chain is a right-handed helix with 3.33 residues per turn, a translation of 1.9 A/residue, and cis amide bonds. In contrast, the

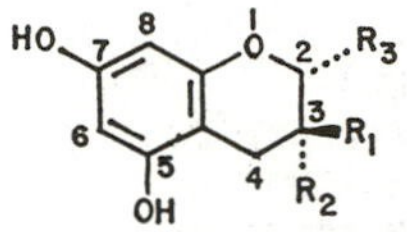

Figure 1. Covalent structures and numbering system for (+)-catechin (R1 = OH, R2 = H, R3 = dihydroxyphenyl) and (-)-epicatechin (R1 = H, R2 = OH, R3 = dihydroxyphenyl).

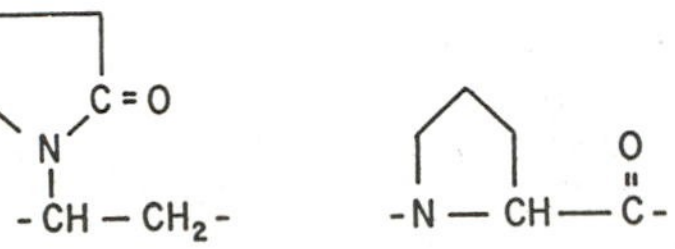

Figure 2. Repeating units in poly(vinylpyrrolidone) and poly(L-proline).

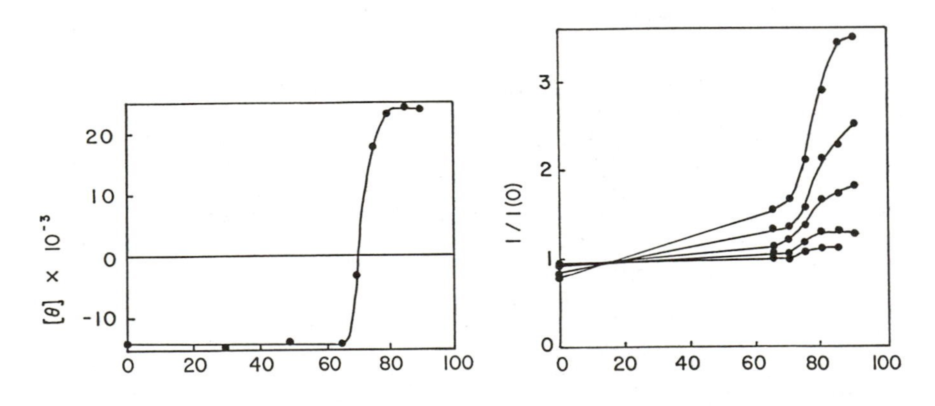


Figure 3. Left panel: Mean residue ellipticity at 210 nm for poly(L-proline) in mixtures of trifluoroethanol and 1-propanol. Right panel: I/I(0) for (+)-catechin in mixtures of trifluoroethanol and 1-propanol that contain poly(L-proline) at a concentration of 0.90 mg/ml. Measurements of I/I(0) were performed at 0, 3, 7, 12, and 20 days after preparation of the solutions. I/I(0) increases with time at the highest 1-propanol concentrations.

emission is slightly quenched if the local structure is a left-handed helix with 3.00 residues per turn, a translation of 3.12 A per residue, and trans amide bonds. The origin of the time dependence of the steady state emission has not yet been identified.

The specificity of the interaction of (+)-catechin with poly(L-proline) is not surprising in view of the biological function of polymers of (+)-catechin and (-)-epicatechin. The presence of these polymers in certain plants makes them an unattractive food source for animals. A molecular basis for this protective action may lie in the ability of polymers of (+)-catechin and (-)-epicatechin to interact strongly with mammalian salivary proteins that have an unusually high content of the L-prolyl residue (10,22).

## ACKNOWLEDGMENTS

This research was supported by National Science Foundation research grant DMB 85-00338. We thank Dr. Richard W. Hemingway for the samples of the dimers.

## LITERATURE CITED

1. Thompson, R. S.; Jacques, D.; Haslam, E.; Tanner, R. J. N. J. Chem. Soc., Perkin Trans.I 1972, 1387.
2. Jacques, D.; Opie, C. T.; Porter, L. J.; Haslam, E. J. Chem. Soc., Perkin Trans. I 1977, 1637.
3. Botha, J. J.; Viviers, P. M.; Ferreira, D.; Roux, D. G. J. Chem. Soc., Perkin Trans. II 1981, 1235.
4. Hemingway, R. W.; Foo, L. Y.; Porter, L. J. J. Chem. Soc., Perkin Trans. II 1982, 1209.
5. Hemingway, R. W.; Karchesy, J. J.; McGraw, G. W.; Wielesek, R. A. Phytochemistry 1983, 22, 275.
6. Mattice, W. L.; Porter, L. J. Phytochemistry 1984, 23, 1309.
7. Porter, L. J. Rev. Latinoamer. Quim. 1984, 15-2, 43.
8. Haslam, E. Biochem. J. 1974, 139, 285.
9. Oh, H. I.; Hoff, J. E.; Armstrong, G. S.; Haff, L. A. J. Agric. Food Chem. 1980, 28, 394.
10. Hagerman, A. E.; Butler, L. G. J. Biol. Chem. 1981, 256, 4494.
11. Porter, L. J.; Woodruffe, J. Phytochemistry 1984, 23, 1255.
12. Bergmann, W. R.; Barkley, M. D.; Mattice, W. L. Polym. Prepr. Div. Polym. Chem., Am. Chem. Soc. 1986, 27(2), 320.
13. Bergmann, W. R. Ph.D. Dissertation, Louisiana State University, Baton Rouge, 1986.
14. Melhuish, W. H. J. Phys. Chem. 1961, 65, 229.
15. Demas, J. N.; Crosby, G. A. J. Phys. Chem. 1971, 75, 991.
16. Traub, W.; Shmueli, U. Nature 1963, 198, 1165.
17. Cowan, P. M.; McGavin, S. Nature 1955, 176, 501.
18. Sasisekharan, V. Acta Crystallogr. 1959, 12, 897.
19. Darsey, J. A.; Mattice, W. L. Macromolecules 1982, 15, 1626.
20. Mattice, W. L.; Mandelkern, L. J. Am. Chem. Soc. 1971, 93, 1769.
21. Gornick, F.; Mandelkern, L.; Diorio, A. F.; Roberts, D. E. J. Am. Chem. Soc. 1964, 86, 2549.
22. Muenzer, J.; Bildstein, C.; Gleason, M.; Carlson, D. M. J. Biol. Chem. 1979, 254, 5629.

RECEIVED March 13, 1987

# EXCIMER PHOTOPHYSICS

Chapter 15

# Excimer Photophysics of Macromolecular Scintillators

D. J. S. Birch[1], A. Dutch[1], R. E. Imhof[1], K. Davidson[2], and I. Soutar[2]

[1]Photophysics Research, University of Strathclyde, Glasgow, G4 ONG, Scotland

[2]Department of Chemistry, Heriot-Watt University, Edinburgh, EH14 4AS, Scotland

The photophysics of intramolecular excimer formation have been studied in vinyl homopolymers labelled with the scintillators BuPBD and PPO. For both systems the fluorescence decay kinetics display unusual features which include the absence of a measureable rate of excimer formation, rate parameters which depend on emission wavelength and time resolved emission spectra which shift to longer wavelengths with increasing time delay. The fluorescence decays are analysed with two alternative kinetic models. One uses three exponentials and the other allows for transient time dependent quenching. Although in some cases both models give good fits they are shown to provide an incomplete description. However, they both provide evidence for there being a number of preformed excimer conformations.

The formation of intramolecular excimers in polymer systems has aroused much interest in recent years (1). Perhaps most notable is the general observation that the reaction kinetics do not obey the accepted Birks scheme for low molecular weight systems (2). In this scheme the fluorescence decay kinetics of the monomer (M) and excimer (D) species can be separated spectrally with fluorescence response functions of the form

$$i_M(t) = B_1 \exp(-t/\tau_1) + B_2 \exp(-t/\tau_2) \qquad (1)$$

0097-6156/87/0358-0170$06.00/0

and $$i_D(t) = B_3[\exp(-t/\tau_1) - \exp(-t/\tau_2)] \quad (2)$$

$\tau_1$ describes the decay of the excimer, which, even in the absence of spectral overlap can appear in the spectral region of monomer emission because of thermally induced reverse dissociation of the excimer producing excited monomer. $\tau_2$ describes the quenched monomer decay time and also appears as a rise time in the excimer fluorescence which is shifted to longer wavelengths with respect to that of the monomer. Because $\tau_1$ and $\tau_2$ are associated with distinct excited state species they should be independent of emission wavelength, although their relative intensities would be expected to change with emission wavelength in acordance with the spectral distributions of the monomer and excimer fluorescence.

The incompleteness of this description when applied to polymer systems has been previously explained by the evidence for isolated monomer sites which do not partake in the excimer forming process and which require the consideration of a third decay component, $\tau_3$, i.e.

$$i_M(t) = B_1\exp(-t/\tau_1) + B_2\exp(-t/\tau_2) + B_3\exp(-t/\tau_3) \quad (3)$$

Such behaviour has been reported in a wide range of polymers including both homo and copolymers labelled with napthalene (3 - 5) and styrene-methyl methacrylate copolymers (6). However, in all these cases no clear evidence for other than a single excimer species has emerged. In addition, a rise time in the fluorescence decay, which can be associated with excimer formation, has also been observed (7).

More recently an alternative explanation of the complex excimer behaviour observed in polymer systems has been proposed (9,10) whereby the close proximity of some fluorophores leads to a time dependent rate of quenching analogous to that predicted by the Einstein-Smoluchowski diffusion theory for low molecular weight systems. This predicts a fluorescence response function of the form

$$i_M(t) = B_1\exp(-t/\tau_1) + B_2\exp(-t/\tau_2-\alpha\sqrt{t}) \quad (4)$$

Where $\tau_1$ describes the excimer dissociation to give excited monomer, $\tau_2$ describes the quenched monomer at longer times and $\alpha$ describes the transient diffusion behaviour (10) which is dependent on the rotational diffusion and collisional properties, though in a manner not so easily described as for small molecules. The same kinetic form has also been used to describe the monomer behaviour when viewed in terms of the energy migration to excimer forming sites (9).

In a previous study (8) we reported the presence of intramolecular excimer formation and evidence for emission from isolated monomer sites in copolymers of vinyl-t-butyl PBD and methyl methacrylate. 2-phenyl-5-(4-biphenyl) - 1,3,4-oxadiazole (PBD) is one of the most widely used scintillator solutes and has long been thought not to form excimers in solution. Its low solubility does not help investigation, but more fundamentally, molecular orbital calculations by Lami and Laustriat (11) predicted that its ground and first excited states are mutually repulsive. However, the slight shift in the fluorescence spectrum to longer wavelengths with increasing solute concentration has been previously attributed to excimer formation in the more soluble derivative, isopropyl PBD (12).

2,5-diphenyl-1, 3-oxazole (PPO) is another important scintillator which, in contrast to PBD, is well known to readily form excimers in solution. Berlman was the first to observe this as early as 1960 (13) but although this and subsequent work (14) included some fluorescence decay measurements the applicability of the expected kinetic scheme (Equations 1 and 2) was not thoroughly tested. This is not really surprising, because although the time resolution of early fluorometers and data analysis methods were adequate for studying monoexponential decays in the nanosecond region, they could not handle complex kinetics other than those on a much longer time-scale, as in the case of the pyrene excimer. The unquenched PBD and PPO lifetimes, which describe the duration over which excimers can be formed, are in the region of 1 ns as compared to ~ 450 ns for pyrene. This means that considerably higher concentrations are required for their excimers to be studied. Covalent bonding to a polymer chain provides a method of ensuring that a large fraction of fluorophores are in close enough proximity to facilitate excimer formation even though the fluorophore solubility may be low. In the light of the work already mentioned with other polymer systems, such a scintillation system would not be expected to obey the conventional low molecular weight kinetics. However, we have recently shown (15) that even in the case of PPO solutions the sub-nanosecond rate of excimer formation leads to deviations from the expected Birks kinetics.

We have thus chosen to pursue these studies using vinyl homopolymers labelled with PBD and PPO with a view to elucidating both the fundamental photophysical processes involved in sub-nanosecond excimer formation and also the effect of segmental motion, energy migration and the constraints imposed by the polymer backbone.

## Experimental

Fluorescence decay and time resolved measurements were performed using an Edinburgh Instruments Model 199 single photon fluorometer. This instrument incorporates an all-metal coaxial flashlamp (16) which enables measurement of the decay components of complex decays down to 200 ps. Reconvolution analysis was performed by fitting over all the fluorescence decay including the rising edge and the kinetic interpretation of the goodness of fit was assessed with regard to errors and limitations in time-resolution recently reviewed (17). Contour spectra were recorded using a Baird Atomic SFR-100 spectrometer interfaced to a PDP 11/04 computer.

All measurements were performed using toluene as solvent unless otherwise stated. Poly (VBuPBD) was prepared by thermally initiated free radical polymerization in benzene solution at 70°C, the mixture being thoroughly degassed and sealed in glass ampules under high vacuum. Purification of the resultant polymer was achieved by multiple reprecipitation from benzene solution into cold methanol ($<$ 263 K) followed by warming to room temperature under constant agitation. Poly (VPPO) was prepared as described by McInally et al (18).

## Results and Discussion

Steady state emission and excitation spectra of the poly (VBuPBD) and poly (VPPO) are consistent with the formation of intramolecular excimers:

(i) The emission spectrum is broadened at low energies relative to the structured monomeric emission band exhibited by the low molar mass analogue (BuPBD and PPO respectively).

(ii) The excitation spectra of polymer and low molar mass analogue are identical and independent of wavelength of analysis indicating the absence of substantial ground state interactions.

Figure 1 illustrates the total fluorescence matrix in contour format for BuPBD and poly (VBuPBD) at 298K. These spectra graphically illustrate the above conclusions and the symmetry of the broadened profile of the polymer emission with respect to the emission/excitation axes is convincing evidence for the excimeric nature of the species responsible for the low energy intensity component. In addition the emission of both polymers reduces to that of the unassociated monomeric chromophore upon dispersal in a glassy (MeTHF) matrix at 77K. The effect is illustrated for poly (VBuPBD) in Figure 2. This observation is consistent with the formation of intramolecular excimers by a conformational sampling mechanism.

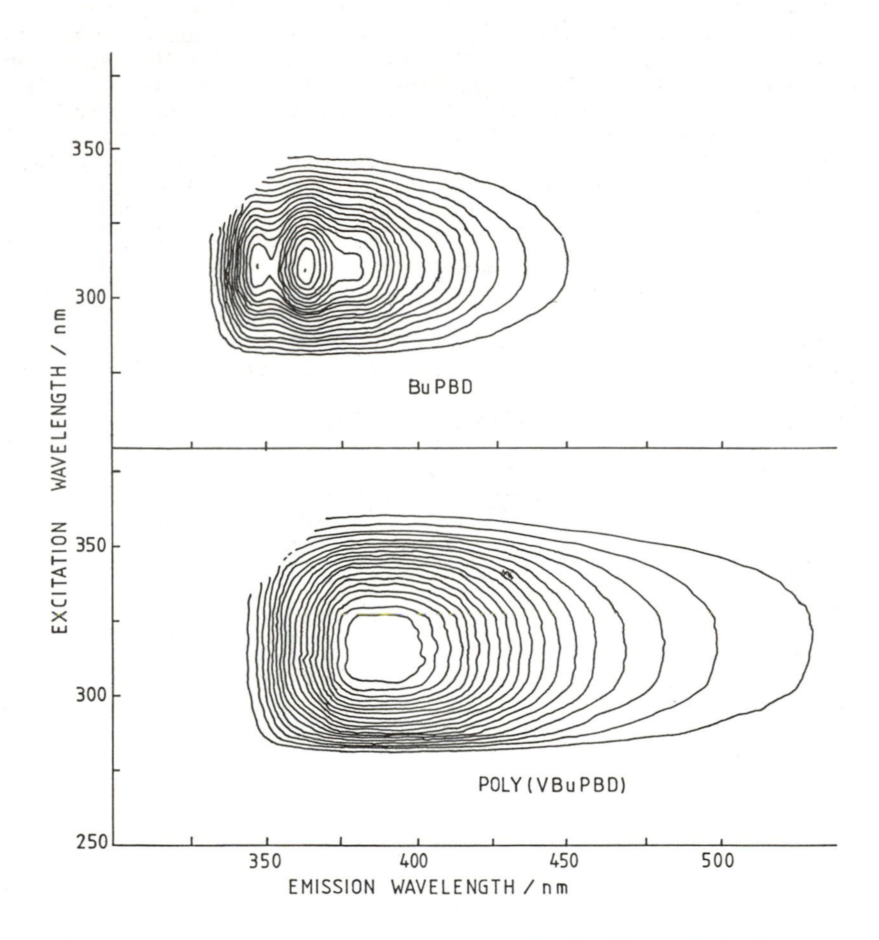


Figure 1 Contour spectra of BuPBD and Poly (VBuPBD)

Fluorescence decays of $10^{-5}$M chromophore solutions of Poly (VBuPBD) and Poly (VPPO) excited at 300 nm were measured over a range of temperatures and emission wavelengths. The low molecular weight solutions of $10^{-5}$M BuPBD and PPO gave good fits to single exponentials with lifetime values of 1.0 and 1.3 ns respectively. Similarly a copolymer of VBuPBD and methyl methacrylate at low aromatic contents (<0.005 mole%) gave a good fit to one decay component indicating that attachment to the polymer does not in itself distort the unimolecular behaviour. The fluorescence decays of the homopolymer are considerably more complex and could not be described by the bi-exponential kinetics of Equations 1 and 2.

Tables I and II show the results of three component fits to Poly(VBuPBD) and Poly(VPPO) respectively at an excitation wavelength of 300 nm. The improvement in goodness of fit when using a three component, rather than two component model is illustrated in Figure 3 for Poly(VBuPBD) at 350 nm and 25°C.

On close inspection the data of tables I and II show a number of inconsistencies with a three component model previously used to describe the excimer, quenched monomer and isolated monomer sites in other polymer systems (3 - 5). For instance, if we consider the 25°C data for Poly(VBuPBD), $\tau_1$ might be considered to represent the excimer, $\tau_2$ the quenched monomer and $\tau_3$ isolated monomer. However, B2 and B3 do not decrease proportionately as measurements are made at wavelengths further displaced from that dominated by monomer emission, though the high degree of monomer and excimer spectral overlap (Figure 1) means that B2 and B3 would not be expected to approach zero. In addition, the decay parameters clearly increase with increasing fluorescence wavelength to the extent that above 400 nm the $\tau_3$ values cannot possibly represent isolated BuPBD sites which would be expected to have a decay time of ~ 1 ns. Moreover, no rise-time was detected even at the longest emission wavelength studied (as would be shown by a negative amplitude component) and a temperature of -60°C, though this might in part be accounted for by the high degree of spectral overlap (15). The Poly(VPPO) data of Table 2 also shows these same anomalies.

Although the three component model provides a good fit in many cases, these discrepancies in the interpretation of the decay parameters led us to investigate if a time dependent quenching model was more appropriate. Indeed, the absence of a measureable rise-time of excimer formation would be consistent with this since it indicates that the excimer conformation is preformed and involves either a rapid reorientation of the fluorophore rather than segmental motion or rapid energy migration to excimer forming sites. Table III shows that fitting Equation 4 to the Poly(VBuPBD) decays

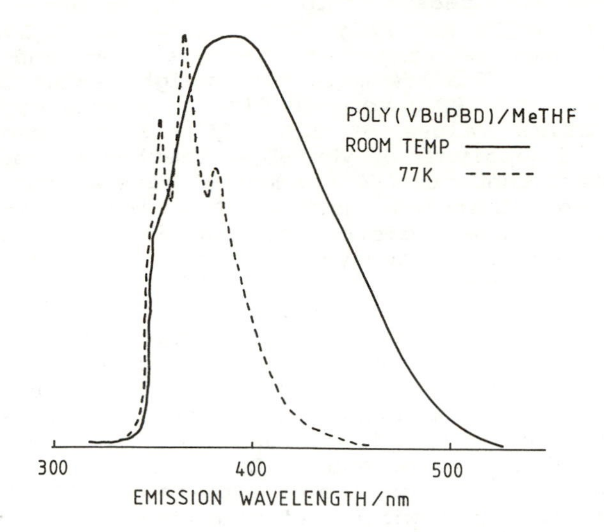


Figure 2 Fluorescence spectra of Poly (VBuPBD) at room temperature (solid line) and 77k (dashed line)

Table I Poly(VBuPBD) Decay Parameters for a Triple Exponential Fit - Equation 3

| $\lambda_F$(nm) | $\tau$1(ns) | $\tau$2 | $\tau$3 | B1(%) | B2 | B3 | $\chi^2$ | T(°C) |
|---|---|---|---|---|---|---|---|---|
| 350 | 5.59 | 0.30 | 1.28 | 12 | 39 | 49 | 1.05 | 25 |
| 400 | 7.13 | 0.53 | 1.79 | 19 | 38 | 43 | 1.18 | 25 |
| 450 | 8.44 | 0.59 | 2.27 | 36 | 22 | 42 | 1.16 | 25 |
| 500 | 12.10 | 1.62 | 4.55 | 16 | 48 | 44 | 1.05 | 25 |
| 350 | 4.99 | 0.29 | 1.50 | 25 | 35 | 40 | 1.1 | 50 |
| 500 | 7.97 | 0.59 | 2.68 | 31 | 21 | 48 | 1.47 | 50 |
| 350 | 5.10 | 0.23 | 1.45 | 8 | 34 | 58 | 1.91 | -60 |
| 400 | 11.10 | 0.65 | 2.29 | 11 | 45 | 44 | 1.36 | -60 |
| 500 | 14.60 | 1.06 | 4.08 | 34 | 30 | 36 | 1.29 | -60 |

Table II Poly(VPPO) Decay Parameters for a Triple Exponential Fit - Equation 3

| $\lambda_F$(nm) | $\tau$1(ns) | $\tau$2 | $\tau$3 | B1(%) | B2 | B3 | $\chi^2$ | T(°C) |
|---|---|---|---|---|---|---|---|---|
| 350 | 7.38 | 0.26 | 1.50 | 36 | 38 | 26 | 1.09 | 25 |
| 400 | 10.10 | 0.49 | 2.81 | 46 | 25 | 29 | 1.49 | 25 |
| 450 | 14.10 | 1.06 | 6.10 | 59 | 9 | 23 | 1.25 | 25 |
| 480 | 13.40 | 1.35 | 5.88 | 71 | 4 | 25 | 1.29 | 25 |
| 500 | 13.60 | 1.10 | 6.16 | 72 | 3 | 25 | 1.24 | 25 |
| 350 | 8.35 | 0.28 | 2.32 | 29 | 34 | 37 | 1.36 | 50 |
| 400 | 9.74 | 0.43 | 2.59 | 41 | 20 | 39 | 1.27 | 50 |
| 500 | 11.50 | 1.28 | 5.56 | 71 | 6 | 23 | 1.28 | 50 |
| 350 | 14.43 | 0.61 | 3.19 | 21 | 43 | 36 | 1.83 | -60 |
| 400 | 17.30 | 0.65 | 4.28 | 34 | 41 | 25 | 1.99 | -60 |
| 500 | 17.71 | 0.38 | 4.08 | 36 | 38 | 26 | 1.35 | -60 |

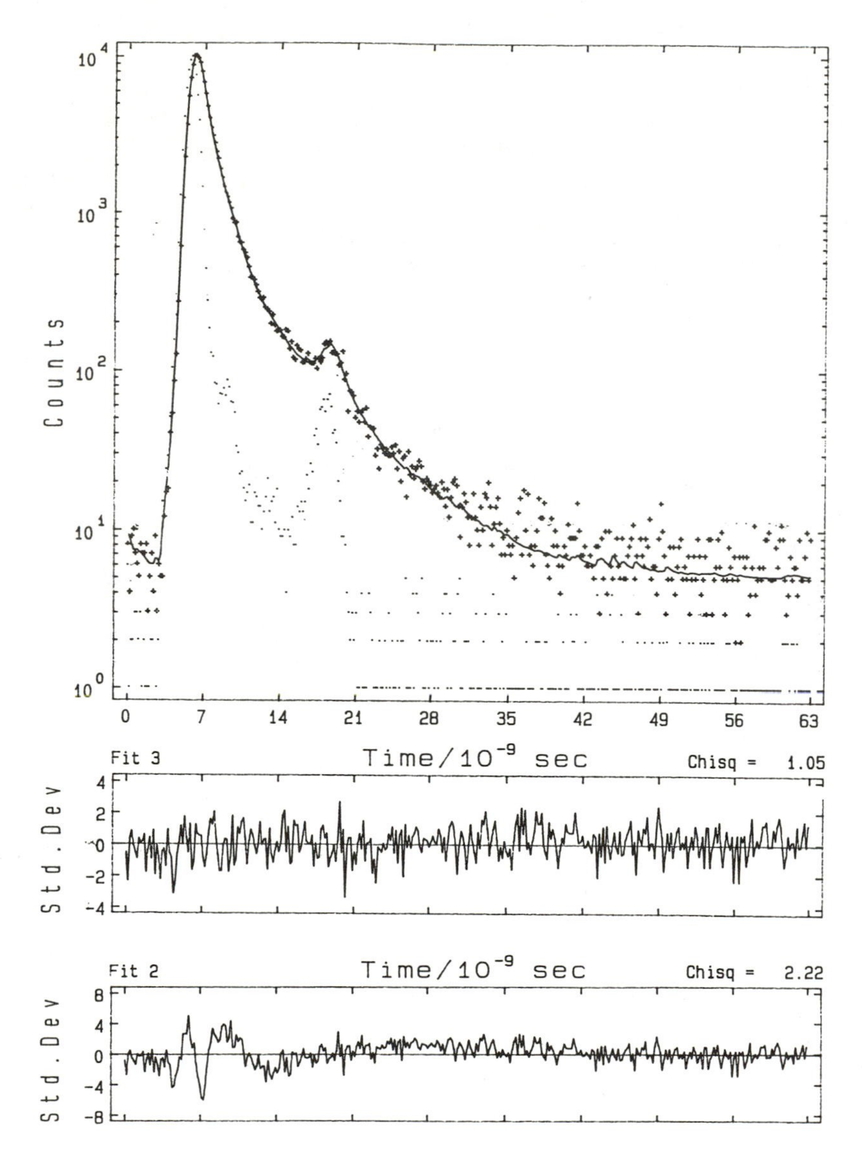


Figure 3 Three component fit and residuals for Poly (VBuPBD) at 350nm and 25°C with residuals for a two component fit also shown.

Table III Poly(VBuPBD) Decay Parameters for Time Dependent Quenching - Equation 4

| $\lambda_F$(nm) | $\tau$1(ns) | B1(%) | $\tau$2 | B2 | $\alpha(ns^{-1/2})$ | $\chi^2$ | T(°C) |
|---|---|---|---|---|---|---|---|
| 350 | 6.37 | 1.5 | 4.50 | 98.5 | 1.71 | 1.03 | 25 |
| 400 | 7.26 | 3.5 | 4.90 | 96.5 | 1.48 | 1.17 | 25 |
| 450 | 9.07 | 5.7 | 8.25 | 94.3 | 1.20 | 1.09 | 25 |
| 500 | 9.46 | 15.6 | 3.99 | 84.4 | 0.51 | 1.19 | 25 |
| 350 | 5.43 | 0.1 | 107.0 | 99.9 | 1.89 | 1.11 | 50 |
| 500 | 7.91 | 0.01 | 6280.0 | 99.99 | 1.27 | 2.01 | 50 |
| 350 | 14.5 | 0.2 | 6.03 | 99.8 | 1.51 | 1.87 | -60 |
| 400 | 11.9 | 1.1 | 8.62 | 98.9 | 1.50 | 1.32 | -60 |
| 500 | 14.6 | 8.2 | 10.40 | 91.8 | 0.98 | 1.31 | -60 |

generally gives a goodness of fit comparable to that obtained using a three component model. Figure 4 illustrates this for the same decay data shown in Figure 3. This equivalence is consistent with the observation made for other polymer systems (10). However, in our case the decay parameters so obtained do not lend themselves to a clear interpretation as the $\tau_2$ values are always significantly greater than the unquenched Poly(VBuPBD) fluorescence lifetime (~ 1 ns). This was particularly so in the analysis of the Poly(VPPO) data such that in most cases the iteration did not converge and predicted very high $\tau 2$ and $\alpha$ values (see also Table III at 50°C). This is in spite of good fits being obtained to this model for solutions of the PPO chromophore itself (19). We are thus led to conclude that although a transient quenching model provides a fair parameterisation, it is no more physically meaningful for these polymer systems. This is emphasised by the fact that the model describes the Poly(VBuPBD) decay equally well at longer wavelengths where it might not be expected to do so. It is also perhaps not surprising that the transient term in Equation 4 is indistinguishable from an extra time independent decay rate. We have also found this frequently to be so in the simpler case of the PPO intermolecular excimer even at a time dispersion of 90 ps/Ch and for decays with 40,000 counts in the peak (19).

Perhaps of most significance when comparing Tables I and III is the common trend for decay parameters to increase with increasing emission wavelength. This behaviour is not predicted by the kinetic models associated with Equations 1 to 4 and we infer from this that more than one excimer conformation is present in both Poly(VBuPBD) and Poly(VPPO). The presence of a second excimeric species involving partial overlap of the two molecules concerned has already been reported for other systems e.g. (20) and (21). Further evidence for this in Poly(VBuPBD) is shown in the time-resolved emission spectrum of Figure 5. At 6 ns following excitation any monomer emission is down to less than 0.1% of its peak intensity. This means that the emission spectra for times equal or greater than this must be due predominantly to the excimer emission. On comparing the spectra at 6 and 9 ns after excitation a shift with increased time delay to longer wavelengths for the peak and red edge of the spectra as a whole is observed. This behaviour is consistent with the energy level of the excimer excited state being dependent on the conformation. The same phenomenon is shown in Figure 6 for Poly(VPPO) where the time resolved spectra continue to shift to longer wavelengths at delay times up to 20 ns. The spectra at longer delay times for both polymers show some evidence of reverse dissociation of the excimer to form excited monomer but in the case of Poly(VPPO) it is considerably less than that of PPO (15).

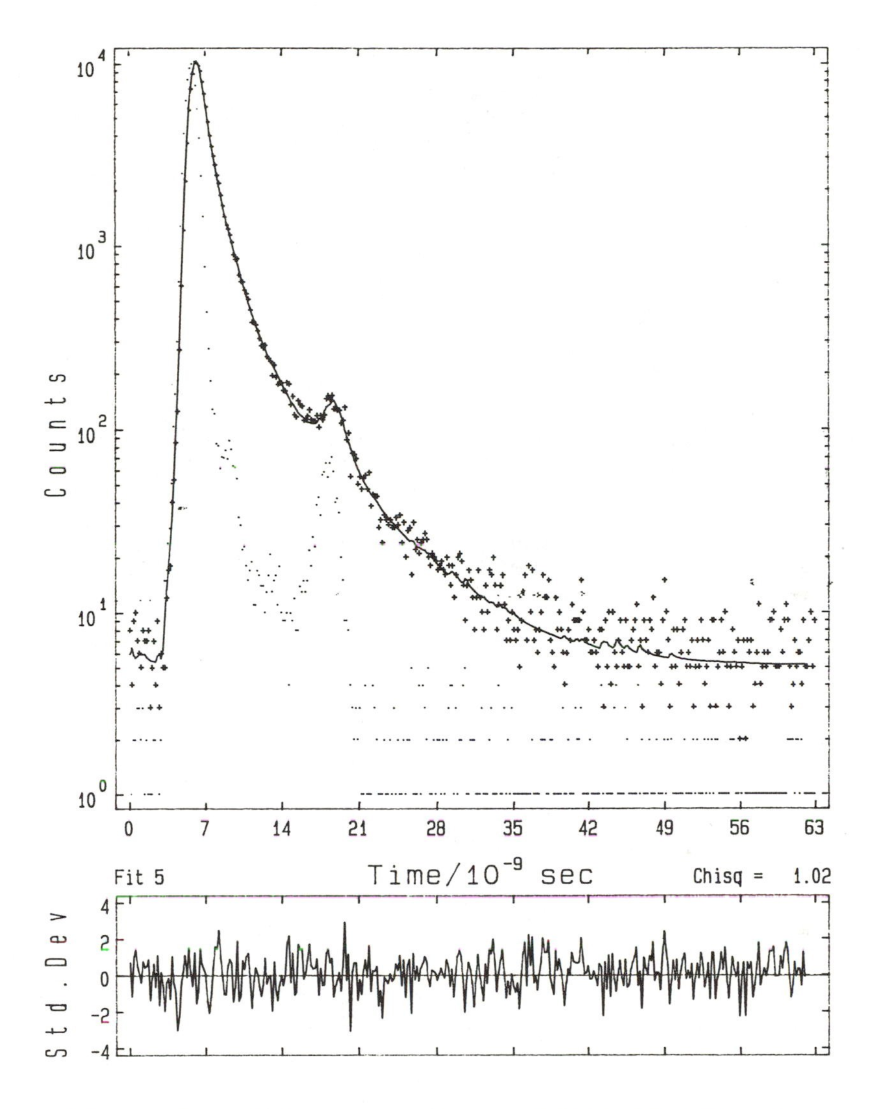


Figure 4 Fit to the decay of Poly(VBuPBD) at 350 nm and 25°C assuming transient quenching and excimer reverse dissociation.

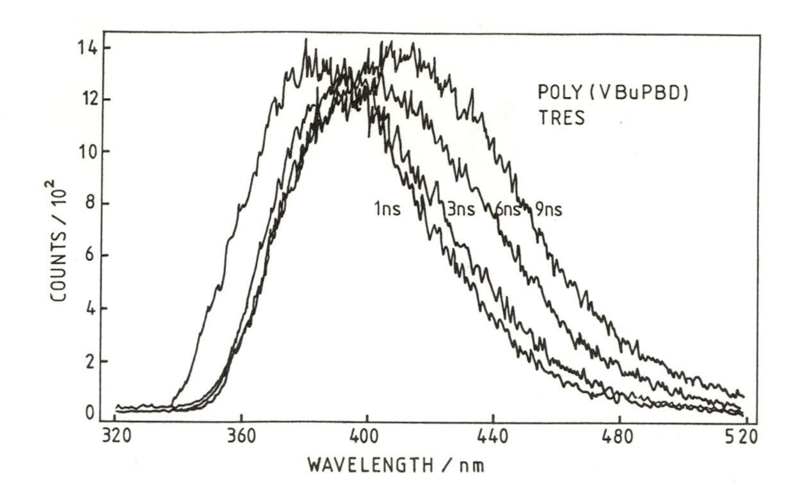


Figure 5 Time-resolved spectra for Poly (VBuPBD) 1, 3, 6 and 9ns after excitation.

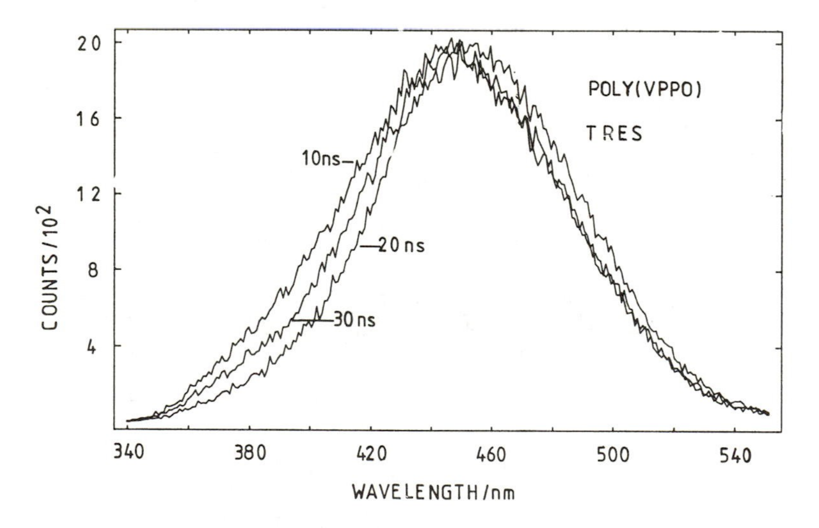


Figure 6 Time-resolved spectra for Poly (VPPO) 10, 20 and 30 ns after excitation.

Given the complexity of excimer behaviour believed to be present in all the decays measured it is perhaps not surprising that there are shortcomings in the different kinetic interpretations offered by Equations 3 and 4. Indeed, particularly for BuPBD which is reluctant to form an intermolecular excimer, it is possible to envisage that there exists in Poly(VBuPBD) a distribution of preformed excimer conformations each with differing overlap, molecular constraint, chromophore separation and decay rate. What is interesting is that these features also look to be apparent in Poly (VPPO) even though PPO itself readily forms an intermolecular excimer. Even if just two excimer conformations are present in these polymers the apparent anomalies of Tables I, II and III now start to appear more rational.

For example, with reference to Poly(VPPO) at 25°C and 350 nm (Table II) the emission can be seen as predominantly (64%) that of decay components more readily attributed to monomer emission ($\tau_3$ = 1.5 ns and $\tau_2$ = 0.26 ns) with a smaller component (36%) of excimer ($\tau_3$ = 7.38 ns). As the emission wavelength is increased the excimer emission intensity (B1) increases with respect to the monomer but, in addition to $\tau_1$, $\tau_3$ now starts to describe the other excimer conformations with $\tau_2$ providing what is probably an incomplete description of the monomer behaviour. This accounts for the relative constancy of $B_3$ as $B_2$ decreases. Quite simply, because Equation 3 contains insufficient decay components the best fit decay parameters are unable to describe the complex decay kinetics in terms of distinct excited state species.

Similar behaviour is evident in the Poly(VBuPBD) data of Table I. The presence of more than one excimer emission would also be expected to distort the transient quenching analysis shown in Table III, leading to anomalously high $\tau_2$ values being obtained. Nevertheless, why the transient quenching model is less appropriate for Poly(VPPO) than it is for Poly(VBuPBD) is still unclear but it is perhaps significant that its excimer decay (both direct emission and reverse dissociation) is considerably more intense than Poly(VBuPBD) which would be expected to more clearly expose the inappropriateness of assuming that only one excimer conformation and decay time is present.

It should be noted that in the case of the PPO intermolecular excimer we have observed no evidence for more than one conformation (19). This fact, when considered with the similarities between the behaviour of the two polymer systems reported here which occurs in spite of marked differences in the susceptibility of their fluorophores to form excimers, is a measure of the dominating influence of torsional constraints imposed by the polymer backbone.

## Conclusion

The excimer photophysics of the two macromolecular scintillators studied has been shown to be too complex to be described by either a kinetic model which includes isolated monomer sites or one which considers a time dependent rate of quenching. This is attributed to the existence in each case of more than one and possibly a distribution of conformations of excimer with an associated spectrum of decay times.

## Acknowledgments

We would like to thank the SERC for research grants and studentships held by A Dutch and K Davidson. We would also like to thank Nuclear Enterprises Ltd. for their support. Acknowledgment is made to the donors of the Petroleum Research Fund, administered by the American Chemical Society, for partial support of this activity.

## Literature Cited

1. Roberts A J and Soutar I. In Polymer Photophysics. Phillips D, Ed.Chapman and Hall, 1985 and references therein.
2. Birks J B Photophysics of Aromatic Molecules; Wiley, 1970.
3. Holden D A, Wang P Y K and Guillet J E, Macromolecules, 1980, 13, 295.
4. Phillips D, Roberts A J and Soutar I, J. Polym. Sci. Polym. Phys. Ed. 1980, 18, 2401.
5. Phillips D, Roberts A J and Soutar I, Polymer, 1981, 22, 427.
6. Soutar I, Phillips D, Roberts A J and Rumbles G, J. Polym. Sci., Polym. Phys. Ed, 1982, 20, 1259.
7. Phillips D, Drake R G and Soutar I (To be published).
8. Anderson R A, Birch D J S, Davidson K, Imhof R E and Soutar I. In Photophysics of Synthetic Polymers; Science Reviews, 1982, 140.
9. Fredrickson G H and Frank C W, Macromolecules. 1983, 16, 572.
10. Itagaki H, Horrie K and Mitka I, Macromolecules 1983, 16, 1395.
11. Lami H and Laustriat G, J. Chem. Phys. 1968, 48, 1832.
12. Horrocks D L, J. Chem. Phys., 1969, 4962, 50.
13. Berlman I B, J. Chem. Phys. 1961, 34, 1083.
14. Yguerabide J and Burton M, J.Chem. Phys. 1962, 37, 1757.
15. Birch D J S, Imhof R E and Dutch A, J Luminescence, 1984, 31, 763.
16. Birch D J S and Imhof R E, Rev. Sci. Instrum. 1981, 52, 1206.

17. Birch D J S and Imhof R E, Analytical Instrum. 1985, 14, 293.
18. McInally I, Soutar I and Steedman W, J Polym. Sci. 1977, 15, 2511.
19. Birch D J S, Dutch A, Imhof R E and Soutar I, J. Photochem. In Press.
20. Nakahira T, Ishizuka S, Iwabuchi S and Kojima K. Macromolecules, 1982, 15, 1217.
21. Johnson G E, J.Chem.Phys. 1975, 62, 4697.

RECEIVED March 13, 1987

# Chapter 16

# Configurational and Conformational Aspects of Intramolecular Excimer Formation

**F. C. De Schryver, P. Collart, R. Goedeweeck, F. Ruttens, F. Lopez Arbelao, and M. VanderAuweraer**

**Department of Chemistry, K.U. Leuven, Celestijnenlaan 200F, B3030 Heverlee, Belgium**

In a polymer the photophysical behavior of the singlet excited state is influenced strongly by the configuration and the conformational distribution within each configuration. This is illustrated in this paper by the excimer formation of 2,4-diarylpentanes and bis(pyrenylalanine)-peptides.

Since the first report of the pyrene excimer by Förster [1] intermolecular excited state interactions have become a key subject in photochemistry and photophysics.

From calculations[2,3] a picture of the excited state complex arises in which the two chromophores, if they are planar, are at an optimum distance of 3.5 Ängstroms in a plane parallel orientation. The stabilization of an excimer is much more dependent on the overlap between the two chromophores than an exciplex in which the coulombic attraction, especially in non polar solvents, is the main stabilizing factor.

Intermolecular excited state complex formation can be described by the kinetic scheme 1 [4]

$$A \xrightarrow{h\nu} A^*$$

$$A^* \xrightarrow{k_1} A + h\nu$$

$$A^* \xrightarrow{k_2} A$$

$$A^* + M \xrightarrow{k_3[M]} (AM)^*$$

$$(AM)^* \xrightarrow{k_4} A^* + M$$

$$(AM)^* \xrightarrow{k_5} A + M + h\nu''$$

$$(AM)^* \xrightarrow{k_6} A + M$$

$$k_7 = k_1 + k_2 \text{ and } k_8 = k_5 + k_6$$

Scheme 1

0097-6156/87/0358-0186$06.00/0

$A^*$ is the locally excited state (LE) and $(AM)^*$ is the excited state complex (E).

The fluorescence spectrum consists of the emission of LE, hv', at higher energies and of the complex E, which emits at lower energies, hv'',due to the stabilisation and repulsion energy terms [5]
In case of a δ excitation the time dependence of the emission can be described by the following equations [6]

$$I_{LE}(t) \approx I_{abs} \frac{k_1}{\beta_2 - \beta_1} [(\beta_2 - X)\exp(-\beta_1 t) + (X - \beta_1)\exp(-\beta_2 t)] \quad \text{eq.1}$$

$$I_E(t) \approx I_{abs} \frac{k_5 k_3 [M]}{\beta_2 - \beta_1} [\exp(-\beta_1 t) - \exp(-\beta_2 t)] \quad \text{eq.2}$$

where $I_{abs}$ represents the number of photons absorbed by A.

$$k_7 + k_3[M] = X \quad \text{eq.3}$$

$$k_8 + k_4 = Y \quad \text{eq.4}$$

$$\frac{X - \beta_1}{\beta_2 - X} = Z \quad \text{eq.5}$$

$$\beta_{2,1} = 1/2\{[(X+Y) \pm (X-Y)^2] + [4k_3k_4]\}^{1/2} \quad \text{eq.6}$$

The quantum yield of fluorescence of the locally excited state, $\Phi_{LE}$, and of the complex, $\Phi_E$, and the ratio thereof can be formulated as follows:

$$\Phi_{LE} = k_1 Y / \beta_1 \beta_2 \quad \text{eq.7}$$

$$\Phi_E = k_5 k_3 [M] / \beta_1 \beta_2 \quad \text{eq.8}$$

$$\frac{\Phi_E}{\Phi_{LE}} = \frac{k_5 k_3 [M]}{k_1 Y} \quad \text{eq.9}$$

At sufficiently low temperature is $k_4 << k_8$, the decay of the locally excited state is monoexponential while the decay of the complex can be described by a difference of two exponentials. When $k_4$ becomes competitive with $k_8$ the decay of the fluorescence of the locally excited state can be described as a sum of two exponentials, the decay of the complex can be described as a difference of two exponentials and the ratio of the preexponential terms should equal -1. An analysis of the decay functions according to Scheme 1 permits the determination of all relevant rate constants[7].

The question we want to addres in this contribution is :If two chromophores are linked by a flexible chain do the confi-gurational and conformational properties of the chain and eventually of the chromophore affect the excimer forma-tion and\or the excimer properties ?

2,4-DIARYLPENTANES and BIS[ 1-(ARYL) ETHYL ] ETHER

The first model study to show a substantial difference in the fluoresecence spectrum of the meso and the racemic diastereoisomer was reported in 1966 by Bovey and Longworth [8]. Based on H-nmr data information concerning the conformational distribution within each diastereoisomer could be obtained[9]. This conformational distribution was also studied for 2,4-diphenylpentanes by ir[10], ultrasonic relaxation[11] and theorethical calculations[12]. These studies indicate that at room temperature the TG conformation and to a very small extent the TT conformation contributes to the ground state of the meso isomer while for the racemic isomer the TT and the GG conformation are the main contributors. The importance of this observation and the consequences on the kinetic aspects of intramolecular excimer formation as well in bichromophoric molecules as in polyvinylaromatic compounds was only fully understood when a systematic study of molecules of the general structure 1 was undertaken.

```
CH3-CH-X-CH-CH3
    |    |
    Y    Y
```

$1_x y$ if meso and $1_x y'$ if racemic

X = O or $CH_2$(indicated by $_c$ or $_o$)
Y = N-carbazoyl = a
= phenyl = b
= 2-pyrenyl = c
= 9-anthryl = d
= 1-pyrenyl = e

Structure 1

As an example the emission spectrum of $1_c c$ is given in fig.1

The Meso Diastereoisomers

Analysis of the fluorescence decay of $1_c a$[13], $1_c b$[14] and $1_c c$[15] led to the conclusion that the excited state complex formation could be described by a monomolecular equivalent of scheme 1 (Scheme 2). Analysis at different wavelenghts within the excimer band further indicated that the decay parameters as well as their contribution did not vary. This clearly supports the formation of only one excimer.

Nmr analysis of the vicinal coupling constants of the methylene protons shows that in the temperature region between -70°C and 30°C these compounds are mainly (>>98%) in the TG conformation. The excimer forming step is one rotation around one bond to form the TT conformation in which the two chromophores overlap extensively.(fig. 2 )

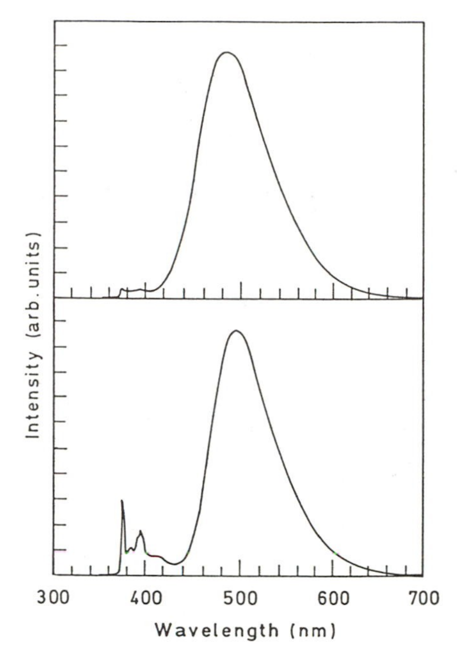


Fig. 1. Corrected emission spectra of $1_cc$ and $1_cc'$ at room temperature in isooctane.

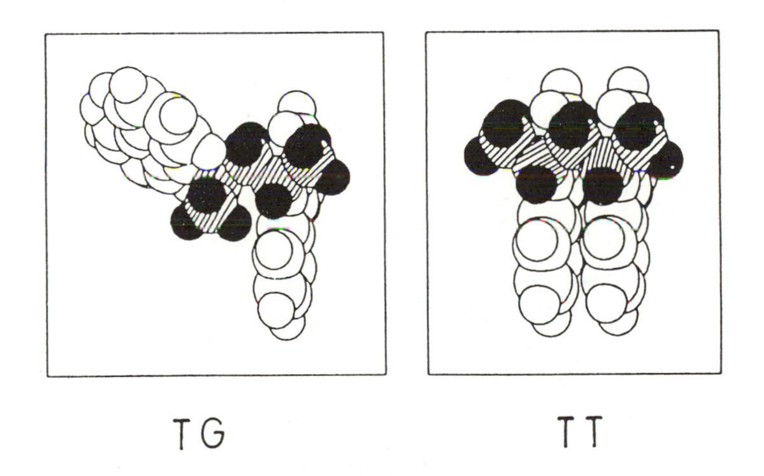


Fig.2.Space filling models representing the TG ground state and $TT^*$ excited state.The actual excimer structure must have the two pyrene groups farther apart for steric reasons.

Scheme 2

$$TG \xrightarrow{h\nu} TG^*$$
$$TG^* \longrightarrow TT^*$$

Important parameters of the photophysics of $1_c$a-c are assembled in table 1.

Table 1 Kinetic and thermodynamic parameters in the excimer formation of $1_c$a-c in isooctane

| parameter | $1_c$a | $1_c$b | $1_c$c |
|---|---|---|---|
| FWMH $cm^{-1}$ a | 3400 | 3900 | 3800 |
| emission max.$^a$nm | 420 | 330 | 485 |
| $E_3$ $kJmol^{-1}$ | 16 | 10 | 17 |
| $k'_3$ $10^{10}$ $s^{-1b}$ | 310 | 10 | 60 |
| $k_8^{-1}$at 20°C in ns | 37 | 22 | 145 |
| $\Delta H^o$ kJ $mol^{-1}$ | -20±3 | - | -16 |
| $\Delta S^o$ $JK^{-1}$ $mol^{-1}$ | -17±5 | -- | -42 |

a)excimer emission b)preexponetial term

Substitution of the central methylene group by an oxygen no longer permits the use of the above mentioned nmr method to obtain information on the conformational distribution for the meso diastereoisomer. An analysis of the luminescence behavior of $1_o$c[15] - a single exponential decay at -30 °C in isooctane when analysed in the locally excited state, an activation energy for excimer formation of 17,6 kJ $mol^{-1}$in isooctane , the presence of only one excimer with a $k_8^{-1}$of 85 ns and a FWHM of 3800 $cm^{-1}$ - suggests a TG ground state conformation.

These results allow the conclusion that in a bichromophoric molecule linked by three carbon-carbon bonds or in the ether analog with a well defined conformation of the chain , a symmetric substitution at the chromophore and in absence of interactions in the ground state excimer formation can be described by a intramolecular versiom of the kinetic scheme 1.

Nmr analysis of $1_c$e[16] indicated that the meso isomer in alkane solvents at room temperature has mainly (>95%) a TG chain conformation while at lower temperatures only the TG conformation is present.Nevertheless the time profile of the fluorescence at -50°C can not be analysed as a single exponential but good fits are obtained upon analysis of the locally excited state as a sum of two exponentials. Spectral information clearly indicates that at this temperature no back dissociation of the complex occurs.[17] A substantial shift of the excimer emission maximum , a broadening of the full width at medium height (FWMH) of the excimer emission and the analysis of the fluorescence decay across the excimer band clearly established the presence of two excimers. Analogous results were obtained for $1_o$e.(table 2)

Table 2 Excited state properties of $1_c$e and $1_o$e in isooctane

| parameter | $1_o$e | $1_c$e |
|---|---|---|
| λ max. 298 K[a] nm | 480 | 472 |
| λ max. 183 K[a] nm | 500 | 492 |
| FWMH(298 K)[a] $cm^{-1}$ | 4300 | 4800 |
| FWMH(183 K)[a] $cm^{-1}$ | 3800 | 4400 |
| $\alpha_1/\alpha_2$ [b]at 450 nm | 1.44 | 1.5 |
| $\alpha_1/\alpha_2$ at 520 nm | 8.5 | 15 |
| $\beta_1^{-1}$ in ns[c] at 450 nm | 80 | 160 |
| $\beta_2^{-1}$ in ns at450 nm | 47 | 80 |
| $\beta_1^{-1}$ in ns at 520 nm | 80 | 161 |
| $\beta_2^{-1}$ in ns at 520 nm | 47 | 80 |
| $k_3'{}_{fo}$($10^{10}$ $s^{-1}$)[d] | 150 | 330 |
| $E_{3,fo}$(kJ $mol^{-1}$)[e] | 21 | 19 |
| $k_3'{}_{po}$($10^{10}$ $s^{-1}$)[f] | 2 | 7 |
| $E_{3\ po}$(kJ $mol^{-1}$)[g] | 12 | 8 |

a) of the excimer emission b) ratio of the preexponential terms of the two exponential decay function describing the time profile of the excimer emission measured at 298 K in isooctane.c) $\beta_{1,2}$ exponential terms of the decay function. Since the amount of back dissociation at this temperature is very small they can be set equal to $k_8^{-1}$.d) preexponetial term of the rate constant for formation of the long lived excimer.e) activation energy of the rate constant for formation of the long lived excimer.f) preexponential term of the rate constant for formation of the short lived excimer.g) activation energy of the rate constant for formation of the short lived excimer.

The formation of two excimers starting from one chain conformation was related to the presence of different rotamers of the non symmetrically substituted 1-pyrenyl group in the TG conformation (fig.3). One of the excimers formed - the longer lived one derived from the $T_1G_1$ conformer - has a $k_8^{-1}$ value at room temperature close to the one observed for respectively meso $1_c$c and meso $1_o$c and emits at longer wavelengths.At low temperature it is the main contributor to the excimer emission.Transient picosecond aborption spectroscopy[18] of $1_o$c and $1_o$e substantiates this interpretation. Since a TT conformation of $1_o$c results in a full overlap of the two pyrene group a similar spatial arrangement can be suggested for this long lived excimer. The second excimer formed from other rotamer(s) will have only partial overlap of the two chromophores. An unequivocal assignment of the decay parameters to the different rotamers forming the two different excimers could be made by the analysis of the decay of the locally excited state and the excimer decay at 205 K where the long lived excimer is the sole contributor and the locally excited state can be correlated with the growing in of this excimer. This suggests that the rate of rotation of the chromophore should be smaller than the rate of excimer formation. Broadening of the nmr signal of the pyrene group at 185K indicates slow rotation of this

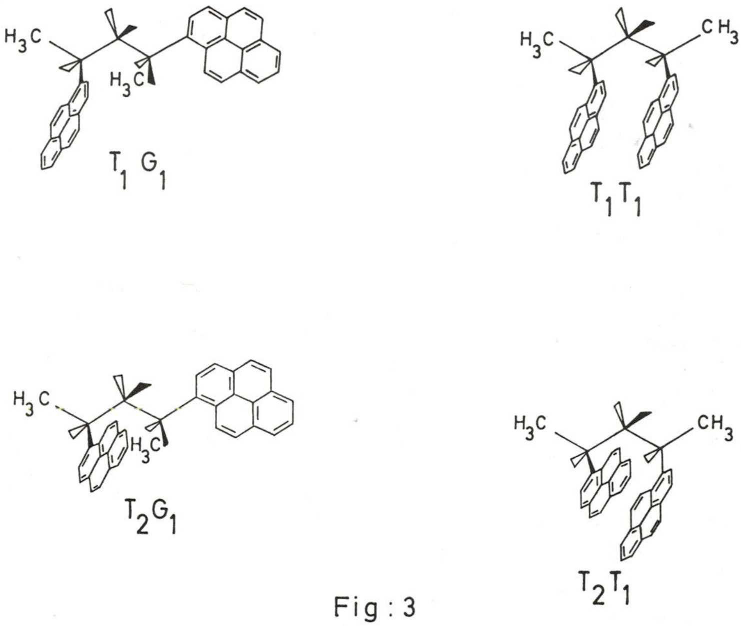

Fig : 3

Fig.3.Schematic representation of two possible ground state conformations leading to two different excimers of $1_c c'$

group at the nmr time scale and hence on the time scale of the photophysical experiment.

The partially overlapping excimer has a lower activation energy of formation due to the fact that a stabilising interaction between the two pyrene groups starts earlier along the reaction coordinate decreasing the activation barrier. The activation energy of formation of the full overlap excimer is in the non symetrically substituted compounds $1_ce$ and $1_ee$, within experimental error, comparable to the activation energies observed for respectively $1_cc$ and $1_oc$

These results indicate that the complexity of the fluorescence decay of meso $1_ce$ and meso $1_oe$, where the pyrene is non symmetrically substituted, is not due to a conformational distribution of the chain but to rotational isomerism around the carbon carbon bond linking the chromophore to the chain backbone providing the possibility to form more than one excimer in these systems.

## The Racemic Diastereoisomers

What are the kinetic consequences if the chromophores are linked by a chain for which more than one chain conformation contributes substantially? This aspect can be considered by a study of the racemic diastereo-isomers of $1_xc$'The racemic diastereisomers $1_xc'$ are systems in which the chain is present in different conformations namely the TT and the GG conformers (fig 4).

In both deuterated cyclohexane and deuterated chloroform, at room temperature, the TT conformation of racemic $1_cc'$[15] is predominant : 80 % TT and 73 % TT respectively. The observed solvent effect on the conformational distribution is similar but smaller than found for $(1_ce')$[17] . These results agree quite well with those obtained on similar compounds[9,10,17,19]

In the NMR data of the racemic diastereoisomers of $1_cc'$ and $1_oc'$ one important difference shows up. The position of the singlet absorption signal of $H_1$ of the pyrene ring is shifted significantly to higher $\delta$ values for $1_oc'$ compared with $1_cc'$ . The position of the $H_1$ protons in the $^1$H-NMR spectrum is a function of the contri-bution of the TT conformation since these protons are shielded by the mutual ring current effect of the pyrene chromophores in the TT conformation. It can therefore be concluded that the equilibrium between TT and GG in the case of $1_oc'$ is shifted more toward the GG conformation compared to $1_cc'$ . This can be explained by comparing the bond lengths of the C-O bond and the C-C bond in the respective compounds. The shorter ether band induces a stronger 1,4 interaction between the methine hydrogens and the pyrene chromophores in the ether compound causing a shift of the conformational equilibrium from TT to GG.

The steady state fluorescence spectra of $1_cc'$ and $1_oc'$ have an excimer fluorescence band that is completely superimposable at all temperatures investigated with that of their respective meso diastereoisomers suggesting an identical geometrical structure of the excimer and hence also the presence of only one excimer.

The fluorescence decay curves of $1_cc'$ and $1_oc'$ monitored at 377 nm and 500 nm can be described by the same decay laws used for the respective meso diastereoisomers.

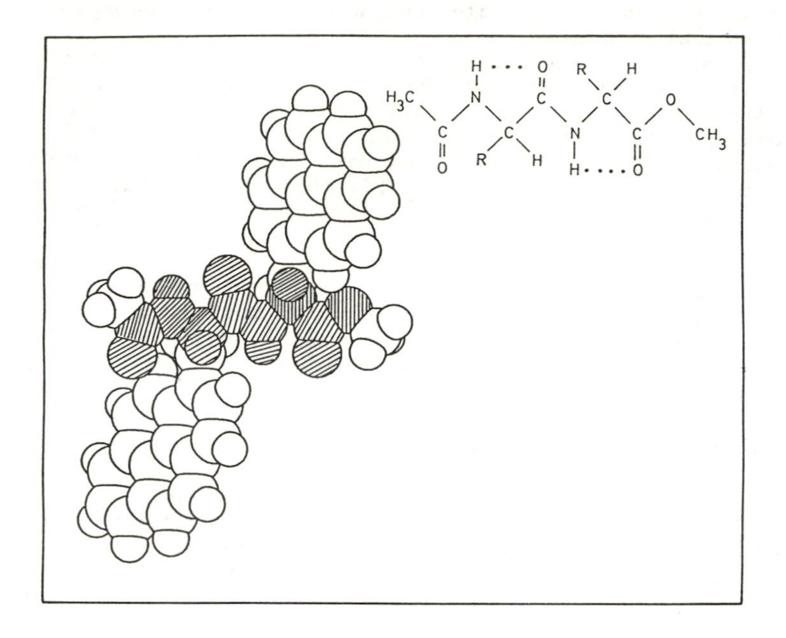

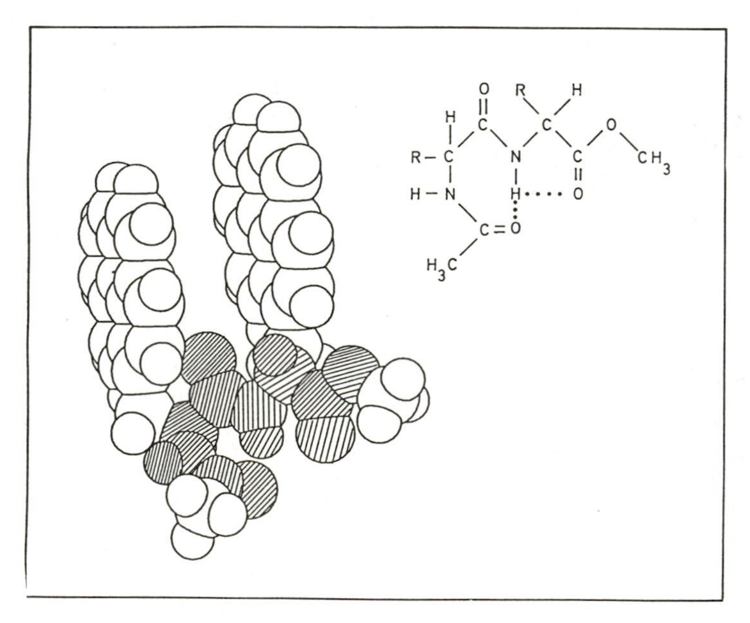

Fig. 4. Space filling representation of the extended ($C_5$) and folded ($C_7$) conformation of 2ae. In the folded conformation the pyrene group are depicted in the excimer geometry.

The fluorescence decays of the locally excited state can be analysed as a two exponential decay function in the temperature domain between 298 K and 233 K . Below this temperature the decays are one exponential. Furthermore excellent agreement between the decay parameters measured in the locally excited and excimer region of the emission spectrum is obtained.

The molecular dynamics of the racemic diastereoisomer upon excitation are however more complicated than that of the meso diastereoisomers due to the presence of two ground state conformations. Since the decay laws used for meso diastereisomers can also be applied to the racemic diastereoisomer, all the rate constants describing the equilibrium between the ground state conformations must be large compared to the rate constant of excimer formation. The question however still remains from which conformation does excimer formation take place. It is not likely that it occurs directly from the GG conformation. This would mean a simultaneous rotation around two bonds and would be associated with a relatively high activation barrier. One possibility is that excimer formation takes place from the TT conformation. This possibility is depicted in scheme 3.

A third route could be excimer formation starting from the TG conformation. This intermediate conformation between the TT and GG conformation has however a relatively high energy content owing to an unfavourable interaction between the methyl group and the pyrene moiety.

Upon taking into account this preequilibrium the rate constants and activation barriers of excimer formation can be rewritten if scheme 3 is considered using the following equations :

$$k_{obs} = f_{TT}\, k_3 \qquad \text{eq 10}$$

$$f_{TT} = \frac{K_1K_2}{1 + K_1 + K_1K_2} \qquad \text{eq 12}$$

$$K_1 = k_a/k_{-a} \qquad \text{eq 13}$$

$$K_2 = k_{-b}/k_b \qquad \text{eq 14}$$

$$E_{obs} = E_3 + \frac{\Delta H°_2 + (1+K_2)\Delta H°_1}{1 + K_1 + K_1K_2} \qquad \text{eq 15}$$

In these equations $k_{obs}$ and $E_{obs}$ are the observed rate constant and activation barrier of excimer formation. The fractions of the TT or TG conformations at a given temperature are represented by $f_{TT}$ and $f_{TG}$..

It could be shown[15] that since the fraction of the TT conformation is considerable and decreases with increasing temperature Scheme 3 should lead to a negative deviation from linearity in the Arrhenius plot of the rate constant of excimer formation.This could be experimentally verified proving the validity of scheme 3

The kinetic and thermodynamic data of $1_c c'$ and $1_o c'$ are summarized in table 3. If the linear part in the Arrhenius plot is

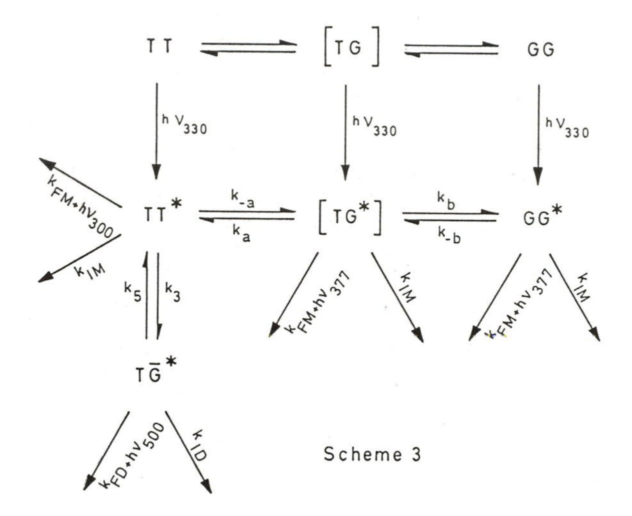
TT
[TG]
GG
hν330
hν330
hν330
TT*
k-a
ka
[TG*]
kb
k-b
GG*
kFM+hν300
kIM
k5
k3
kFM+hν377
kIM
kFM+hν377
kIM
TḠ*
kFD+hν500
kID

Scheme 3

extrapolated to higher temperatures, an estimate of $f_{TT}$ can be made. This assumes that the extrapolated rate constants at all given temperatures are equal to the value of $k_3$. The variation of $f_{TT}$ with temperature in the high temperature region (303 K-343 K) can then be used to determine $\Delta H_o$ and $\Delta S_o$ for the conformational equilibrium betwee TT and GG of the racemic diastereoisomer. In the case of $1_cc'$ the respective values are 12,2 $kJmol^{-1}$ and 35 $Jmol^{-1}K^{-1}$. In the case of $1_oc'$ these values are 5 kJ $mol^{-1}$ and 12 $Jmol^{-1}K^{-1}$ respectively. The conformational distribution between TT and GG can then be calculated and equals 70 % TT/ 30% GG for $1_cc'$ and 61 % TT/ 39 % GG for $1_oc'$ at room temperature.

Table 3. Kinetic and thermodynamic parameters $1_cc'$ and $1_oc'$ in iso-octane (a) : pre-exponential factor in the Arrhenius equation; (b) determined in the linear portion of the Arrhenius plot at low temperature (243 K-203 K)

| | $1_cc'$ | $1_oc'$ |
|---|---|---|
| $E_{obs}$ ($kJ\ mol^{-1}$) (b) | 20 (± 2) | 22 (± 2) |
| $k^o_{obs}$ ($s^{-1}$) (a) | 1.8 (±0.2)x$10^{11}$ | 3.3 (±0.3)x$10^{11}$ |
| $E_5$ ($kJ\ mol^{-1}$) | 34 (± 5) | 40 (± 8) |
| $k^o_5$ ($s^{-1}$) (a) | 5.0 (± 1)x$10^{13}$ | 3 (± 2)x$10^{14}$ |
| $\Delta H_o$ ($kJ\ mol^{-1}$) | -14 (± 7) | -18 (± 10) |
| $\Delta S_o$($Jmol^{-1}\ K^{-1}$) | -47 (± 12) | -57 (± 30) |

The results obtained for the racemic diastereoisomers $1_cc'$ and $1_oc'$ indicate that even when the rate of conformational change between the different chain conformations is much faster than the rate of excimer formation , resulting in fluorescence decays similar to those of the meso diastereoisomers it is possible to extract some information on the conformational distribution of the chain from the obtained data.

BIS(PYRENYLALANINE)PEPTIDES

By choosing another link between the two chromophpores it is possible to slow down the rate of conformational interchange as to be in the same order of magnitude as the inverse of the fluorescence lifetime of the chromophore. This was realized in bis(pyrenyl-alanine)peptides of the general structure 2.

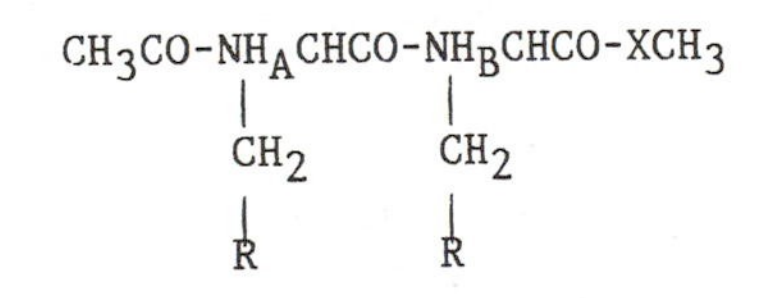


2

if $H_A$,$H_B$ is replaced by a methylgroup

2($ME_a$or $ME_b$)

R = 1-pyrenyl = a

R = 2-pyrenyl = b

Structure 2

N-acetyl-bis(1-pyrenyl)alaninemethylester 2a can occur in two diastereoisomeric forms- threo(t) and erythro(e)- which could be separated by chromatography[20] . In inert solvents,not capable of forming hydrogen bonds,the ratio of excimer emission over emission from the locally excited state is substantially higher than in solvents which can accept hydrogen bonds. The decrease of the emission ratio can be correlated with the Taft[21] basicity parameter..These observations and the fact that 3ae always has a higher efficiency of excimer formation than 3at were explained by a consecutive kinetic scheme involving the locally excited state of pyrene in two different conformations of the peptide chain: an extended chain conformation, $C_5$, with alternate side chains(fig.4) and a folded conformation,$C_7$,with quasi parallel side chains(fig.4) and supported by an intramolecular hydrogen bond between the carbonyl function of the acetyl protecting group and the amine function of the second pyrenylalanine residue[22].

Only the $C_7$ conformation allows rotation of the aromatic side groups to a partially overlapping excimer geometry within the lifetime of the excited pyrene moeiety. This point could be proven by the substitution of the hydrogen involved in the hydrogen bond formation in the $C_7$ conformation by a methyl group[22] 2($ME_b$)a resulting in the dissapearence of the excimer band in the emission spectrum.

The fluorescence decay of the locally excited state at temperatures were the excimer does not disociate back could be analysed as a sum of two exponentials[23] .That this is not due to roational isomerism of the 1-pyrenyl group could be proven by the analysis of 2b which showed analogous behavior.The analysis of the fluorescence decay according to scheme 4 permits the determination of the ratio of $C_7/C_5$.

Scheme 4.

$$C_5^{LE*} \underset{k^{-fol}}{\overset{k^{fol}}{\rightleftharpoons}} C_7^{LE*} \underset{k^{-ex}}{\overset{k^{ex}}{\rightleftharpoons}} C_7^{EX*}$$

In inert solvents the molecule is stabilised by folding leading to a high $C_7$ population and hence a more intense excimer emission than in hydrogen bonding solvents that shift the conformational distribution in the ground state more to the $C_5$ conformer.When compared in the same solvent the threo diastereoisomer has a lower folded population then the erythro in part due to increased steric hinderance and in part due to the absence of a stabilising N-H pyrene interaction

The solvent induced shift of the conformational equilibrium was confirmed,in toluene at -20°C a ratio of 3 and 0.75 was calculated for 2ae and 2at respectively while in ethylacetate these values decrease to 0.8 and 0.4.

CONCLUSIONS

In this contribution we attempt to evaluate the influence of conformational and configurational aspects ,both on excimer formation rates and on excimer properties.From the results presented, we con-clude that the difference in photophysical behavior of two different configurations can be related to the conformational distribution of each diastereoisomer. If only one conformation is present complex fluorescence decay still can arise by the non symmetrical substitution of the chromophore linked to the chain.

If more then one chain conformation is present will this lead to complexity in the fluorescence decay only if the rate of conformational change is comparable to the rate of excimer formation a in the dipeptides. In absence of this complication it is however necessary to interpret the obtained results taking into account the fast preequilibrium between the different conformers.

The nature of the excimer formed and its properties will for a given chromophore depend on the diastereoisomer. The binding energy of the excimer will be affected by stabilising or destabelising effects of the chain. Furthermore if rotational isomerism of the chromophore is possible and interconversion between the rotamers is slow compared to excimer formation the occurece of more then one excited state complex is possible and depends on the respective stabilisation of the diffetrent complexes.The relative contribution of each at a given temperature will depend on the rotational distribution and the respective rate of excimer formation

## Acknowledgments

The authors wish to thank the Belgian National Science Foundation for continous support to the laboratory.Dr.M.VdA is a Reseach Associate of the NFWO.Dr.Boens is thanked for his essential contribution to the development of the software used in the analysis of the fluorescence decays obtained by time correlated single photon counting.

## REFERENCES

(1)Förster,T. and Kaspar,K.,Z.Phys.Chem.(N.F.),**1**,275,(1954).
(2)a.Azumi,T. and McGlynn,S.P.,J.Chem.Phys.,**39**,1186(1963)
b.Azumi,T. and McGlynn,S.P.,J.Chem.Phys.,**41**,3131(1964)
c.Azumi,T.,Armstrong,A.T.,and McGlynn,S.P. J.Chem.Phys,**41**, 839,(1964)
d.Förster,T.,Pure and Appl.Chem.,**4**,121(1962)
e.Murell,J.N., and Tanaka,J.,J.Mol.Phys.,**7**,363,(1964)
(3)a.Azumi,T. and Azumi,H.,Bull.Chem.Soc.Jpn.,**39**,2317,(1966)
b.Chandra,A.K. and Lim,E.C.,J.phys.Chem.,**48**,2589(1968)
c.Post,M.F.M.,Langelaar,J. and Van Voorst ,J.D.W., Chem.Phys.,**15**,445(1976)
d.Padma Malar,E.J. and Chandra,A.K., Theor.Chim.Acta, **55**,153(1980)
(4)a.Döller,E. and Förster,.,Z.Phys.Chem.(N.F.),**34**,132(1962)
b.BirksJ.B.,"Photophyics of aromatic molecules",ch.7, Wiley- Interscience,London,1970
(5)Stevens,B. and Ban,M.I.,Trans.Far.Soc.II ,**60**,1515(9164)

(6)a.ref.5b p.309
b.O'Connor,D. and Ware,W.R.,J.Am.Chem.Soc.,**101**,121,(1979)
(7)Palmans,J.P.,Van der Auweraer,M.Swinnen,A.M.and De Schryver,F.,J.Am.Chem.Soc.,**106**,7721(1984)
(8)Longworth,J.W. and Bovey,F.A.,Biopolymers ,**4**,1115(1965)
(9)Bovey,F.A.,Hood III,F.P.,Anderson,E.W. and Snyder,L.C., J.Chem.Phys.,**42**,3900(1965)
(16)S. Ito,M. Yamamoto, Y. Nishijima, Bull.Chem.Soc.Jap 55,(1982) 363.
(10)Jasse,B.,Lety,A. and Monnerie,L.,J.Mol.Struct., **18**,413(1973)
(11)Froelich,B.,Noel,C.,Jasse,B.and Monnerie,L., Chem.Phys.Lett.,**44**,159(1976)
(12)a.Gorin,S. and Monnerie,L.,J.Chim.Phys. Physicochim. Biol.,**67**,869(1970)
b.Yoon,D.Y.,Sundararajan,P.R. and Flory,P.J., Macromolecules,**8**,776(1975)
c.McMahon,P.E. and Tincher,W.C.,J.Mol.Spectrosc., **15**,180(1965)
(13)a.De Schryver,F.C.,Vandendriessche,J., Toppet,S., Demeyer,K. and Boens,N.?Macromolecules,**15**,406(1982)
b.Vandendriessche,J.,Palmans,P.,Toppet,S.,Boens,N,De Schryver,F.C. and Masuhara,H.,J.Am.Chem.Soc. **106**,8057(1984)
(14)Monnerie,L.,Bokobza,L.,De Schryver,F.C.,Moens,L.,Van derAuweraer,M. and Boens,N.,Macromolecules,**15**,64(1982)
(15)Collart,P.,Toppet,S.,and De Schryver,F.C.,Macromolecules in press
(16)Collart,P.,Toppet,S.,Zhou,Q.F.,,Boens,N. and De Schryver,F.C.,Macromolecules,**18**,1026(1985)
(17)Collart,P.Demeyer,K.,Toppet,S. and De Schryver,F.C., Macromolecules,**16**,1390(1983)
(18)a.Masuhara,H.,Tanaka,J.A.,Mataga,N.,De Schryver,F.C. and Collart,P.,Polym.J.,**15**,915(1983)
b.Masuhara,H.,Tamai,T.,Mataga,N.,Collart,P. and De Schryver,F.C.,submitted
(19)Ito,S.,Yamamoto,M. and Nishijima,Y.,Bull.Chem.Soc.Jpn., **55**,363(1982)
(20)Goedeweeck,R. and De Schryver,F.C.,Photochem.Photobiol **39**,515(1984)
(21)Mortimer,J.K.,Abbaud,J.L.,Abraham,M.H. and Taft,R.W., J.Org.Chem.,**48**,2877(1983)
(22)Goedeweeck,R.,Ruttens,F.,Lopez Arbelao,F. and De Schryver,F.C.,submitted
(23)Goedeweeck,R.,Van der Auweraer,M. and De Schryver,F.C., J.Am.Chem.Soc.,**107**,2334(1985)

RECEIVED September 14, 1987

# Chapter 17

# Photophysics of 1,5-Naphthalene Diisocyanate-Based Polyurethanes

**Charles E. Hoyle and Kyu-Jun Kim**

**Department of Polymer Science, University of Southern Mississippi, Hattiesburg, MS 39406–0076**

Using both steady-state and transient fluorescence decay spectroscopy, the formation of intramolecular excimers in dilute solution of a naphthalene diisocyanate based polyurethane is identified. Investigation of an appropriate model compound leads to the conclusion that hydrogen bonding is a key factor in stabilizing excimers formed from naphthyl carbamates. While the decay kinetics of the model naphthyl carbamate are described by a typical Birks excimer scheme involving a single excited species in dynamic equilibrium with the excimer, the polymer decay kinetics can only be adequately interpreted by an "isolated monomer" scheme involving both an interactive (excimer forming) and non-interactive (isolated monomer) excited naphthyl carbamate moiety. The extent of excimer formation is dependent on the ability of the solvent to solvate the polyurethane, i.e., excimer formation is increased in poor solvents. In addition, excimer formation in solid polyurethane films is quite high where hydrogen bonding, as identified by the shift in the carbonyl stretching frequency, is prevalent.

Excimers, which are simply excited state complexes formed from equivalent chemical species, one of which is excited prior to complexation, were first reported in small molecule systems. However, during the past 20 years one of the more vigorous research areas in photophysics has been the investigation of excimers formed in polymers (1). In most of the cases reported to date, the excimer studies in polymer systems have been conducted on polymers bearing pendant aromatic chromophores. Only a few papers have been published on excimers formed from polymers with the two species participating in excimer formation spaced at relatively large

0097–6156/87/0358–0201$06.00/0

intervals from each other along the backbone (2-5). Herein, we report on intramolecular excimers between naphthyl carbamate groups spaced periodically in polyurethanes made from 1,5-naphthalene diisocyanate. It is found that even in very dilute solutions, well below the concentrations required for intermolecular interaction, excimer emission can be quite strong, depending on the nature of the solvating system employed. The driving mechanism for excimer formation of the naphthyl carbamate groups is shown to be based, at least in part, on hydrogen bonding.

## EXPERIMENTAL SECTION

### Materials

Dichloromethane, dimethylformamide (DMF), and benzene were obtained from Burdick and Jackson and used without further purification. 2,3-Butanediol and 1-butanol (Aldrich) were used as received. Propyl benzene (Aldrich) was distilled before use. Deionized water was used.

### Equipment

Emission spectra and absorption spectra were recorded on a Perkin-Elmer 650-10S Fluorescence Spectrophotometer and a Perkin-Elmer 320 UV Spectrophotometer, respectively. Fluorescence decay data were obtained on a single-photon-counting apparatus from Photochemical Research Associates. The samples were bubbled with nitrogen for the steady-state fluorescence spectra and the fluorescence decay measurements. In some cases, front face spectra were taken. The data were analyzed by a software package from PRA based on the iterative convolution method. $^{13}C$ NMR spectra were obtained on a JEOL FX90Q, and FTIR spectra were recorded on a Nicolet 5DX. The elemental analyses were conducted by M-H-W Laboratories of Phoenix, AZ.

### Synthesis of Model Compounds

1,5-Naphthalene diisocyanate (NDI). To a stirring solution of p-dioxane (50 mL) containing the 1,5-diaminonaphthalene (Fluka, 5.1 g) was added trichloromethyl chloroformate (Fluka, 17 g) in p-dioxane (15 mL) through an addition funnel under a nitrogen stream. A white precipitate was immediately observed. After addition, the temperature was increased to reflux. The forming HCl was removed by passing through water. After 1 hour, the solution turned clear and was allowed to react for another 3 hours. The p-dioxane was evaporated under reduced pressure and the resulting solid was vacuum sublimed twice to give colorless crystals in 71% yield: mp 126-128°C (lit[6]. mp 129.5-131°C); IR 3022, 2300, 1600, 1500 $cm^{-1}$; $^{13}C$ NMR 130.9, 128.7, 127.7, 124.2, 122.2 ppm (benzene); Anal. $C_{12}H_6N_2O_2$ Calc. C, 68.57; H, 2.88; N, 13.33; Found C, 68.62; H, 2.91; N, 13.32.

Propyl N-(1-naphthyl) carbamate (PNC). Into 50 mL of an ethyl acetate (distilled and dried) solution of 1-naphthyl isocyanate (Aldrich, distilled, 10 g) was added 1-propanol (Baker, distilled and dried, 7.1 g) through an addition funnel. The temperature was increased to the reflux temperature of ethyl acetate and the mixture was allowed to react for 8 hours under a nitrogen stream. The precipitate was removed by filtration and ethyl acetate was evaporated under reduced pressure. The forming product was purified by recrystallization from $CH_3CN$: mp 72°C; IR 3325, 2950, 1675, 1600, 1540, 1500 $cm^{-1}$; $^{13}C$ NMR 156.1, 135.3, 129.1, 126.8, 126.6, 125.5, 123.4, 120.9, 67.1, 23.2, 10.9 ppm (DMF); Anal. $C_{14}H_{15}NO_2$ Calc. C, 75.24; H, 6.59; N, 611; Found C, 75.36; H, 6.51; N, 6.18.

Propyl N-methyl N-(1-naphthyl) carbamate (PNMNC). For the preparation of this compound, all reactions were carried out in an ice bath. NaH (Alfa, 50% in oil, 1.2 g) was dispersed in DMF (Burdick and Jackson, dried, 5 mL) and the solution was cooled in an ice bath. The propyl 1-(N-naphthyl) carbamate (3.3 g) in DMF (15 mL) was added to a reactor through an addition funnel dropwise under a nitrogen stream. Immediately, hydrogen was generated and the reaction mixture turned green. The stirring was continued for 1.5 hours. $CH_3I$ (Aldrich, distilled, 4.0 g) in DMF (10 mL) was added dropwise and the reaction mixture turned colorless. The solution was allowed to react for 2 hours. When the reaction was complete, the precipitates were removed by filtration and DMF was evaporated under reduced pressure. The residue was redissolved in diethyl ether and the insolubles were removed by filtration. After evaporation of solvent, the resulting liquid was vacuum distilled. A colorless liquid was obtained; bp 120-125°C at 0.5 mm Hg; IR 3060, 2950, 1700, 1600, 1500 $cm^{-1}$; $^{13}C$ NMR 158.6, 142.9, 137.4, 133.3, 131.2, 130.4, 129.3, 128.9, 128.5, 127.7, 125.5, 69.5, 40.7, 25.0, 12.7 ppm (neat); Anal. $C_{15}H_{19}NO_2$ Calc. C, 74.05; H, 7.04; N, 5.75; Found C, 74.16; H, 6.95; N, 5.69.

Dipropyl N,N'-naphthalene-1,5-diylbiscarbamate (1,5-DNB). To a p-dioxane solution containing 1-propanol (1.71 g) and dibutyltin dilaurate (Polysciences, 0.2 g) was added 1,5-naphthalene diisocyanate (1.5 g) at once. The reaction mixture was heated to 80° C and allowed to react for 4 hours with stirring under a nitrogen stream. After evaporation of p-dioxane, the forming product was purified by recrystallization from $CH_3CN$/DMF (50/50); mp 203°C; IR 3290, 3000, 1685, 1540, 1500 $cm^{-1}$; Anal. $C_{18}H_{22}N_2O_2$ Calc. C, 65.44; H, 6.71; N, 8.48; Found C, 65.38; H, 6.63; N, 8.59.

Synthesis of NDI-Based Polyurethanes (NDI-650 and NDI-2000). To 20 mL of 1,1,2,2-tetrachloroethane (Baker, distilled and dried) containing polytetramethylene ether glycol (Polysciences, Average MW 650, 2.17 g of average MW 2000, 6.68 g), dibutyltin dilaurate (Polysciences, 0.11 g), and Dabco (Aldrich, 0.08 g) was added 1,5-naphthalene diisocyanate (0.7 g) at once. The mixture was heated to 100° C and allowed to polymerize for 2.5 hours with stirring under a nitrogen stream. The 1,1,2,2-tetrachloroethane was evaporated under reduced pressure and the products were dissolved in $CH_2Cl_2$. The $CH_2Cl_2$ solution was poured dropwise into 500 mL of cyclohexane. The precipitated polymers were collected and dried; IR 3310, 2930, 1740, 1695, 1540, 1500 $cm^{-1}$; $^{13}C$ NMR 157.2, 136.2, 130.4, 128.4, 122.3, 120.3, 73.1, 67.9, 29.1, 28.6 ppm ($CH_2Cl_2$); Anal. Calc. C, 65.71; H, 9.15; N, 3.26; Found C, 65.99 H, 9.32; N, 3.29. The molecular weights of NDI-650 and NDI-2000 were 51,000 and 63,000 respectively by GPC (obtained at peak maxima).

**RESULTS AND DISCUSSION**

The results of our studies will be presented in three sections. The first deals with the fluorescence properties of model naphthyl carbmates. This first section is particularly important in furnishing the background necessary to interpret the results in the next two sections which are concerned with the fluorescence of naphthalene diisocyanate (NDI) based polyurethanes in solution and film.

Fluorescence of Naphthyl Carbamate Models. Two compounds (shown below) based on carbamate substituted naphthalene will be considered as models for the naphthalene diisocyanate based polyurethanes under investigation in this paper. The propyl N-(1-naphthyl) carbamate (PNC) and dipropyl N,N'-naphthalene-1,5-diylbiscarbamate (1,5-DNB) model compounds were made by reacting the requisite mono- or diisocyanate with 1-propanol.

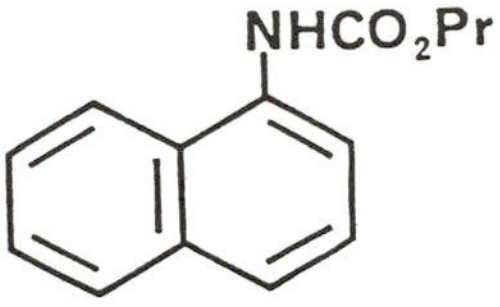

$NHCO_2Pr$

PNC

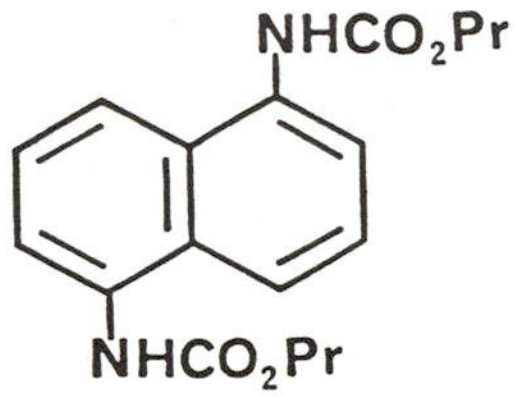

$NHCO_2Pr$

$NHCO_2Pr$

1,5-DNB

Figure 1 shows the fluorescence spectra of dilute solutions of PNC ($4.4 \times 10^{-4}$ M) and 1,5-DNB ($1.5 \times 10^{-4}$ M) in dichloromethane. Each exhibits a single exponential decay ($\tau_1$ in Table I) with the biscarbamate 1,5-DNB having the shortest lifetime (1.84 ns). (In each case, the decay curves were single exponential and invariant when measured at any emission wavelength from 330 nm to 400 nm). Unfortunately, 1,5-DNB is only sparingly soluble in most organic solvents and attempts to obtain concentrated solutions were limited to PNC. As the concentration of PNC increases (Figure 2), a new red-shifted emission appears. The red-shifted emission spectrum is characteristic of naphthalene substituted compounds and, in keeping with traditional interpretation of such phenomena, are assigned to excimers between naphthyl carbamate molecules. In accordance with this interpretation, the fluorescence decay curves were recorded in concentrated solutions of PNC at a series of wavelengths spanning the range from 330 nm to 500 nm. Figure 3 shows the decay curve for PNC (1.2 M) recorded at a representative wavelength (330 nm) in the monomer emission region. The 330 nm decay curve is readily fit to a double exponential decay function with decay parameters of 0.33 ns

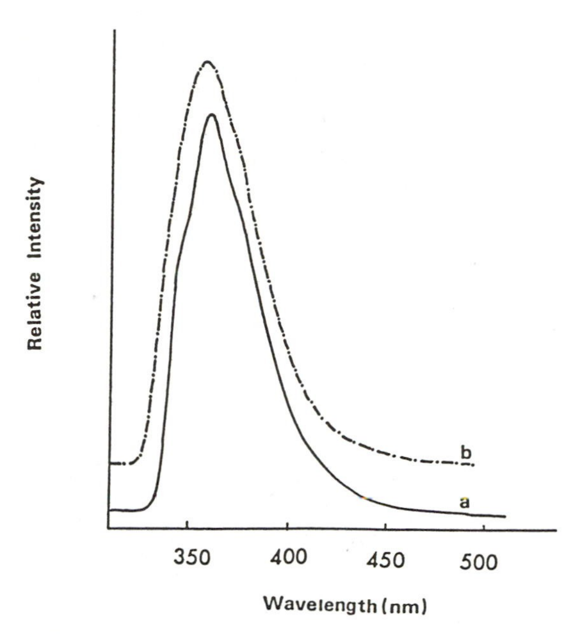


Figure 1. Steady state fluorescence spectra ($\lambda_{ex}$ = 300 nm) of 1,5-DNB (curve a, 1.5 X $10^{-4}$ M) and PNC (curve b, 4.4 X $10^{-4}$ M) in dichloromethane.

**Table I.** Lifetime data for model and polymer systems (in $CH_2Cl_2$ and 1-butanol)

| Compound | $\tau_1$[f] (ns) | $\tau_2$[g] (ns) | $\tau_3$[h] (ns) |
|---|---|---|---|
| 1,5-DNB (dilute)[a] | 1.84 | ---- | ---- |
| 1,5-DNB (dilute)[b] | 2.48 | ---- | ---- |
| PNC (dilute)[c] | 3.53 | ---- | ---- |
| PNC (concentrated)[d] | ---- | 0.33 | 15.3 |
| (15.4)[i] | | | |
| NDI-650 (dilute)[e] | 2.04 | 1.20 | 19.2 (21.4)[i] |
| NDI-650 (dilute)[j] | 2.41 | 1.22 | 22.9 (22.3)[i] |

a. 3.0 X $10^{-4}$ M in $CH_2Cl_2$
b. 3.0 X $10^{-4}$ M in 1-butanol
c. 4.4 X $10^{-4}$ M in $CH_2Cl_2$
d. 1.2 M in $CH_2Cl_2$
e. 0.01 g/dL in $CH_2Cl_2$
f. Unquenched monomer lifetime
g. Quenched monomer lifetime
h. Excimer lifetime
i. Lifetime of long-lived component monitored at 480 nm
j. 0.01 g/dL in 1-butanol

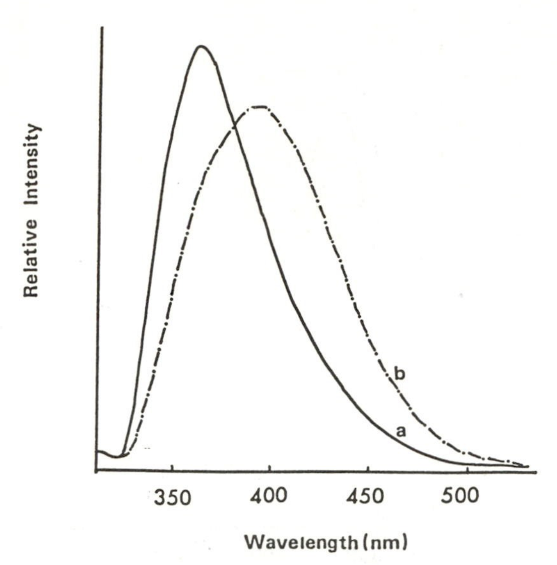


Figure 2. Steady state fluorescence spectra ($\lambda_{ex}$ = 300 nm) of PNC in dichloromethane at concentrations of 4.4 X $10^{-4}$ M (curve a) and 2.2 M (curve b).

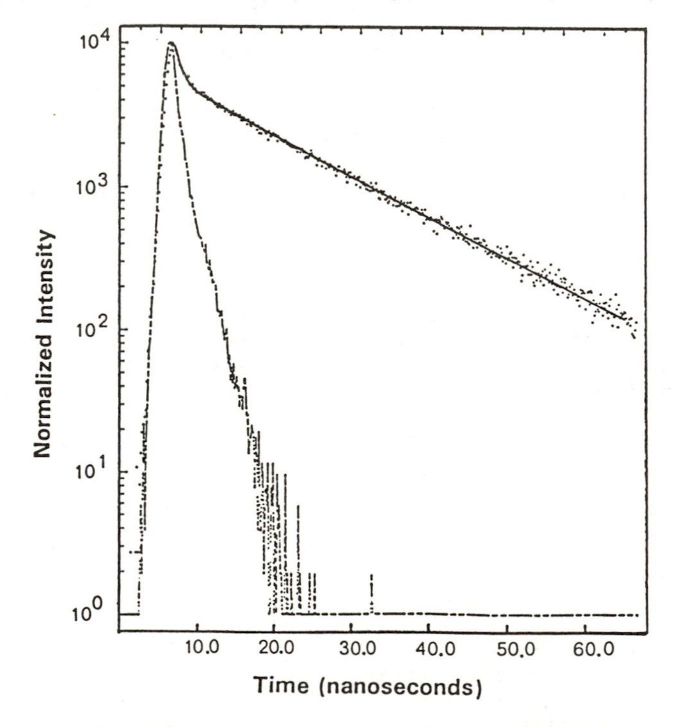


Figure 3. Fluorescence decay curve ($\lambda_{ex}$ = 300 nm, $\lambda_{em}$ = 330 nm) of PNC (1.2 M) in dichloromethane.

($\tau_2$) and 15.3 ns ($\tau_3$). By monitoring at 480 nm in the excimer region and fitting the long-lived portion of the decay curve to a single exponential decay function, a lifetime of 15.4 ns (Table I) was recorded. Comparison of the short-lived component (0.33 ns--$\tau_2$) of the monomer decay in the concentrated solution with the lifetime (3.53 ns--$\tau$) in the dilute solution leads to the conclusion that the monomer emission in the concentrated solution is significantly quenched. Both the large decrease of the fluorescence lifetime of the monomer naphthyl carbamate and the presence of the long-lived component in the monomer emission region suggests that the monomer exists in dynamic equilibrium with the excimer as represented by the classical scheme (Scheme I) for excimer kinetics. The value for $k_{MD}$ has a finite value which makes the dissociation of the excimer E* into its component species M and M* a viable process.

At this point, it is appropriate to consider the factors which may enter into stabilization of the excimer in the PNC solutions. First, it is worth noting that at least some excimer formation, as monitored by the recording of fluorescence decay curves in the monomer region, occurs at concentrations of PNC as little as 0.005 M in organic solvents. However, it is only above 0.5 M that appreciable build up of excimer emission is recorded.

It has long been recognized that polyurethanes, which contain carbamate chromophores, are characterized by hydrogen bonding between the hydrogen attached to the nitrogen of the carbamate group and the carbamate carbonyl (or an ether group if present). This contributes to the unique physical properties of polyurethanes. Unsurprising, the PNC model system also shows appreciable hydrogen bonding which is prevalent at higher concentrations between carbamate chromophores. This is illustrated in Figure 4 by the IR spectra of PNC recorded at two concentrations. In the dilute solution (0.2 M) the IR shows a single peak in the N-H stretching region. This peak at 3416 $cm^{-1}$ results from a non-bonded (or free) N-H stretching in the carbamate moiety. At the higher concentration (1.5 M) a new band due to a hydrogen-bonded N-H stretching appears at 3320 $cm^{-1}$. It is in this concentration region that the higher degree of excimer formation is recorded in Figure 2. Thus, it is reasonable to assume that hydrogen bonding between the PNC molecules contributes to the stabilization of the excimer. In support of this supposition, propyl N-methyl N-(1-naphthyl) carbamate (PNMNC) with a methyl group substituted on the nitrogen carbamate was synthesized. The methyl group prohibits hydrogen bonding. Figure 5 shows that little or no excimer emission is recorded for PNMNC up to concentrations of 2.2 M in dichloromethane. Only at concentrations above 3 M can any excimer emission be observed. Consequently, it can be concluded that hydrogen bonding is indeed an important factor in the excimer formation of PNC.

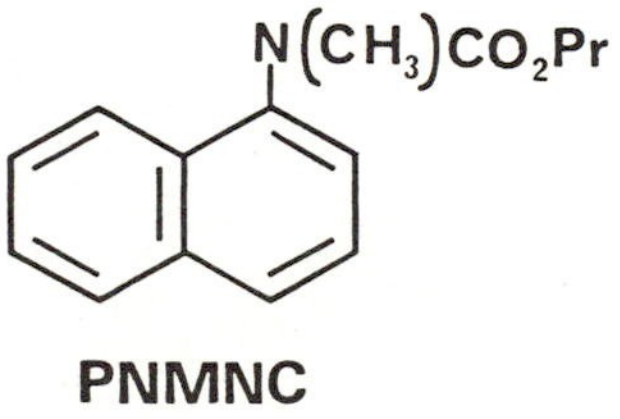

$N(CH_3)CO_2Pr$

PNMNC

Scheme I

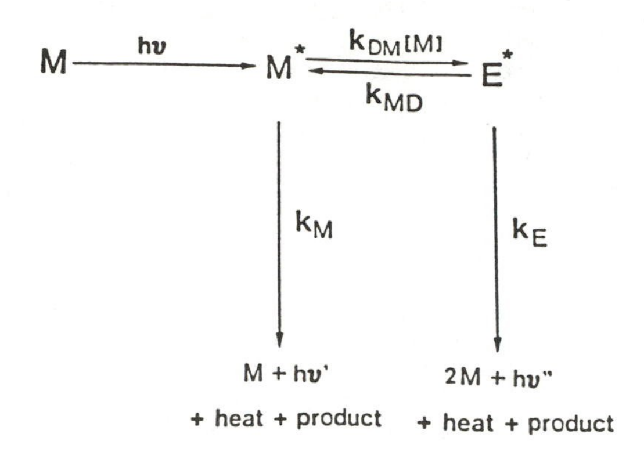


$k_M$ = non-radiative plus radiative rate constant for excited PNC monomer $M^*$.

$k_{DM}$ = rate constant for excimer formation between M and $M^*$.

$k_{MD}$ = rate constant for dissociation of excimer E into component species M and $M^*$.

$k_E$ = non radiative plus radiative rate constant for excimer $E^*$.

M = ground state PNC.

$M^*$ = excited state PNC.

$E^*$ = PNC excimer.

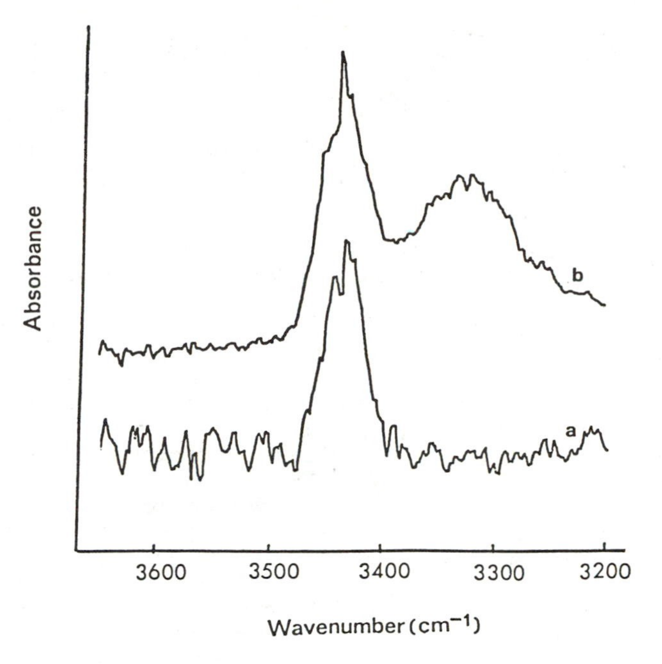


Figure 4. Infrared spectra of PNC at concentrations of 0.2 M (curve a) and 1.5 M (curve b) in dichloromethane.

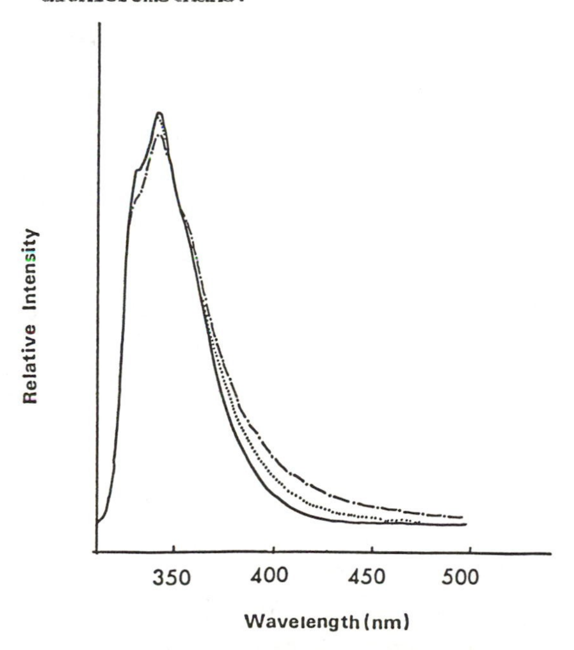


Figure 5. Steady state fluorescence spectra ($\lambda_{ex}$ = 300 nm) of PNMNC at concentrations of 4.1 X $10^{-4}$ M (————), 2.2 M (··········) and 3.3 M (—·—·—·—).

Photophysics of a 1,5-Naphthalene Diisocyanate Based Polyurethane in Solution. Having established the basic features of excimer formation in naphthyl carbamates and realizing that hydrogen bonding plays a significant role in excimer formation, we turn to consideration of a 1,5-naphthalene diisocyanate (NDI) based polyurethane which has the naphthyl carbamate groups periodically spaced as an integral part of the polymer backbone. The polymer, designated as NDI-650, is based on NDI and polytetramethylene ether glycol (average molecular weight of 650) and has a molecular weight of 51,000. The steady-state fluorescence spectrum of a very dilute dichloromethane solution of NDI-650 is shown in Figure 6. Although the basic structure of the fluorescence spectrum of NDI-650 is similar to the fluorescence of biscarbamate 1,5-DNB in Figure 1 (curve a), there is a distinctive red-shifted tail above 400 nm. Moreover the fluorescence decay curves in dilute solution (monitored at any wavelength from 330 nm to 500 nm) cannot be fit to a single exponential decay function. These results indicate intramolecular excimer formation in the dilute NDI-650 solution.

To establish the nature of the excimer formation as either intramolecular or intermolecular, an extremely dilute solution ($1 \times 10^{-4}$ g/dL) of NDI-650 was prepared. Even in this ultra-dilute solution excimer emission was present. Figure 7 shows a plot of the excimer (430 nm) to monomer (350 nm) intensity ratio as a function of polymer concentration. Not until a concentration of 1.0 g/dL is reached does intermolecular excimer formation between naphthyl carbamate groups on different polymer backbones become important.

To identify the nature of the absorbing chromophore in NDI-650 which leads to the emission spectra recorded in Figure 6, excitation spectra were recorded at a variety of emission wavelength settings. The excitation spectrum in Figure 8 is exemplary of the curves recorded. Also included in Figure 8 is the absorption spectrum of NDI-650 for comparison. Except for minor differences due to recording on two instruments (fluorescence excitation spectra are uncorrected for wavelength response) the excitation spectrum is identical to the absorption spectrum. Although not shown, the absorption and excitation spectra of the model 1,5-DNB are identical to those for NDI-650. In general, it can be safely concluded that the species responsible for the initial absorption of light leading to excimer formation in dilute solution is a single ground state naphthyl biscarbamate moiety in the backbone of NDI-650.

In the discussion of the fluorescence spectra of NDI-650, it was noted that the emission decay curves are not single exponential. Additionally, unlike for concentrated solutions of PNC, the decay curves recorded in the monomer emission region (anywhere from 330 nm to about 370 nm) could not be fit to a double exponential decay function. However, the decay curve (monitored at 330 nm in the monomer emission region) obtained experimentally (Figure 9) could be fit to a triple exponential decay function with decay parameters of 1.20 ns, 2.04 ns, and 19.2 ns. In addition, fitting the long-lived component of the decay curves monitored at 480 nm to a single exponential decay function gave a value of 21.4 ns for NDI-650, very similar to the value for the long-lived component of the monomer decay. This data is summarized in Table I along with the lifetime data previously obtained for PNC and 1,5-DNB.

By analyzing the results from the decay curves compiled in Table

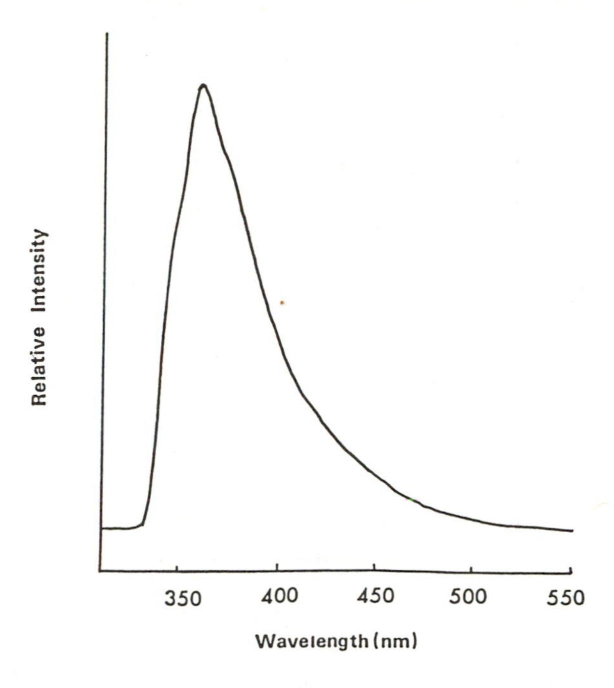


Figure 6. Steady state fluorescence spectrum of NDI-650 (0.01 g/dL) in dichloromethane ($\lambda_{ex}$ = 310 nm).

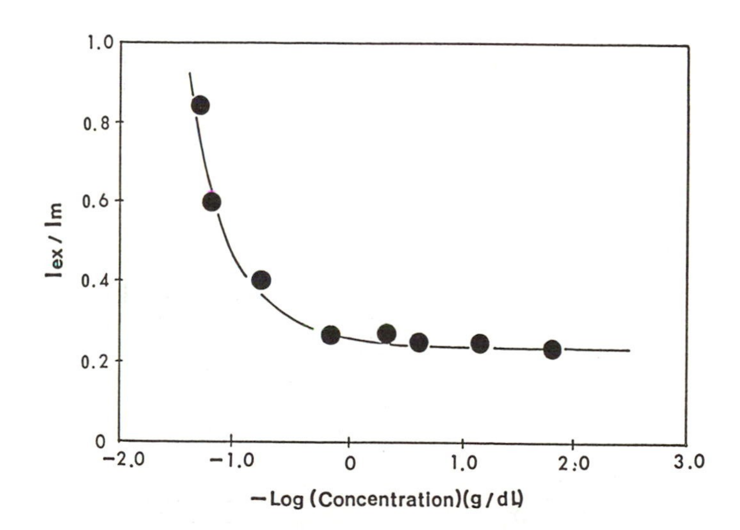


Figure 7. Ratio of excimer ($I_{ex}$) to monomer ($I_m$) fluorescence intensity ($\lambda_{ex}$ = 310 nm) as a function of NDI-650 concentration in dichloromethane.

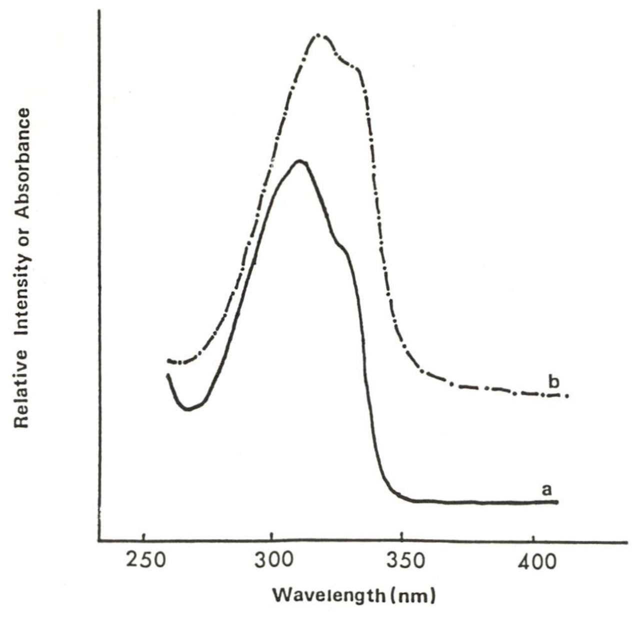


Figure 8. Absorption (curve a, 0.01 g/dL) and excitation spectra (curve b, $\lambda_{em}$ = 460 nm, 1 X $10^{-3}$ g/dL) of NDI-650 in dichloromethane.

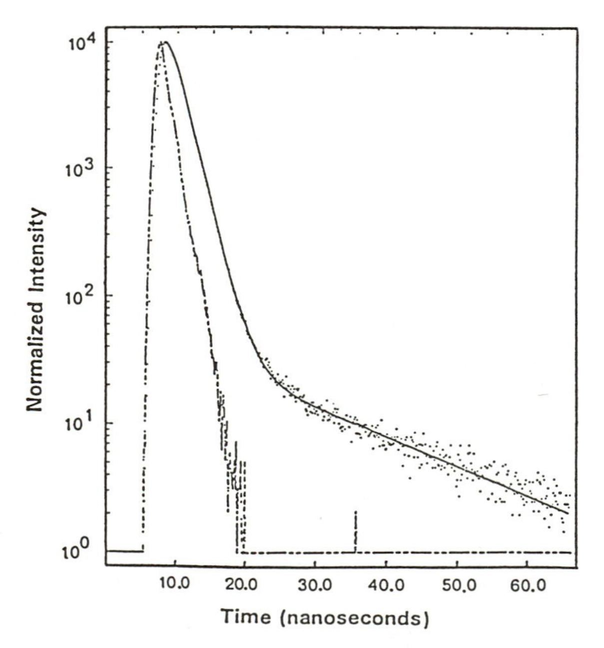


Figure 9. Fluorescence decay curve ($\lambda_{ex}$ = 310 nm, $\lambda_{em}$ = 330 nm) of NDI-650 (0.01 g/dL) in dichloromethane.

I, one can infer from the long-lived decay of the monomer emission ($\lambda_{em}$ = 330 nm) that the excimer is reversibly formed and exists in dynamic equilibrium with an excited state monomer species. This is in reasonable agreement with the reported (7,8) reversibly formed excimers between naphthyl chromophores in poly (vinyl naphthalene). The two short lived components are attributed (after Phillips et al. 7,8) to monomeric species, one which is quenched and in a dynamic equilibrium with the excimer and another which is an "isolated monomer" unquenched by interaction to form an excimer. The longer lived component ($\tau_1$ = 2.04 ns) of the two fast components corresponds to the lifetime of the model biscarbamate DNB (1.84 ns), and probably results from the isolated monomer. The shorter lifetime ($\tau_2$ = 1.20 ns) can be assigned to the quenched monomer emission in equilibrium with the excimer. The "isolated monomer" concept, as recently reported by Phillips et al (7,8) and Holden et al (9) to account for the excimer kinetics in polymers with pendant naphthalene groups, adequately accounts for the excimer kinetics of dilute solutions of NDI-650 in dichloromethane. The basic mechanism for the "isolated monomer" is given in Scheme II with the parameters as defined.

In the case of NDI-650, since the 2.04 ns lifetime ($\tau_1$) for the "isolated monomer" is essentially identical, within our error limits, to the 1,5-DNB lifetime of 1.84 nsec, we conclude that the rate constant $k_{12}$ for conversion of the excited isolated monomer $M_1$* into the excited excimer forming monomer $M_2$* is low compared to the rate constant $k_{M1}$ for deactivation of $M_1$*.

In a recent report, Holden et al (9) postulated that $k_{12}$ was low compared to $k_{M1}$ for polymers with naphthyl chromophores separated by greater than three carbon atoms. The naphthyl groups in the backbone of NDI-650 are certainly separated by more than three carbon atoms and thus our results are quite consistent with those of Holden et al (9).

To provide confirming evidence for existence of the isolated monomer, the fluorescence decay curves were analyzed for 1,5-DNB and NDI-650 in a second solvent system (1-butanol) in which the 1,5-DNB lifetime increased to 2.48 ns. As shown in Table I, the decay parameters for the triple exponential fit to the decay curve of NDI-650 in 1-butanol lead to the same conclusion as the results in dichloromethane. Particularly important to note is the agreement between $\tau_1$ for 1,5-DNB (dilute) in 1-butanol and $\tau_1$ obtained for NDI-650 in 1-butanol. In short, the $\tau_1$ value is altered by the solvent both for the 1,5-DNB small molecule model and NDI-650. The correlation of the $\tau_1$ values in the two solvents provides additional evidence for the "isolated monomer" scheme.

Having established the existence of the excimer emission of NDI based polyurethanes in solution, and realizing that the intramolecular excimer forming naphthyl carbamate groups are located on the backbone of the polymer, it becomes apparent that an excellent opportunity exists for chain conformational studies as a function of solvent. Figure 10 shows the steady-state fluorescence spectra of NDI-650 in four solvents with distinctively different solvating power. In each case (curves a-d) both monomer and excimer emission are observed; however, the ratios of excimer to monomer emission reflect conformational differences of the NDI-650 polymer in the solvent employed. The excimer to monomer intensity ratio

Scheme II

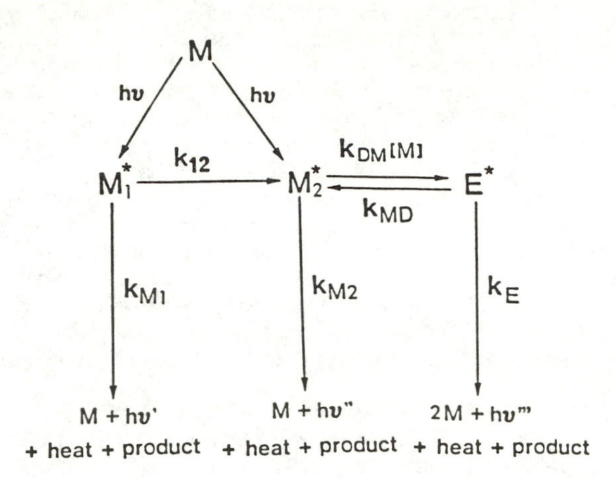


$k_{M1}$ = non-radiative plus radiative rate constant for excited isolated naphthyl monomer $M_1^*$.

$k_{M2}$ = non-radiative plus radiative rate constant for excited excimer forming naphthyl monomer $M_2^*$.

$k_{DM}$ = rate constant for excimer formation between $M_2^*$ and M.

$k_{MD}$ = rate constant for dissociation of excimer $E^*$ into component species $M_2^*$ and M.

$k_E$ = non-radiative rate constant for excimer $E^*$.

$k_{12}$ = rate constant for formation of excited excimer forming naphthyl monomer $M_2^*$ from the excited isolated naphthyl monomer $M_1^*$.

M = ground state naphthyl monomer.

$M_1^*$ = excited isolated naphthyl monomer.

$M_2^*$ = excited excimer forming naphthyl monomer.

($I_{ex}/I_m$) recorded at 430 nm ($I_{ex}$) and 355 nm ($I_m$) in Table II for the four pure solvent systems under consideration shows an increase in the order: $I_{ex}/I_m$ (2,3-butanediol) < $I_{ex}/I_m$ (dichloromethane) < $I_{ex}/I_m$ (benzene) < $I_{ex}/I_m$ (propyl benzene). This parallels the decrease in solubility parameters for each solvent and reflects the ability of the solvent to effectively solvate the polyurethane chain. In other words, propyl benzene is a fairly "poor" solvent for NDI-650 and results in a more compact polymer chain with higher degree of intramolecular excimer formation than benzene, dichloromethane, and 2,3-butanediol. The ratio of $I_{ex}/I_m$ parallels the decrease in the solubility parameter (Table II). If indeed our supposition concerning the relationship between the solubility parameter and $I_{ex}/I_m$ is true, then we should be able to find a solvent system with an extremely high solubility parameter which also has a large value for $I_{ex}/I_m$. In order to attain such a solvent system, DMF and water were mixed in a 70/30 ratio (vol %) and the photophysical parameters recorded (Table II). For the DMF/water (70/30) system, the $I_{ex}/I_m$ ratio is quite high indicating a compact coil with a significant degree of intramolecular excimer formation. Apparently, from the data in Table II, both propyl benzene with a low solubility parameter and the DMF/water (70/30) solvent system with a high solubility parameter are relatively "poor" solvents for the NDI-650 polymer while dichloromethane and 2,3-butanediol are relatively "good" solvents.

Photophysics of NDI-650 and NDI-2000 polyurethane Films. The fluorescence spectrum of an NDI-650 polyurethane film ($\lambda_{ex}$ = 300 nm) is dominated by excimer emission (Figure 11). A preliminary multiexponential analysis of the excimer and monomer emission decay curves indicates a complicated photophysical system. However, it can be reported that the long-lived component of the excimer decay curve (taken at 500 nm) does yield a lifetime of ~22 ns for excimer emission, close to that obtained in $CH_2Cl_2$.

In comparing to the results for the NDI-650 polyurethane, the fluorescence spectrum of a polyurethane film of NDI-2000 [based on NDI and poly (tetramethylene ether glycol) with average molecular weight of 2,000] shows appreciable emission from the monomer component (Figure 11). Although the excimer and monomer decay curves are complex and difficult to analyze, the long-lived component of the excimer decay curve (recorded at 500 nm) is about 21 ns, in close agreement with the lifetime obtained for the NDI-650 film.

In contrasting the results for the NDI-650 and NDI-2000 films, it is apparent that the extent of excimer formation is greater for the NDI-650 film. It may be argued that the increased relative concentration of naphthyl carbamate chromophores in the NDI-650 film leads to increased excimer formation. It should, however, be pointed out that there is a distinct and perhaps critical structural difference between the NDI-650 and NDI-2000 polyurethane films which may manifest itself in increased excimer formation. In the NDI-650 polyurethane film, there is a significant degree of hydrogen bonding to the carbonyl on the urethane moiety. This is exemplified by the infrared spectra of the NDI-650 and NDI-2000 films given in Figure 12 which indicates that NDI-650 has primarily hydrogen bonded carbonyls (1695 $cm^{-1}$) while the NDI-2000 film has a high content of

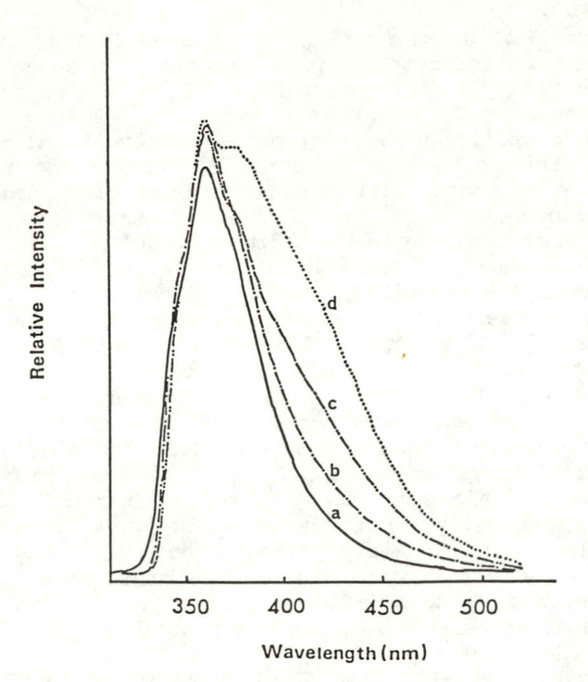


Figure 10. Steady state fluorescence spectra ($\lambda_{ex}$ = 310 nm) of NDI-650 in 2,3-butanediol (curve a, < 0.01 g/dL), dichloromethane (curve b, 0.01 g/dL), benzene (curve c, 0.01 g/dL) and propyl benzene (curve d, 5 X $10^{-3}$ g/dL).

**Table II.** Excimer to monomer intensity ratios for NDI-650 in several solvents

| Solvent | $I_{430}/I_{355}$ | Solubility Parameter $(J/m^3)^{1/2}$ X $10^{-3}$ |
|---|---|---|
| Propyl benzene[a] | 0.55 | 17.6 |
| Benzene[b] | 0.35 | 18.8 |
| Dichloromethane[b] | 0.21 | 19.8 |
| 2,3-Butanediol[c] | 0.11 | 22.7 |
| DMF/$H_2O$(70/30)[d] | 0.91 | 31.7 |

a. Concentration of NDI-650 is 5 X $10^{-3}$ g/dL.
b. Concentration of NDI-650 is 0.01 g/dL.
c. Concentration of NDI-650 is < 0.01 g/dL. Fluorescence emission spectrum was measured for a filtered solution after heating at about 60° C for 30 minutes.
d. Concentration of NDI-650 is 1.5 X $10^{-3}$ g/dL.

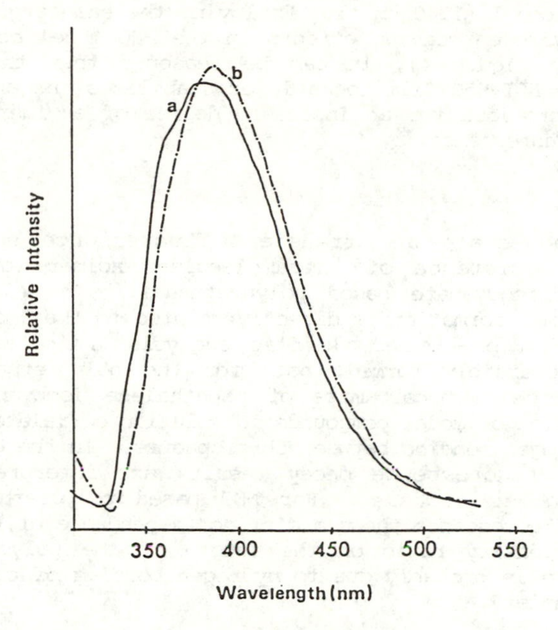


Figure 11. Steady state fluorescence spectra of NDI-2000 (curve a) and NDI-650 (curve b) films ($\lambda_{ex}$ = 300 nm).

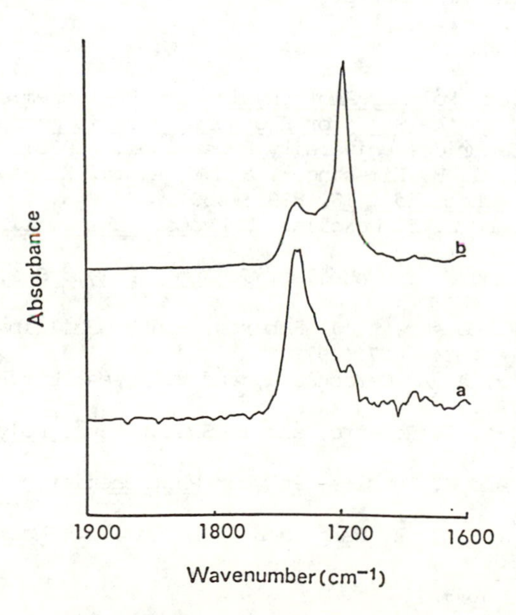


Figure 12. Infrared spectra of NDI-2000 (curve a) and NDI-650 (curve b) films.

non-bonded carbonyl (1740 $cm^{-1}$). Employing the same arguments used to describe hydrogen bonding effects in the PNC model compounds in solution (see Figure 4), it can be reasoned that the hydrogen bonding in the NDI-650 film locks in or stabilizes the excimer site geometry and provides for an increased degree of excimer, relative to monomer, fluorescence.

## CONCLUSION

Both steady-state and transient fluorescence spectroscopy indicate the existence of intramolecular excimer formation in naphthalene diisocyanate based polyurethanes. In solution, the extent of excimer formation is directly related to the "poorness" of the solvent. A preliminary kinetic analysis indicates that the excimer is reversibly formed, both for the polyurethane and an appropriate model monocarbamate of naphthalene isocyanate. The excimer formation of model compounds in solution correlates with the degree of hydrogen bonding between chromophores. In the case of the polymer, the fluorescence decay results are interpreted by an "isolated monomer" scheme. For NDI based polyurethane films, steady-state fluorescence shows a distinct dependence of the excimer to monomer intensity ratio on the length of the polyol segment. This dependence is probably due to hydrogen bonding of carbonyls in the urethane moiety.

ACKNOWLEDGMENTS. This work was sponsored by the Office of Naval Research. We thank Chia-Hu Chang for helpful discussion during the preparation of this manuscript.

## LITERATURE CITED

1. J. E. Guillet, Polymer Photophysics and Photochemistry: An Introduction to the Study of Photoprocesses in Macromolecules, Cambridge, Cambridge University Press, 391, (1985).
2. J. A. Ibemsi, J. B. Kimssinger, and M. Ashraf El-Bayoumi, J. Polym. Sci., Chem. Ed., 18, 879 (1980).
3. N. S. Allen and J. F. McKellar, Makromol. Chem., 179, 523 (1978).
4. N. S. Allen and J. F. McKellar, J. Appl. Polym. Sci., 22, 2085 (1978).
5. M. Graley, A. Reiser, A. J. Roberts, and D. Phillips, Macromolecules, 14, 427 (1978).
6. A. Barbalata, A. A. Caraculacu, and V. Iurea, Eur.Poly. J., 14, 427 (1978).
7. D. Phillips, A. J. Roberts, and I. Soutar, Eur. Poly. J., 17, 101 (1981).
8. D. Phillips and G. Rumbles, Polymer Photochemistry, 5, 153 (1984).
9. D. A. Holden, P. Y. K. Wang, and J. E. Guillet, Macromolecules, 13, 295 (1980).

RECEIVED August 4, 1987

# ENERGY MIGRATION

## Chapter 18

# Electronic Energy Relaxation in Aromatic Vinyl Homopolymers

**H. F. Kauffmann, W.-D. Weixelbaumer, J. Bürbaumer, and B. Mollay**

**Institut für Physikalische Chemie, Universität Wien, Wahringerstrasse 42, A-1090 Wien, Austria**

Transient fluorescence spectroscopy on nanosecond and picosecond time scales has been used in an attempt to study the pathways of excitation energy transport (EET) and trapping for poly-(1-vinylnaphthalene), p-VN, and poly-(N-vinylc arbazole),p-N-VCz, in dilute fluid solution. Excited state localization has been probed by collecting the transient fluorescence patterns of excimer-forming-sites (EFS) in typical trapping experiments. Fluorescence rise-profiles of both mobile and static excimers have been analyzed on the premises of low-dimensional transport topology and "effective" **diagonal** disorder of aromatic hopping sites. A trapping function of the form $k(t) = b + ct^{-1/2}$, a distribution of transport states $\{X_1\}$ and a small ensemble of energy-relaxed, monomeric tail-states have been processed in kinetic schemes. The calculated profiles are **nonexponential** and can recover satisfactorily the experimental curves. The rationale behind - a **time-dependent** excited-state random walk among energy-dispersive chromophores - has been discussed.

Electronic excitations in molecular aggregates have received considerable attention from theoretists and experimentalists in recent years. Much of the work has focused on excitonic transport in molecular crystals (1), in substitutionally disordered mixed crystals (2) and in amorphous structures (3). Furthermore, excitation energy transfer (EET) has been studied in fluorescence concentration depolarization of donor molecules in solution (4) as well as in excited-state energy transfer of donor - acceptor dyes in condensed phase systems (5).

NOTE: Dedicated to Professor Adolf Neckel on the occasion of his 60th birthday.

0097-6156/87/0358-0220$06.50/0

The effect of disorder on excitation energy transport in polychromophoric arrays is a formidable theoretical problem and has been investigated extensively. Special emphasis has been given to the dynamical nature of migrative motion (6). For infinite systems of donor ensembles, randomly distributed and energetically equivalent (positional disorder), incoherent evolution of electronic transport has been formulated and analyzed in terms of symmetric Pauli master equations (PME) (7). From diagrammatic (8) and perturbative approaches (9), but also from continuous time random walk analyses (CTRW) (10) and fractal arguments (11) there is agreement now that electronic transport is **nondiffusive**, i. e. **time-variant**, and becomes diffusive, i. e. time-independent, after a relatively long time comparable to an average hopping time of excitation. Intrinsic excited state **dispersion** of random, positional disorder - evidenced by typical **nonexponential** transient fluorescence - has been experimentally verified in dye solutions (12-14), where energetic fluctuation of molecular self energies are smeared out by thermal activation.

The situation is even more complex for molecular aggregates for which spatial, random fluctuations of intermolecular potentials significantly exceed kT. Such systems are said to posess **diagonal** disorder where positional disorder causes a pronounced spread of site energies and, thus, in addition, **energetic** disorder (15). Dispersion as a consequence of distribution of self-energies is well documented by the inhomogeneous (Gaussian) broadening of optical $S_1 \leftarrow S_0$ transitions in vapor condensed organic $\pi$-systems (16). Further experimental manifestations of energetic disorder are fluorescence line narrowing (17), the identification of typical mobility edges (18, 19) as well as energetic narrowing of time gated, transient emission spectra (20).

The spread of site energies in systems of diagonal disorder is of considerable impact on the space-time evolution of an elementary excitation. Energy-dispersive excited-state walks are involved in a hierarchy of sequential steps of gradually decreasing event times which have been recognized to produce nondiffusive, i. e., time-dependent phenomena, in general. A transient observable is expected, therefore, to show a **nonexponential** relaxation profile on relevant time scales. Nonexponential decay analysis in random systems with energetic disorder has been achieved, in particular, for triplet-excitations (21, 22) which are - because of their reduced hopping times and their large inherent lifetimes- conveniently accessible to experimental investigations.

For an assembly of donor molecules **covalently fixed** in close, local proximity to a macromolecular environment, similar successful theoretical treatments - as outlined above for randomized arrays of independent donors (8-11)- have not been performed, so far. Among polychromohoric systems set up by correlated donor sites **aromatic vinyl-**

**homopolymers** are a special class of synthetic, linear macromolecules in which the aromatic subsystem, such as benzene, naphthalene or carbazole is chemically linked to the main chain and periodically separated by three saturated carbon atoms. Poly-(styrene), p-S, poly-(vinylnaphthalenes), p-VN, or poly-(N-vinylcarbazole), p-N-VCz, are prominent examples. Because of the incorporation of dyads of different tacticity these polymers are stereochemically not pure. In toto, they represent predominantly disordered finite-size, many-particle systems superimposed by interfering local domains of slight short-range order. Consequently, this results in significant positional heterogeneity leading to spatial distribution of chromophores which is neither random nor periodical for the majority of experimental conditions.

Therefore, for a single excitation statistically created amidst a swarm of polymer-bound (identical) aromatic sites in a homopolymer, geometric and energetic inhomogeneity must induce even more complex transport dynamics as compared to random, amorphous materials, and there does not yet exist - for low dimensional transport media, in particular, a comprehensive understanding on a microscopic level. Statistical correlation effects, local semi-coherent evolutions, a distribution of various transport dimensionalities, energetic disorder and non-equilibrium as well as the finite volume for polymers, in general, are serious problems which make the analyses of stochastic master equations practically impossible. So far, only isoenergetic hopping processes based on approximants of GAF- and LAF- expansions (8) have been analyzed in polymeric materials, provided the chromophores are random and chromophore density is small (23-26).

## TRANSIENT FLUORESCENCE IN AROMATIC VINYLPOLYMERS

While microscopic theories designed to model the dynamics of electronic transport in the high density limit of polymer-bound chromophores are rather underdeveloped until recently, the phenomenological aspects of electronic energy diffusion and excitonic states in aromatic vinylpolymers have now been a topic of fundamental interest over a period of twenty years (27). Excitation transfer was claimed from studies of fluorescence depolarization (28) and it was hypothetisized from experiments based upon energy transfer (29)), dynamical quenching (30) and excimer formation (31). From literature of polymer photophysics the nature of polymer-inherent traps has been discussed almost exclusively in terms of well established stereochemical concepts of physical-organic chemistry. In the limit of bichromophoric interaction an electronic trap can be considered to be a localized state of an excimer or an excimer-forming- site, EFS, (32). EFS refer to polymer-inherent, intrinsic conformations of distinct pairs of chromophores, with different geometries and density numbers. They are immobilizing the hopping process by absorbing the excited state energy in their potential sinks, and afterwards, they are dissipating the excitation via significant red shifted fluorescence in a subsequent step. Because of its migrative nature the trapping process is a very efficient event in an aromatic homopolymer. Since the quantum yields of radiation are considerably high for excimers, the dominant fluorescence in a CW-experiment is mostly excimeric, in nature.

**Pulsed** excitation techniques have been applied to synthetic polymers over a period of 10 years now with the objective to study the reponse-laws of monomer- and excimer states. Special emphasis has been given to time-resolved fluorescence of aromatic polymers in **dilute, fluid** solution. The primary aspect of early investigations was concerned with the evaluation of **segmental rotation** on the basis of transient Birks kinetics. For the major part of these polymers, however, a more accurate data analysis has revealed a more complex excited-state dynamics. It is generally observed in these polymers, that the low-intensity, monomer fluorescence can be fitted more satisfactorily by a sum of **three** exponentials, while for the consecutive profiles of excimer fluorescence, quite obviously, **more than two** exponentials are required to recover the data. This is indicative of a discrepancy between experiment and kinetic scheme which, quite obviously, shows that - even in the absence of energy migration - rotational motion between correlated chromophores cannot be modeled in analogy to collisionally induced, translatory diffusion of independent and randomly distributed, low-molecular chromophores. The significant deviations from conventional Birks kinetics have led to an increased discussion as to the nature of polymeric $S_1$-states and their intrinsic relaxation channels. In the meantime, these kinetic complications have been partially overcome by strategies which correlate the number of exponentials numerically extracted from reconvolution procedures (33) to the number of kinetic transients in linear coupled schemes. The role of physical transients have been discussed in terms of a) partial overlap high-energy excimers (34), b) rotational isomeric states of bichromophoric, excited conformations (35) and c) monomer-like excited states in statistical triads of aromatic copolymers (36).

Typical multiexponential fluorescence convolutes studied, more recently, in our laboratory are given in Figures 1a-1c. Figure 1a refers to the low-energy excimer fluorescence of p-1-VN ($5.10^{-4}$ base molar in benzene, 25°C) at emission wavelengths around $\lambda_{em}$ = 470 nm. The ns profile has been acumulated by means of conventional flash-lamp excitation ($\lambda_{exc}$ = 295 nm, typical pulsewidth of 1.5 ns FWHM) and single-photon-timing detection (SPT). Numerical reconvolution yields a best-fit **three** exponential form according to

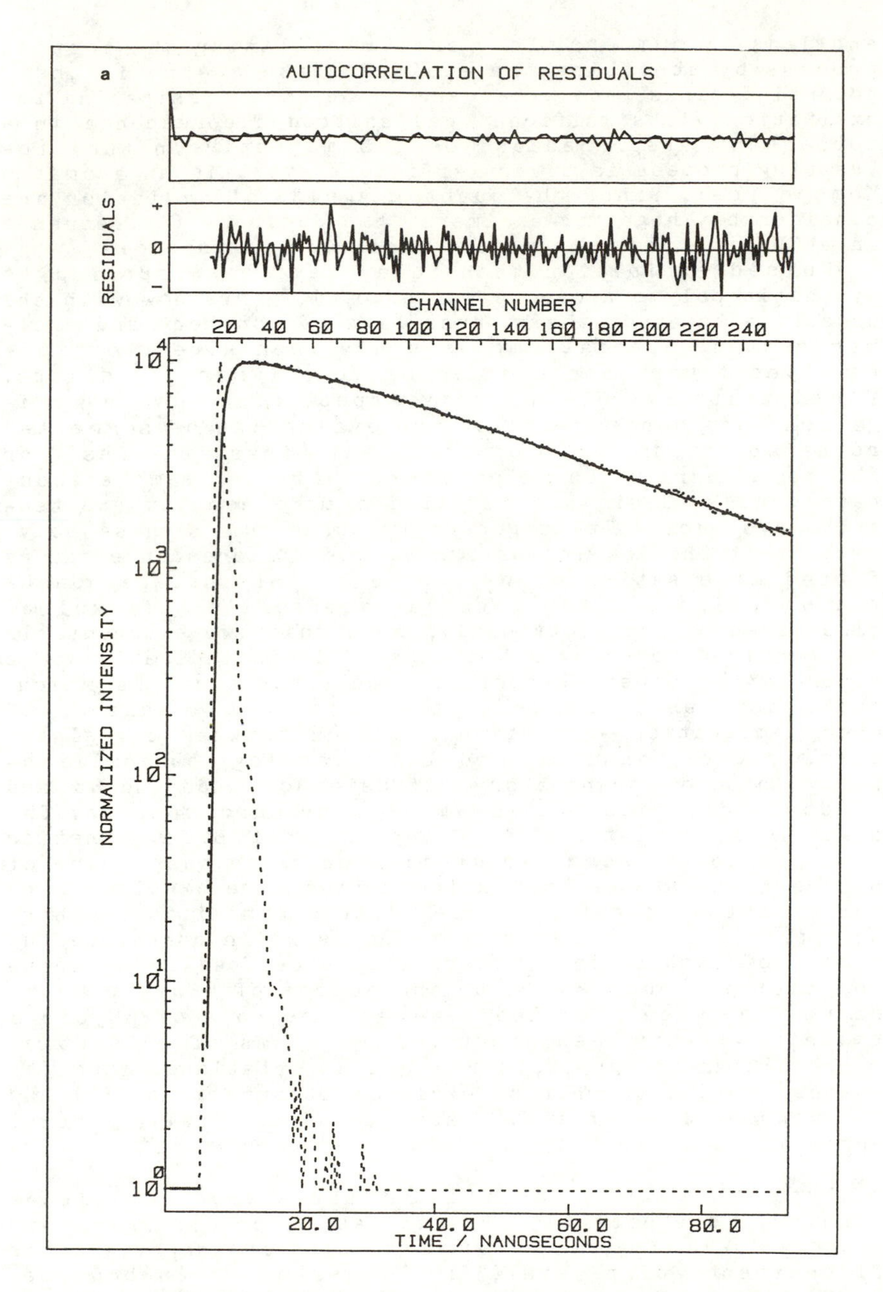


Fig. 1a. Nanosecond time-resolved fluorescence profile of p-VN at $\lambda_{em}$ = 470 nm (low-energy regime); broken curve (---), lamp; dots (....), experimental data; smooth solid line (——), best-fit three exponential form, according to Equation I.

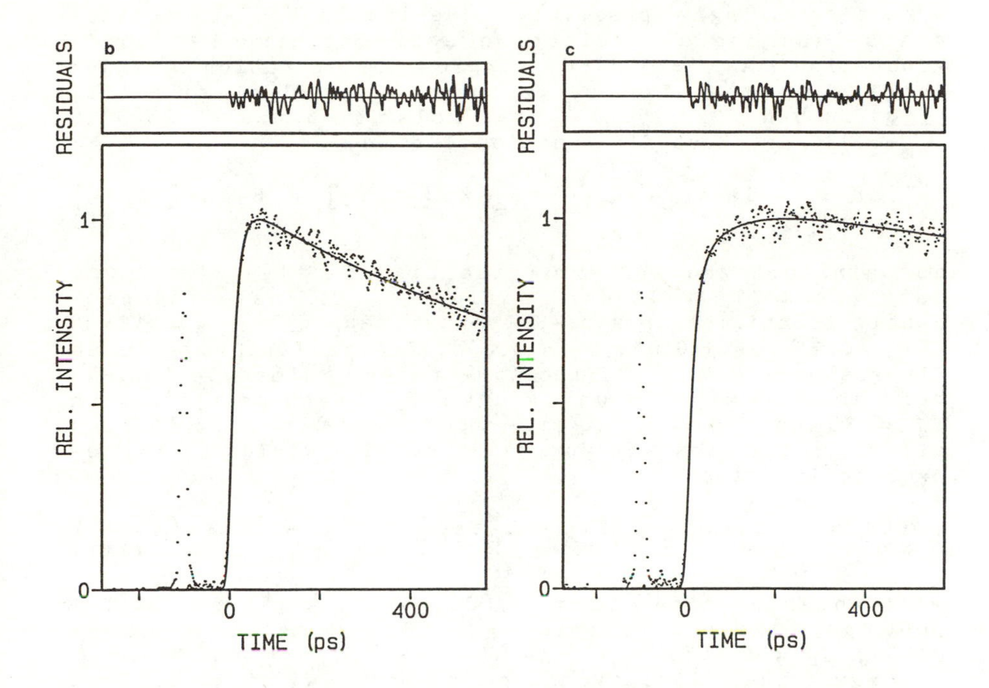


Fig. 1b. Picosecond time-resolved fluorescence of p-N-VCz at $\lambda_{em}$ = 380 nm ($E_1$); dotted spike (....), apparatus function (prepulse); dotted pattern (....), experimental data convolute; solid curve (——), three exponential fit, according to Equation II.

Fig. 1c. Picosecond fluorescence rise of p-N-VCz at $\lambda_{em}$ = 460 nm ($E_2$); dotted spike (....), instrumental response (prepulse); dotted pattern (....), experimental fluorescence; solid line (——), best-fit to triple-exponential rise curve, according to Equation III. (Reproduced with permission from Ref. 38. Copyright 1986: American Institute of Physics)

$$F_E(t) = -A_2\exp[-t/T_2] - A_3\exp[-t/T_3] + A_4\exp[-t/T_4] \qquad (I)$$

showing a **double**-exponential build up with apparent rise times $T_2$ = 0.36 ns, $T_3$ = 13.4 ns, and a single decay time $T_3$ = 41.9 ns ($A_2$ = 0.15, $A_3$ = 0.05, and $A_4$ = 0.21). Figures 1b and 1c correspond to fluorescence patterns on a ps time scale. They have been obtained by streak camera detection using a passively mode-locked $Nd^{3+}$-phosphate-glass laser as an excitation source (third harmonics generation, $\lambda_{exc}$ = 351 nm, single-shot operation). Figure 1b displays the high-energy, partial overlap excimer $E_1$ (34a) in p-N-VCz (5.10$^{-5}$ base molar in benzene, 25$^o$) at $\lambda_{em}$ = 380 nm. Best-fit functional form

$$F_{E_1}(t) = -B_1\exp[-t/T_1] + B_2\exp[-t/T_2] + B_3\exp[-t/T_3] \qquad (II)$$

contains **one growth** term (rise time $T_1 \approx$ 14 ps) and one decay term ($T_2 \approx$ 400 ps), while $T_3$ = 2.6 ns - independently identified in a ns-SPT experiment (37) - was fixed ($B_1$ = 0.12, $B_2$ = 0.04, $B_3$ = 0.09). Figure 1c shows the ps fluorescence profile of the low-energy excimer $E_2$ in p-N-VCz at the emission wavelength $\lambda_{em}$ = 460 nm (38). With $T_4$ = 35 ns fixed (c.f. ns-experiment in Ref.(37)) to allow for the flat maximum reconvolution yields a **triple**-exponential rise

$$F_{E_2}(t) = -C_1\exp[-t/T_1] - C_2\exp[-t/T_2] - C_3\exp[-t/T_3] \qquad (III)$$

with apparent rise times $T_1 \approx$ 20 ps, $T_2 \approx$ 75 ps and $T_3 \approx$ 3000 ps, respectively ($C_1$ = 0.096, $C_2$ = 0.009, $C_3$ = 0.002, $C_4$ = 0.107).

From a phenomenological point of view, it is natural to interpret the multiexponential curves in Equations I, II, and III on the basis of energy cascading models. Such schemes assume - parallel to a single monomer and excimer state - additional electronic dwell-stations to be involved in serial energy relaxation processes. In a quantitative treatment, one has to diagonalize, then, rate equations of the form

$$\dot{\mathbf{x}} = \mathbf{A}\ \mathbf{x} \qquad (1)$$

Here, $\dot{\mathbf{x}}$ and $\mathbf{x}$ are column vectors containing the components of interacting transients $X_i$ and $\mathbf{A}$ is the physical reaction matrix. For the transient pattern of the low-energy excimer $E_2$ in p-N-VCz a detailed analysis based upon a trap controlled rate constant formalism between coupled excited states has been given recently (37). In analogy, the four exponential form of $E_2$ (Figure 1c, Equation III) might be correlated to the number of four kinetic transients. Again, such a procedure should allow, in principle, to recover both the time constants and the sign of the pre-exponentials for plausible data sets of kinetic fluorescence parameters. A similar treatment will, presumably, hold for modeling the biexponential growth of low-energy excimer fluorescence in p-VN (Figure 1a, Equation I).

Nevertheless, we must note, that the rigorous application of multiexponential trial functions in polymer fluorescence analyses must be discussed with some reservations, in general. First, the majority of kinetic entities proposed from transient measurements and processed in linear kinetic schemes must be formulated within the rather unsharp limits of phantom-states as in fact nothing is known with respect to their geometries, their mutual energetical positions and their emission bandwidths. Second, the number of exponentials reconvolved from experimental raw data is strongly dependent on the quality of data and, in addition, on the time scale over which the collection of photons has been performed. It might be argued, therefore, that the number of extractable exponentials has no real physical relevancy other than as a lower limit to the actual number. Thus, the corresponding time constants $T_i$ and amplitudes $A_{ij}$ have little meaning unless the experiment has been able to resolve all the exponential terms. Third, - and this point has been fully addressed in the introductory section -, EET per se, may induce the possibility of **nonexponential** relaxation kinetics, thus, compromising any analysis of polymer fluorescence in terms of a sum of exponentials, no matter what the quality of the data is. On these premises, multiexponential reconvolution must be considered to be only a useful tool for a simple curve parametrization, while the number of exponentials extractable stands for a lower bound of experimentally accessible terms in an exponential series approach of a typically nonexponential profile. Time-dependent migration and nonclassical trapping dynamics in low dimensional aromatic polymers have been formulated, first, by Fredrickson and Frank (39) in their critical paper and, in the meantime, nonconventional relaxation of polymer fluorescence has become a subject of increasing interest, both theoretically (23, 24, 40) and experimentally (25, 26, 38). Dispersive first-order kinetics of serial hopping events and its effect on the observables of monomer-and excimer fluorescence in a transient experiment will be discussed in the next section.

## NONEXPONENTIAL MODELS

As briefly outlined in the introductory section, configurational and ensemble averages for an adequate description of EET in an aromatic homopolymer have not been solved so far. From a more qualitative point of view, however, the concepts of **diagonal** disorder and **dispersive** excitation energy transport - as applied to condensed disordered matter (15) - have turned out to be quite appropriate strategies for aromatic polymers as well. Accordingly, these polymers may be treated, in its wider connotation, as spatially disordered organic "wide gap" materials for which the strong fluctuations of aromatic site positions produce, in addition, pronounced **energetic** disorder. For low-temperature, bulk p-N-VCz, which is a good model for a 3-D disordered, random solid with a Gaussian-type distribution of self-energies (15, 41, 42), the situation is schematically sketched in Figure 2 (situation a). In the electronic $S_1$-state the disorder forms up a) a density-of-state distribution $N(\epsilon)$ of energetically different and localized $\{S_1\}$ states and b) a significant off-diagonal broadening of interchromophoric coupling energies. An elementary excitation, (*), created at random within the distribution of carbazole states is, therefore, subject to a) energetic fluctuations of hopping sites and b) variation of intersite transition rates. Since both the hopping rates and the number of sites decrease as a function of intersite distance, dissipative energy relaxation of an excited-state walker is expected to undergo a systematic deceleration while traversing through the distribution (↓). Therefore, as time proceeds and relaxation evolves, the waiting times gradually increase and finally, they formally tend to infinite as transport becomes, more and more, immobilized, either in energy-relaxed, monomeric tail states , or in traps and EFS outside the distribution of bulk states. Excitonic hopping motion across energy-dispersive carbazole sites must, therefore, proceed with a distribution of jump-frequencies $\phi(\nu)$, decreasing in time. Consequently, the net effect for an excited-state carrier is a nondiffusive random walk with a hopping frequency to correspond to a time dependent stochastic variable (cf. last section). A time-dependent, ensemble-averaged hopping rate function of the form

$$k(\mathbf{t}) = b' + c'\mathbf{t}^{\alpha - 1} \qquad (2)$$

has been confirmed by Monte-Carlo simulation (15) and by analytical techniques (43). b' corresponds to the isotropic transfer rate of a diffusive exciton in a (hypothetical) periodical array of energetically equivalent aromatic sites and $\alpha$ is the dispersion parameter ( $0 < \alpha < 1$) which is related to the halfwidth of the Gaussian density-of-state distribution (15). The term $c't^{\alpha-1}$ takes into account the nondiffusive, i. e. time-dependent, transport regime where c' is a complex physical quantity of a hopping frequency affected by disorder and dispersion. Clearly, formulation of a rate function of the form given in Equation 2 must induce non-autonomous coefficients in network equations of kinetic transients $X_i$. In vector notation the equations of evolution read

$$\dot{\mathbf{x}} = \mathbf{A}(t)\ \mathbf{x} \qquad (3)$$

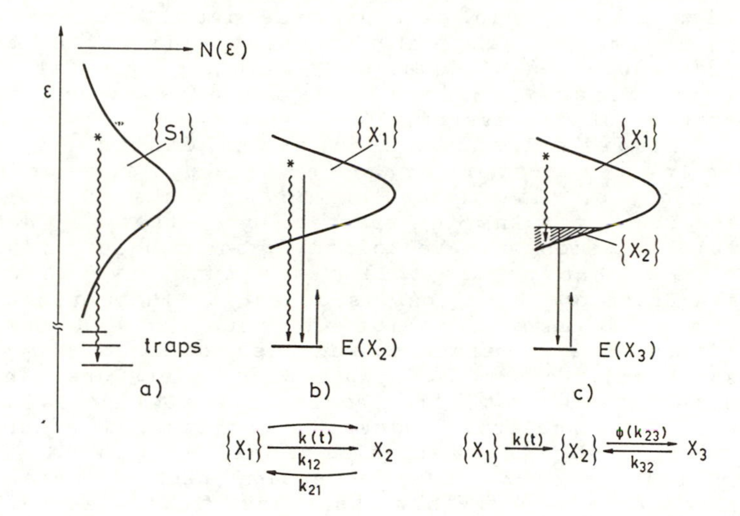


Fig. 2. Diagonal disorder and density-of-states distribution in aromatic vinyl-homopolymers; a) 3 D, random solid state, b) and c) low dimensional, non-random topology in fluid solution. See text for details.

Here $\dot{\mathbf{x}}$ and $\mathbf{x}$ are column vectors of the components $\dot{x}_i(t)$ and $x_i(t)$, respectively, and $\mathbf{A}(t)$ is now a physical reaction matrix containing time-dependent elements. Fluorescence of kinetic transients, e. g., the relaxation profiles of monomer- or excimer fluorescence are, therefore, strictly **nonexponential** for which closed form, analytical solutions can be found in few cases, only. A convincing manifestation of nonexponential trapping in low-temperature, solid state p-N-VCz is a recent analysis by Bässler et al. (42). With the use of rate function in Equation 2, the transient ps-rise profile of the low-energy excimer $E_2$ has been satisfactorily fitted to the numerical solution of Equation 3 with a single-fit dispersion parameter $\alpha$ between 0.2 and 0.8 depending on the temperature of the system.

In a **liquid** solvent an aromatic homopolymer is not a static low temperature array for EET, but it represents an assembly of time dependent arrangements of localized states. Thus, for the interpretation of trapping data in in **fluid** solution we have to modify the concept, inasmuch as solvent-chromophore collisions must have an intervening effect on EET. First, stochastic fluctuations must show up in segmental motion and thus, in a pronounced time-dependence of space-coordinates of aromatic sites. Transport and trapping are proceeding, therefore, through the interplay of both electronic intersite coupling and activation controlled, segmental rotation. Furthermore, rotational sampling will interfere, in general, with immobilization of EET. Consequently, back transfer (detrapping) from imperfectly absorbing sinks - as a result of thermal release and conformative motion, will lead to reversible transport pathways and, thus, to an increased probability of revisiting a site. Second, solvent scatter tends to narrow the energetic spread of aromatic sites. The effect of the heat-bath on EET is, therefore, to smear out the distribution of density-of-states, thus promoting, last not least, diffusive random walks across isoenergetic hopping sites.

The capacity of solvent-chromophore perturbation to effect motional narrowing and to overcome static site energy disorder in aromatic polymers depends on the magnitude and time scale of these interactions and the **width** of the static distribution of chromophores. This width, we believe, is presumably different for correlated chromophores in concentrated and dilute solutions. Thus, the annihilation of diagonal disorder by motional narrowing is expected to be important in typical 3-D morphologies of concentrated solutions where positional fluctuation of aromatic side groups are nearly random and small. However, for polymer morphologies under consideration here - p-N-VCz and p-VN in dilute, fluid solution of a 'good' solvent - the effect of cancellation of self-energy distribution is expected to be comparatively small. The cause for imperfect narrowing in the low chromophore density limit is, presumably, a kind of remaining static, diagonal disorder which is evidenced by inhomogeneous broadening of the absorption line shape (44). This residue of a dispersive contribution arises from anomalous fluctuations and local heterogeneities of chromophores, quite typical for the quasi-1-D dimensional contour of an extended and dilute aromatic homopolymer. Such fluctuations are inherently ill-behaved and their statistics are, predominantly, non-Gaussian. As a consequence, the density of state-profile and the dispersion parameter are rather undefined. Despite these further complications, we may expect the low dimensionality in tandem with an 'effective" diagonal disorder of chromophores to cause a profoundly pathological behavior of migrating singlet-excitations in these polymers. Therefore, a trapping function of the form

$$k(t) = b + ct^{-1/2} \qquad (4)$$

reflecting low dimensionality, disorder and inhomogeneity, and formally equivalent to Equation 2 for $\alpha = 1/2$ - should provide a quite satisfactory approach. In Equation 4 the square root law corresponds to the time-dependent transport regime which has become a theoretically sound result for one-dimensional-walks, in appropriate limits (11, 45, 46). c is a hopping frequency, again, affected by disorder, dispersion, and low-dimensionality. b represents the asymptotic, discrete hopping rate allowing for the admixture of isotropic, diffusive transport.

A simple model case of excimer population via low dimensional, dispersive transport and rotational sampling in fluid solution is illustrated in Figure 2 (situation b). Here, $\{X_1\}$ corresponds to a density-of-states distribution (arbitrary) which is intended to show significant difference from the Gaussian shape of a 3D-disordererd chromophore ensemble in situation a. $X_2$ denotes the excimer state, which is discrete and energetically well below the transport states. The wavy line (⌇) displays the serial events of EET through the transport band, the arrow pointing at downwards (↓) shall indicate rotational sampling, while the arrow in upward direction (↑) shall illustrate the dissociative behavior of the excimer state (37). In kinetic formulation the situation displayed corresponds to a monomer-excimer pair in presence of EET (phenomenological rate function: k(t)), rotational sampling (rate constant: $k_{12}$, simplified assumption, cf. below) and detrapping (rate constant: $k_{21}$). Because of the reversibility of the problem exact, analytical solutions cannot be found. Under the condition of weak coupling ($k_{21} \ll (k(t) + k_{12})$) an asymptotic WKB approach proves useful for the description of the transient long-time behavior of a fluorescence observable (40). With k(t) from Equation 4 the fluorescence response functions of monomer $x_1(t)$ and excimer $x_2(t)$ have the approximate form

$$x_1(t) \simeq A_{11}(t)\exp[-(2ct)^{1/2}]\exp[-t/T_1] + A_{12}(t)\exp[-t/T_2] \quad (5)$$

$$x_2(t) \simeq A_{21}(t)\exp[-(2ct)^{1/2}]\exp[-t/T_1] + A_{22}(t)\exp[-t/T_2] \quad (6)$$

with $1/T_1 = k_1 + k_{12} + b$, $1/T_2 = k_2 + k_{21}$ and rate constants $k_1$ and $k_2$ which are composed quantities of radiative and nonradiative deactivation for monomer and excimer, respectively. $A_{11}(t)$, $A_{12}(t)$, $A_{21}(t)$ and $A_{22}(t)$ are time dependent, but slowly varying amplitudes depending on the lower bound $t_o$ within the WKB-limit. Clearly, the equations of relaxation are nonexponential and dominated by the typical, **stretched** exponential form $\exp[-(2ct)^{\alpha}]$ for $\alpha = 1/2$ (cf. next section). Within the validity of the WKB-approximation, the lower bound $t_o$ (short-time limit) can be as short as several hundred picoseconds which depend on the EET parameters b and c. For moderate transfer Equations 5 and 6 appear to cover a time scale from the late sub-ns up to the early ns regime in a time resolved fluorescence experiment. A preliminary result of nonexponential curve fitting to the sub-ns rise profile of the low-energy excimer in p-1-VN is shown in Figure 3a. The noisy points (symbol: o) correspond to the experimental data reconvolved from the raw data (Figure 1a) by means of multiexponential curve parametrization (see, Equation I). The smooth full curve is an initial fit based upon Equation 6. Fit-parameters are given in caption of Figure 3. The agreement between experimental data and theoretical trial function is satisfactory, for the present, and, furthermore, it demonstrates that a nonexponential functional form, such as Equation 6 can replace a sum of two exponentials.

However, the WKB-equations 5 and 6 derived for relatively long times cannot hold, in general, when analyzing ps profiles. In the case of weak coupling ($k_{21}$  $k(t)$) and under conditions of predominantly migrative population ($k_{12} = 0$) the short-time solution to the situation b in Figure 2 has the form

$$x_2(t) = B_1\left\{\exp[-t/T_1]\exp[-(2Ct)^{1/2}] - \exp[-t/T_2]\right\} + B_2\left\{\mathrm{erf}\left[(1/T_1 - 1/T_2)^{1/2}\, t^{1/2} + a\right] - \mathrm{erf}\, a\right\}\exp[-t/T_2] \quad (7)$$

with $1/T_1 = k_1 + k_{12} + b$, $1/T_2 = k_2$, and the composed quantity $a = c(1/T_1 - 1/T_2)^{-1/2}$. $B_1$ and $B_2$ are constants and erf m is the error function. We suggest that this type of trapping function already applied to quasi-1D molecular crystals (45) might be applicable to the fluorescence of the high-energy, partial overlap excimer in p-N-VCz which is a "preformed" trap (34a) and thus, presumably, subject to a direct, migrative population from $\{X_1\}$. Initial attempts of fitting Equation (7) to

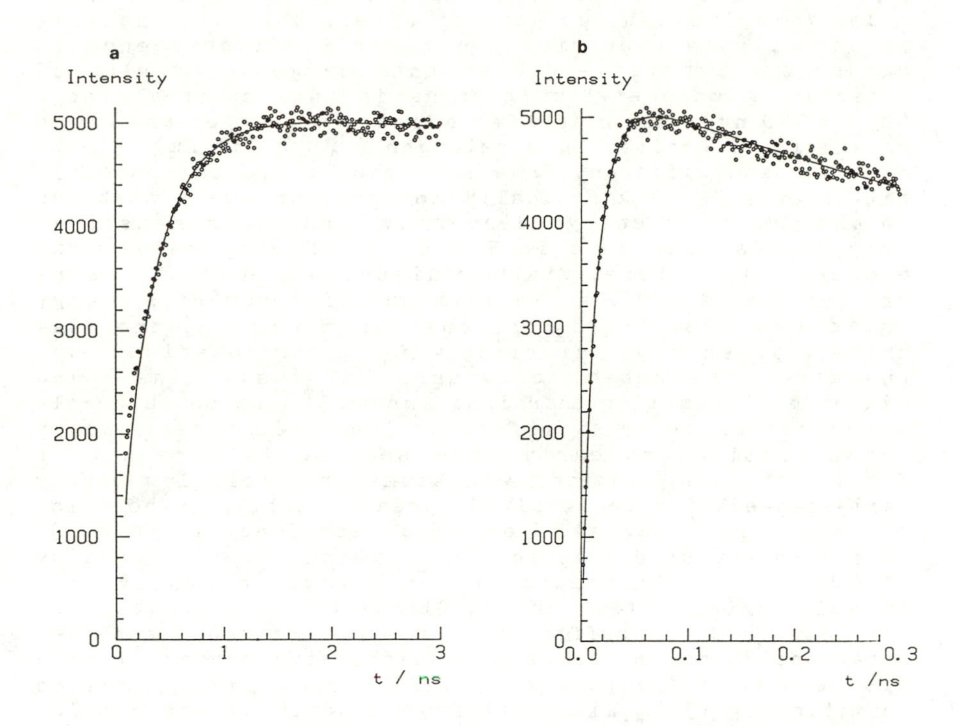


Fig. 3. Nonexponential curve fitting based upon fluorescence data convolutions in Figures 1a and 1b, respectively. Fig. 3a: dots (symbol o), experimental (artificial) pattern, according to Equation I; smooth solid curve (——), preliminary fit to Equation 6 with parameters $k_1 = 0.02\ ns^{-1}$, $k_2 = 0.015\ ns^{-1}$, and $k_{21} = 0.01\ ns^{-1}$ fixed; adjusted parameters: $b = 2\ ns^{-1}$ and $c = 0.45\ ns^{-1/2}$. Fig. 3b: dots (o), experimental (synthetic) data according to Equation II; smooth solid curve (——), fit to Equation 7; parameters: $k_1 = 0.083\ ns^{-1}$ and $k_2 = 0.05\ ns^{-1}$ (fixed), $b = 57\ ns^{-1}$ and $c = 1.5\ ns^{-1/2}$.

the parametrized experimental data (see, Equation II) based upon the experimental data convolute of Figure 1b is shown in Figure 3b. It becomes evident, that Equation (7) can recover the experimental rise, quite satisfactorily. Nevertheless, we have to stress that the analysis of the ultrafast ps rising edge of $E_1$-fluorescence is rather problematic - due to the energetic overlap of interfering monomer-like photons in this spectral range ($\lambda_{em}$ = 380 nm) and to scatter effects caused by the close proximity of excitation wavelength ( $\lambda_{em}$= 351 nm).

A version different from situation b treats monomeric tail states $X_2$ as kinetically independent dwell stations in the course of energy-dispersive and low-dimensional trapping (situation c in Figure 2). Energy relaxation proceeds, therefore, via 1) a distribution of host-singlet states $\{X_1\}$, 2) a small amount of monomeric, energy relaxed tail states $\{X_2\}$ (hatched area ) and 3) the discrete, low-energy excimer $E_2$ - $X_3$ in our notation. Furthermore, the scheme considers EET (wavy line ) and migrational sampling (straight arrows, $\downarrow\uparrow$ ) to be separable processes which is satisfied at least in case of bulky aromatic sites at early time scales. As a result of factorization of motion, we assume therefore low-energy tail-states $\{X_2\}$ to be involved, predominantly, in rotational sampling processes ( distribution function: $\Phi(k_{23})$; rate constant of detrapping: $k_{32}$) which are triggered by the density of transport states $\{X_1\}$ via migrative relaxation ($k(t)$). If we model, first, the distribution of tail states to be sufficiently narrow and thus, consider rotational sampling to be a nondispersive process with a discrete rate constant $k_{23}$, the problem can be treated analytically (38). With $k(t)$ from Equation 4 the $\delta$-pulse solution to the excimer $X_3$ reads

$$x_3(t) = A_{31}(t)\exp[-t/T_1] + A_{32}(t)\exp[-t/T_2] + A_{33}(t)\exp[-t/T_3] \quad (8)$$

$1/T_1$ equals $k_1$ + b ($k_1$ = $k_2$, reciprocal lifetime of isolated chromophore), while $1/T_2$ and $1/T_3$, respectively, refer to the well-known eigenvalues for the coefficient determinant of a reversible monomer-excimer pair (47). $A_{31}(t)$, $A_{32}(t)$, and $A_{33}(t)$ correspond to time dependent amplitudes which contain nonexponential contributions in terms of stretched exponentials, error-functions and Dawson integrals (38). In Figure 4a Equation 8 has been applied to the ps rise profile of $E_2$-fluorescence in p-N-VCz (Figure 1c). The noisy pattern (symbol: 0, artificial noise!) corresponds to the parametrized four exponentials of the experimental data (Equation III), the smooth full curve is a preliminary fit based upon Equation 8 with three fit-parameters b, c, and $k_{23}$ given in caption of Figure 4a. Agreement between experimental and theoretical profile is satisfactory, but, quite obviously, not perfect, as indicated by the systematic deviations within the range from 50 ps up to 90 ps. Undoubtedly, a more

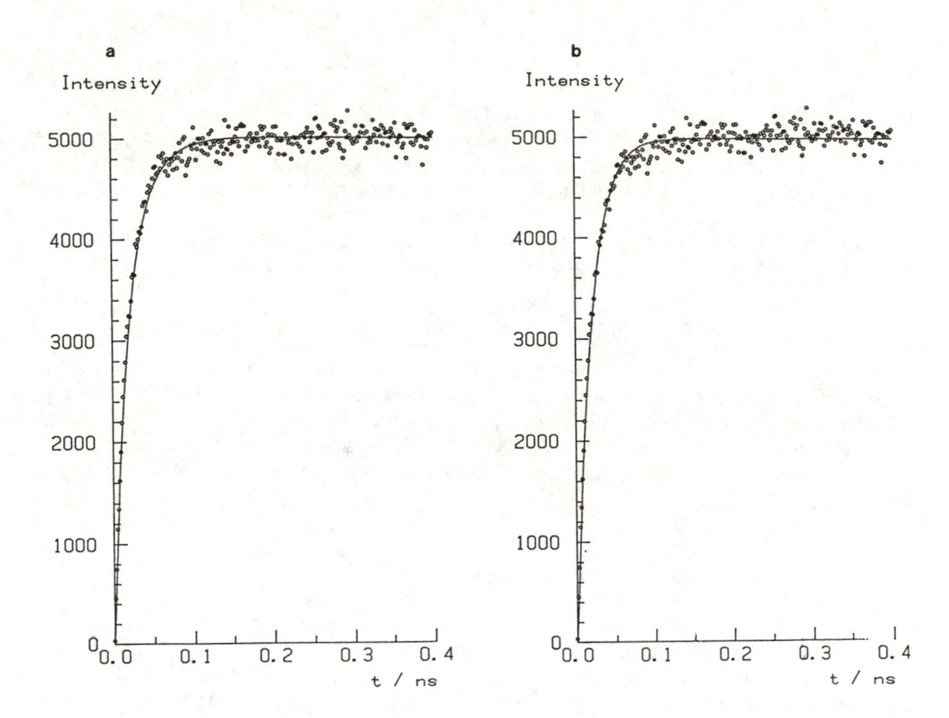


Fig. 4. Nonexponential curve fitting to data convolute in Fig. 1c: dots(o), (synthetic) experimental data from parametrized three exponentials of Equation III; smooth solid curves (———), preliminary fits based upon Equation 8 (Fig.4a) and the numerical solution with $k_{23}(t)$ from Equation 9 (Fig.4b). Fixed parameters: $k_1 = k_2 = 0.083\ ns^{-1}$, $k_3 = 0.02\ ns^{-1}$, and $k_{32} = 0.009\ ns^{-1}$. Adjusted parameters: $b = 10\ ns^{-1}$, $c = 50\ ns^{-1/2}$, and $k_{23} = 50\ ns^{-1}$ (Fig.4a); $b = 50\ ns^{-1}$, $c = 6\ ns^{-1}$, $\beta = 18\ ns^{-1}$, and $\gamma = 5.5\ ns^{-1/2}$ (Fig.4b). (Fig. 4a reproduced with permission from Ref. 38. Copyright 1986: American Institute of Physics)

realistic model of rotational sampling must account for a distribution of rotational frequencies $\phi(k_{23})$. Since the tail states $\{X_2\}$ corresponding to bichromophoric starting conformations are subject to an energetical width (see, sketch in c), they are expected to start their relaxation (toward excimer formation) under locally and energetically varying kinetic conditions. Therefore, a distribution of unimolecular rotational steps, each occuring in both parallel and serial processes should cause **dispersion**, again, with a time-dependent, average rate function $k_{23}(t)$ reflecting the dynamic correlations of microscopic rotational diffusion on a macroscopic scale (cf. next section). Clearly, a rate function of the form

$$k_{23}(t) = \beta + \gamma t^{-1/2} \tag{9}$$

- suggested previously (48) and phenomenologically similar to Equation 4 - cannot yield a closed form solution to the $\delta$-pulse behavior of the excimer $X_3$, but it has turned out that numerical Runge-Kutta computer solutions can recover the experimental data quite well (cf. Figure 4b).

## SUMMARY AND CONCLUDING REMARKS

In this work the main aspect has been concerned with the problem of electronic energy relaxation in polychromophoric ensembles of aromatic homopolymers in dilute, **fluid** solution of a "good" solvent. In this morphological situation microscopic EET and trapping along the contour of an expanded and mobile coil must be expected to induce rather complex rate processes, as they proceed in typically low-dimensional, nonuniform, and finite-size disordered matter. A macroscopic transport observable, i.e., **excimer fluorescence**, must be interpreted, therefore, as an ensemble and configurational average over a convolute of individual disordered dynamical systems in a series of sequential relaxation steps. As a consequence, transient fluorescence profiles should exhibit a more complicated behavior, as it can be modelled, on the other hand, on the basis of linear rate equations and multiexponential reconvolution analysis.

Thus, the question of central concern raised in our contribution has been the macroscopic formulation of EET and its relation to the experimental observable of excimer fluorescence in a time-resolved experiment. EET has been discussed, here, as a dispersive, i.e., time-dependent process in deterministic monomer-excimer models which had been the subject of a detailed kinetic analysis in recent work (38, 40). With the use of rate function $k(t)$ (Equation 4) it is natural to yield typical nonexponential intensity-time profiles, either in form of an asymptotic approach (Equations 5,6), or in closed form analytical solutions (Equations 7,8). The physics emerging from these formulas, or more precisely, the rationale behind k(t) rests on the hypothesis of an "effective" diagonal disorder. This sort of a quasi-static energetic disorder arises from the pathologically large spread of self-energies in low-dimensional aromatic vinylpolymers which cannot be smeared out by the relatively slow motions of bulky aromatic groups via Brownian rotational sampling. Although the shape of density-of-states function is predominantly non-Gaussian - due to the low-dimensionality and the local inhomogeneities -, we may, nevertheless, expect the dissipative excited-state relaxation to undergo a profoundly dispersive dynamics in these polymers. Since both the off-diagonal elements controlling the probability of nonradiative transition and the density numbers of hopping sites decrease as the intersite separation increases, an elementary excitation is subject to a systematic deceleration in the course of its energy relaxation. After executing, first, the rapid jumps, the hopping exciton will gradually be slowed and, finally, as time goes on, arrive at monomeric tail states, where excitonic residence times are large and thus, rotational sampling becomes more and more likely to compete with multipolar transitions.

A final note regarding the deceleration of energy-dispersive EET in these polymers appears in order. Quite obviously, excited-state relaxation of disordered polymers falls into a general class of hierarchically constrained dynamics where in a series of sequential, correlated processes a faster event initializes a slower one. A simple model class of hierarchically constraint successive and irreversible steps has been shown to give a typical "Williams-Watts" stretched exponential relaxation law (49). Even if a complete connection to microscopic constraints in real systems cannot yet be built, there are, nevertheless, good reasons to assume that a similar stretched exponential form might emerge from energy relaxation of distributed $\{X_1\}$-states in aromatic polymers

$$G(t) = \int_0^t \phi(\tau)\ \exp[-t/\tau]\ d\tau \xrightarrow{\ ?\ } \exp[-(t/\tau)^{\alpha}]\ ,\quad 0<\alpha<1 \qquad (10)$$

Here, the excitation function of $\{X_1\}$-states, G(t), specifies the configurational and ensemble average of a time dependent probability according to which the polymer remains excited as a whole at a monomeric site. $\tau$ is the single hopping time for diffusive excitonic motion in (hypothetical) regular arrays of isoenergetic aromatic groups and $\phi(\tau)$ is the (unknown) distribution of stepping times caused by the stochastic constraints of serial events in the energy cascade. On these premises, the power law term in Equations 2 and 4, respectively, would arise naturally, therefore, from a sequence of irreversible relaxation processes. Since these considerations are valid in the absence of any reverse constraint (dissociative traps), only, its applicability to reversible, deterministic schemes, as discussed here, must be restricted to **weakly** coupled systems, where detrapping is only a small perturbation to the overall trapping process (38, 40). The power law was proposed, originally, for the simulation of charge carrier transport (50), but it has turned out, in the meantime, to reflect, quite generally, the stochastic nature of transient relaxation phenomena in disordered materials. Thus, it has been demonstrated that the existence of a stretched exponential in Equation (10) is by no means limited to serial (constrained) dynamics. The law can be obtained as a consequence of independent, parallel (unconstrained) relaxation events as well (51,52) for which dielectric dipole relaxation (53), direct energy transfer (54), photochemical conversion (55) and spin-relaxation (56) are typical examples. However, the generality and widespread use of the law in different fields shows that the macroscopic law is rather unspecific and, quite obviously, insensitive with respect to different mechanisms of stochastic relaxation dynamics, at least at distinct time scales (21) . Therefore, a rate function of the form suggested in Equation (9) for modeling a distribution of constrained and unconstrained event times in rotational sampling might be a plausible assumption.

In summary, we have shown that excimer rise profiles in aromatic homopolymers (Figures 1a-1c) can be rationalized in terms of a stochastic model which has nothing to do with the interaction of states in linear kinetic schemes. Even if a rising edge analysis of an excimer in fluorescence reconvolution must be viewed with some reservations (38), our results, in principle, demonstrate that a macroscopic time-dependent term (Equations 2, 4) - representative for a microscopic dispersion of event times - can emulate two or three exponentials. Moreover, the rate function presented (Equation 4) has the advantage of yielding analytical solutions ( Equations 5-8 ) which allow one to analyze the relevancy of nonexponential contributions to ps and ns time scales in synthetic data simulation (38, 40). Nevertheless, we have to note that time-dependent trapping functions show up in increasing mathematical complexity, while, on the other hand, the procedures for testing acceptability of these functions are very crude. Undoubtedly, the results of parameter fitting need further proof in order to eliminate cross correlations and convolution errors. Therefore, tedious computational work is necessary in forthcoming transient experiments with regard to a direct implementation of raw data to arbitrary nonexponential trial functions in an iterative convolute-and- compare algorithm. Apart from the uncertainities in nonexponential data analysis, the EET-parameters b and c in the algebraic power law of Equation 4 are not easy to discuss, for the moment (cf. order of magnitude fits in Figures 3, 4), as they contain - parallel to the static properties of low dimensionality, disorder and heterogeneity - the dynamical effect of motional narrowing caused by solvent scatter and rotational sampling. As long as systematic transient experiments down to lower temperatures have not been done, it is therefore yet not possible to allow drawing order of estimates for the average hopping time of a polymer exciton in fluid solution.

To close this paper, we believe that both the theoretical and experimental aspects of excited-state relaxation in aromatic polymers will continue to be the subject of lively debate in the near future. Thus, the analyses of non-equilibrium transport based upon asymmetric energy-space master equations (43, 57) as well as theoretical models for a description of EET **and** rotational sampling are challenging many-particle problems in polymer photophysics. From an experimental standpoint of view, the time resolution of fluorescence system-configurations requires further improvement in order to test these concepts. Moreover, site-selective pulse-and-probe techniques should help to reveal transient excited-state distributions, energy relaxation and trapping on sub-ps time scales in forthcoming measurements.

ACKNOWLEDGMENTS

This work has been supported by the Fonds zur Förderung der wissenschaftlichen Forschung, Wien, Austria (Projekt No. P 4309 and P 6101C) which is gratefully acknowdledged. H. F. K. wishes to thank Österreichische Forschungsgemeinschaft, Wien, Austria, for providing funds from the program "Internationale Kommunikation". Furthermore, acknowledgement is made to the Donors of the Petroleum Research Fund, administered by the American Chemical Society, for partial support of this activity.

Literature Cited

1. Agranovich, V. M.; Galanin, M. D. Electronic Excitation Energy Transfer in Condensed Matter; North- Holland: Amsterdam, 1982;
2. Kopelman, R. In Spectroscopy and Excitation Dynamics of Condensed Molecular Systems; Agranovich, V. M.; Hochstrasser, R. M., Eds.; North-Holland: Amsterdam, 1983; p 139.
3. Yen, W. M.; Selzer, P. M., Eds. Laser Spectroscopy of Solids; Topics in Applied Physics Vol. 49; Springer: New York, 1981;
4. Lutz, D. R.; Nelson, K. A.; Gochanour, C. R.; Fayer, M. D. Chem. Phys. 1981, 58, 325.
5. Miller, D. P.; Robbins, R. J.; Zewail, A. H. J. Chem. Phys. 1981, 75, 3649.
6. Silbey, R. In Ref. 2; p 1.
7. Haan, S. W.; Zwanzig, R. J. Chem. Phys. 1978, 68, 1879.
8. a) Gochanour, C. R.; Andersen, H. C.; Fayer, M. D.

ibid. 1979, 70, 4254; b) Loring, R. F.; Andersen, H. C.; Fayer, M. D. ibid.1982, 76, 2015.
9. a) Huber, D. L. In Ref. 3, p 83; b) Sahimi, M.; Hughes, B. D.; Scriven, L. E.; Davis, H. T. J. Chem. Phys., 1983, 78, 6849.
10. Klafter, J.; Silbey, R. ibid. 1981, 74, 3510.
11. Klafter, J.; Blumen, A. ibid. 1984, 80, 875.
12. Gochanour, C. R.; Fayer, M. D. J. Phys. Chem. 1981, 85, 1989.
13. Miller, R. J. D.; Pierre, M.; Fayer, M. D. J. Chem. Phys. 1983, 78, 5138.
14. Even, U.; Rademan, K.; Jortner, J.; Manor, N.; Reisfeld, R. Phys. Rev. Lett., 1984, 42, 2164
15. Bässler, H. Phys. Status Solidi B, 1981, 107, 9.
16. Jankowiak, R.; Rockwitz, K.-D.; Bässler, H. J. Phys. Chem. 1983, 87, 552.
17. For Example, Selzer, P. M. In Ref. 3, p 113
18. Jankowiak, R; Bässler, H. Chem. Phys. 1983, 79, 57
19. Morgan, J. R.; El-Sayed, M.A. J. Phys. Chem. 1983, 87, 383.
20. Göbel, E. O.; Graudszus, W. Phys. Rev. Lett. 1982, 48, 1277
21. Richert, R.; Bässler, H. J. Chem. Phys. 1986, 84, 3567.
22. Anacker, L. W.; Parson, R. P.; Kopelman, R. J. Phys. Chem. 1985, 89, 4758.
23. Fredrickson, G. H.; Andersen, H. C.; Frank, C. W. Macromolecules 1983, 16, 1456; ibid. 1984, 17, 54.
24. Ediger, M. D.; Fayer, M. D. J. Chem. Phys. 1983, 78, 2518; Macromolecules 1983, 16, 1839; J. Phys. Chem. 1984, 88, 6108.
25. Ediger, M. D.; Domingue, R. P.; Petersen, K. A.; Fayer, M. D. Macromolecules 1985, 18, 1182;
26. Fredrickson, G. H.; Andersen, H. C.; Frank, C. W. J. Polym. Sci. Polymer Phys. Ed. 1985, 23 591.
27. For Example, Klöpffer, W.; In Electronic Properties of Polymers; Mort, J.; Pfister, G.; Eds.; Wiley: New York, 1982; p 171.
28. a) Schneider, F. Z. Naturforsch. A 1969, 24A, 863; b) Reid, R. F.; Soutar, I. J. Polym. Sci., Polym. Phys. Ed. 1978, 16, 231;
29. North, A. M.; Treadaway, M. F. Eur. Polym. J. 1973, 9, 609.
30. Webber, S. E.; Avots-Avotins, P. E.; Deumie, M. Macromolecules 1981, 14, 105.
31. Fitzgibbon, P. D.; Frank, C. W. Macromolecules 1982, 15, 733; Gelles, R.; Frank, C. W. ibid. 1982, 15, 741; 15, 747.
32. Semerak, S. N.; Frank, C. W. Adv. Polym. Sci. 1984, 54, 33.
33. O'Connor, D. V.; Ware, W. R.; Andre, J. C. J. Phys. Chem. 1979, 83, 1333.
34. a) Johnson, G. E. J. Chem. Phys. 1975, 62, 4697; b) Gupta, A.; Liang, R.; Mocanin, D.; Kliger, D.; Goldenbeck, R.; Horwitz, J.; Miskowski, V. M. Eur. Polym. J.

1981, 17, 485; c) Itagaki, H.; Okamoto, A.; Horie, K.; Mita, I. Eur. Polym. J. 1982, 18, 885.
35. De Schryver, F. C.; Demeyer,; Van der Auweraer, M.; Quanten, M.; Ann. N. Y. Acad. Sci. 1981, 366, 93.
36. Phillips, D.; Roberts, A. J.; Soutar, I. J. Polym. Sci., Polym. Phys. Ed. 1982, 20, 411.
37. Kauffmann, H.; F.; Weixelbaumer, W.-D.; Bürbaumer, J.; Schmoltner, A.-M.; Olaj, O. F. Macromolecules 1985, 18, 104.
38. Kauffmann, H. F.; Mollay, B.; Weixelbaumer, W.-D.; Bürbaumer, J.; Riegler, M.; Meisterhofer, E.; Aussenegg, F. R. J. Chem. Phys. 1986, 85, 3566
39. Fredrickson, G. H.; Frank, C. W. Macromolecules 1983, 16, 572
40. Weixelbaumer, W.-D.; Bürbaumer, J.; Kauffmann, H. F. J. Chem. Phys. 1985, 83, 1980.
41. Rockwitz, K.-D.; Bässler, H. Chem. Phys. 1982, 70, 307.
42. Peter, G.; Bässler, H.; Schrof, W.; Port, H. ibid. 1985, 94, 445
43. Grünewald, M.; Pohlmann, B.; Movaghar, B.; Würtz, D. Philos. Mag. B 1984, 49, 341;
44. a) Hesse, R.; (unpublished); see Ref. 15: p 25; b) Buchberger, E.; Kauffmann, H. F. (unpublished)
45. Wieting, R. D.; Fayer, M. D.; Dlott, D. D. J. Chem. Phys. 1978, 69, 1996.
46. Kenkre, V. M. In Exciton Dynamics in Molecular Crystals and Aggregates; Höhler, G., Ed.; Springer Tracts in Modern Physics, Vol. 94; Springer: Berlin, 1982; p 1.
47. Birks, J. B. Photophysics of Aromatic Molecules; Wiley-Interscience: New York, 1970; p 304.
48. Itagaki, H.; Horie, K.; Mita, I. Macromolecules 1983, 16, 1526.
49. Palmer, R. G.; Stein, D. L.; Abrahams, E.; Anderson, P. W. Phys. Rev. Lett. 1984, 53, 958.
50. Scher, H., Montroll, E. W. Phys. Rev. B 1975, 12, 2455.
51. Klafter, J.; Blumen, A. Chem. Phys. Lett. 1985, 119, 377.
52. Bendler, J. T.; Shlesinger, M. F. Macromolecules 1985, 18, 591; Montroll, E. W.; Bendler, J. T. J. Stat. Phys. 1984, 34, 129
53. Williams, G.; Watts, D. C. Trans. Faraday Soc. 1970, 66, 80.
54. Blumen, A. Nuovo Cimento 1981, 63 B, 50
55. Richert, R.; Bässler, H. Chem. Phys. Lett. 1985, 116, 302.
56. Chamberlin, R. V.; Mozurkewich, G.; Orbach, R. Phys. Rev. Lett. 1984, 52, 867
57. Parson, R. P.; Kopelman, R. J. Chem. Phys. 1985, 82, 3692

RECEIVED March 13, 1987

# Chapter 19

# Influence of Chromophore Organization on Triplet Energy Migration in Amorphous Polymer Solids

Richard D. Burkhart and Norris J. Caldwell

Department of Chemistry, University of Nevada, Reno, NV 89557

A study of the comparative rates of triplet exciton migration has been carried out on molecularly doped polymers and on vinyl aromatic polymers. In both cases specific rate constants for triplet exciton migration were estimated from rate constants for triplet-triplet annihilation. The rate data were obtained by using a laser pulse-optical probe method to determine triplet concentrations directly by triplet-triplet absorption. It is found that triplet exciton migration rates for polymers are ten-fold to one-hundred-fold larger than those for doped polymer matrices probably due to the more dense local chromophore concentrations in the former.

Natural chemical and physical processes occurring in living organisms have traditionally been a source of curiosity to practicing scientists and have often provided the jumping-off place for fundamental studies. Photochemists, in particular, have been led to fertile fields for investigation after contemplating the various light-induced biological processes such as photosynthesis and the visual process. One of the very clear lessons which these investigations have brought to light is the importance of proper organization of molecules or chromophoric groups in order to accomplish the task at hand. Somehow the organism manages to achieve the proper structural arrangement taking advantage of chemical and physical laws and its own evolutionary propensities. Chloroplasts in the leaves of green plants provide the classical example.

The challenge which these naturally occurring processes presents, of course, involves the laboratory construction of organizational units which also are able to perform some pre-determined, photo-initiated task. It seems to be widely accepted that a necessary first step to this end requires a clear understanding of the interaction between the mechanism of electronic energy migration and the role played by the positioning of

0097-6156/87/0358-0242$06.00/0

chromophoric groups(1). Most of the examples to be discussed here and, in fact, the majority of experimental studies have focussed upon organic chromophores which exist in the ground state with zero electron spin angular momentum. Thus, spin allowed singlet-singlet and triplet-triplet transitions and spin forbidden singlet-triplet transitions are the primary ones to be considered.

In the present work a major emphasis has been placed on those processes involving excited triplet states. A partial justification for this particular choice is that many different experimental methods exist for evaluating the rate of triplet exciton migration. These include the use of spatially intermittent excitation (also called the transient grating method)(2), time-dependent optical anisotropy(3), luminescence quenching(4), time dependent spectral shifts(5), and rates of triplet-triplet annihilation(6). In addition, the relatively long lifetime of triplet states places a lighter burden of performance on detection equipment and makes the direct observation of these species a relatively simple task.

A severe and sometimes critical weakness associated with the study of triplet states is that it is usually difficult to prepare them by direct photon absorption. Usually an intermediate excited singlet state must be formed followed by intersystem crossing or else a sensitization technique is used. In either case, one must be careful to distinguish between energy migration among triplets on the one hand and among precursor states on the other.

## Doped Polymer Matrices

**The Secluded Pair.** An example of the effects of energy migration involving excited electronic states which precede triplet formation is provided by experiments on time-dependent optical anisotropy(3). For example, when polystyrene matrices doped with 1,2-benzanthracene (BZN) are photoexcited using a vertically polarized beam, the resulting phosphorescence polarization increases in absolute magnitude with increasing time over a period of 50 msec as does the delayed fluorescence polarization. The fact that these delayed luminescence polarizations are near zero when observed at fifty μsec after the excitation pulse is understandable if rapid scrambling of the transition dipole vector has occurred among the precursor excited singlet states. But why, then, does the polarization not maintain this zero value at longer times?

A plausible answer to this question leads to a first glimpse of some rudimentary structural features appearing even in these molecularly doped matrices. In any solute-solvent system, even an ideal one, there will be some distribution function describing average intermolecular separation distances. Thus, if the distribution is frozen at some instant in time there will exist certain regions which are more densely populated than others. The model adopted here for the doped polymer matrix is essentially that of a fluid solution, but not necessarily an ideal one, which has been instantaneously frozen so as to immobilize both solute and solvent molecules. In densely populated regions, triplet exciton migration will give rise to relatively rapid triplet-triplet annihilation and a faster than average rate of triplet migration and decay. It is these same regions in which there will be the most complete scrambling of the transition dipole both for precursor singlet states and triplets. As the emission intensity decreases the triplet states which remain at longer times tend to be those in less populated regions with few neighbors properly situated to be an energy acceptor. Thus, at long times, triplet emission comes primarily from those molecules which are the same ones excited by the initial excitation pulse.

Upon closer examination it is found that the delayed fluorescence (DF) polarization from these BNZ/polystyrene matrices approaches a value near 0.25 which is one-half the theoretical maximum of 0.5. This fact suggests the existence of another type of structural feature, a so-called secluded pair, in which two molecules are significantly closer to each other than to any third species(7). A triplet exciton localized at such a pair would be expected to spend, on average, fifty percent of its time at the molecule which originally absorbed the exciting photon and fifty percent at the partner site. Thus, triplet-triplet annihilation involving interaction of a second exciton with this pair-trapped species will, with equal probability, produce a DF photon nearly co-linear with the excitation polarization or else randomly oriented with respect to it.

Thus, BNZ-doped polystyrene appears to possess two structural features which influence the characteristics of the delayed luminescence. One of these involves clusters of BNZ molecules consisting of relatively densely populated regions along with others which are much less densely populated. The other consists of molecular pairs which are capable of trapping excitons by virtue of kinetic barriers as opposed to energetic ones. Although it might seem that features of this sort would be common to all molecularly doped polymers, a search for them in N-ethylcarbazole (NEC) doped polystyrene revealed no indication of their presence. In particular, there was found to be no time dependence of the delayed luminescence polarization on the millisecond time scale(8). With NEC, however, one does find a dependence of the delayed luminescence polarization on dopant concentration. That is, the polarization decreases monotonically as the concentration increases.

Triplet Decay and Exciton Migration. The rate of triplet state decay following a photoexcitation pulse is conveniently followed by monitoring the time dependence of the delayed fluorescence. A limitation of this approach is that absolute triplet concentrations cannot usually be evaluated and so rate constants for processes having a kinetic order greater than unity cannot be determined.

More information about these important higher order decay processes is available by direct measurements of absolute triplet concentrations. For example, in the case of BNZ-doped polystyrene it was possible to obtain $k_2$ in the rate equation

$$-dT/dt = k_1T + k_2T^2 \tag{1}$$

by measuring the time dependence of T using optical absorbance of this species(9). In this equation T is the triplet concentration, $k_1$ is the rate constant for the first order decay of triplets and $k_2$ is the specific rate for triplet-triplet annihilation.

The cluster model which was proposed to explain the time

dependent increase of the absolute delayed luminescence polarization would also predict that apparent $k_2$ values should decrease monotonically with time. Triplet-triplet absorption studies on BNZ-doped polystyrene, in fact, were carried out to test this very point. It was indeed observed that in the first 25 msec following excitation there is a steady decline in $k_2$ and that the final asymptotic value of $k_2$ increased with increasing solute concentration(9). Such experiments do not, of course, prove that the suggested model is correct but are, at least, consistent with the model's predictions and provide evidence for the power of this type of technique.

An additional benefit to be derived from direct measurements of $k_2$ involves its relation to rates of exciton migration using the theory of diffusion controlled processes. The rate constant for a purely diffusion controlled reaction may be written in the form (10)

$$k_{df} = 4\pi rDN/1000 \tag{2}$$

where r is the encounter diameter, D is the diffusion coefficient and N is Avogadro's number. If D is expressed in $cm^2/sec$ and r in cm, then $k_{df}$ has the units of $M^{-1} sec^{-1}$. An estimate of 15 A was suggested by Birks(11) for the encounter diameter for triplet-triplet annihilation. The rate constants $k_{df}$ and $k_2$ are linked by an efficiency factor, that is, $k_2=qk_{df}$ where $q^2$ has upper and lower limits of 2 and 0 respectively(12). Thus, measurements of $k_2$ yield experimental values of qD. Measurements of this sort have been made for a number of different polymer/dopant systems. These have been collected over the last several years in these laboratories and are summarized in Table I.

Table I. Second Order Rate Constants for Triplet Decay Found in Various Polstyrene/Dopant Systems

| Dopant Molecule | Concentration (M) | T (K) | $k_2$x$10^{-3}$ + ($M^{-1}$ $sec^{-1}$) |
|---|---|---|---|
| N-carboethoxycarbozole | 0.0944 | 77 | 1.8 |
| 1,2-benzanthracene | 0.230 | 298 | 11.1 |
| 1,2-benzanthracene* | 0.114 | 298 | 29.2 |
| naphthalene | 0.047 | 298 | |

*Using poly( -methylstyrene) as the matrix
+Long time asymptotic value of $k_2$

For the polystyrene matrices it is clear that $k_2$ values on the order of $10^3$ to $10^4$ seem to be typical. Values of qD were calculated from these rate constants using as encounter radii either the 15 A suggested by Birks or the average intermolecular separation distance calculated from the equation derived by Chandrasekhar (13), whichever was smaller. The results are presented in Table II. Again very little individuality among the dopant molecules is noted.

Table II. Values of qD for Various Dopant/Polystyrene Systems

| Dopant Molecule | r* (A) | Concentration (M) | $qD \times 10^{12}$ ($cm^2$/sec) |
|---|---|---|---|
| N-carboethoxycarbazole | 14.4 | 0.0944 | 1.8 |
| 1,2-benzanthracene | 12.8 | 0.134 | 1.6 |
| naphthalene | 18.2 | 0.047 | 1.6 |

*Calculated average intermolecular separation distances

As a final step in this analysis it may be assumed that these exciton migrations occur by a sequence of random flights for which the mean square displacement in unit time is equal to 6D. The mean square displacement is the product of the frequency of migratory hops (f) and the square of the length of each hop ($\ell^2$). Thus, $f\ell^2 = 6D$ and, using the mean separation distance of solute molecules as $\ell$ and abitrarily setting q = 1, one finds f values between 500 $sec^{-1}$ and 600 $sec^{-1}$ for the three systems of Table II.

These results may well be typical of most polymer/dopant systems. Certainly the behavior of these systems with respect to triplet exciton migration seems quite similar in spite of their rather distinctive molecular architecture. Let us now turn to an examination of systems in which chromophore units are regularly arranged by virtue of their being bonded to the backbone of a polymer chain. In this way, it may be possible to assess the effects of chromophore organization on the mechanism of triplet exciton migration and decay.

## Pure Polymer Systems

General Observations. The triplet photophysical behavior of most polymeric systems is quite dependent on the particular physical state. It is important to distinguish, for example, whether one is dealing with neat polymer films or with solutions. If the latter, then the fluidity of the solution will also be important.

One reason for this preoccupation with the physical state concerns the ease with which triplet excimer formation occurs. Phosphorescence from triplet excimers, for example, is common in solid polymeric films but much less common in rigid solutions. For this reason, the interpretation of rate processes involving triplet states tends to be simpler to handle and more susceptible to quantitative treatment for rigid polymer solutions.

An interesting feature of delayed luminescence from polymers which sets them apart from their monomeric analogues is the very prominent delayed fluorescence (DF) emission which is usually observed. This, in fact, may be taken as a primary feature of photophysical behavior which may be directly linked to chromophore organization. For many different polymers it has been shown that the process of triplet-triplet annihilation is responsible for the prominent DF emission observed. In fact, for dilute polymer solutions the annihilative process appears to be almost exclusively intramolecular in character(14,15).

Since it will be necessary to distinguish between triplet excitons which are mobile from those which are trapped at an excimer forming site, the symbol $T_m$ will be used to represent molar concentrations of the former and $T_t$ for the latter. If more than one type of trapped triplet is present, an additional numerical subscript will also be used, i.e. $T_{t1}$, $T_{t2}$ or, in general, $T_{tj}$.

Neat Polymer Films. It is interesting to compare the triplet photophysical properties of poly(N-vinylcarbazole) (PVCA)(16) on the one hand and poly(1-vinylnaphthalene) (P1VN)(17) on the other when each is examined as a pure polymer film. Both polymers exhibit a prominent excimer phosphorescence band as well as a distinct delayed fluorescence emission. In addition, the delayed fluorescence arises by a process of triplet-triplet annihilation for both polymers. Furthermore, the luminescence decay kinetics suggest that equilibria of the type

$$T_t \rightleftarrows T_m + {}^1E^0 \tag{3}$$

exist where ${}^1E^0$ is an excimer forming site. That is, excimer formation is a reversible process. One manifestation of this equilibrium is the shift of the center of gravity of the phosphorescence band to longer wavelengths as the temperature is raised. Presumably this occurs because of the selective loss of the highest energy excimer species by detrapping as the temperature rises. The mobile excitons produced are then free to probe available trap sites and, of course, will tend to populate the lower energy ones selectively. Thus, the average energy of populated excimer sites decreases with increasing temperature.

An important difference in photophysical behavior between the naphthalenic and carbazole polymers is the lifetime of the mobile exciton compared with that of the triplet excimer. For PVCA the mobile exciton is much shorter lived than the excimeric species but for P1VN just the reverse is true. In both polymers the primary mode of delayed fluorescence production involves a hetero-annihilation of the type

$$T_m + T_t \longrightarrow {}^1E^* + {}^1M^0 \tag{4}$$

Thus, the delayed fluorescence lifetime is essentially equal to that of $T_m$ for PVCA and to $T_t$ for P1VN. The latter is easy to prove since excimer phosphorescence and delayed fluorescence lifetimes are the same for P1VN. For PVCA at 77 K $\tau_{DF} \ll \tau_{phos}$ but no direct and independent measure of the lifetime of $T_m$ in PVCA has been accomplished at this time.

Delayed fluorescence lifetimes for these two polymers indicate that there is more than one type of excimeric species present. It had been pointed out by Siebrand(18) that when this situation obtains and when the lifetime of mobile the exciton is less than that of the excimeric species, then one should observe that the delayed fluorescence intensity goes through a maximum when plotted as a function of the absolute temperature. This predicted maximum is, in fact, found for solid films of PVCA and, as expected, is not observed for solid films of P1VN.

Polymers in Rigid Solution. The emission spectrum of PCVA in 2-methyltetrahydrofuran (MTHF) at 77 K consists of prominent delayed fluorescence and phosphorescence bands(19). For this reason it was decided to investigate the rate of triplet exciton decay in these rigid solutions and to treat the data in terms of concurrent first and second order processes. For systems in which an equilibrium distribution of potential reactants may be assumed, eq 1 may be employed for data analysis. It is not clear, however, that such a distribution is valid for polymer solutions especially in light of evidence suggesting that T-T annihilations occur principally by intra-coil processes(14-15).

A recent investigation was carried out to find a valid method for analyzing data on the rate of triplet state disappearance using models involving only intramolecular migration of triplet excitons (Burkhart, R. D. Chem. Phys. Lett., in press). A one-dimensional model allowing only nearest neighbor migrations was compared with a three-dimensional random flight model. Physically realistic results were obtained only with the three-dimensional model.

Furthermore, the diffusion coefficients for triplet exciton migration extracted from this three-dimensional intramolecular model were nearly the same as those obtained using the conventional kinetic equation (i.e. eq 1). The hopping frequencies for triplet exciton migration in PVCA for these three models are summarized in Table III(20). Neither the electron exchange mechanism(21) nor the Förster

Table III. Frequencies of Triplet Exciton Migration for Rigid Solutions of PVCA in MTHF at 77 K Using Various Models for Data Analysis

| Assumed Model | Migration Frequency ($cm^{-1}$) |
|---|---|
| Conventional | $3x10^5$ |
| Intra-coil, one dimensional | 67 |
| Intra-coil, three dimensional | $2x10^4$ |

dipole-dipole mechanism(22) can account for the one-dimensional model. For the other two models, however, the frequencies are on the order of ten-fold larger than those found for doped polystyrene matrices.

Thus, for the polymer systems investigated to this point, there are only modest enhancements of the exciton migration frequency compared with molecularly doped polystyrene. Furthermore, the enhancements which are found may be due simply to a higher local density of chromophores in the case of the polymeric systems.

Time-Dependent Phosphorescence Spectra of Polymers. Although a strong similarity exists between phosphorescence spectra of vinyl aromatic polymers and the corresponding monomeric analogues, it is interesting to focus attention upon the differences which exist between these spectra. A sample comparison is provided in Figure 1 between NEC and PVCA both as dilute solutions in MTHF at 77 K (23). Both spectra are recorded using comparable conditions and instrument parameters.

In the spectrum of PVCA it seems that the 0-0 emission band is present only as an unresolved shoulder and that the remaining structural features of this spectrum are somewhat broadened compared with those of NEC. An attempt was made to achieve better resolution of the 0-0 band of PVCA using narrow monochromator slits and multiple scans of the emission band. In addition, spectra were recorded using variable delay times following excitation. These results are presented in Figure 2.

Two observations about the 0-0 phosphorescence band of PVCA emerge from these experiments. In the first place a clear resolution of this band is, in fact, achieved. In addition, the intensity of this band decreases relative to longer wavelength components by using long delay times on the order of several hundred milliseconds. An obvious corollary to this effect is that apparent phosphorescence decay times will depend upon the wavelength chosen for the measurement. Such effects are not large but they are readily measurable.

Evidently these interesting triplet state phenomena are recorded here for the first time for a polymer system. Similar observations have been made, however, by Bässler and coworkers on amorphous benzophenone(5). The interpretation put forward to account for the effect involves triplet exciton migration in the inhomogeneously broadened profile. As indicated above for solid film spectra, the tendency is for the migrating excitons to seek out the lowest energy sites. Therefore, excitons at relatively high energy sites may not only relax to the ground state but may also migrate to lower energy sites releasing some fraction of their energy to the lattice. The rate of disappearance of such excitons will clearly be greater than that of lower energy ones which have fewer channels available for energy dissipation.

In order to demonstrate this effect to best advantage it was necessary to choose a PVCA sample having a relatively low molecular weight. In this way interference of the phosphorescence emission by delayed fluorescence is minimized. These are provacative results because they indicate that there may be no well defined lowest triplet state in vinyl aromatic polymers unless special steric or electronic effects are present which nullify inter-chromophore interactions. On the other hand, they may provide an additional tool with which to investigate rates of energy migration in polymers and in some polymer/dopant systems as well.

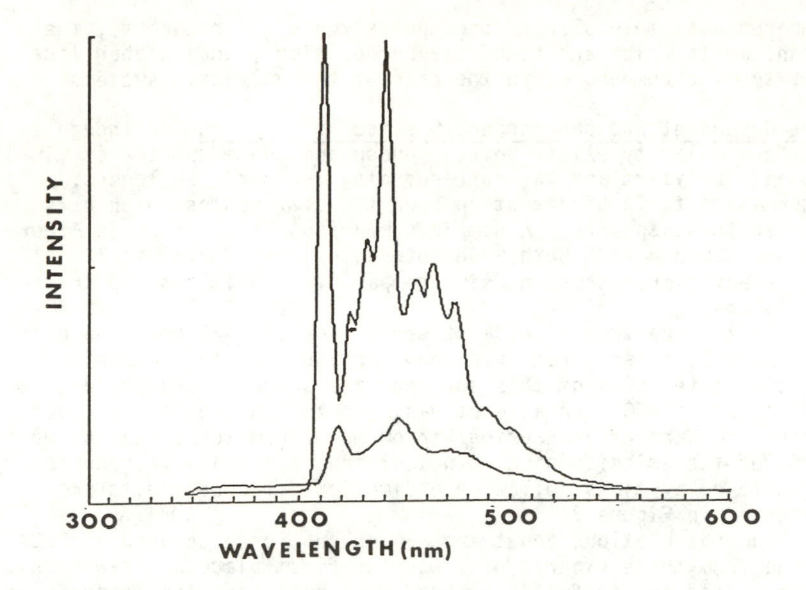


Figure 1. Phosphorescence spectra of NEC (upper) and PVCA (lower) in solution at 77 K. Both spectra are recorded at 50 msec following the excitation pulse using a monochromator band pass of 6 nm.

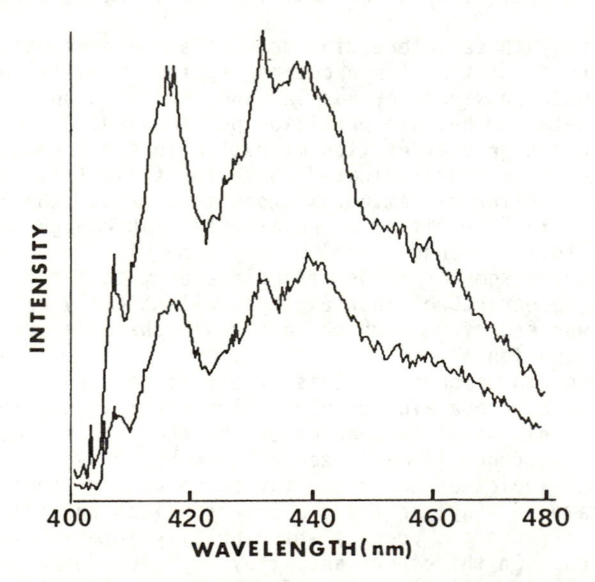


Figure 2. Time resolved phosphorescence spectra of PVCA in solution at 77 K using a monochromator band pass of 2 nm. Delay times of 400 msec (upper) and 800 msec (lower) were used.

## Acknowledgment

This work was supported by the U.S. Department of Energy under Grant Number DE-FG08-84ER45107.

## Literature Cited

1. Guillet, J. Polymer Photophysics and Photochemistry; Cambridge University Press: Cambridge, 1985; pp 241-251.
2. (a) Meyer, E. G.; Nickel, B. Z. Naturforsch. 1980, 35A, 503. (B) Burkhart, R. D. J. Am. Chem. Soc. 1974, 96 6276.
3. Burkhart, R. D.; Abia, A. A. J. Phys. Chem. 1982, 86, 468.
4. David, C.; Demarteau, W.; Geuskens, G. Eur. Polym. J. 1970, 6, 537.
5. Richert, R.; Ries, B.; Bässler, H. Phil. Magazine B. 1984, 49, L25.
6. Caldwell, N. J.; Burkhart, R.D. Macromolecules 1986, 19, 1653.
7. These pair structures had earlier been postulated on the basis of luminescence decay kinetics. For details see Burkhart, R. D. Chem. Phys. 1980, 46, 11.
8. Abia, A. A. Ph.D. Thesis, University of Nevada, Reno, 1984.
9. Burkhart, R. D. J. Phys. Chem. 1983, 87, 1566.
10. Noyes, R. M. Progr. Reaction Kinetics 1961, 1, 130.
11. Birks, J. B. Photophysics of Aromatic Molecules; Wiley: New York, 1970; p 390.
12. Values of this efficiency factor have been evaluated for only a few molecules. For a detailed discussion see, Saltiel, J.; Marchand, G. R.; Smothers, W. K.; Stout, S. A.; Charlton, J. L. J. Am. Chem. Soc. 1981, 103, 7159.
13. Chandrasekhar, S. Rev. Mod. Phys. 1943, 15, 1.
14. Pasch, N. F; Webber, S. E. Chem. Phys. 1976, 16, 361.
15. Klöpffer, W.; Fischer, D.; Naundorf, F. Macromolecules 1977, 10, 450.
16. Burkhart, R. D.; Avilés, R. G. Macromolecules 1979, 12, 173.
17. Burkhart, R. D.; Avilés, R. G.; Magrini, K. Macromolecules 1981, 14, 91.
18. Siebrand, W.; J. Chem. Phys. 1965, 9, 234.
19. Klöpffer, W.; Fischer, D. J. Polymer Sci., Part C 1973, 40, 43.
20. The rate constants providing the basis for this analysis were taken from the data of reference 6.
21. Inokuti, M.; Hirayama, F. J. Chem. Phys. 1965, 43 1978.
22. Förster, T. Z. Naturforsch. 1946, 4a, 321.
23. The data of Figures 1 and 2 were obtained in this laboratory by Gregory Haggquist using a locally constructed phosphorimeter described in reference 6.

RECEIVED April 27, 1987

## Chapter 20

# Triplet Antenna Effect in Poly(acetonaphthyl methacrylate)

**David A. Holden and Ali Safarzadeh-Amiri**

**Guelph–Waterloo Centre for Graduate Work in Chemistry, Department of Chemistry, University of Waterloo, Waterloo, Ontario N2L 3G1, Canada**

The different types of experiments used to study migration and transfer of triplet excitation energy within individual polymer chains are discussed. Certain advantages are offered by a method involving analysis of the quenching of phosphorescence of the donor chromophores by copolymerized phosphorescent traps. This approach is illustrated with new copolymers containing 1-aceto-2-naphthyl methacrylate as the donor and 2,4-diaceto-1-naphthyl methacrylate as the acceptor. The triplet states in poly(acetonaphthyl methacrylate) are mobile and a hopping frequency of $3x10^3$ $s^{-1}$ between individual donors was determined for poly(1-aceto-2-naphthyl methacrylate) in dilute solution in 1:1 THF/2-methyl THF glass at 77K.

The "triplet antenna effect" is the hopping or migration of triplet excitation between donor chromophores until transfer to a triplet quencher can occur. Singlet and triplet electronic energy migration have captured the imagination of many polymer photophysicists for several reasons (1-3). These processes represent the reproduction in a single macromolecule of the same "antenna effect" which feeds electronic excitation into the reaction centre of plant chloroplasts (4,5). The chloroplast is arguably the most important of all photochemical systems, and yet the antenna effect has been largely ignored by both biophysicists and classical photochemists. Here is a chance for polymer photophysicists to contribute to the larger sphere of science. Energy migration may also have an immediate practical consequence in the photochemistry of certain polymers (6). It leads to enhanced quenching, both by unreactive chromophores (stabilizers) and by photolabile chromophores (sensitizers). What are the overall effects on the useful lifetimes of new plastics?

0097-6156/87/0358-0252$06.00/0

## Review of the Literature

Estimates of the rate and range of triplet energy migration as measured by different techniques vary enormously. For this reason it is worthwhile to summarize the literature on triplet energy migration in polymers and to point out some assumptions in the different experimental approaches.

The major experimental approaches to detecting triplet migration are:

1. By studies of delayed fluorescence at low temperature.
2. By analysis of the quenching of phosphorescence by small molecules at low temperature.
3. By observing the quenching of photochemical reactions of the triplet state, both by small-molecule quenchers and by copolymerized quenchers in fluid solution.

At low temperature triplet lifetimes in polymers such as poly(N-vinyl carbazole) and poly(2-vinyl naphthalene) are sufficiently long that it is possible to have more than one excited state per chain. Interaction of two migrating triplets leads to triplet-triplet annihilation and delayed fluorescence. Thus the observation of delayed fluorescence by itself indicates triplet energy migration (7-9) . Quantitative measurements of the rate of hopping $\omega$ of the triplet excitation are obtained by observing the quenching of triplet phosphorescence and delayed fluorescence by added small-molecule quenchers such as 1,3-pentadiene or biacetyl (10,11) or by observing the effects of molecular weight on the ratio of intensities of delayed fluorescence and phosphorescence (12-15). The difficulty with interpretation of quenching by small molecules is determination of the true number of contacts between an individual chain and the quencher. Among the assumptions made in arriving at a value for $\omega$ from small-molecule quenching studies are an isotropic quencher distribution and an educated guess of the interaction distance between donors. Values of $\omega$ obtained for naphthalene-containing polymers by the two methods differ substantially, from $\omega = 100\text{-}300\ s^{-1}$ for the quenching experiments (10,11,16,17) to $\omega \simeq 10^5\text{-}10^6\ s^{-1}$ from molecular weight effects (10,15).

Different problems are encountered when quenching studies are carried out on fluid solutions at room temperature (2,18). Analysis of the quenching effect is usually made using the Stern-Volmer equation:

$$\phi_0/\phi = \tau_0/\tau = 1 + k_Q\tau_0[Q] \qquad (1)$$

where [Q] is the quencher concentration, $k_Q$ is a rate constant for bimolecular quenching, and $\phi_0$ and $\tau_0$ are the quantum yield and lifetime of the triplet state in the absence of quencher. The quenching phenomenon can be studied both by observing a reduction in the quantum yield $\phi$ of a photoreaction proceeding from the triplet state, or by direct determination of the triplet decay time by transient absorption spectroscopy. The problem begins with the interpretation by the rate constant $k_Q$, which contains contributions both from energy migration and diffusion. It is necessary to compare $k_Q$ with the value of $k_Q$ obtained for the same chromophore in a system where there can be no energy migration. Such a system can be a small molecule model compound containing only a single chromophore or a copolymer containing only a low mole fraction of chromophore-containing units. In the former case problems arise because the diffusion coefficient D of the polymer is smaller than that of the small molecule, so that the approximation is frequently made that D = 0 for the polymer. In the second approach there may be differences between homopolymer and copolymer in the accessibility of the chromophore to the quencher, depending on the relative size of adjacent monomer units.

Webber and co-workers concluded that the mobility of the triplet state in poly(2-vinyl naphthalene) during its μs excited state lifetime was negligible at room temperature (19). On the other hand Scaiano's group conducted a thorough study of quenching of polymers of substituted acrylophenones by small molecules (20), and observed that $k_Q$ for the homopolymer was always less than $k_Q$ for the dilute copolymer, a result which would also indicate the absence of significant triplet energy migration in fluid solution. However, because this group believed they had evidence from another type of experiment (see below) for very fast triplet migration ($\omega \simeq 10^{11}$ $s^{-1}$) in the homopolymers, they concluded instead that the effect of energy migration in polymers was to decrease the efficiency of quenching by small molecules. It is difficult to visualize how a random process such as energy migration could consistently carry the excitation away from a potential quencher, and in our opinion, the body of results compiled by Scaiano's group actually constitutes strong evidence against efficient triplet migration in poly(acrylophenone) and similar polymers at room temperature.

A further class of experiments to determine the role of triplet energy migration involves studies on polymers containing copolymerized quenchers. For example Scaiano and co-workers observed that the quantum yield of chain scission in poly(acrylophenone) decreased with irradiation time. The authors concluded that terminal acrylophenone chromophores created by the Norrish Type II reaction were acting as quenchers, and developed a model incorporating energy migration between chromophores on the chain to end groups. This analysis gave values of $\omega$ of up to $4x10^{11}$ $s^{-1}$ (21,22). However, the same group examined copolymers of phenyl vinyl ketone with o-tolyl vinyl ketone and found that at 10% of the 2nd monomer the quantum yield of chain scission was reduced by only 50% (23). Such an efficiency of quenching is inconsistent with a high hopping frequency. Other experiments on copolymerized quenchers have also failed to uncover evidence for significant triplet energy migration on the time scale of the excited state lifetime (24,25). It must be stressed that Scaiano's argument for efficient triplet energy migration also relies on a quenching experiment - and therefore should be subject to the same conclusion reached by that group in their studies of the quenching by small molecules, namely that energy migration reduces the efficiency of quenching rather than enhancing it.

The preceding analysis and review of the literature indicate a need for additional types of experiments to study triplet mobility in polymers. One experiment which has been particularly useful in studies of singlet energy migration in polymers, involves determination of the quenching of donor emission by a known mole fraction of a copolymerized luminescent quencher (26). We have extended this approach to the study of triplet states. The polymers chosen for study are homopolymers of isomeric acetonaphthyl methacrylates (aceto-NMA, **1**). A related monomer, 2,4-diaceto-1-naphthyl

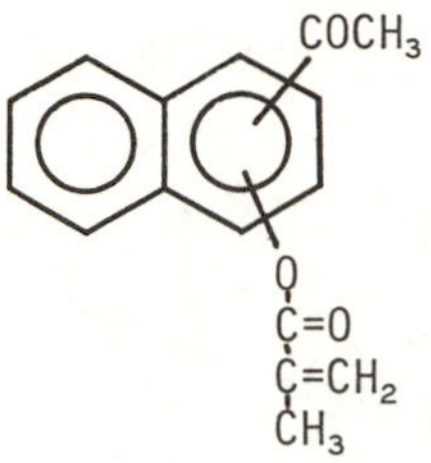

1 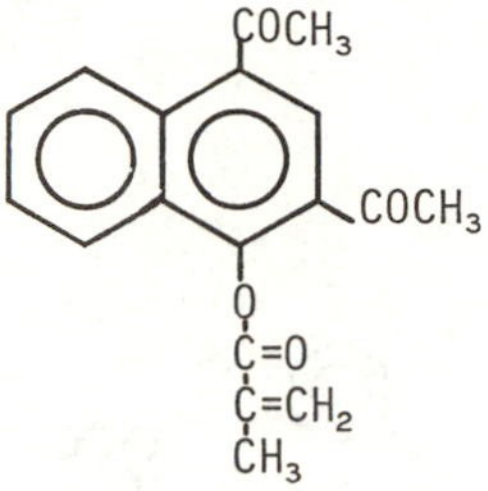 2

methacrylate **2** functions as a copolymerizable triplet quencher. These polymers were chosen for several reasons. The phosphorescence of the acetonaphthyl ester chromophore is intense at 77K, so that spectra of high quality are obtained. At the same time the quantum yield of intersystem crossing from $S_1$ to $T_1$ is 1.0, so that interference from singlet energy migration is not a problem. Finally, the triplet states have $\pi\pi^*$ character and correspondingly long lifetimes at 77K. As a result the effects of triplet energy migration are expected to be particularly evident.

## Experimental

Materials Full details of the preparation, purification and spectroscopic characterization of the monomers and polymers will be published elsewhere (Holden, D.A.; Safarzadeh-Amiri, A. Macromolecules, submitted). Four isomeric acetonaphthols and 2,4-diaceto-1-naphthol were synthesized by known reactions (27), as summarized in Scheme I. Purification of these compounds was achieved by a combination of recrystallization, distillation and sublimation, and took advantage of the large differences in boiling point and solubility between intramolecularly H-bonded isomers such as 1-aceto-2-naphthol and intermolecularly H-bonded isomers such as 4-aceto-1-naphthol. The methacrylate esters were synthesized by reaction with methacryloyl chloride in $CH_2Cl_2$ at 0-20°C in the presence of 1 equivalent of pyridine. The monomers are crystalline solids and their structures were verified by IR, NMR and elemental analysis.

Polymers were prepared by AIBN-initiated free-radical polymerization in degassed toluene at 60°C. With the exception of poly(4-aceto-1-naphthyl methacrylate), the polymers precipitated on formation. Polymers were purified by reprecipitation from $CHCl_3$ into methanol. Copolymer compositions were determined by UV spectroscopy using the extinction coefficients of the corresponding homopolymers and of 2,4-diaceto-1-naphthyl methacrylate monomer as references. Tables I and II list the properties of the homopolymers and copolymers used in the present study.

Spectroscopic Techniques Luminescence spectra were recorded on a Perkin-Elmer MPF-2A spectrofluorimeter with phosphorescence and low temperature accessories. Sample cells consisted of 3 mm o.d. tubes of low-fluorescent quartz connected to a 6 mm diameter stem, and all samples were deoxygenated with high purity nitrogen introduced through NMR septa via a pair of long stainless steel needles.

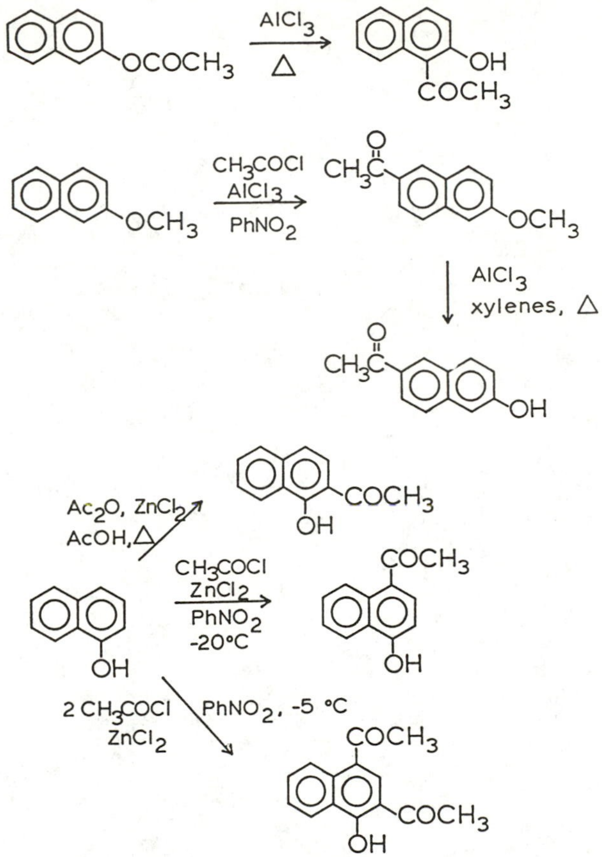

Scheme 1. Synthesis of isomeric acetonaphthols.

Table I. Homopolymers from isomeric acetonaphthyl methacrylates

| Monomer | % Conversion | $M_n(x10^{-4})$ |
|---|---|---|
| 1-aceto-2-NMA | 70 | 12.1 |
| 6-aceto-2-NMA | 15 | 11.3 |
| 2-aceto-1-NMA | 19 | 1.2 |
| 4-aceto-1-NMA | 15 | 4.3 |

Solutions were prepared having optical densities at the excitation wavelength of 0.1 to 0.2 in a 1 mm cell. 2-Methyltetrahydrofuran (2-MeTHF) and THF were purified by distillation from $LiAlH_4$ under nitrogen and spectra were recorded of polymers dissolved in 1:1 THF/2-MeTHF. This mixture gives a glass of the same high optical quality as 2-MeTHF. Relative quantum yields of phosphorescence were obtained by comparison of emission intensities of samples having the same optical densities at the excitation wavelength. Phosphorescence decays were recorded using a recording oscilloscope and camera and cutting out the excitation beam using an electromechanical shutter.

## Results and Discussion

The isomeric poly(acetonaphthyl methacrylates) are white powders soluble in THF, $CH_2Cl_2$ and $CHCl_3$. Figure 1 is the infrared spectrum of a film of poly(6-aceto-2-NMA) and shows clearly the ester and aromatic ketone carbonyl bands of equal intensity. The most remarkable feature of these polymers is the degree to which they retain static electricity. This property is much more marked for the poly(aceto-NMA) samples than for any poly(N-vinyl carbazole) sample prepared in our laboratory. It causes the polymer to disperse over the inside surfaces of glass sample vials and makes transfer of the dry powders very difficult. Figure 2 compares the UV absorption spectra of the four homopolyers. Strongly marked differences exist between the isomeric polymers depending on whether the dipole moments of the two substituents on the naphthalene ring reinforce or cancel each other.

Luminescence Behaviour The photophysics of poly(acetonaphthyl methacrylate) is dominated by the triplet state. The absence of measurable fluorescence means that the quantum yield for intersystem crossing from $S_1$ to $T_1$ is essentially unity. Poly(aceto-NMA) is strongly phosphorescent at low temperature, as illustrated for one isomer in Figure 3. The presence of the C=O substituent on the aromatic ring could also increase the rate of $T_1$ to $S_0$ internal conversion, with a corresponding decrease in the phosphorescence quantum yield and lifetime. This does not occur, however, as the decay curves of Figure 4 show. Enhancement of $S_1$-$T_1$ intersystem crossing without an increase in the rate of $T_1$-$S_0$ intersystem crossing is typical of $\pi\pi^*$ lowest triplets (28).

Studies of Triplet Antenna Effect When copolymers of 1-aceto-2-NMA with small amounts of 2,4-diaceto-1-NMA are excited at a wavelength where almost all of the light is absorbed by the first monomer the phosphorescence spectrum consists of contributions from both

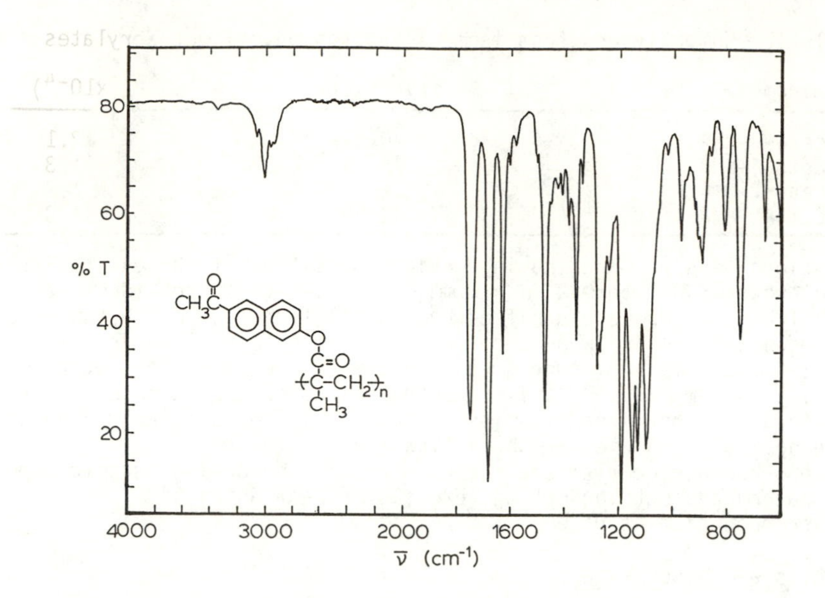


Figure 1. IR spectrum of a film of poly(6-aceto-2-naphthyl methacrylate).

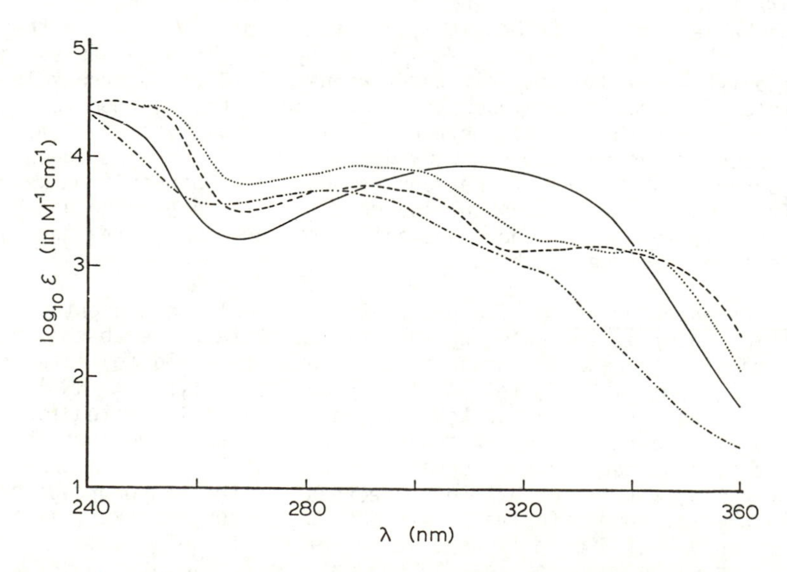


Figure 2. UV spectra in $CH_2Cl_2$ of isomeric poly(acetonaphthyl methacrylates). (-••-••) poly(1-aceto-2-NMA); (••••••) poly(6-aceto-2-NMA); (----) poly(2-aceto-1-NMA); (———) poly(4-aceto-1-NMA).

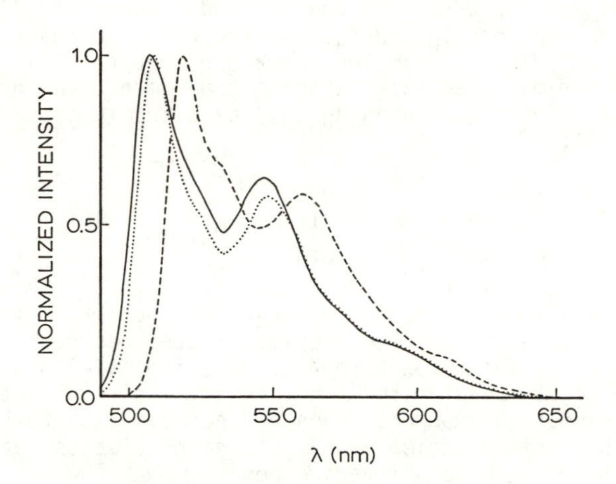


Figure 3. Phosphorescence spectra in 1:1 THF/2-MeTHF glasses at 77K. (——) 4-aceto-1-NMA; (•••••) poly(methyl methacrylate-co-1.5% 4-aceto-1-NMA; (----) poly(4-aceto-1-NMA).

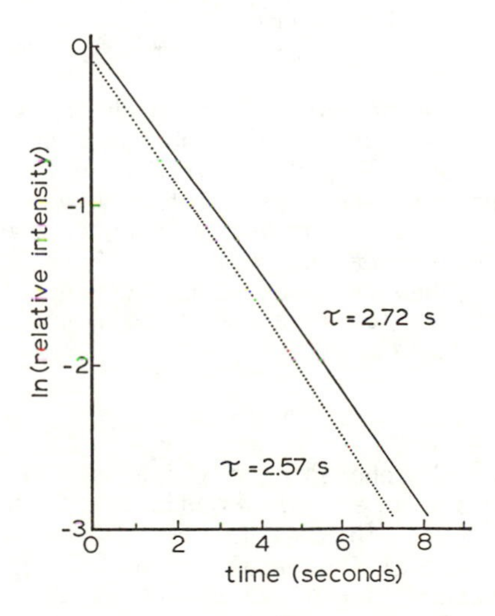


Figure 4. Phosphorescence decays in 1:1 THF/2-MeTHF glasses at 77K. (••••) poly(methyl methacrylate-co-0.51%-1-aceto-2-NMA); (——) poly(methyl methacrylate-co-5.6% 2-naphthyl methacrylate)

chromophores. 2,4-Diaceto-1-NMA thus functions as a highly efficient triplet quencher. Typical phosphorescence spectra are shown in Fig. 5. These spectra can be readily separated into donor and acceptor components $I_D$ and $I_A$. If the relative quantum yields $\phi_D/\phi_A$ of emission from the donor and acceptor chromophores are measured in a separate experiment, the quantum efficiency of triplet quenching $\phi_{DA}$ can be determined using equation 2 (26):

$$\frac{\phi_{DA}}{1 - \phi_{DA}} = \frac{I_A}{I_D} \cdot \frac{\phi_D}{\phi_A} \tag{2}$$

Table II lists values of $\phi_{DA}$ for a series of copolymers. At high mole fractions of quencher, the overlap of quenched regions on the polymer causes $\phi_{DA}$ to approach 100% asymptotically with increasing mole fraction f of 2,4-diaceto-1-NMA. What is significant is the limiting number of chromophores quenched per 2,4-diaceto-1-NMA unit as the quencher mole fraction approaches zero. Because of the difficulty of measuring copolymer compositions at low 2,4-diaceto-1-NMA content, we were only able to go to 1% as the lower limit. At this copolymer composition $\phi_{DA}$=0.40 and the average number of chromophores $\overline{n}$ quenched per 2,4-diaceto-1-NMA unit, given by equation 3, is 40 ± 4.

$$\overline{n} = \lim_{f \to 0} \frac{\phi_{DA}}{f} \tag{3}$$

This kind of quenching efficiency is too high for single-step transfer from donor to acceptor and establishes the existence of triplet energy migration. This value of $\overline{n}$ is obtained directly from experimental data and so represents a better way of expressing data for triplet energy migration than the hopping frequency $\omega$, a quantity derived using several assumptions. For comparison with the results of others, however, we can calculate $\omega$ from $\overline{n}$ assuming that triplet energy migration is a random walk process:

$$\omega = \overline{n}^2/\tau_0 \tag{4}$$

where $\tau_0$ is the donor phosphorescence lifetime with no quencher present. For $\overline{n}$ = 40 and $\tau_0$ = 0.51 seconds the calculated hopping frequency is $3 \times 10^3$ $s^{-1}$. This value lies between those determined for poly(vinyl naphthalene) from analysis of quenching by small molecules and from observation of molecular weight effects on the relative intensities of delayed fluorescence and phosphorescence (10, 13, 29). It is many orders of magnitude less than those determined by Scaiano's group from interpretation of chain-scission behaviour in poly(acrylophenones) in fluid solution (21,22). Energy migration does not have a high activation energy. In the case of poly(acrylophenones), the triplet lifetime at room temperature is shortened from its 77K value by several orders of magnitude because of the onset of efficient deactivation processes, including

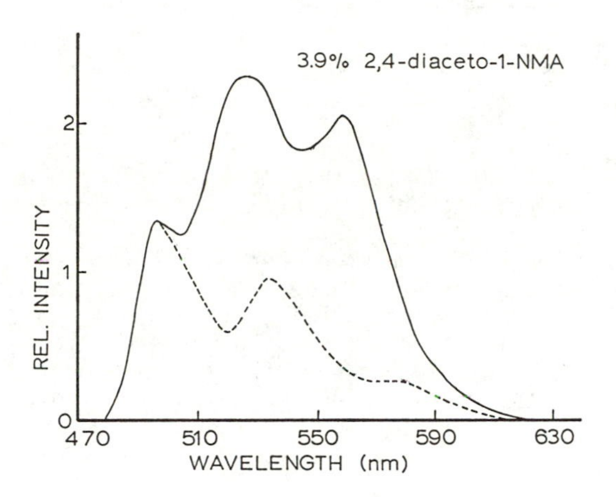


Figure 5. Phosphorescence spectra of polymers in 1:1 THF/2-MeTHF glasses at 77K. (----) poly(1-aceto-2-NMA); (——) poly(1-aceto-2-NMA-co-3.9% 1-aceto-2-NMA, $\lambda_{ex}$ 292 nm. The spectra are normalized to the same intensity at 496 nm.

Table II Quantum Yields $\phi_{DA}$ of Triplet Energy Transfer in Copolymers of 1-Aceto-2-NMA With 2,4-Diaceto-1-NMA Traps

| Sample | Mol % 2,4-diaceto-1-NMA in feed | Mol % 2,4-diaceto-1-NMA in copolymer | % conversion | $\overline{M}_n$ [a] | $\phi_{DA}$ [b] |
|---|---|---|---|---|---|
| 1 | 1.15 | 1.0 | 58 | 84000 | 0.40 |
| 2 | 2.36 | 1.8 | 60 | 109000 | 0.62 |
| 3 | 4.80 | 3.9 | 65 | 119000 | 0.85 |
| 4 | 7.21 | 6.6 | 58 | 62000 | 0.88 |
| 5 | 9.74 | 8.7 | 54 | 75000 | 0.91 |

[a] By membrane osmometry
[b] In 1:1 THF/2-MeTHF glass at 77K.

photochemical reactivity of the triplet state. One would therefore expect triplet energy migration to be an insignificant process in such polymers at room temperature.

In summary the use of polymers containing copolymerized luminescent quenchers has allowed determination of the mobility of the triplet state in individual chains of poly(1-aceto-2-naphthyl methacrylate). Triplet energy migration was shown to be a significant process in polymers when the lifetime of the migrating excited state is unshortened by photochemical reaction or quenching by oxygen.

## Acknowledgment

This research is supported by the Natural Sciences and Engineering Research Council of Canada.

## Literature Cited

1. Guillet, J. Polymer Photophysics and Photochemistry Cambridge University Press, Cambridge, UK, 1985, Chapter 9.
2. Webber, S.E. in Polymer Photophysics; D. Phillips, editor; Chapman and Hall, London, UK, 1985, Chapter 2.
3. Iinuma, F.; Mikawa, H.; Shirota, Y. Mol. Cryst. Liq. Cryst. 1981, 73, 309.
4. Porter, G.; Archer, M.D. Interdiscip. Sci. Rev. 1976, 1, 119.
5. Porter, G. Proc. Roy. Soc. (London), Ser. A 1978, 362, 281.
6. Holden, D.A.; Jordan, K.; Safarzadeh-Amiri, A. Macromolecules 1986, 19, 895.
7. Klöpffer, W. Spectrosc. Lett. 1978, 11, 863.
8. Kim, N.; Webber, S.E. Macromolecules 1980, 13, 1233.
9. Burkhart, R.D.; Avilés, R.G.; Magrini, K. Macromolecules 1981, 14, 91.
10. Pasch, N.F.; McKenzie, R.E.; Webber, S.E. Macromolecules 1978, 11, 733.
11. Webber, S.E.; Avots-Avotins, P.E. J. Chem. Phys. 1980, 72, 3773.

12. Yokoyama, M.; Tamamura, T.; Nakano, T.; Mikawa, H. J. Chem. Phys. 1976, 65, 272.
13. Pasch, N.F.; Webber, S.E. Chem. Phys. 1976, 16, 361.
14. Klöpffer, W.; Fischer, D.; Naundorf, G. Macromolecules 1977, 10, 450.
15. Pasch, N.F.; Webber, S.E. Macromolecules 1978, 11, 727.
16. David, C.; Demarteau, W.; Geuskens, G. Eur. Polym. J. 1970, 6, 537.
17. Cozzens, R.F.; Fox, R.B. J. Chem. Phys. 1969, 50, 1532.
18. Holden, D.A.; Ng, D. S-K.; Guillet, J.E. Br. Polym. J. 1982, 14, 159.
19. Pratte, J.F.; Noyes, W.A., Jr.; Webber, S.E. Polym. Photochem. 1981, 1, 3.
20. Scaiano, J.C.; Lissi, E.A.; Stewart, L.A. J. Am. Chem. Soc. 1984, 106, 1539.
21. Encinas, M.V.; Funabashi, K.; Scaiano, J.C. Macromolecules 1979, 12, 1167.
22. Scaiano, J.C.; Selwyn, J.C. Macromolecules 1981, 14, 1723.
23. Bays, J.P.; Encinas, M.V.; Scaiano, J.C. Macromolecules 1979, 12, 348.
24. Sanchez, G.; Knoesel, R.; Weill, G. Eur. Polym. J. 1978, 14, 485.
25. Leigh, W.J.; Scaiano, J.C.; Paraskevopoulos, C.I.; Charette, G.M.; Sugamori, S.E. Macromolecules 1985, 18, 2148.
26. Holden, D.A.; Guillet, J.E. Macromolecules 1980, 13, 289.
27. Merle-Aubry, L.; Holden, D.A.; Merle, Y.; Guillet, J.E. Macromolecules 1980, 13, 1138.
28. Birks, J.B. Photophysics of Aromatic Molecules, Wiley-Interscience, New York, NY, 1970.
29. David, C.; Lempereur, M.; Geuskens, G. Eur. Polym. J. 1972, 8, 417.

RECEIVED March 13, 1987

# Chapter 21

# Aromatic Polymers in Dilute Solid Solutions

## Electronic Energy Transfer and Trapping

Walter Klöpffer

Battelle-Institute e.V., Am Romerhof 35, D-6000 Frankfurt am Main, West Germany

Energy transfer on isolated macromolecules in rigid environment is charaterized by finite volume (length), one dimensional exciton hopping, exciton trapping at monomeric trapping sites of unknown structure and the absence of (sandwich) excimer emissions. The methods used to elucidate the mechanism of energy transfer comprise UV-absorption, fluorescence and phosphorescence spectra and quenching, delayed fluorescence caused by triplet-triplet annihilation and dynamic measurements.

Energy transfer in polymers has been studied in the pure solid state, in heterogeneous systems (e.g. polymer blends), in liquid solutions and in solid solutions. The last case, which will be considered here, provides relatively simple and clear experimental conditions since interactions between the macromolecules can be excluded by dilution and molecular movement is severely restricted by low temperature and rigid environment. Thus, excitonic energy transfer can be studied without competing molecular movement. The luminescence of dilute, solid solutions of aromatic polymers is not dominated by excimers - in sharp contrast to the other modes of observation - so that side group fluorescence and phosphorescence can be observed. This does not mean, however, that exciton trapping processes are absent in these systems.

If the exciton motion is considered to be due to random hopping between the aromatic side groups along the chain, this movement should be quasi-one-dimensional (1). The term "exciton", if used for energy migration in amorphous polymer systems, always refers to the exciton hopping model as opposed to the exciton as a crystal state (exciton band model). The disadvantage of lower order in these systems compared to molecular crystals is counterbalanced by a higher degree of "isolation" in dilute solutions of macromolecules, jumps between different chains being very unlikely. Another unique fea-

0097-6156/87/0358-0264$06.50/0

ture, which links the experiments discussed here with photosynthesis (2) and recent theories (3) is the finite volume of polymer coils or finite length of the polymer chains. This property causes the most unusual effect of the molar mass dependence of triplet-triplet annihilation (4-7).

There are, however, numerous other less showy observations which give insight into energy transfer of aromatic macromolecules. Some of these will be discussed in the following three sections, followed by a brief account on triplet-triplet annihilation.

## Ultraviolet Absorption

UV-Absorption spectroscopy would deserve more attention in polymer research than it actually receives (8). The absorption bands of aromatic groups in polymers have nearly the same spectral position as the corresponding bands of monomeric models, provided that the backbone is not of the $\pi$-conjugated type. From that and from the absence of any additional absorption bands in pure polymers it is concluded that the ground-state interaction between the aromatic groups is small. The polymer bands are broader than the corresponding monomer bands due to the fact that spectroscopically there is no truly dilute polymer solution (8), the side groups being tightly connected with the backbone and - at least in case of homo polymers - in close neighborhood with their fellows.

This broadening of the bands partly compensates the obvious decrease of the molar absorption coefficient ($\varepsilon$) at the band maxima. It is not true, however, that the compensation is complete, as was roughly estimated in an earlier paper (9). Perfect compensation would mean that the total absorption strength of a particular transition ($n \rightarrow m$), as measured by the oscillator strength (f) and related to the integrated absorption band (Equation 1), is the same in the polymer and in a suitable monomeric model.

$$f_{nm} = 4.319 \times 10^{-9} \int \varepsilon (\nu')d\nu' \qquad (1)$$

$\varepsilon$ is given in $l\ mol^{-1}\ cm^{-1}$, $\nu'$ in $cm^{-1}$

Recent quantitative work by J. Kowal (10) involving careful band separation showed that the hypochromy, i.e. a true decrease of $f_{nm}$, in Poly(N-vinylcarbazole) (PVCA) may reach the order of the strong hypochromy observed in solutions of helical DNA (11). In THF solution, e.g., the oscillator strength of the first two transitions ($^1A \rightarrow {}^1L_b$, 29,000 $cm^{-1}$ and $^1A \rightarrow {}^1L_a$, 34,000 $cm^{-1}$) amounts to f = 0.050 and 0.14 for N-Ethylcarbazole, but only f = 0.037 and 0.085 for PVCA (10).

There is no significant difference between radically and cationically prepared PVCA samples although the tacticity is reported to be different (12) and other spectral properties differ drastically (13-15). The results obtained on PVCA confirm earlier work by Okamoto et al. (16) reporting hypochromy data on several aromatic polymers in liquid solution obtained by graphical band separation.

UV-absorption studies on aromatic polymers at low temperature solid glasses seem to be scarce. In Figure 1, the low temperature spectra of PVCA and its monomeric model N-Methyl-carbazole in MTHF glass at 77 K are compared (17). The monomer clearly shows better resolution due to sharper absorption bands in the two lowest transitions but again the hypochromic effect in the polymer is evident.

In order to explain the hypochromy in macromolecules, helical regions have been postulated theoretically (18) and evidenced in the case of DNA (11) whose absorption strength returns to a normal value in helix breaking solvents. A small hypochromic effect has indeed been measured in isotactic relative to atactic Polystyrene by Vala and Rice (18). In the case of PVCA, however, neither PVCA rad nor PVCA cat is purely syndiotactic or isotactic, respectively (12) and no significant difference between PVCA samples of different origin has been observed by Kowal (10) and Okamoto et al. (16). Alternatively, therefore, the tight packing of bulky aromatic side groups in wormlike chains, even without perfect ordering, should be considered as a possible cause of the large hypochromic effects observed.

## Fluorescence and Singlet Energy Transfer

In contrast to film- and liquid solution fluorescence, which in most cases is excimeric (9,19-22), solid solutions are mostly of the monomeric type (here, as throughout this article, "monomer" does not designate the chemical monomer but rather a suitable model substance of the basic unit).

Figure 2 shows typical solid solution spectra of cationically prepared PVCA and 1,3-Biscarbazolylpropane (BCP). The main difference of this polymer spectrum relative to the low molar mass model is a small red shift which is also observed in other aromatic polymers (Table I), with exception of the more open-structured Poly(2-naphthylmethylacrylate). In the case of P2VN no monomer data for exactly the same solvent and temperature could be found in literature, the blue shift indicated may therefore be fortuitous. The reason for the red shift, which is not observed in absorption (Figure 1), may either be a larger Stokes shift of all side groups, due to the different electronic environment felt by the side groups, or the fluorescence may originate from a few side groups acting as traps which are populated by energy transfer. Again, the traps may have a discrete structure or represent low energy tail states observed in dispersive excitation transport (30-35) which is likely in all amorphous molecular systems.

A distinction between the different possibilities can be achieved by time-resolved fluorescence and quenching experiments which have not yet been performed for aromatic polymers in solid solution. A more detailed discussion on exciton trapping in pseudo one-dimensional systems is presented in the next section, since much more information is available about phosphorescence quenching and triplet excitons than about fluorescence.

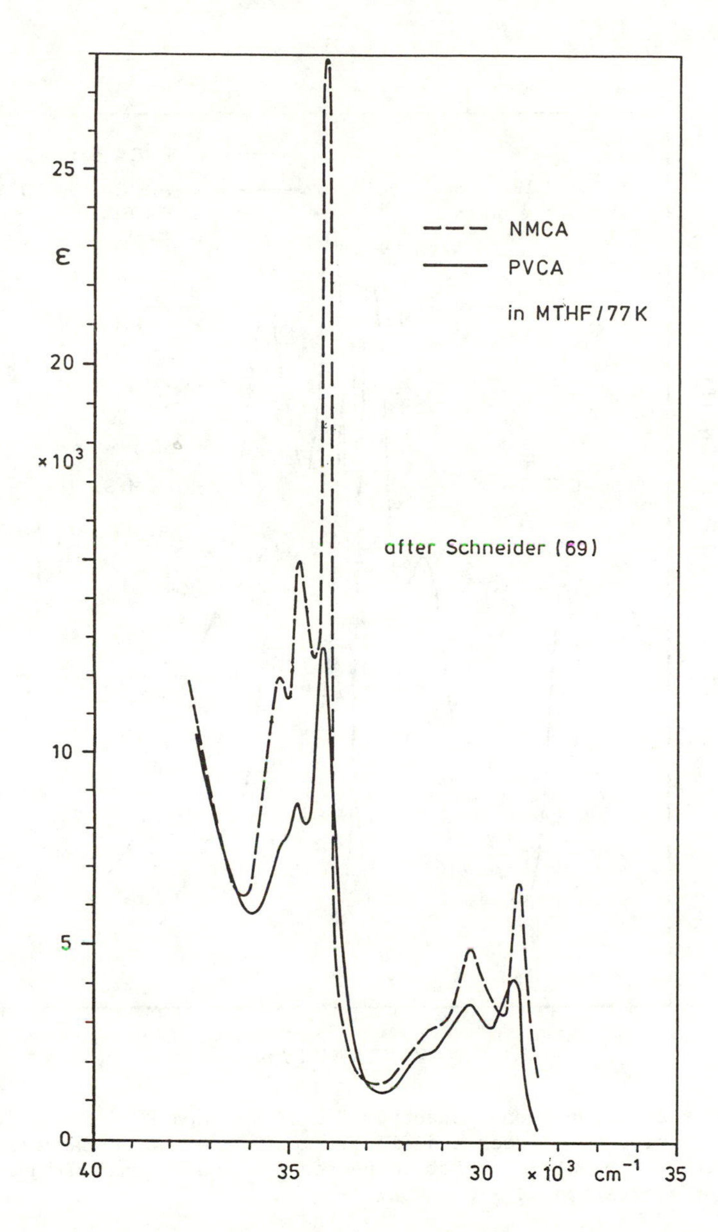


Figure 1. UV-absorption spectra of PVCA and monomeric model N-methylcarbazole; replotted after Schneider (17)

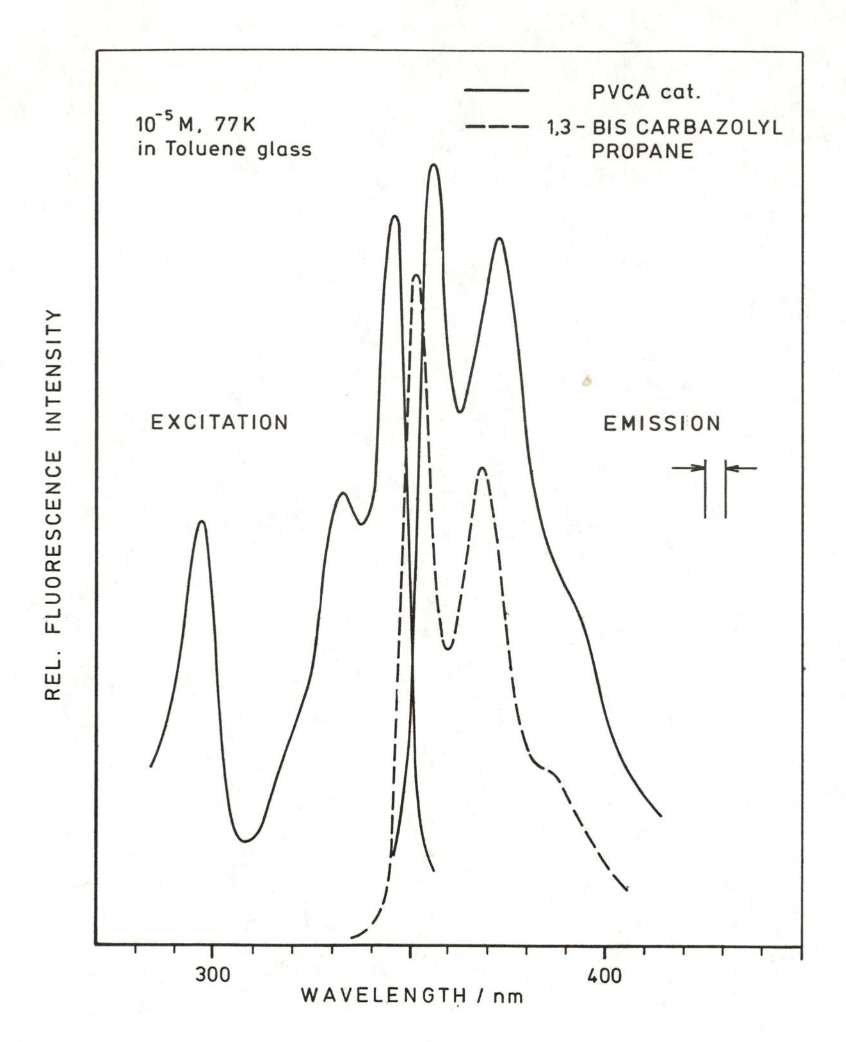


Figure 2. Fluorescence spectra of PVCA cat and PCP. The emission intensities of polymer and dimeric model are not comparable; the excitation spectrum is not corrected for equal quantum intensity; excitation at 330 ± 4 nm

Table I: Polymer Redshift Observed in Fluorescence

| Polymer | $S_1/cm^{-1}$ (a) | Lit. |
|---|---|---|
| Poly(1-vinylnaphthalene) (P1VN) | 200 (b) | (19) |
| Poly(2-vinylnaphthalene) (P2VN) | blue shift? (c) | (23-25) |
| Poly(acenaphthylene) (PACN) | 560 | (26) |
| Poly(N-vinylcarbazole) (PVCA) | 300-360 (d) | (14,27,28) |
| Poly(2-naphthylmethyl-methacrylate) | 0 | (29) |

(a) Wavenumber of $S_1 \rightarrow S_0$ of monomeric model compound minus $S_1 \rightarrow S_0$ of polymer, measured at 0-0 or 0-1 etc. vibronic peak of the fluorescence spectra; the fluorescence spectra of polymer and model have been measured in the same solvent at 77 K if not indicated otherwise.
(b) The spectra of monomeric model and polymer have slightly different shapes.
(c) 2 EtN at -80 °C, THF: 0-3 peak at 29690 $cm^{-1}$; P2VN at 77 K, MTHF and THF/Diethylether: 0-3 peak at 30.000 $cm^{-1}$
(d) The higher value has been measured by Itaya et al. (14) relative to oligomers of PVCA.

An indication for energy transfer along the chain is obtained from radically prepared PVCA (Figure 3) whose fluorescence spectrum is a superposition of monomer type emission (the shoulder at 28,000 $cm^{-1}$ $\hat{=}$ 357 nm corresponds to the 0-0 peak at 356 nm in Figure 2) and a new band at 26,900 $cm^{-1}$ $\hat{=}$ 372 nm which exactly coincides with the 0-1 vibrational peak in the cationic PVCA.

There can be little doubt that this new band, which is always more prominent in the radically prepared, predominantly syndiotactic PVCA samples (12-14,28) is due to the "second excimer", "high energy excimer" or "Trap II", which first was detected in solid PVCA at low temperature (4,15,36). The correctness of this interpretation has recently been conformed in a study of the meso- and rac-2,4-di-N-carbazolyl-pentanes as models of isotactic and syndiotactic sequences of PVCA (37). It is now generally accepted that syndiotactic sequences of PVCA tend to form a partial overlap excimer with low binding energy and, hence, high frequency fluorescence. This model of Trap II has been proposed by Itaya et al. for the first time (14). The complementary preference of isotactic sequences to form the "true excimer", "low energy excimer" or "full overlap (sandwich)

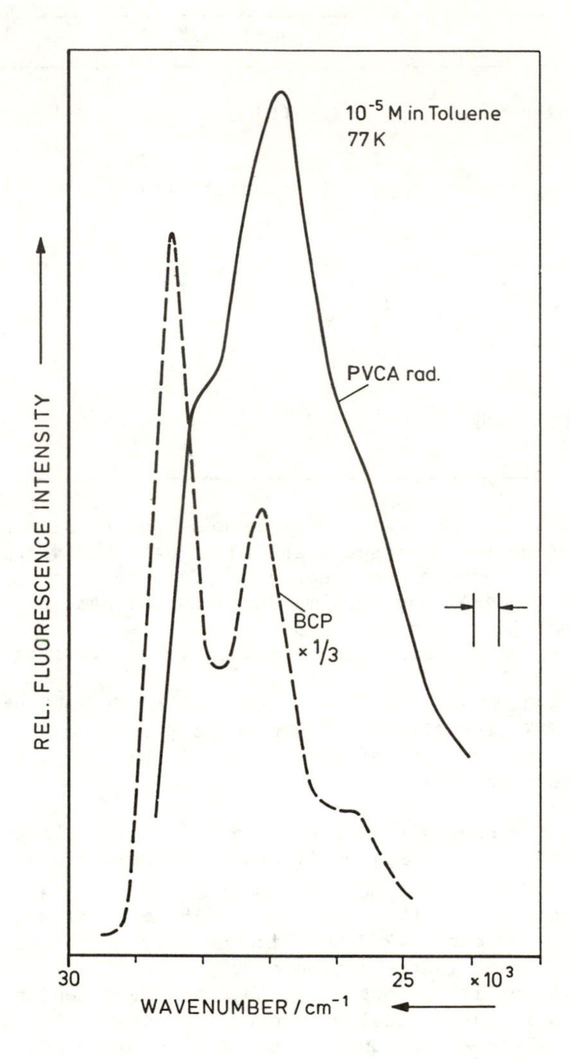


Figure 3. Fluorescence spectra of PVCA rad and BCP; emission slit width indicated; free of reabsorption, corrected for equal quantum intensity

excimer" is also evident from the studies of stereoregular models (37). It was first demonstrated for models of polystyrene by Bokobza et al. (38).

This excimer, which is easily formed in most aromatic polymers, is not observed in glassy solutions. Why, then, can the high energy excimer be observed in solid solutions of PVCA rad? G.E. Johnson (40) pointed out that the excimer-forming site corresponding to this excimer - now known to involve the overlap of only one aromatic ring (14) - needs only a small activation energy, if any. They are present, therefore, even at low temperature and frozen in during solidification of the solvent. It should furthermore be noted that this process is not completely absent in cationically prepared PVCA in qualitative agreement with its only partial isotactic character. In some PVCA cat samples, the 0-1 vibronic peak appears to be much stronger than shown in Figure 2, indicating some Trap II emission (28).

The absence of low energy excimer in solid solutions of PVCA and other aromatic polymers can easily be explained by the need of thermal activation to form a sandwich-pair (excimer-forming site) of two neighboring bulky aromatic groups. This has been shown by Frank for P2VN and Poly(4-vinylbiphenyl) (21,39).

It can furthermore be concluded that during cooling (in most cases just putting the sample tube into liquid nitrogen) the macromolecules have enough time to disperse the (low energy-)excimer-forming sites which must be present at room temperature as can be concluded from the films formed at this or higher temperature (9, 21, 39).

Since it is not likely that (nearly) each sidegroup forms a "Trap II" even in radically prepared PVCA, energy transfer has to be involved from the randomly excited groups to the high energy-excimer-forming sites. Since Förster-type long range transfer can safely be excluded if the acceptor is an excimer-forming site (9), only exciton hopping can explain this energy transfer. This has independently been shown by fluorescence depolarization of solid solutions of PVCA by Schneider (17).

## Phosphorescence and Triplet Energy Transfer

Phosphorescence, the spin forbidden, radiative $T_1 \rightarrow S_o$ transition can easily be observed in most solid solutions of aromatic polymers. For reasons to be discussed in the next section, low excitation intensity and low molar mass of the polymers favor phosphorescence relative to delayed fluorescence. Excitation is done in most cases via the singlet system, but especially naphthalene based polymers can be sensitized conveniently due to the large $S_1 - T_1$ gap (11,000 $cm^{-1}$) of this chromophore (1). If $S_o \rightarrow S_n$ absorption is used for creating the triplets, intersystem crossing may occur from the singlet exciton, as it moves along the chain or from the trapped singlet states discussed in the preceeding section.

There can be no doubt that mobile triplets do exist in solid solutions of aromatic polymers, but this does not mean that the phosphorescences observed are necessarily emitted from (hopping) excitons (41).

A typical phosphorescence spectrum of atactic PS has been published by Vala, Haebig and Rice (19) together with the monomeric model ethylbenzene. The two spectra do not show substantial differences. PVCA is compared with N-isopropyl-carbazole and BCP in Figure 4. The loss of vibronic resolution, due to inhomogeneous broadening of the polymer phosphorescence and a bathochromic shift of a few hundred $cm^{-1}$ relative to monomeric or oligomeric models is characteristic for all polymers investigated (Table II). Although the shifts observed are smaller than typical binding energies of excimers (which again are not observed) they considerably surpass the thermal energy kT of 53.5 $cm^{-1}$ at 77 K.

Table II: Polymer Red Shift Observed in Phosphorescence

| Polymer | $T_1/cm^{-1}$ (a) | Lit. |
|---|---|---|
| Poly(styrene) (PS) | 250 (b) | (19) |
| Poly(1-vinylnaphthalene) (P1VN) | 410 | (42) |
| Poly(2-vinylnaphthalene) (P2VN) | 480 (c) | (6,43) |
| Poly(N-vinylcarbazole) (PVCA) | 410-510 | (4,28) |
| Poly(4-vinylbenzophenon) (PVB) | 370 (d) | (44) |
| Polyriboadenylic acid | 500 (e) | (45) |

(a) Wavenumber of $T_1 \rightarrow S_0$ of monomeric model compound minus $T_1 \rightarrow S_0$ of polymer, measured at 0-0 or 0-1 etc. vibronic peak of the phosphorescence spectra; the phosphorescence spectra of polymer and model have been measured in the same solvent at 77 K if not indicated otherwise
(b) Spectra of monomeric model and PS have slightly different resolution; the value given is an average of the differences of 3 corresponding peaks
(c) Nishijima's polymer was a copolymer of 2VN (60 %) with styrene (43)
(d) Copolymer of 4-vinylbenzophenone (70 mol%) with styrene measured against a VB/S copolymer with only 0.7 % VB
(e) A helical oligomer showed a red shift of 350 $cm^{-1}$ vs. monomeric model

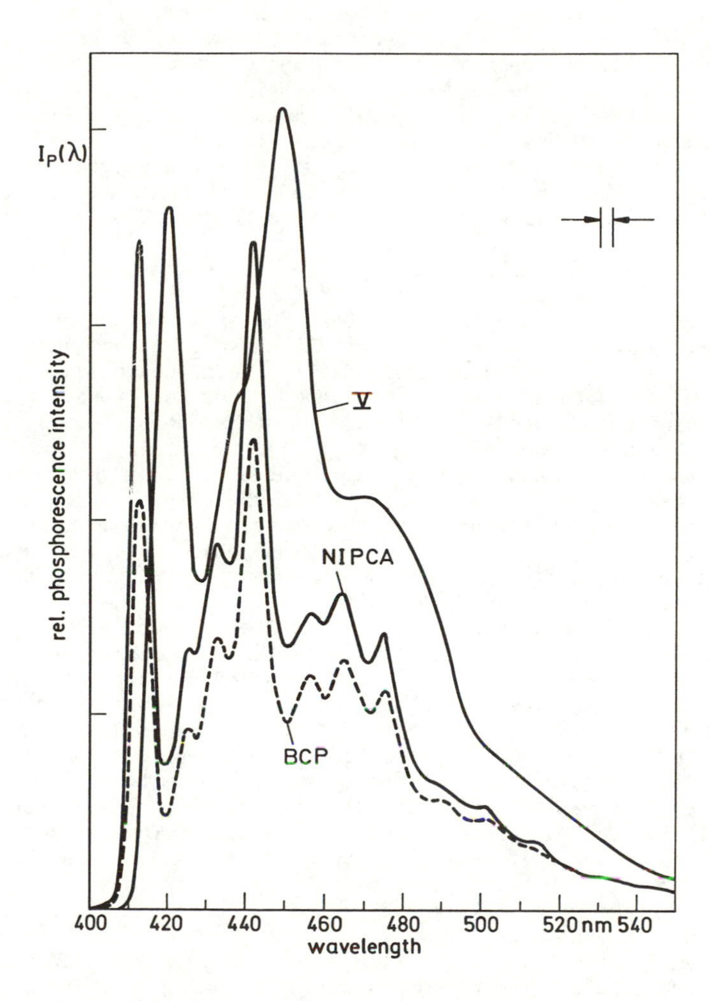


Figure 4. Phosphorescence spectra of radically prepared PVCA(V), isopropylcarbazole (NIPCA) and 1,3-bis-N-carbazolylpropane in MTHF glass at 77 K; the intensities are not comparable; emission slit width indicated. Reproduced with permission from Ref. /10/. Copyright 1986 Hüthig & Wepf Verlag, Basel, Heidelberg, New York

One method of gaining insight into the mechanism of energy transfer is quenching, using the luminescence (in this case phosphorescence) as a probe. The quenching molecules are chosen according to their energy levels, solubilities, emission characteristics, etc. Two types of experiments have been performed on several aromatic polymers, which will be treated here in an analogous way:

- Sensitized phosphorescence of copolymers consisting of suitable host/quencher pairs, the quenching molecules being the minority components and have the lower triplet levels
- Ermolaev-type quenching experiments (46) where suitable quenchers (phosphorescent as well as non phosphorescent ones) are added in varying concentrations to the solution and frozen in.

In both methods, the polymer or copolymer to be studied is in dilute, solid solution, typically $10^{-4}$ to $10^{-3}$ mol base unit $l^{-1}$. The first method has the advantage that the ratio mol quencher/mol basic unit of polymer is reasonably well known from copolymer composition. The copolymer should be of the random (not block) type and the molar mass should be known in order to know the average number of quenching groups per macromolecule (if this number is $\lesssim 1$, "ordinary" kinetics are suspected to fail).

In the second method, the volume concentration is known (where in most cases the remarkable shrinking of the sample volume upon cooling is neglected) rather than the above mole ratio. The problem is that we do not know exactly how many quenching molecules are near enough to one of the polymer chains in order to efficiently trap the excitons moving along the chain.

In the usual evaluation procedure a Stern-Volmer plot ($I_0/I$ or "quenching factor" $Q = (I_0-I)/I$ vs. molar concentration of quencher) is interpreted according to a theory of Voltz et al. (47) developed for quenching of singlets in solution by diffusion (both molecular and excitonic) plus Förster's dipol transfer. In the following, the one dimensional character of exciton diffusion is emphasized using the $^1$D-random walk model by H.B. Rosenstock (48). This model assumes a majority of equal lattice points, a small fraction being replaced by "absorbing points" (quenchers). The exciton moves around in a random next neighbor hopping process and has a spontaneous "emission probability" per unit time. This emission probability is in the real world actually the sum of radiative and non-radiative rate constants.

In an $^1$D-random walk, the high probability of multiple visits to individual lattice points makes energy transfer less effective compared to the $^3$D-case where the probability of return to the starting point is included by F = 0.34 in Equation (2)

$$Q = (1-F) \cdot n \cdot c \qquad (2)$$

describing the dependence of quenching factor Q on n, the number of exciton jumps per lifetime and c, the ratio mol quencher/mol host (49-51). Equation (2) expresses the linear dependence of Q on c which is characteristic for Stern-Volmer plots (for small c, the mole ratio is porportional to the molar concentration used in SV plots). The main result of Rosenstock's work (48) is shown graphically in Figures 5 and 6. Figure 5 is a straightforward transcription

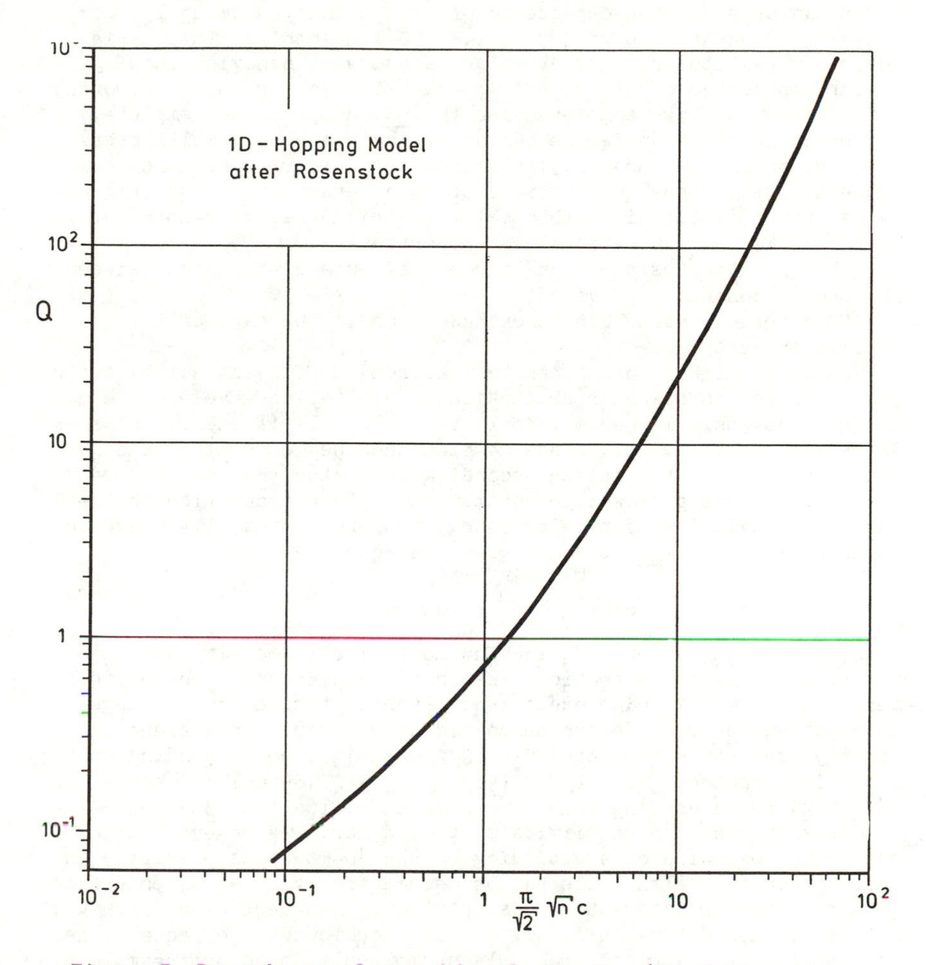


Figure 5. Dependence of quenching factor $Q = (I_0-I)/I$ on the number of jumps during exciton lifetime (n) and ratio mol quencher/mol basic unit (c) in a $^1$D-random walk (48)

of Rosenstock's Figure 4 (48) into the notation more familiar in luminescence quenching. This diagram allows the calculation of n if Q has been measured at least for one mole ratio c. Before using this evaluation one has to be sure that the system behaves as a one dimensional one. In Figure 6, Q(c) is calculated for three values of n. As can be seen, the dependence of Q on c changes smoothly from linear to quadratic above $Q \approx 1$, i.e. 50 % quenching. This is in sharp contrast to ordinary SV or 3D-random walk behavior showing a linear dependence of Q(c). It is in qualitative agreement, however, with Förster's long range transfer (51-53) despite the completely different energy transfer mechanisms (single step vs. multi step) involved in the two models. The reason for this unexpected coincidence may be the great importance of quenchers near to the originally excited molecules in both models, favoring energy transfer at high quencher concentrations overproportionally.

The Q(c) curves shown in Figure 6 allow a distinction between 3D- and 1D-behavior, provided

- Förster's transfer can be excluded (as in the case of T-T-transfer)
- Q-values higher than 1 can be measured (sufficient solubility of quencher, no quencher absorption at excitation wavelength etc.)

In practice, this method will only be applicable if energy transfer is relatively efficient, since at high quencher concentrations single step triplet transfer according to Dexter (84) and Ermolaev (46) will always be possible and may simulate a transition to the quadratic region. This type of energy transfer is usually evaluated using F. Perrin's (54) formula given in Equation (3).

$$I/I_o = \exp.\ (-C/C_o) \tag{3}$$

C is the concentration of quenching molecules in mol $l^{-1}$ and $C_o$ is the critical concentration. The critical concentration for this mechanism is mostly high due to short range of electron exchange interaction (84). For benzophenone (energy donor)/naphthalene (energy acceptor = quencher) $C_o = 0.20$ mol $l^{-1}$, corresponding to a critical radius $r_o = 1.27$ nm (46) or $C_o = 0.186$ mol $l^{-1}$ and $r_o = 1.29$ nm, according to a more refined evaluation (55).

Keeping in mind the limits mentioned above, a re-evaluation of published quenching data according to the 1D-model is possible. In order to calculate the mole ratios needed from volume concentrations [Qu] reported in literature, a simple model has been used which will not be discussed here in detail. It uses cubic cells of equal size, each cell containing either a solvent molecule, a quenching molecule or a basic unit of the polymer. It is furthermore assumed that 4 places around each basic unit are available to quenching molecules to which there is no specific interaction - neither attractive nor repulsive. Under these assumptions we obtain Equation (4):

$$c = 0.4[Qu] \tag{4}$$

c is the ratio mol (effective) quencher/mol basic unit and [Qu] is the volume concentration of the quencher in mol $l^{-1}$.

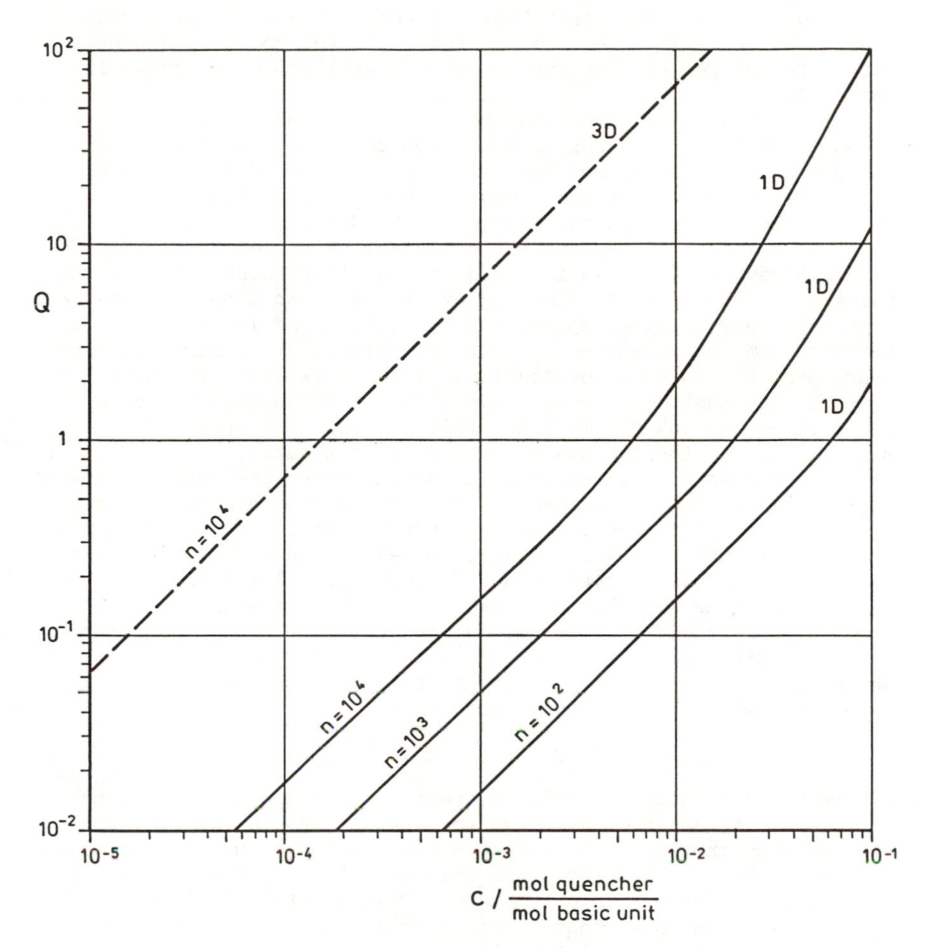


Figure 6. $^1$D-quenching curves calculated from Figure 5 assuming different numbers of jumps during exciton lifetime (n), as indicated. For comparison, one $^3$D-quenching curve after Equation (2) is shown for the highest value of n

The product of c x P, where P = M (Pol)/M (basic unit) is the average degree of polymerization, gives the average number of quenching molecules per macromolecule. In case of copolymers with a quenching minority component ($c \ll 1$), c is molar composition in mol quenching/mol main component. The influence of end groups is considered to be negligible in all cases, i.e. the model is not well suited for oligomers. The results of the evaluation are shown in Table III.

The "upward" curvature in the SV plot (in our case the equivalent plot Q vs. c) which is also "not uncommon" in liquid solution and probably an indication for down chain exciton diffusion according to Webber (62), has been found in Nos. 5-8, 10,11 and 13. Only in No. 12,Q(c) is definitely linear up to Q=6 (74).

The most convincing quadratic dependence and an indication for linear behaviour at low c is found in the first quantitative work on polymer phosphorescence quenching by Eisinger and Schulman (61), see Figure 7. They followed up reports by Behrson and Isenberg (63) on the quenching of phosphorescence of nucleic acids by paramagnetic ions, what was taken as evidence for triplet excitons by these authors. The number of exciton jumps, n, is calculated to be in the order of $10^4$ to $10^5$ for P2VN and PVCA and about an order of magnitude lower for the other polymers investigated.

There remain, however, some questions about the real meaning of these data and the correctness of the evaluation procedure. First of all, the data presented in Table II suggest efficient trapping of the triplet excitons by intrinsic, not excimeric traps of unknown structure. This is corroborated by reports of all authors who performed quenching and phosphoresence decay measurements that the phosphorescence lifetime is not shortened by adding the quencher. This is a clear proof that it is primarily not the free triplet which is observed in phosphorescence but rather a trapped species, since otherwise Equation (5) should apply.

$$Q = (I_0-I)/I = (\tau_0-\tau)/\tau \quad (5)$$

A more discriminating picture of the role of traps in triplet exciton mobility can be drawn from the work by Webber (64, 65) who studied the kinetics of triplet quenching using a quencher molecule with a short lived triplet state (biacetyl) as a probe of the triplet exciton lifetime. In a study on P2VN (64), PACN (64) and PVCA (65) it has been shown that trapping is more efficient in PVCA compared to P2VN where triplet excitons exist long after the decay of delayed fluorescence (next section).

PACN has been interpreted as slow exciton macromolecule in accordance with this work (No. 4 in Table III) and with the microstructure which prevents any close overlap of nearby naphthalene groups needed for efficient electron exchange interaction.

Coming back to the interpretation of the results presented in Table III we have to admit that a satisfactory theory including the role of traps does not yet exist. The following considerations, however, do show nevertheless that the exercise was useful:

1) In the region of high mol ratios of quencher to basic unit, the intrinsic traps, being in the minority, should play a minor role. The $^1$D-random walk theory should therefore hold in the quadratic part of the Q(c) plot.

Table III: Evaluation of Phosphorescence Quenching According to the 1D-Hopping Model

| No. | Polymer | Quencher | c(Q=1) | CxP | n | Lit. |
|---|---|---|---|---|---|---|
| 1 | PS | Piperylene | $2x10^{-4}$ (Q = 0.6) (a) | ? | | (56) |
| 2 | P1VN | -"- | | 1 | | (1) |
| 3 | P2VN | -"- | $2.6x10^{-3}$ | 8-32 | $5.5x10^{4}$ | (57) |
| 4 | PACN | -"- | $8.9x10^{-3}$ | 5(c) | $4.7x10^{3}$ | (58) |
| 5 | PVCA | Naphthalene | $3.1x10^{-3}$ | ? | $3.8x10^{4}$ | (59) |
| 6 | PVCA-co-alt. fumaronitril | -"- | | 1 | | (59) |
| 7 | PVCA (rad.) | -"- | $1.8x10^{-3}$ | 1-23 | $6.5-11x10^{4}$ | (13) |
| 8 | PVCA (cat.) | Piperylene | $3.4x10^{-3}$ | 1-10 | $3.2x10^{4}$ | (13) |
| 9 | Poly(acrylo-phenone) (isotactic) | Naphthalene | $10^{-2}$ | 12-45 | 3700 | (60) |
| 10 | Poly(acrylo phenone) (atactic) | -"- | $1.3x10^{-2}$ | 3-39 | 2200 | (60) |
| 11 | Poly(vinyl-benzophenone-co-styrene)(77% VB) | -"- | $1.5x10^{-2}$ | ? | 1600 | (44) |
| 12 | Poly(naphthyl)-methacrylate) | Piperylene | $5.4x10^{-3}$ | ? | (d) | (74) |
| 13 | Poly(adenylic acid) | $Mn^{++}$ $Co^{++}$ $Ni^{++}$ | $5.6x10^{-3}$ | ? | $1.35x10^{4}$ (e) | (62) |

(a) Highest Q measured
(b) 3 experimental point
(c) At Q=1; linear correlation up to Q=2
(d) No transition to quadratic region up to Q=6, [Qu] = 0.08 mol $l^{-1}$
(e) Average from
Q=1 n=11800; Q=10 n=13200; Q=100 n=15700

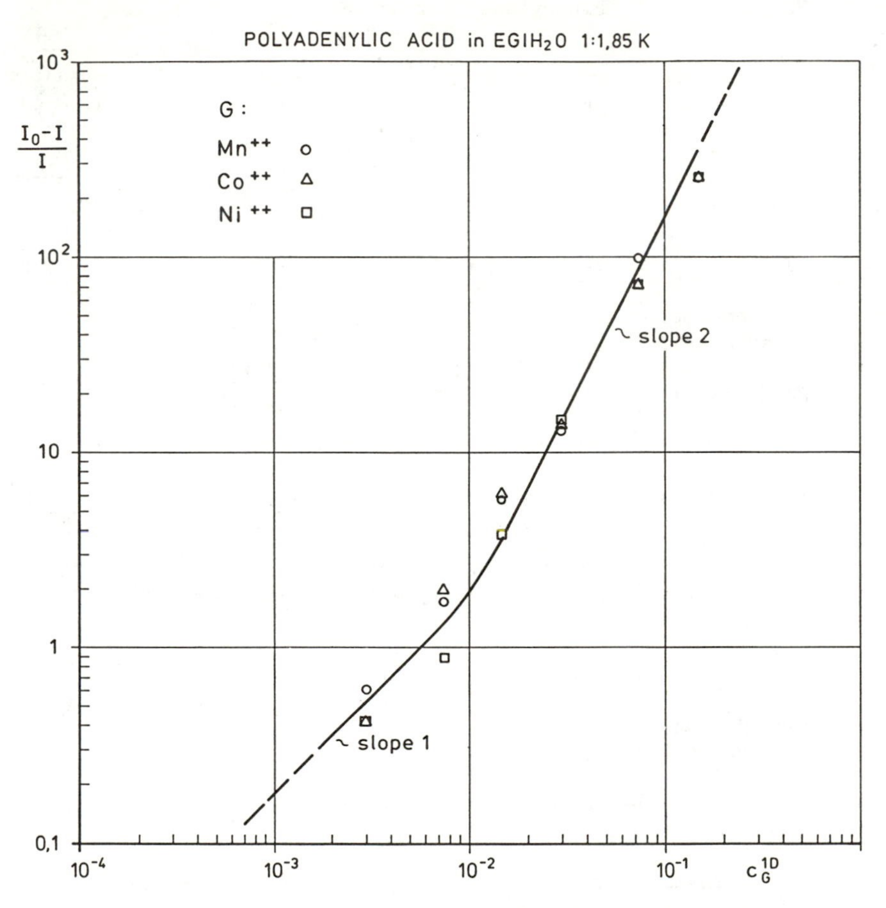


Figure 7. Phosphorescence quenching in polyadenylic acid, data recalculated after Eisinger und Schulman (61)

2) The mole ratio of intrinsic trap to basic unit should roughtly be euqal to c at Q=1 (50 percent quenching), see 4th column in Table III.
3) A first indication of the true dimensionality of exciton diffusion can be gained, at the basis of which more elaborate kinetic experiments and theories (3,30-34,66-71) can be put into action.
4) The hopping times estimated from the data of Table III ($t_h = 1/k_h = \tau/n$) are more realistic than data presented earlier. The measured phosphorescence decay (in the order of several seconds for aromatic polymers) is not suitable as triplet exciton lifetime ($\tau$), since this figure characterizes the trap rather than the exciton. The mean triplet exciton lifetime is mostly found in the order of 10 ms. Taking this value and $n = 10^4$ to $10^5$ we obtain hopping times in the order of 0.1 to 1 µs, corresponding to triplet hopping rates of $10^6$ to $10^7$ $s^{-1}$ for aromatic macromolecules with efficient triplet transfer. They should not be mistaken as time-independent constants (although this is assumed in the simple random walk model) since exciton migration in unordered systems is dispersive, i.e. time-dependent (30).

Finally it should be pointed out that a 1D-model similar to the one used in this paper has been applied by Slobodyanik et al. (72) to singlet transfer in solid solutions of PVCA copolymers with varying degrees of energy trapping nitrocarbazole groups. The experimental data excellently fit a Q(c) curve with a linear and a quadratic part, yielding $n \approx 3000$ in agreement with the value given by Slobodyanik (72). Care must be taken, however, due to the possibility of Förster's single step transfer in this system, which could result in a Q(c) dependence indistinguishable from 1D exciton hopping.

## Delayed Fluorescence and Triplet-Triplet Annihilation

In the previous section, delayed fluorescence has been disregarded although it constitutes an additional deactivation channel for triplets in P1VN (1,56), P2VN (6,57,73), PVCA (4,5,7,13,59,65), P2NMA (74,75) but not in PACN (64).

Delayed fluorescence in a rigid matrix proves triplet excitons if other delayed $S_1 \rightarrow S_o$ emissions, as high temperature phosphorescence or ionisation followed by ion-electron recombination can be excluded (15). The delayed fluorescence observed in solid solutions of Polyriboadenylic acid (45) seems to be due to this latter process (78).

For P2VN, Avakian et al. (79) showed that indeed T-T annihilation causes the delayed fluorescence using magnetic field modulation of DF in solid solutions of this polymer. The basic processes in generating delayed fluorescence by triplet-triplet annihilation (80) are shown in Equation (6).

$$T_1(\text{mobile}) + T_1\ (\text{mobile or trapped}) \rightarrow S_n + S_o \qquad (6)$$

$$S_n \longrightarrow S_1 \rightarrow S_o + DF$$

In the case of polymers with high trapping efficiency, the heterofusion (81) between mobile triplets (excitons) and trapped ones is likely to prevail. The delayed singlets can therefore be created as excitons or as trapped states.

The main facts about T-T annihilation in solid solutions of aromatic polymers may be summarized as follows:

1. Delayed fluorescence (DF) can be observed in several aromatic polymers in addition to phosphorescence after excitation via the singlet system or using triplet sensitizers.
2. The DF can be quenched by suitable triplet quenching molecules (e.g. those listed in Table III).
3. The non exponential decay of DF is always much faster (in the 1 to 100 ms range) than the phosphorescence decay (in the 0.1 to 10 s range for $\pi\pi^*$ states) which is mostly close to or even identical with the exponential decay of monomeric models.
4. The DF spectra are very similar to those of prompt fluorescence.
5. The intensity of DF increases, that of phosphorescence decreases with increasing molar mass (M) of the polymer, all other conditions being kept constant. The slope of this increase differs strongly for different polymers. At high M a saturation region can be observed.
6. The relative intensity of DF increases, that of phosphorescence decreases with increasing excitation intensity, if all other conditions including M are kept constant.

Two extreme cases can easily be understood, at least qualitatively: low M + low excitation intensity and high M + high excitation intensity. In the first case, only a few polymer chains are occupied by a triplet (mobile or trapped), the majority being unoccupied. In this case, the bimolecular process (Equation 6) cannot take place due to the sequestration of the triplets. In the second case, ordinary "macroscopic" kinetics may already apply, albeit modified by the $^1$D-system.

It is the intermediate range which is not yet fully understood. Clearly, the distribution of the triplets among the chains should be governed by Poisson statistics which qualitatively predict the M- and intensity effects observed. Quantitatively, however, Poisson statistics fail to explain the effects, since M influences the DF and P intensity even in a region where there should be simultaneously many excitons per chain so that, statistically, there should be little differences between the individual macromolecules. Yokoyama et al. proposed (59) that triplet traps at the chain ends, supposed not to contribute to T-T annihilation, could explain the M dependence of DF. If this is true, there remains the question, why heterofusion, so common in other systems, should not work between triplet excitons moving along the chain and trapped triplets at the chain ends.

Webber and Swenberg (82) worked out a master equation theory for excitonic annihilation processes in low dimensional, finite lattices. This theory seems to describe well the process for polymers with few or shallow traps, as P2VN, but not for those containing many relatively deep traps, as PVCA. Finally, a list of those features which at the present time prevent a full understanding of T-T annihilation in dilute, solid solutions of aromatic polymers should give us an appreciation of the problems to be solved in the future:

1. [1]D (or near so) exciton hopping and possible transition to higher dimensionality in tightly packed coils.
2. Disordered systems likely to show dispersive exciton hopping.
3. Presence of trap sites of still unknown structure, leading to heterofusion.
4. Possible pecularities of [1]D T-T annihilation (83).
5. Finite volume or, better, finite length effects, and Poisson distribution.
6. Possible influences of the end groups.

As a first step to solve part of these problems, better quantitative data, using well defined polymers in narrow molar mass fractions are needed, combining steady state and time-resolved experiments.

Acknowledgement is made to the donors of the Petroleum Research Fund, administered by the American Chemical Society, for partial support of this activity.

## LITERATURE CITED

1. Cozzens, R.F.; Fox,R.B.: J. Chem. Phys. 1969, 50, 1532.
2. Chlorophyll Organization and Energy Transfer in Photosynthesis, Ciba Foundation. Symposium 61 (new series), Excerpta Medica, Amsterdam, 1979.
3. Ediger, M.D.; Fayer, M.D.; J. Phys. Chem. 1984, 88, 6108.
4. Klöpffer, W.; Fischer, D.:J. Pol. Sci. Pol. Symposium 40 1973, 43.
5. Yokoyama, M.; Nakano, T.; Tamamura, T.; Mikawa, H.: Chem. Letters (Tokyo), 1973, 509.
6. Pasch, N.F.; Webber, S.E.: Chem. Phys. 1976, 16, ,361.
7. Klöpffer, W.; Fischer, D.; Naundorf, G.: Macromolecules 1977, 10, 450.
8. Klöpffer, W.: Introduction to Polymer Spectroscopy; Polymers, Properties and Applications, Vol. 7; Springer, Berlin, 1984.
9. Klöpffer, W.: J. Chem. Phys. 1969, 50, 2337.
10. Klöpffer,W.; Rippen, G.; Kowal, J.: Makromol. Chem., Macromol. Symp., 1986, 5, 187.
11. Thomas, R.: Biochem. Biophys. Acta, 1954, 14, 231.
12. Okamoto, K.; Yamada, M.; Itya, A.; Kimura, T.; Kusabayashi, S.: Macromolecules, 1976, 9, 645.
13. Itaya, A.; Okamoto, K.; Kusabayashi, S.: Bull. Chem. Soc. Japan, 1976, 49, 2037.
14. Itaya, A.; Okamoto, K.; Kusabayashi, S.: Bull. Chem. Soc. Japan, 1976, 49, 2082.
15. Klöpffer, W. in Electronic Properties of Polymers; Mort, J.; Pfister, G., Eds.: Wiley, New York, 1982,pp. 161.
16. Okamoto, K.; Itaya,A.; Kusabayashi, S.: Chemistry Letters (Japan), 1974, 1164.
17. Schneider, F.: Z. Naturforsch. 1969, 24a, 863.
18. Vala, M.T., Jr.; Rice, S.A.: J. Chem. Phys. 1963, 39, 2348.
19. Vala, M.T., Jr.; Haebig, J.; Rice, S.A.: J. Chem. Phys. 1965, 43, 886.
20. Hirayama, F.: J. Chem. Phys. 1965, 42, 3163.
21. Frank, C.W.: J. Chem. Phys. 1974, 61, 2015.

22. Pillips, D., Ed.: Polymer Photophysics, Chapman and Hall: London, 1985.
23. Fox, R.B.; Price, T.R.; Cozzens, R.F.; McDonald, J.R.: J. Chem. Phys. 1972, 57, 534.
24. Holden, D.A.; Ren, X.-X.; Guillet, J.E.: Macromolecules, 1984, 17, 1500.
25. Ito, S.; Yamamoto, M.; Nishijima, Y.: Reports Progr. Pol. Phys. Japan, 1979, 22, 453.
26. David, C.; Baeyens-Volant, D.; Piens, M.: Europ. Pol. J. 1980, 16, 413.
27. Rippen, G.; Klöpffer, W.: Ber. Bunsenges. Phys. Chem. 1979, 83, 437.
28. Rippen, G.; Kaufmann, G.; Klöpffer, W.: Chem. Phys. 1980, 52, 165.
29. Nakahira, T.; Ishizuka, S.; Iwabuchi, S.; Kojima, K.: Makromol. Chem. Rapid Commun. 1980, 1, 759.
30. Richert, R.; Bässler, H.: J. Chem. Phys. 1986, 84, 3567.
31. Jankowiak, R.; Bässler, H.: Chem. Phys. 1983, 79, 57.
32. Schönherr, G.; Eiermann, R.; Bässler, H.; Silver, M.: Chem. Phys. 1980, 52, 287.
33. Rockwitz, K.-D.; Bässler, H.: Chem. Phys. 1982, 70, 307.
34. Bässler, H.: phys. stat. sol. (b) 1981, 107, 9.
35. Peter, G.; Bässler, H.; Schrof, W.; Port, H.: Chem. Phys. 1985, 94, 445.
36. Johnson, P.C.; Offen, H.W.: J. Chem. Phys. 1971, 55, 2945.
37. Vandendriessche, J.; Palmans, P.; Toppet, S.; Boens, N.; De Schryver, F.C.; Masuhara, H.: J. Am. Chem. Soc. 1984, 106, 8057.
38. Bokobza, L.; Jasse, B.; Monnerie, L.: Eur. Pol. J. 1977, 13, 921.
39. Frank, C.W.; Harrah, L.A.: J. Chem. Phys. 1974, 61, 1526.
40. Johnson, G.E.: J. Chem. Phys. 1975, 62, 4697.
41. Klöpffer, W.: Spectroscopy Letters 1978, 11, 863.
42. Fox, R.B.; Cozzens, R.F.: Macromolecules, 1969, 2, 181.
43. Ito, S.; Nishino, S.; Yamamoto, M.; Nishijima, Y.: Reports Progr. Pol. Phys. Japan 1981, 24, 481.
44. David, C.; Naegelen, V.; Piret, W.; Geuskens, G.: Eur. Pol. J. 1975, 11, 569.
45. Hélène, C.; Longworth, J.W.: J. Chem. Phys. 1972, 57, 399.
46. Ermolaev, V.L.: Sov. Phys. Uspekki 1963, 80, 333 (Engl. Translation).
47. Voltz, R.; Laustriat, G.; Coche, A.: J. Chim. Phys. (Paris) 1966, 63, 1253.
48. Rosenstock, H.B.: J. Soc. Indust. Appl. Math. 1961, 9, 169.
49. Rosenstock, H.B.: J. Chem. Phys. 1968, 48, 532.
50. Rudemo, M.: J. Soc. Indust. Appl. Math. 1966, 14, 1293.
51. Klöpffer, W.: J. Chem. Phys. 1969, 50, 1689.
52. Förster, Th.: Z. Naturforsch. 1949, 4a, 321.
53. Förster, Th.: Discussions Faraday Soc. 1959, 27, 7.
54. Perrin, F.: Compt. Rend. Acad. Sci. (Paris) 1924, 178, 1978.
55. Inokuti, M.; Hirayama, F.: J. Chem. Phys. 1965, 43, 1978.
56. Fox, R.B.; Price, T.R.; Cozzens, R.F.: J. Chem. Phys. 1971, 54, 79.
57. Pasch, N.F.; McKenzie, R.D.; Webber, S.E.: Macromolecules, 1978, 11, 733.

58. David, C.; Lempereur, M.; Geuskens, G.: Eur. Pol. J. 1972, 8, 417.
59. Yokoyama, M.; Tamamura, T.; Nakano, T.; Mikawa, H.: J. Chem. Phys. 1976, 65, 272.
60. Kilp, T.; Guillet, J.E.: Macromolecules 1981, 14, 1680.
61. Eisinger, J.; Schulman, R.G.: Proc. Natl. Acad. Sci. USA 1966, 55, 1387.
62. Hargreaves, J.S.; Webber, S.E.: Macromolecules 1984, 17, 235.
63. Behrson, R.; Isenberg, I.: J. Chem. Phys. 1964, 40, 3175.
64. Webber, S.E.; Avots-Avotin, P.E.: Macromolecules 1979, 12, 708.
65. Webber, S.E.; Avots-Avotin, P.E.: J. Chem. Phys. 1980, 72, 3773.
66. Klafter, J.; Silbey, R.: J. Chem. Phys. 1980, 72, 849.
67. Blumen, A.; Zumofen, G.: Chem. Phys. Lett. 1980, 70, 387.
68. Fitzgibbon, P.D.; Frank, C.W.: Macromolecules 1982, 15, 733.
69. Blumen, A.; Klafter, J.; Zumofen, G.: Phys. Rev. B. Rap. Commun. 1983, 28, 6112.
70. Fredrickson, G.H.; Frank, C.W.: Macromolecules 1983, 16, 1198.
71. Klafter, J.; Blumen, A.: J. Chem. Phys. 1984, 80, 875.
72. Slobodyanik, V.V.; Faidysh, A.N.; Yaschuk, V.N.; Fedorova, L.N.: Prikladnoi Spektroskopii 1982, 36, 309.
73. Pratte, J.F.; Webber, S.E.: J. Phys. Chem. 1983, 87, 449.
74. Somersall, A.C.; Guillet, J.E.: Macromolecules 1973, 6, 218.
75. Pasch,N.F.; Webber, S.E.: Macromolecules 1978, 11, 727.
76. Faidysh, A.N.; Slobodyanik, V.V.; Yaschuk, V.N.; Naidenov, V.P.: J. Opt. i Spectrosk. (USSR) 1979, 47, 510.
77. Faidysh, A.N.; Slobodyanik, V.V.; Yaschuk, V.N.: J. Luminescence 1979, 21, 85.
78. Bazin, M.; Santus, R.; Hélène, C.: Chem. Phys. 1973, 2, 119.
79. Avakian, P.; Groff, R.P.; Suna, A.; Cripps, H.N.: Chem. Phys. Letters 1975, 32, 466.
80. Pope, M.; Swenberg, C.E.: Electronic Processes in Organic Crystals, Clarendon Press: Oxford, 1982.
81. Tedder, S.H.; Webber, S.E.: Chem. Phys. Letters 1975, 31, 611.
82. Webber, S.E.; Swenberg, C.E.: Chem. Phys. 1980, 49, 231.
83. Suna, A.: Phys. Rev. 1970, B1, 1716.
84. Dexter, D.L.: J. Chem. Phys. 1953, 21, 836.

RECEIVED April 29, 1987

Chapter 22

# Excited-State Singlet Energy Transport in Polystyrene

**Daniel R. Coulter[1], Amitava Gupta[1], Vincent M. Miskowski[1], and Gary W. Scott[2]**

[1]Jet Propulsion Laboratory, Chemical and Mechanical Systems Division, California Institute of Technology, Pasadena, CA 91109
[2]Department of Chemistry, University of California, Riverside, CA 92521

Photophysical properties of polystyrene and copolymers of styrene and a chemically attachable quencher have been determined in dilute solution and in solid films. In dilute solution, it is found that energy migration is not extensive. In room temperature solid films, quenching studies shown effective long range energy transport. Rough estimates of the excimer forming site concentration and hopping rate have been reported. Temperature dependent studies on films down to ∿20K have shown that both monomer and excimer emission come from intrinsic traps. Steady state and time resolved fluorescence measurements as a function of temperature have allowed determination of the activation energy for monomer detrapping and have placed an upper limit on the activation energy of migration.

Singlet excited state energy migration in aromatic polymers has been proposed to be important for many years (1). There remains, however, considerable disagreement (1,2) as to the extent of excitation mobility, and little information (3) is available as to the time-scale of energy migration.

Our current work (1,4) in this field attempts to directly address these questions. We here report the results and interpretation of photophysical experiments involving two different types of perturbation of the phototypical aromatic polymer, polystyrene. First, we have prepared a series of co-polymers of styrene with an excitation quencher, 2-(2'-hydroxy-5'-vinylphenyl)-2H-benzotriazole, abbreviated 2H5V. Secondly, we have looked at the effect of temperature on the photophysical properties of these polymers.

## Experimental

Preparation and characterization of the polymers has been described in detail (4). Results are summarized in Table I.

Fluorescence measurements employed a Perkin-Elmer MPF-66 spectrofluorimeter with an excitation wavelength of 260 nm and excita-

0097-6156/87/0358-0286$06.00/0

tion and emission spectral slit widths of 2 nm. A band pass filter was used to further isolate 260 nm excitation light.

The excitation was chopped at a frequency of 34Hz and phase-sensitive detection was used. In this way, the long-lived (seconds at 20K) phosphorescence and delayed fluorescence were selectively excluded from these data, and only the "prompt" fluorescence was analyzed. Data including the long-lived signals were measured in separate experiments, and will be reported elsewhere.

Time-resolved studies employed previously-described (4) equipment but were improved upon as follows. Part of the excitation Nd-YAG laser beam was split off and delivered to the phototube, serving as a marker pulse to trigger our Biomation 6500 for data acquisition; the sample emission signal was timed by a laser pulse delay line to arrive ∿15 ns after the marker pulse. Laser jitter was thereby minimized, and computer signal averaging more precise.

Variable-temperature experiments employed a Cryogenic Technology Model 21 closed-cycle helium refrigerator. Samples were deposited as films upon quartz disks from $CH_2Cl_2$ solution, with minimal exposure to light. They were mounted on a copper sample holder bolted to the cold stage of the refrigerator, with copper grease used to give good thermal contact of all surfaces. A Chromel-Au (0.07%Fe) thermocouple was mounted on the opposite side of the sample holder, at the same height as the sample. A copper heat shield was mounted over the cold stage/sample, a shroud with quartz windows was emplaced and the sample space was pumped on for ∿24 hours in order to remove $O_2$ and $CH_2Cl_2$ dissolved in the polymer. Samples prepared in this way were extremely photostable.

## Background: Aromatic Excited States and Excimers

Excimers, that is complexes of excited state molecules with corresponding ground state molecules, are (5) characteristic of aromatic $\pi\rightarrow\pi^*$ excited states. For our present purposes we wish to consider the lowest $\pi\rightarrow\pi^*$ excited state of benzene and substituted derivatives together with the corresponding excimer.

The lowest singlet excited state ($\pi\rightarrow\pi^*$) of, for example, ethyl benzene, shows absorption and emission maxima at, respectively (1), ∿260 and ∿280 nm. The temperature dependence of the emission in dilute polar organic solution has been investigated (1b), and it was found that there is a temperature dependent non-radiative rate component that follows an Arrhenius law $Ae^{-\Delta E/kT}$ with an activation energy $\Delta E \sim 2300\ cm^{-1}$ and $A \sim 10^{12} s^{-1}$. The ratio of the emission quantum yield ($\Phi_E$) and lifetime($\tau$) is temperature independent, consistent with a temperature independent radiative rate constant $k_v$, and limiting low temperature values of $\Phi_E$ and $\tau$, achieved by ∿250K, are 0.12 and 21 ns, respectively, in dichloroethane solution.

Excimer formation of ethylbenzene in excited state has been studied in concentrated (.1-1M) methylcyclohexane solution at -78°C by Hirayama, et al (6). They found, in addition to the "monomer" emission, a broad concentration dependent excimer emission with $\lambda_{max}$ = 322 nm. From the concentration dependence they were able to extract "intrinsic" monomer and excimer yields, that is, the quantum yields if all excited states were present as either monomer or excimer; these amounted to, respectively, 0.30 and 0.021 under the given conditions.

The intrinsic weakness of benzene excimer emission derives from its extreme dipole forbiddeness (7). Its weakness relative to monomer emission under dilute solution conditions additionally results from the diffusion limit upon excimer formation rates, along with thermally induced excimer dissociation. These latter considerations are largely irrelevant to polymers, vide infra.

## Polystyrene in Dilute Solution

In Figure 1 we compare emission spectra for polystyrene in dilute solution and as a solid film, and for a model monomer, ethylbenzene, in dilute solution. Polystyrene in solution exhibits, in addition to a monomer-like emission, a broad excimer emission maximizing at ∿330 nm. This emission spectrum is not unique to high molecular weight polymer. Indeed, 1,3-diphenylpropane exhibits (1a,8) very similar total emission spectra. The excimer emission lifetime is (4) 12.5 ns in $CH_2Cl_2$ at room temperature, while monomer-like emission decay and excimer emission rise times are reported (3) to be of the order of a nanosecond in cyclohexane solution.

Importantly, emission spectra of polystyrene and of model compounds in rigid dilute solution (1), e.g., 1:1 diethylether-tetrahydrofuran at 77K, are all essentially identical, with no trace of excimer emission.

The model which has been developed (1,3) to fit these results involves formation of excimers via thermally activated phenyl group motion. Such motion is restricted in a rigid matrix, and "pre-formed" excimer sites, where little phenyl motion would be required for collapse to the excimer, are evidently rare.

However, Itagaki, et al (3) propose that energy migration along polymer chains is still important, as a result of a detailed analysis of molecular weight effects on photophysical parameters, with the average number of phenyl rings covered by singlet energy migration estimated to be ∿7-8 in the high molecular weight polymers; the characteristic timescales of energy hopping and of polymer internal rotation to an excimer-forming conformation were proposed to be, respectively, ∿30 ps and ∿7 ns. The relatively short 1 ns monomer decay lifetime and excimer rise time thus was inferred to result from excitation sampling along the chain to find favorable conformations, with the additional restriction that excimer dissociation to monomer was negligibly slow on the timescale of the excimer lifetime. The latter restriction is characteristic of aromatic polymers in general.

We determined photophysical properties for our copolymers in dilute solution, and found no effects (9) upon either total emission spectra (Figure 2) or emission lifetimes. However, because the mole % quencher in our copolymers is quite small, this result is completely consistent with the interpretation of Itagaki, et al (3). All we can say from our results is that energy migration is not extensive in dilute solution.

## Solid Polystyrene

The emission spectrum of solid polystyrene at room temperature has already been shown in Figure 1. It shows a broad excimer emission at 325 nm and only a trace of monomer-like emission. The excimer and

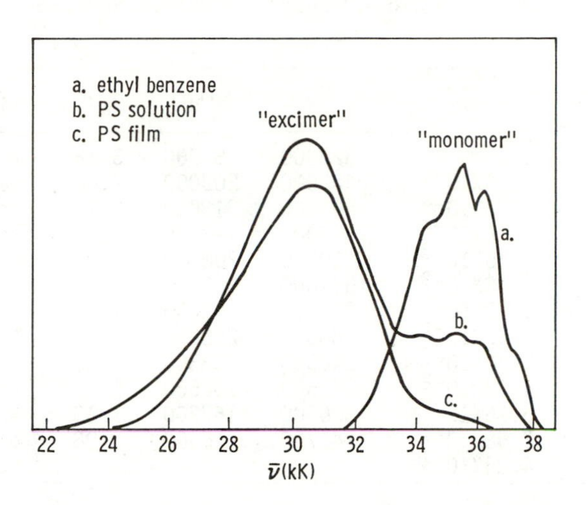


Figure 1. Emission spectra of ethylbenzene and polystyrene.

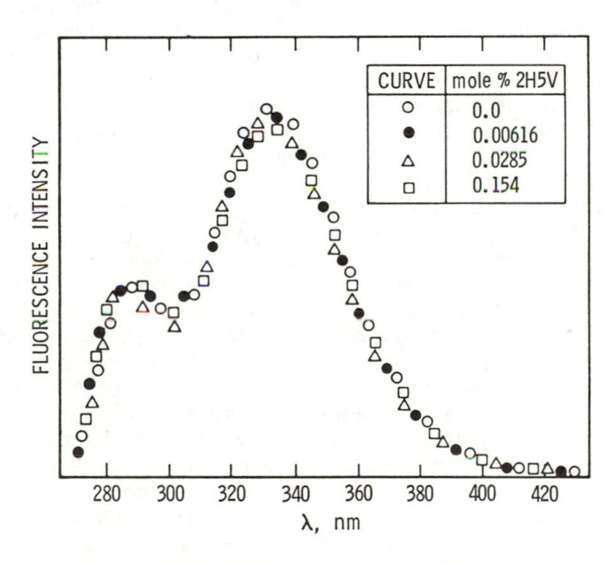


Figure 2. Emission spectra for $CH_2Cl_2$ solutions of poly(styrene-Co-2H5V) copolymers. The phenyl group concentration is constant at 5 X $10^{-3}$M.

Table I. COMPOSITION AND MOLECULAR WEIGHT DATA FOR POLY(STYRENE-CO-2H5V) COPOLYMER SAMPLES

| 2H5V, mol % | 2H5V, mol/L | $\overline{M}_w$ | $\overline{M}_n$ | $\overline{M}_w/\overline{M}_n$ | initiator |
|---|---|---|---|---|---|
| 0.0 | 0.0 | 328000 | 87000 | 3.78 | Polysciences |
| 0.00155 | $1.55X10^{-4}$ | 402000 | 202000 | 2.06 | peroxide |
| 0.00273 | $2.73X10^{-4}$ | 463000 | 218000 | 2.12 | peroxide |
| 0.00616 | $6.16X10^{-4}$ | 364000 | 192000 | 1.89 | peroxide |
| 0.0114 | $1.15X10^{-3}$ | 410000 | 200000 | 2.05 | peroxide |
| 0.0285 | $2.88X10^{-3}$ | 511000 | 302000 | 1.69 | AIBN |
| 0.527 | $5.32X10^{-3}$ | 511000 | 302000 | 1.69 | AIBN |
| 0.154 | $1.55X10^{-2}$ | 480000 | 280000 | 1.71 | AIBN |
| 0.268 | $2.70X10^{-2}$ | 240000 | 133300 | 1.80 | AIBN |
| 0.483 | $4.88X10^{-2}$ | 242600 | 132500 | 1.83 | AIBN |
| 0.570 | $5.76X10^{-2}$ | 306000 | 167200 | 1.83 | AIBN |
| 2.55 | $2.58X10^{-1}$ | 241700 | 116400 | 2.08 | peroxide |
| 4.78 | $4.83X10^{-1}$ | | | | peroxide |

monomer-like emission decay lifetimes are respectively 21.5 ns and 1 ns. We have not been able to observe an excimer rise-time to date and conclude that it must be <1 ns. The absence of correlation of monomer decay and excimer rise is intriguing. In contrast to our solution results, the copolymers show large emission quenching effects as solids, Figure 3. For the 0.0527 mole % 2H5V sample, for which there is ∿70% emission quenching and also samples with lower 2H5V content, we found no change in emission decay lifetime, which strongly suggests that an excimer precursor is being quenched. Curiously, the monomer-like emission lifetime was also unaffected.

For higher 2H5V contents, decreases in emission lifetimes could be observed, Figure 4. The decays were now no longer exponential but could be fit to the Forster quenching equation (4). An analysis of the time resolved data then allowed us to deconvolute the total emission quenching into that due to (presumed) precursor quenching and that due to Forster quenching of the excimer, Figure 5. From the relative efficiency of trapping of mobile excitation by 2H5V quencher and by excimer traps, we could conclude that excimer sites occurred only about 1 per 950 monomers. This is a factor of 10 lower than an estimate for polystyrene by Franck and Harrah (1d) but very similar to an estimate by Kloppfer (10) for poly (vinylcarbazole). It is furthermore possible to obtain a crude estimate for a site-to-site hopping rate if 3-D isotropic diffusion is assumed. At 50% quenching (0.025 mole % 2H5V), the rate for excitation quenching, $k_Q$, should be equal to the mobile excitation decay lifetime in the absence of added quenchers, $\tau_M{}^0)^{-1}$. Then, in the 3-D case (11)

$$k_Q = ( M^0)^{-1} = X_Q P k_{MM} \qquad (1)$$

where $X_Q$ is the molar fraction of quenchers, P is a lattice-dependent constant ∿0.7 for cubic lattices, and $k_{MM}$ is the single-step excitation hopping rate. Thus, we have Equation 2.

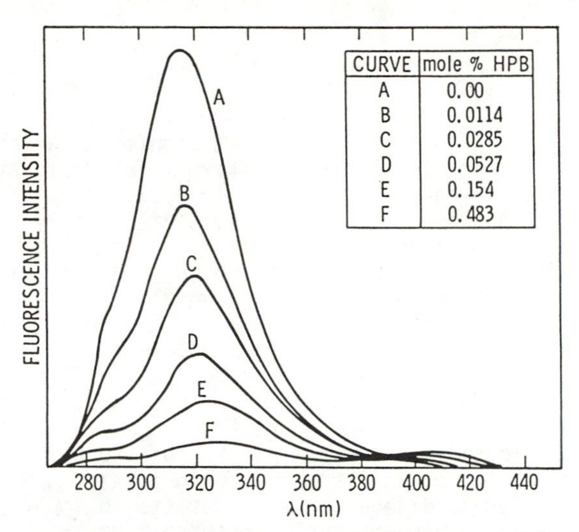


Figure 3. Emission spectra for solid films of poly(styrene-Co-2H5V) copolymers at room temperature.

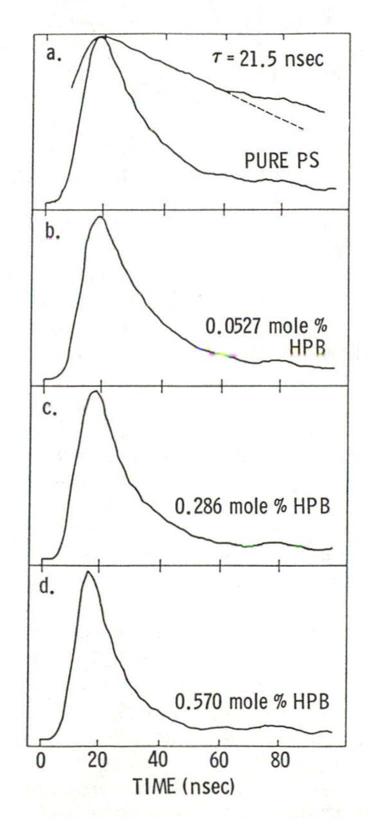


Figure 4. Time resolved excimer emission for poly(styrene-Co-2H5V) copolymers at room temperature.

$$k_{mm} \sim (5.7 \times 10^3)(\tau_M{}^0)^{-1} \quad (2)$$

If $\tau_M{}^0$ is about 500 ps (which is allowed by the data presently available) then $k_{MM} \sim 10^{13}S^{-1}$, two orders of magnitude faster than estimated by Itagaki, et al (3) for polystyrene in dilute solution. A similar estimate of hopping frequency for poly (vinyl carbazole) in the solid state has been made by Bassler, et al. (12). This is not entirely surprising, since the solid polymer should on the average present several phenyl groups close enough (6Å) for dipole-dipole energy transfer. It should be noted however, that the turning over of the precursor quenching at high mole % 2H5V in Figure 5 is inconsistent with an isotropic model, e.g., apparently about 15-20% of the initial excitation is unavailable for quenching via singlet migration, possibly because it is trapped in restricted regions (excimer-rich or quencher-poor) of the inherently inhomogeneous polymer. Concurrently, the readily quenchable portion of the precursor excitation need not necessarily be well-treated by an isotropic model.

We now decided to extend our measurements to low temperatures, in the hope of slowing down excitation transport to an accessible time regime. Emission spectra for pure polystyrene at low temperatures are shown in Figure 6. At the lowest temperatures a very strong monomer-like emission is evident. As the temperature is increased, this emission rapidly decreases, while excimer emission increases, resulting in a clean isoemissive point at 327 nm (13).

At 150K, nearly all monomer-like emission has disappeared and the significant change above this temperature is that the excimer emission slightly broadens with loss of the isoemissive point.

The presence of an isoemissive point is symptomatic of an interconversion between two emissive species. However it is noteworthy that the excimer emissive is not extrapolating to zero at the lowest temperatures. Because of the well-defined isoemissive point, we were able to deconvolute the total emission into monomer excimer emissions using the T>150K excimer emission as one limit. The resulting monomer emissions are shown in Figure 7; the spectral profile compares well with that for dilute ethylbenzene Figure 1 and (1b). The temperature dependence of the integrated intensities is shown in Figure 8, and an Arrhenius plot of the monomer emission data is shown in Figure 9.

We also determined the temperature dependence of the emission lifetimes, as shown in Figure 10. It is remarkable that the excimer decay lifetime is almost completely temperature independent, showing no detectable rise time, while the monomer lifetime decreases rapidly with temperature. An Arrhenius plot of the monomer lifetime data is shown in Figure 11. It gives, within experimental error, the same activation energy as that for the monomer intensity, Figure 9.

Finally, we performed similar experiments for the various copolymers. As shown in Figure 12 for one of the copolymers, despite a large (70%) decrease in excimer intensity, the thermal behavior is qualitatively identical, and an Arrhenius plot of the monomer intensity data, Figure 13, gives the same activation energy, $\sim$130-150 $cm^{-1}$.

As indicated in Figure 8, the limiting (at presently accessible temperatures) low-temperature excimer emission intensity is only about 20% lower than the room-temperature value and no rise-time consistent with the monomer decay can be discerned. The limiting low

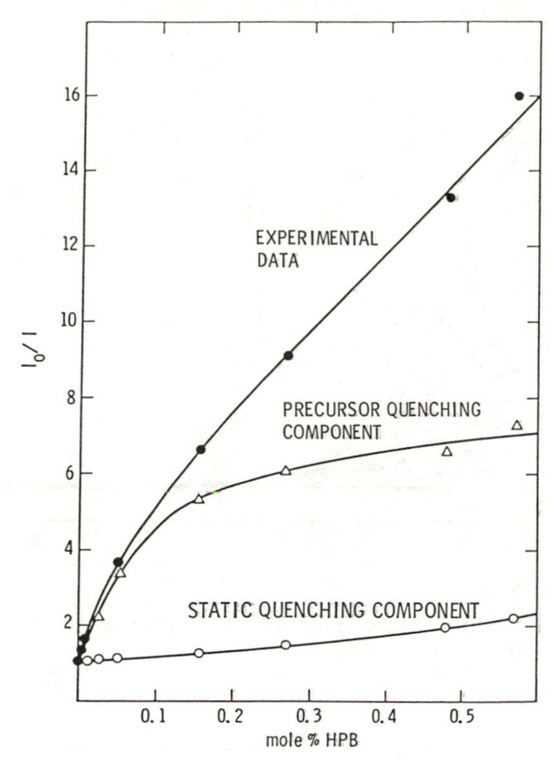


Figure 5. Excimer emission intensity as a function of composition for copolymers at room temperature.

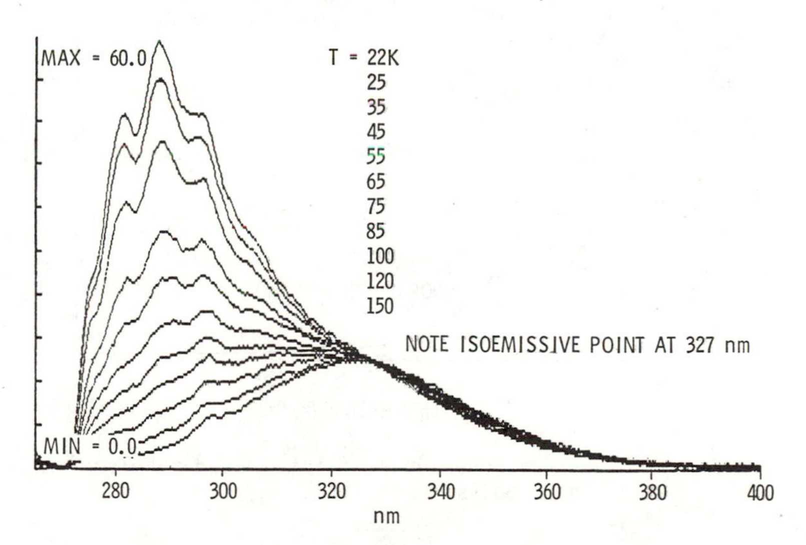


Figure 6. Temperature dependent fluorescence for a film of pure polystyrene.

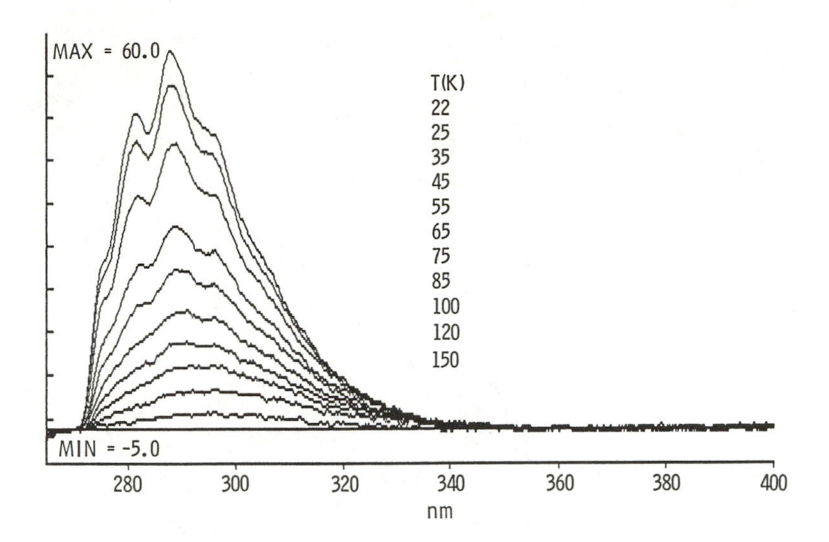


Figure 7. Data of Figure 6 after subtraction of excimer emission. See text.

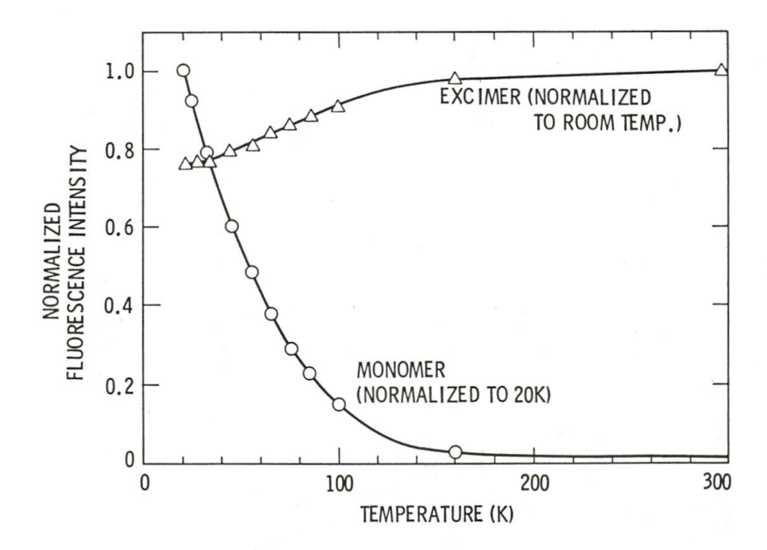


Figure 8. Temperature dependence of integrated resolved emission for a polystyrene film.

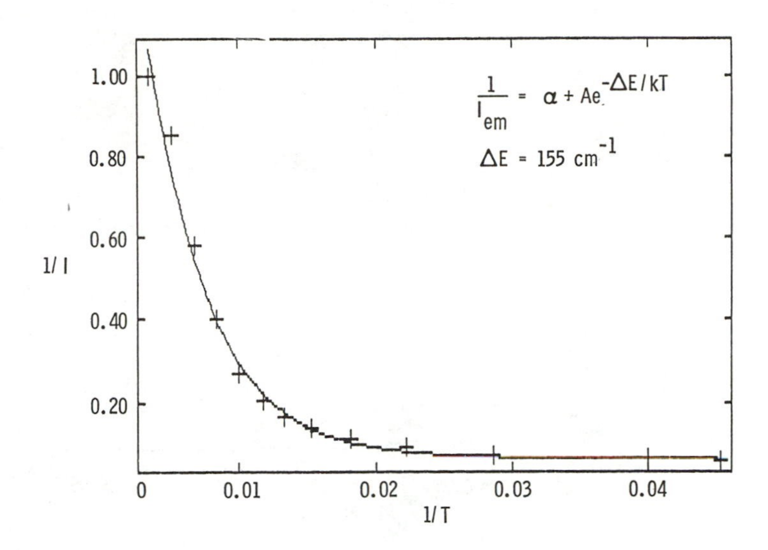


Figure 9. Arrhenius fit to the monomer intensity data of Figure 8.

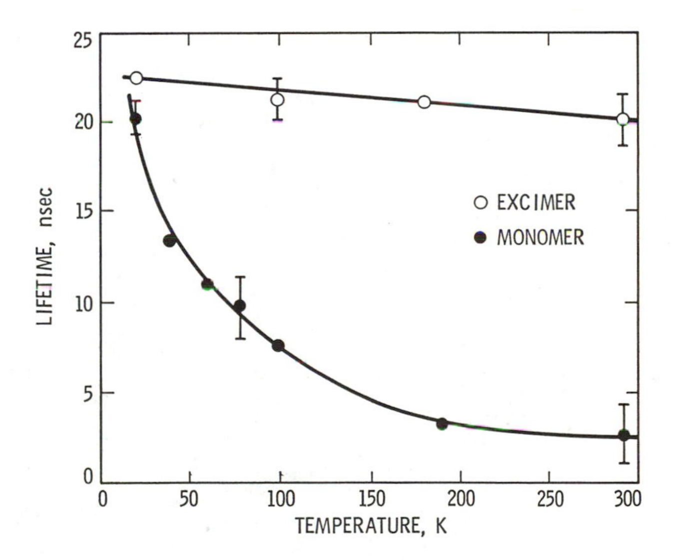


Figure 10. Temperature dependence of fluorescence decay lifetimes of a polystyrene film.

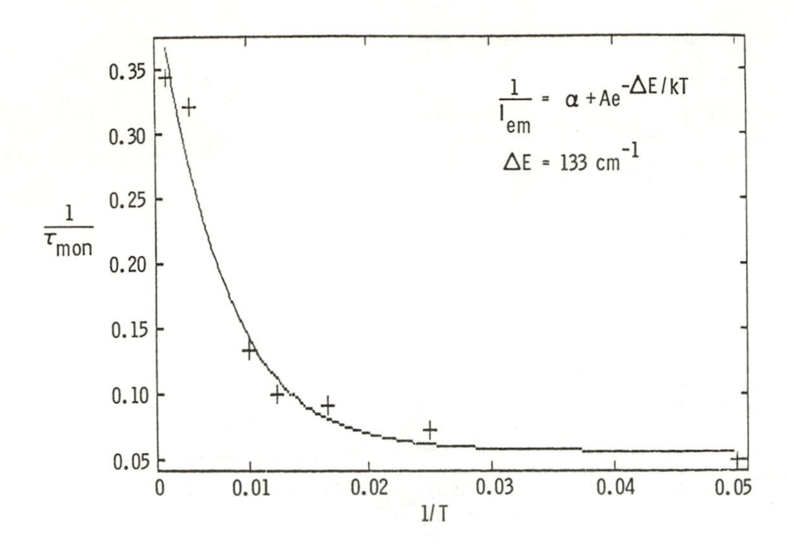


Figure 11. Arrhenius fit to the monomer lifetime data of Figure 10.

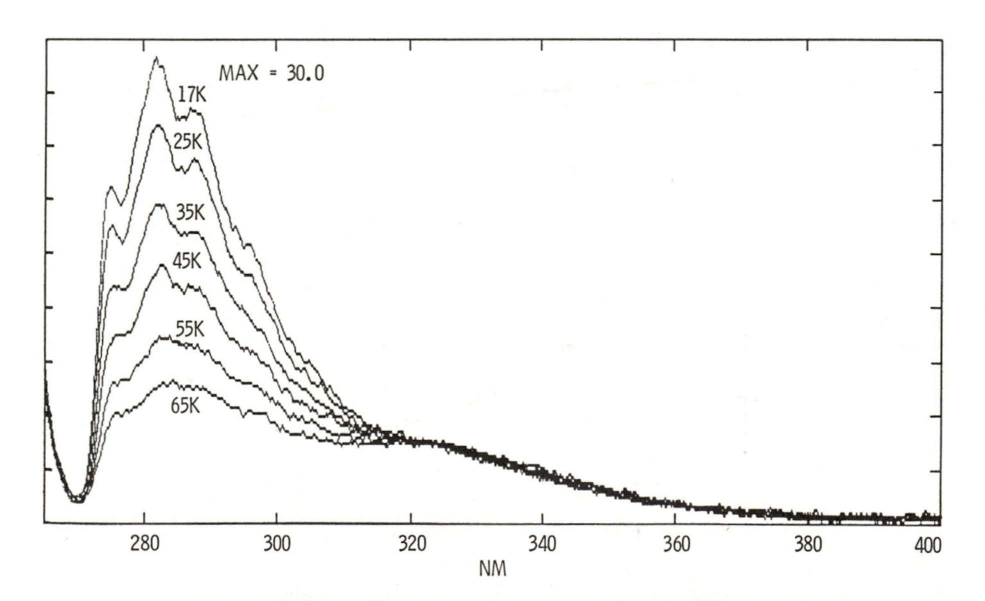


Figure 12. Temperature dependent fluorescence for a poly(styrene-Co-2H5V) copolymer, 0.0527 mole % 2H5V.

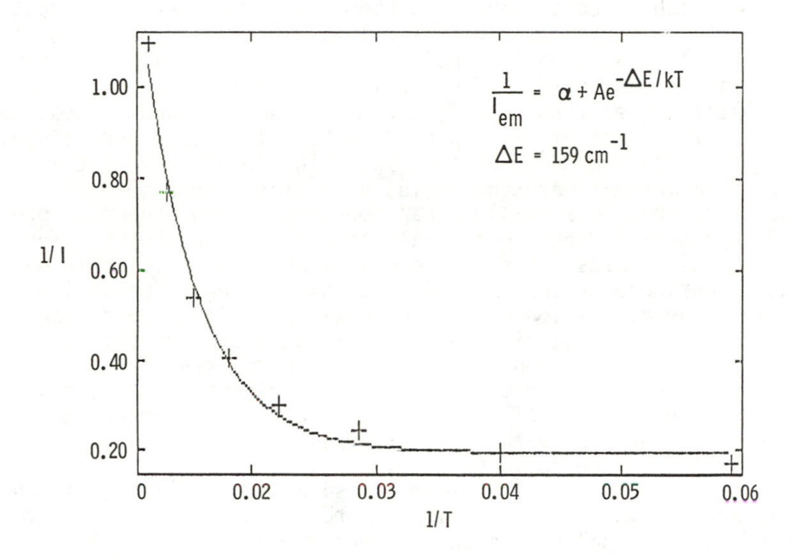


Figure 13. Arrhenius fit to the monomer intensity data of Figure 12.

temperature monomer lifetime (∿22 ns) is very similar to that of ethylbenzene in dilute solution (16). We therefore conclude that the emissive monomer is predominantly not the source of low-temperature excimers, but that at higher temperatures it can be detrapped in a thermally activated process, activation energy ∿140cm$^{-1}$, to yield excimer additional to the low-T limit. Note that the limiting low-temperature relative emission intensities are consistent with this hypothesis. Hirayama's (6) "intrinsic" monomer and excimer quantum yields of, respectively, 0.30 and 0.021, for ethylbenzene predict that if the emissive monomer comprises 20% of the total initial excitation that monomer emission should be ∿3 times as intense as excimer emission, which is consistent with Figure 6. Moreover, the hypothesis that the emissive monomer is a relatively small fraction of the total initial excitation is consistent with our inability to measure excimer rise times corresponding to monomer decay as simulations of the biphasic curves resulting from a ∿20% rise component convoluted with the normal excimer decay proved to be effectively indistinguishable from single component decays (15). This would not be true if the emissive monomer were the exclusive excimer precursor.

We think that this shallow trap monomer is predominantly "preformed" and achieved by the initial absorption of a photon. Thus, comparison of Figures 6 and 12 shows that monomer emission is decreasing considerably less rapidly than excimer emission at low temperature. The activation energy is consistent with expectation (3, 1d,16) for a phenyl rotation and the activated process is likely then such a rotation into a conformation where energy-transfer to a neighboring phenyl group can occur. We have not attempted to correlate the population of the shallow trap with chain dyads because we suspect that interactions with phenyl groups of adjacent chains are much more important in the solid polymer.

On the other hand, the excimer emission because it is 80% noncorrelated with monomer trap emission and because it is effectively quenched in the copolymers even at low temperatures, must largely arise from a mobile precursor. The activation energy for hopping of this precursor is implied to be <10 cm$^{-1}$. This is not unreasonably low(12,17), and indeed, the zero-point energy of the phenyl chromophore could in principle allow completely activationless hopping (tunneling) at reasonable rates. Determination of the true situation will require measurements at still lower temperatures, which are now in progress. We note that the polystyrene emission spectrum at 4.2K reported in (1d) indicates a monomer/excimer intensity ratio nearly the same as our 20K spectra.

Before we leave this section we should note that our explanation of the monomer emission as due to a special pre-formed shallow trap suggests that the low-temperature monomer/excimer emission intensity should be morphology dependent, hence dependent upon sample history, molecular weight, etc. This is indeed the case and we have found that different batches of polymer, aged samples, or samples prepared in different ways than described, may give slightly different intensity ratios at 20K. However, our results for samples prepared as described are completely reproducible and the temperature dependences remain qualitatively the same for deviant samples.

## CONCLUSIONS

It is clear from our results that the initial excitation in solid polystyrene must be extremely mobile in contrast to the case of the polymer in solution. However, it is now clear that available experimental results do not directly examine the mobile excitation. In particular, observed "monomer-like" emission is due to a shallow trap, which vitiates a previous analysis (1d,18) that assumed it to be due to the mobile excitation. Direct determination of mobile excitation dynamics will require extension of photophysical measurements to the picosecond time regime and efforts by us to accomplish this goal are now in progress.

## Acknowledgments

The research described in this paper was performed by the Jet Propulsion Laboratory, California Institute of Technology, under contract with the National Aeronautics and Space Administration.

## Literature Cited

1. a) Vala, Jr., M. T.; Haebig, J.; Rice, S. A. J. Chem. Phys. 1965 43, 886. b) Heisel, F.; Laustriat, G., J. Chem. Phys. Harrah, L. A., Frank, C. W. J. Chem. Phys. 1974, 61, 1526, d) Gupta, M. C.; Gupta, A.; Horwitz, J., Kliger, D. Macromolecules 1982, 15, 1372.
2. MacCallum, J. R. Eur. Polym. J., 1981, 17, 209.
3. Itagaki, H.; Horie, K.; Mita, I.; Washio, M.; Tagawa, S.; Tabata, Y. J. Chem. Phys. 1983, 79, 3996.
4. Coulter, D. R.; Gupta, A.; Yavrouian, A.; Scott, G. W.; O'Connor, D.; Vogl, O.; Li, S.-C. Macromolecules 1986, 19, 1227.
5. Birks, J. B. Photophysics of Aromatic Molecules; Wiley: New York 1970.
6. Hirayama, F.; Lipsky, S. J. Chem. Phys. 1969, 51, 1939.
7. a) Vala, Jr., M. T.; Hillier, I. H.; Rice, S. A.; Jortner, J. J. Chem. Phys. 1966, 44, 23, b) Ron, A.; Noble, M.; Lee, E.K.C. Chem. PHys. 1984, 83, 215, c) Glass, L.; Hillier, I. H.; Rice, S. A. J. Chem. Phys. 1965, 45, 3886.
8. Hirayama, F. J. Chem. Phys. 1965, 42, 3163.
9. We do see some quenching for very concentrated solutions, see (4).
10. Klopffer, W. In Electronic Properties of Polymers; Mort, J.; Pfister, G., Eds.; Wiley: New York, 1982; p. 161.
11. Montroll, E. W.; Weiss, G. H. J. Math. Phys. (N.Y.) 1956, 6, 167.
12. Peter, G.; Bassler, H.; Schrof, W.; Port, H. Chem. Phys., 1985, 94, 445.
13. In previous work (4), we also saw a weak emission near 400-450 nm, which increases intensity upon photo-oxidation (14). Interestingly, in the present work, involving rigorously degassed samples, little or no trace of this emission could be detected at any temperature, suggesting that it was entirely due to photo-oxidation in the previous work. For one sample, an extremely weak (<$10^{-2}$ times as intense as intrinsic fluorescence) emission was seen in the 400-450 nm region, which at 17K was resolved in-

to three sharp features at 396, 425, and 457 nm. The 1700 $cm^{-1}$ spacing of these features is attricutable to vibronic structure in ν(CO) and the frequency strongly suggests a backbone ketone (14). The wavelength of this emission is in fact reasonable for such a species.
14. Gueskens, G.; Baeyens-Volant, D.; Delaunois, G;, Lu-Vinh, Q.; Piret, W.; David, C. Euro. Polym. J. 1978, 14, 291.
15. Calculations assumed the two observed decay constants and the kinetic scheme described in the text.
16. Tanabe, Y., J. Polym. Sci. Polym. Phys. Ed. 1985, 23, 601.
17. Rose, T. S.; Righini, R.; Fayer, M. D. Chem. Phys. Lett. 1984, 106, 13.
18. Semerak, S. N.; Frank, C. W. Can. J. Chem., 1985, 63, 1328.

RECEIVED March 13, 1987

Chapter 23

# Significance of Energy Migration in the Photophysics of Polystyrene

J. R. MacCallum

Chemistry Department, University of St. Andrews, St. Andrews, Scotland

The photophysical behavior of poly(vinyl aromatics) has stimulated much interest from both practical and theoretical points of view. The proposal that such molecules exhibit electronic energy migration, e.e.m., resulted in, among other things, development of the concept of synthetic antennae macromolecules (1). The object of this paper is to introduce a note of caution to interpretation of observed photophysical phenomena. The bulk of the discussion is concerned with polystyrene, although the behaviour of poly(vinyl naphthalene) will also be considered.

The basic feature of the concept of e.e.m. is the proposal that excitation energy can be transferred from one aromatic group to another until the energy is irreversibly trapped by either an excimeric structure or an additive. The starting point of my review is an examination of data reported for small molecule analogues which have very similar characteristics such as absorption/emission, overlap, quantum yield, and lifetime, as the repeat units of the macromolecules under consideration. In fact rotational and translational freedom is higher in the small molecule analogues. Since the possibility of e.e.m. in fluid solutions of benzene and its methylated derivatives, (2) and naphthalene (3) had been investigated and rejected it seemed there is a good case for examining the evidence available to support the thesis that energy migration is a common phenomenon in macromolecules.

## Solutions

Quenching Studies. A number of papers have been published (4,5,6) reporting mesurement of both monomer and excimer quenching for polystyrene. The evidence adduced for the occurrence of e.e.m. is that obtained by comparing the measured rate constant for monomer quenching with that calculated for diffusion limited

0097-6156/87/0358-0301$06.00/0

contacts. The following comments are relevant to this procedure:

(a) an accurate monomer decay life-time is required of the quenching rate constant.
(b) uniform distribution of the quencher between the polymer coil and the bulk of the solution is assumed.
(c) correction must be made for the overlap of monomer and excimer emission.

A number of publications have appeared in which migration coefficients for polystyrene are determined but which do not fulfill (a) and (c). Correction for non-random distribution of quencher, (b), is complex, if even possible. For PVN the complexity of the decay profile for the monomer emission makes quenching studies rather difficult to interpret (7).

The most important condition which should be tested experimentally is that of overlap of emission bands. It has been shown that excimer overlap with monomer emission is significant for polystyrene (5) thus any determination of relative excimer/monomer yields must make allowance for this factor. This correction is large when the area of excimer peak is much greater than that of the monomer, a situation which prevails for solutions of polystyrene. Examination of emission spectra of naphthalene containing polymers indicates that when excimer is formed the emission overlaps to a large extent with monomer emission. A further experimental problem can arise when the wave-length of excitation radiation is close to that of emission, for this situation Raman scattering can contribute to the observed intensities.

Polarisation. In the event that e.e.m. takes place then the emission from the trap, excimer in the case of pure polymers, should be completely depolarised. Some time ago data was published on the emission of excimer for both polystyrene and poly($\alpha$-methyl styrene) indicating polarised emission (8) and therefore little e.e.m. More recently Phillips questioned the validity of the data and reported mesurements which suggest the excimer emission is depolarised (9). An experimental difference between the two sets of data is apparent - Phillips' solutions were more dilute than these used in ref. 8. The range of concentrations has subsequently been extended with the results shown in Figure 1. A possible explanation for the effect of concentrations and molecular weights on the extent of polarisation of excimer emission is that the rate of rotational relaxation of this bulky entity becomes slower than the emission life-time as these two parameters increase. It is difficult to reconcile the concept of e.e.m. with polarised excimer emission.

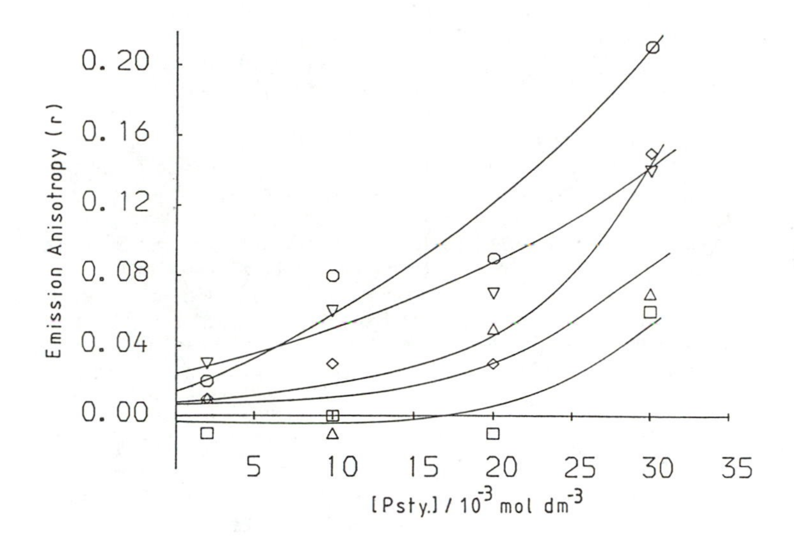


Figure 1. Emission anistropy (r) of excimer for solutions of polystyrene of varying concentration and molar mass.

□ - MW 3000 △ - MW 20,400

◇ - MW 195,000 ▽ - MW 490,000

○ - MW 15,000,000

Decay Profile. It has been proposed that the monomer emission decay is predominantly exponential with an additional minor late exponential component attributed to excimer dissociation (10,11). The contribution of the late component is less than 1% of the population of emitting species. Regardless of mechanism, e.e.m. is a diffusive process, one to three dimensional, in that an exciton migrates into a trap. Such a process must inevitably result in a non-exponential decay of the migrating exciton, i.e. the excited monomer. The observed exponentiality of the monomer emission can only be rationalised on the basis that (a) e.e.m. is very rapid and is complete before the decay measurement can be made, that is within 500 ps of excitation, or (b) no significant amount of e.e.m. takes place.

The theoretical implications of (a) are fascinating since single step transfer as part of extensive e.e.m. would occur on a time-scale much faster than would be predicted by exchange of dipole transfer and would be of the magnitude typical of crystals which exhibit ground-state interaction. Furthermore, if (a) were the case then all excimer traps would be populated on this very fast time basis. However the observed behaviour shows a definite rise-time of the order of a nanosecond for the excimer state.

Trapping Experiments. A possible means for demonstrating e.e.m. is by chemical incorporation of very small number of traps which compete with the excimer traps for the migrating exciton. Efficient transfer to the trap as measured by steady-state or transient decay behaviour would suggest e.e.m. played a role in populating the trap. The key factor in this type of experiment is that all molecules must contain a trap. Non-uniform distribution of traps makes data interpretation very difficult. Recently trapping experiments have been performed and it has been claimed that the results prove the occurrence of e.e.m. in polystyrene (12). However, on examining the experimental details it is apparent that for the very low concentration of trap used a number of macromolecules do not contain a trap and consequently the sample is non-uniform. With no correction made for this fact the interpretation of the observed behaviour is open to question. Use of traps is not a good procedure since the actual spatial distribution of the trap must be known and must be uniform before reliable analysis of experimental data can be made.

## Mechanism

The photophysical behaviour of polystyrene in solution can be explained by a Birks-type scheme in which $k_{DM}$ is an average rotational constant whereby a 1:3 intramolecular excimer is formed. On this basis, assuming excimer dissociation is relatively unimportant we have

$$\frac{\text{Intensity excimer } (I_E)}{\text{Intensity monomer } (I_M)} = \frac{k_{DM} \quad k_{FD}}{k_{FM}(k_{FD} + k_{ID})}$$

The only temperature dependent parameter is $k_{DM}$ and the activation energy for solutions has been measured as 4.1 kcal $mol^{-1}$ (4). Temperature variation of the relaxation of polystyrene main chain bonds has been investigated (13) giving an activation energy of 4.1 kcal $mol^{-1}$ and a room temperature value of $4 \times 10^9$ $s^{-1}$. The agreement between these two different experimental measurements suggests they are both concerned with the same process.

## Solid State

Data available on polarisation measurements is in conflict. Two major difficulties are found for measurements on solid state samples; they are (1) scatter of excitation and emission radiation, and (2) strain resulting in birefringence in the samples. However, data other than polarisation are available for examination.

On cooling films of pure polystyrene to 77K the ratio $I_E/I_M$ goes to almost zero (14). Assuming that excimer traps are preformed in the film when cast, the population of such traps will remain constant, once the film is formed. Neither the exchange nor the dipole-dipole mechanism of energy transfer is particularly sensitive to temperature and thus it is difficult to see why the traps should not be populated at very low temperatures, if indeed the mechanism of population is by e.e.m. into the traps.

The disposition of phenyl groups in the solid amorphous polymer is of relevance to the photophysical properties. Following refractive index measurements Prest has proposed (15) that the phenyl units are oriented in the plane of the film and indeed a thin layer of highly strained polymer can be produced by solvent casting. Since the bulk of excitation radiation is absorbed in the surface of a sample the major component of the emission can occur from a birefringent surface layer, so that the observed polarisation/depolarisation can be dominated by this influence rather than by e.e.m., and therefore by sample preparation.

The structure of amorphous polystyrene has been examined using X-rays (16) and the analysis suggests that phenyl-phenyl contacts dominate with both inter and intra molecular contacts resulting in stacking effects. Such an interaction is consistent with the stability derived from face to face configurations (17,18). Thus it is reasonable to propose that excimer formation in polymer films results from a liberational motion of the face-to-face phenyl units following vibrational relaxation of the excited monomer unit. This mechanism would account for the non-observation of excimer at very low temperatures at which the vibrational energy is insufficient to attain the required conformation.

Blends of polystyrene with poly(vinyl methyl ether) have been examined by Gelles and Frank (19) and the results analysed using sophisticated statistical methods involving e.e.m. A number of important assumptions are made:

(a) rotational sampling is not important.
(b) intermolecular excimer occurs in concentrated but not dilute solutions.
(c) the concentration of chain conformations is independent of molecular weight.
(d) the probability of hopping is the same for all steps.

It would be reasonable to confirm some of these assumptions by other experimental means before citing this work as unequivocal evidence for e.e.m. Furthermore the authors made no correction for monomer/excimer band overlap.

Solid state behavior of poly(vinyl naphthalene) is somewhat similar to that of polystyrene. The contribution from excimer is much diminished at low temperatures. Indeed in frozen dilute solutions of copolymers containing added traps excimer disappears totally (20).

In a number of ways the photophysical behaviour of small molecules and polymers are similar. The rejection of e.e.m. among the low molecular weight analogues of polystyrene and poly(vinyl naphthalene) requires the proposition that the macromolecules have some special property which promotes e.e.m., if that phenomenon does in fact occur. One very obvious factor is the 1,3 disposition of the aromatic groups along the main chain. However it is not apparent why this factor should enhance chromphore to chromophore energy transfer. A definite consequence of the placing of chromophores along a polymer backbone is the enhanced possibility of forming excimer by rotational relaxation. Conformation of the molecules, both large and small, is important. Polymer molecule/solvent interactions have been demonstrated by variation of solvent but an early observation by Hirayama (21) is worth highlighting. It was shown that $I_E/I_M$ for 1,3 diphenylpropane changed markedly on changing solvent indicating that phenyl/phenyl and phenyl/solvent interactions played an important role in determining the efficiency of forming the excimeric structure.

## Literature Cited


1. Polymer Photophysics and Photochemistry, J. E. Guillet, Cambridge University Press, 1985.
2. P. K. Ludwig and C. D. Amata, J. Phys. Chem., 72, 3725 (1968).
3. J. B. Birks and M. S. S. C. P. Leite, J. Phys. B. Atom. Molec. Phys., 3, 513 (1970).
4. T. Ishii, T. Handa, and S. Matsunaga, Macromolecules, 11, 40 (1978).
5. T. Ishii, T. Handa, and S. Matsunaga, Makromol. Chem., 178, 2351 (1977).
6. B. R. M. Kyle and T. Kilp, Polymer, 25, 989 (1984).
7. D. Phillips, A. J. Roberts, and I. Soutar, Polymer, 22, 427 (1981).
8. J. R. MacCallum, Polymer, 23, 175 (1982).
9. L. Gardette and D. Phillips, Polymer, 25, 336 (1984).
10. D. Phillips, A. J. Roberts, G. Rumbles, and I. Soutar, J. Polym. Sci., Polym. Phys. Ed., 20, 1739 (1982).
11. K. Horie, I. Mita, S. Tagawa, Y. Tabata, and M. Washio, J. Chem. Phys., 79, 3996 (1983).
12. D. Phillips, A. J. Roberts, and I. Soutar, Macromolecules, 16, 1593 (1983).
13. C. Friedrich, F. Laupretre, C. Noel, and L. Monnerie, Macromolecules, 14, 1119 (1981).
14. C. W. Frank and L. A. Harrah, J. Chem. Phys., 61, 1526 (1974).
15. W. J. Prest, Jr., and D. J. Luca, J. Appl. Phys., 50, 6067 (1979).
16. G. R. Mitchell and A. H. Windle, Polymer, 25, 906 (1985).
17. D. Y. Yoon, P. R. Sundararajam, and P. F. J. Flory, Macromolecules, 8, 776 (1975).
18. P. C. Hagele and L. Beck, Macromolecules, 10, 213 (1977).
19. R. Gelles and C.W. Frank, Macromolecules, 15, 741 (1982).
20. D. Ng, K. Yoshiki, and J. E. Guillet, Macromolecules, 16, 568 (1983).
21. F. Hirayama, J. Chem. Phys., 42, 3163 (1985)


RECEIVED September 2, 1987

Chapter 24

# Complex Decay of Fluorescence in Synthetic Polymers

**David Phillips**

**The Royal Institution, 21 Albemarle Street, London, W1X 4BS, England**

Some simple fluorescence decay laws are given which result from models concerned with relaxation phenomena, heterogeneity, molecular motion and excitation migration in synthetic polymers exhibiting fluorescence. Application of such laws to decays from a poly(diacetylene) in rigid crystal form; to polystyrene in fluid solution, and to vinyl naphthalene co-polymers in fluid solution are considered briefly.

The great sensitivity of fluorescence spectral, intensity, decay and anisotropy measurements has led to their widespread use in synthetic polymer systems, where interpretations of results are based upon order, molecular motion, and electronic energy migration (1). Time-resolved methods down to picosecond time-resolution using a variety of detection methods but principally that of time-correlated single photon counting, can in principle, probe these processes in much finer detail than steady-state techniques, but the complexity of most synthetic polymers poses severe problems in interpretation of results.

The single or dual exponential fluorescence decay encountered in simple small molecules in fluid or rigid media, is not often normally expected in synthetic polymers. There are, in general, several causes which have been reviewed (1,['Non-exponential kinetics.' Phillips, D. in NATO ASI 'Excited-state Probes in Biochemistry and Biology' Szabo, A.G.and Masotti, L. Eds. Plenum Press, (in press.] ) of the deviations from simple mathematical forms of decay law. Stated briefly, these are heterogeneity, motion, complex formation, and energy transfer and migration.

## HETEROGENEITY

For more than one simultaneously excited,non-interacting species, the decay of total fluorescence will be described in principle by 'Equation 1'. The situation with two non-interacting species is fairly common, but as the number of species increases, interactions such as energy transfer are bound to become more probable, complicating the kinetics.

0097–6156/87/0358–0308$06.00/0

$$I(t) = \sum_i A_i e^{-t/\tau_i} \tag{1}$$

In the extreme of a large number of non-interacting sites, such as molecules adsorbed on a solid surface, in defects in molecular crystals or in some polymeric species, the decay may be better described by a distribution of decay times, suitably weighted about some mean value.

A recent treatment by Albery et al (3) gives a rate-parameter k as a distribution represented by

$$k = \bar{k} \exp(\gamma x) \tag{2}$$

Thus the decay of concentration C of a species from initial concentration $C_0$ is given by

$$\frac{C}{C_0} = \frac{\int_{-\infty}^{+\infty} \exp(-x^2) \exp[-\tau \exp(\gamma x] dx}{\int_{-\infty}^{+\infty} \exp(-x^2) dx} \tag{3}$$

where

$$\tau = \bar{k}t, \text{ and } \int_{-\infty}^{+\infty} \exp(-x^2) dx = \pi^{1/2}$$

Since any sample of polymer is characterised by a distribution of molecular weights and the fluorescence of a chromophore is in principle environmentally sensitive, even for non-interacting chromophores it would not be surprising if a distribution of decay times were observed in these situations, which however, corresponds to that observed in a free chromophore in solution, and the decay should be modelled adequately by a rate constant. This is certainly not the case for interacting chromophores, where the local environment will be critical in determining the decay rate of any particular fluorophore. In a homopolymer, the principal cause of heterogeneity will be the tacticity of the polymer, isotactic, syndiotactic and atactic polymers being expected to behave very differently. In cases where nominally 'atactic' polymers consist of isotactic and syndiotactic sequences, the decay may in favourable simple cases be interpretable in terms of a summation of exponential decays of two kinetically distinct species. For a wide distribution of sites, a kinetic model recognising this heterogeneity may be more appropriate, although this yields information of limited usefulness.

In co-polymers, heterogeneity of environment of a chromophore by virtue of composition becomes of overriding concern. In our earlier work on co-polymers of vinyl naphthalene (4-10) and polystyrene (11-13), the models adopted to explain results deliberately emphasised heterogeneity at the expense of, say, energy migration.

## RELAXATION

In many synthetic polymers, particularly the vinyl aromatic type, excimer formation is significant. In the simplest case of excimer formation in free molecules in fluid solution, the decays of uncomplexed chromophore (monomer) and excimer are of the form

$$I_M(t) = A_1e^{-\lambda_1 t} + A_1e^{-\lambda_2 t} \quad (4)$$

$$I_E(t) = A_3(e^{-\lambda_1 t} - e^{-\lambda_2 t}) \quad (5)$$

In synthetic polymers, the interpretation is necessarily more difficult. The form of 'Equation 4' and 'Equation 5' requires that the kinetics of formation and decay of complexes are modelled adequately by rate-constants and that they take place in a homogeneous medium. If, as in synthetic polymers, the population of excimer trap sites, may occur through energy migration or rotational diffusion, a rate-constant may not be an adequate representation of the process, some time-dependent parameter being required (see below.) Heterogeneity may also play an important role. Thus in earlier work the fluorescence decay of excimer-forming polymers was modelled adequately by a scheme based upon simple excimer kinetics to which had been added terms to account for the occurrence in co-polymers of monomer sites which, by their isolation, could not form excimers (4-10). For polymers which contain isotactic and syndiotactic sequences, or rather, are made up of meso and racemic triads (14), the kinetics may be similarly a superimposition of simple schemes appropriate for the different sequences.

## MOTION

Translational diffusion results in a decay law of the form (15) of 'Equation 6'. Itagaki (16) et al have proposed a decay law for restricted rotational motion of similar type.

$$I(t) = \exp(-At-2Bt^{1/2}) \quad (6)$$

They propose that in excimer-forming vinyl aromatic polymers in fluid solution, the

$$I(t) = A \exp\{-(at + bt^{1/2})\} + B \exp(-ct) \quad (7)$$

rate-determining step in excimer formation is rotational diffusion, leading to use of a term such as 'Equation 7' to model the fluorescence. Data from other workers data were claimed to be compatible with this model, although the normal stringent tests of acceptability were not applied in this case. In general, fluorescence anisotropy measurements are of more importance in determining motion in synthetic polymers in fluid solutions but such measurements are beyond the scope of this paper.

## ENERGY MIGRATION

In cases where energy migration is a dominant feature of luminescence, as in molecular crystals, various forms of decay are expected depending upon circumstances, but relying upon solutions, usually complex, to the basic rate equations where E(t) is the time-dependent population of the initially excited (exciton) state, T(t) the population of the trap state, $k_E$ the decay rate constant for band states, $k_T$ the decay rate constant for the trap states, and $k_L(t)$ the time-dependent

$$\dot{E}(t) = -[k_E + k_L(t)]\, E(t) \tag{8}$$

$$\dot{T}(t) = -k_T T(t) + k_L(t)\, E(t) \tag{9}$$

trapping rate functions, the form of which depends upon the effective transport topology. For a strictly one-dimensional transport, Fayer has given the form of $k_L(t)$ as'Equation 10'.

$$k_L(t) = At^{-1/2} \tag{10}$$

For quasi-one-dimensional, two-dimensional, and three-dimensional diffusional processes, other forms are appropriate (17-21). Thus very extensive theoretical and picosecond experimental work on electronic excited state transport in finite volumes of randomly distributed molecules has been reported, which shows that there are significant deviations in the behaviour of finite volume systems compared with the infinite volume systems considered above. The treatment is mathematically complex and the results will not be given here explicitly. Frederickson and Frank (22) have used this treatment to suggest possible forms for the decay of monomer and growth and decay of excimer fluorescence in vinyl aromatic polymers where electronic energy migration might be a dominant process. In this work the suggestion was made that complex mathematical forms resulting from this treatment might be simulated by multiple exponentials, and thus care was needed in attaching physical significance to parameters derived from this procedure. A simplification of the Frederickson and Frank approach gave as the decay law for a monomeric species in a polymer exhibiting energy migration the expression.

$$I(t) = A \exp\{-(at+bt^{1/2})\} + B \exp(-ct) \tag{11}$$

An exact solution to the 'Equation 8' and'Equation 9' for one-dimensional and higher dimensionality cases of diffusion has been developed. However this is beyond the scope of the present paper, although results in one case are discussed below.

The concept of 'fractal' dimensionality is clearly of use in interpreting results of energy transfer experiments; thus for example for a two-dimensional donor-acceptor system, the fluorescence decay function of the donor is given by

$$I(t) = \exp\,[-t/\tau_D - \gamma_A (t/\tau_D)^{\beta}] \tag{12}$$

$$\text{where, } \beta = \bar{d}/s,\ \gamma - x_A\,(d/\bar{d}) V_d R^d\, \Gamma\ (1-\beta) \tag{13}$$

$\tau_D$ is the lifetime of the donor, s is the order of the multipolar interaction, $x_A$ is the fraction of fractal sites occupied by the acceptor, d is the Euclidean dimension, $\bar{d}$ the fractal dimension, $R_o$ the critical transfer distance and $V_d$ the volume of a unit sphere in d dimension.

## TESTING OF MODELS

It is clear from the above discussion that selection of a mathematical model which will describe uniquely the fluorescence decay of a complex polymer system is at best a hazardous procedure. We describe below three areas of such fitting: one successful, another possibly so, a third in which recent experiments may require that previous conclusions be revised.

POLYDIACETYLENE ONE-DIMENSIONAL SYSTEM. 1OH is a monomer which leads to the production of orientated fibrous polymer (PDA-1OH) on solid-state polymerization (27). Detailed methods of preparation of this polymer can be found elsewhere (28,29) the structure being given below, where R and R' are $-CH_3$ and $-CH_2OH$ respectively for 1OH and are, so far, the smallest substituents capable of providing a unique polymerization direction and a reactive lattice packing, primarily as a virtue of hydrogen-bonding occuring between the $-CH_2OH$ groups. On polymerization fibrillation occurs as a result of the considerable strain built up. The latter is due to the difference of the packing nature of the monomer molecules (0.41 nm apart) and the polymer repeat unit length (0.49 nm.) The weak Van der Waals interactions of the $-CH_3$ groups are therefore disrupted. This produces fibres having good crystallinity with a high degree of chain alignment as shown by the optical dichroism. The polymer chain axes lie along the fibre axis as revealed by electron diffraction studies. Typically, the fibres are more than 10 μm in length and about 60 nm in diameter, as obtained from small-angle-X-ray scattering studies (27). This polymer is clearly less perfect than other PDA single crystals, and as such luminesces. Local defects such as chain kinks, dislocations, cross-links (unlikely), etc. therefore alter the dimensionality to $\hat{d} \geq 1$. In other words these systems are quasi-one-dimensional. Misaligned chains have been observed by electron microscopy. The effect of disorder (present only to a small extent) should have important consequences on the diffusivity and drift velocity of carriers or excitations in the material. The exact 1D decay law for excitons has been derived (25) as follows. The survival fraction of random walkers in 1D $n(\eta = 0,t)$ for the case of physical interest ($\eta \leq x$) where x is the concentration of deep traps, $\eta$ is a dimensionless field parameter, and W is the zero field jump rate between neighbours, is given by

$$n_1(t) = \frac{8}{\pi^2} \int_0^{\infty} ds \, \frac{s \exp(-\pi^2 x^2 Wt/s^2)}{(e^s - e^{-s})} \qquad (12)$$

The 1D decay law from photoconduction experiments is known to be of the form

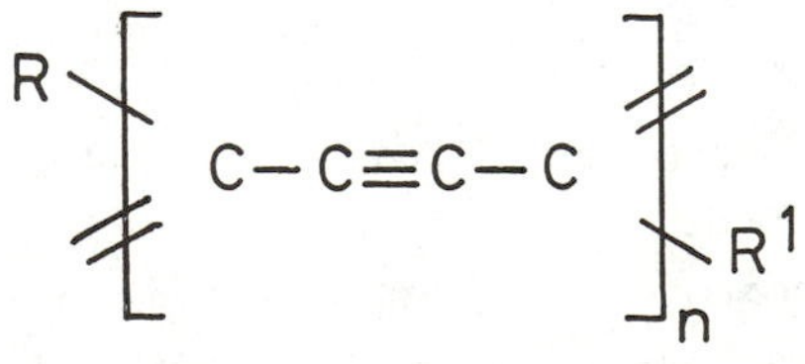

$$\left[ \begin{matrix} R \\ \\ \end{matrix} \; C-C\equiv C-C \; \begin{matrix} \\ R^1 \end{matrix} \right]_n$$

$$E(t) = n_1(t)\, e^{-K_E t} \qquad (13)$$

Substitution of these equations gives for the time-dependence of T

$$T(t) = \frac{8}{\pi^2} e^{-Yt'} \int \frac{S\,ds}{(e^s - e^{-s})\gamma} [1-e^{-\gamma t'/s}] \qquad (14)$$

where $\gamma = 1+s^2Z$, $t' = W_1 t = \frac{\pi^3 (PA) t}{16}$, $Z = \frac{K_E - K_T}{W_1}$,

$PA = \frac{16\pi^2 x^2 W}{\pi^3}$, $W_1 = \pi^2 x^2 W$, $Y = \frac{K_T}{W_1}$

This equation can be solved numerically, and in normalised form the function can be represented on a log vs $t^\alpha$ plot as a straight line when $\alpha = 0.45$ (Figure 1.) The actual fluorescence decay of 1OH at room temperature is shown in Figure 2.

Detailed analysis of several decay and pump curves revealed good fits for two exponential terms. However, the time constants ($\tau_i$) obtained crucially depended on the sampling-window of the decay profile chosen. Longer lifetime values were obtained at longer times; t-values changing from ca. 20 ps - 2.25 ns. This suggests that the decay is neither single-nor double-exponential but involves a distribution of decay times. It must be appreciated that very lengthy programming effort would be required to fit these data, after *either* deconvolution of the data *or* reconvolution of the fitting function, to the equations derived from the MPW-theory (25) for 1D-exciton migration. To get over this problem the data were fitted to six exponentials to deconvolve from the instrument response function and then replotted on a log intensity vs $t^\alpha$ plot. A linear plot was obtained for $\alpha = 0.42$ (Figure 3), in good agreement with theory. The same treatment for data obtained at 4.2K yielded the best fit for $\alpha = 0.45$, as shown in Figure 4.

We conclude therefore that in the case of this highly-ordered, rigid polymer, time-resolved fluorescence measurements of excitation migration yield results which are compatible with an exact solution for one-dimensional diffusion.

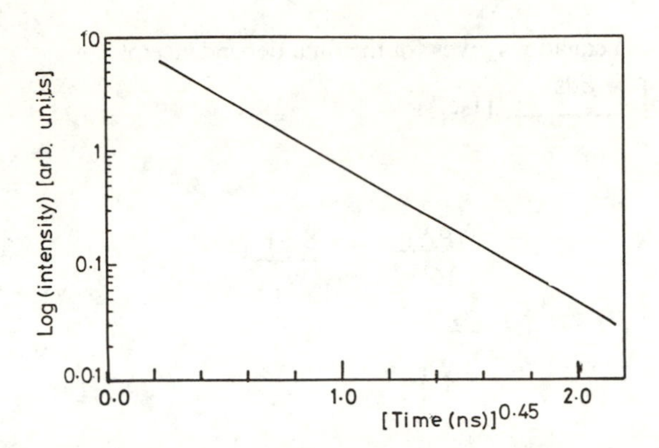


Figure 1.Simulated (1D exact) decay based upon MPW theory[25]. Plot of intensity vs. $t^{\alpha}$, $\alpha = 0.45$ (from Ref. 27.)

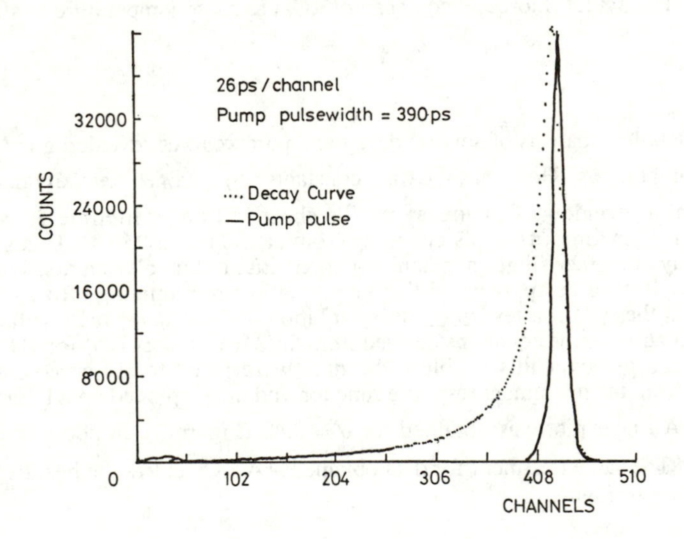


Figure 2. Fluorescence decay of PDA - 1OH (300K) Raw data $\lambda_e = 476.5$ nm, $\lambda_r = 565$ nm (from Ref. 27.)

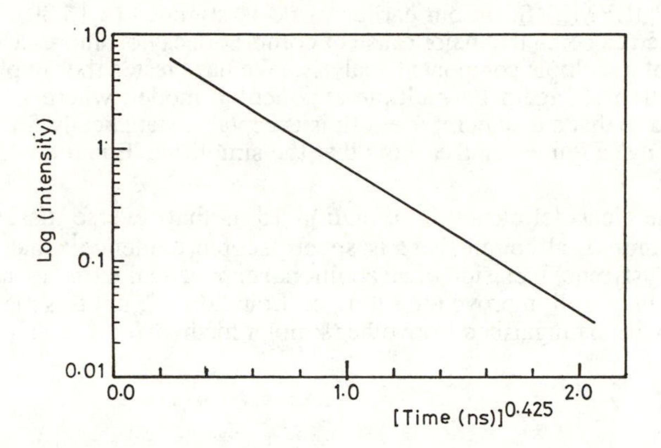


Figure 3. Experimental fit of log intensity vs $t^{\alpha}$ for PDA 1OH fluorescence at 300K, with $\alpha = 0.425$.

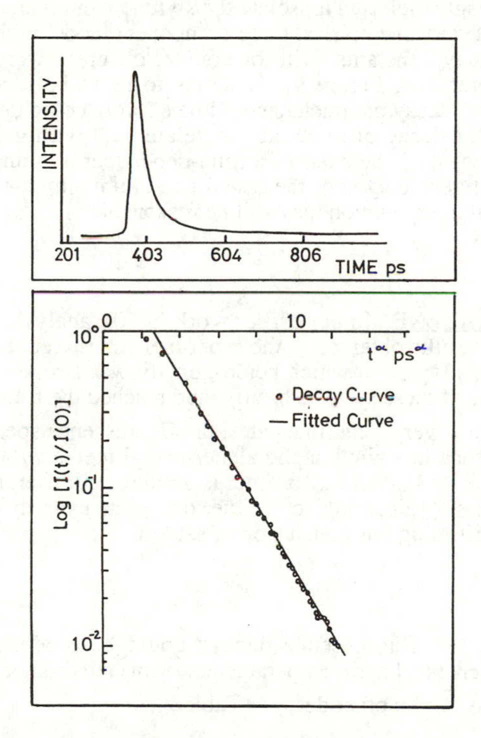


Figure 4. Experimental fit of log intensity vs $t^{\alpha}$ for PDA 1OH fluorescence at 4.2K, with $\alpha = 0.45$.

STYRENE POLYMERS. In our earlier work on styrene (12,13,30), heterogeneity was emphasised as being the major cause of complex decay of fluorescence, leading to the adoption of a multiple component analysis. We have tested the simplest alternative model, 'Equation 15' against a multiple exponential model, where we have shown, (Figures 5 and 6) that a dual component fit is acceptable statistically for homopolymer monomer decay. Figures 7and 8 show that the simplified 'Equation 15' is certainly unacceptable.

One clear deficiency in 'Equation 15' is that reverse dissociation of the excimer is omitted, although there is spectroscopic evidence[12] that it does occur slightly in polystyrene. Inclusion of an additional exponential term to account for this phenomenon may well improve the fitting of 'Equation 15', but this method will only with difficulty be distinguished from other simpler methods.

$$I(t) = \exp[-(at + bt^{1/2})] \tag{15}$$

That simple models suffice in polystyrene polymers is further illustrated by work we have carried out on polystyrene copolymerised with a small concentration of

phenyl oxazole POS (11). In this work, careful study of the decay characteristics of emission at wavelengths selected to isolate the styrene monomer, styrene excimer, and POS trap, when fitted to an empirical three-component model without constraint, gave the same three decay constants, with of course, different weighing factors in the different spectral regions, Figure 9. It seems to us to be inconceivable that the recovery of identical decay parameters could be entirely coincidental. Moreover, the values recovered for decay of monomer correlate well with grow-in of trap. This would not be expected on the basis of a time-dependent trapping function. We feel that in the case of styrene polymers, the case for careful multiple-component fitting and subsequent physical interpretation may still be reasonable.

VINYL NAPHTHALENE. In our earlier work (4,10), analysis of fluorescence was based solely upon results obtained in the monomer fluorescence spectral region, and again a model was adopted in which heterogeneity was stressed. We have recently reinvestigated some of these systems briefly, and reached the following conclusions.

(1) Even at very long time gates, the fluorescence spectrum in the excimer region of fluorescence in a vinyl naphthalene/methyl methacrylate co-polymer is not time-dependent, and thus either the excimer is unique, or , different excimer sites have identical decay times. 'Tail-fitting' to excimer decays at a variety of wavelengths gave the same result confirming this conclusion. (Table 1.)

(2) Excimer fluorescence decays could be modelled adequately by a triple-component scheme, but decay parameters recovered did not correlate well with those from analysis of monomer decay. (Table 2.)

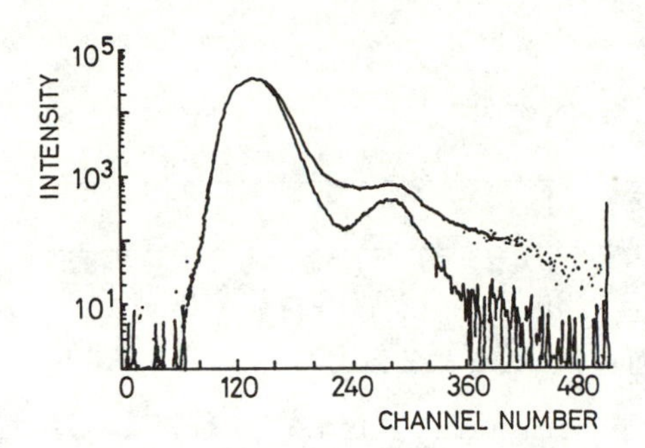


Figure 5. Double-exponential fit to polystyrene decay measured at 270 nm with excitation at 257.25 nm.

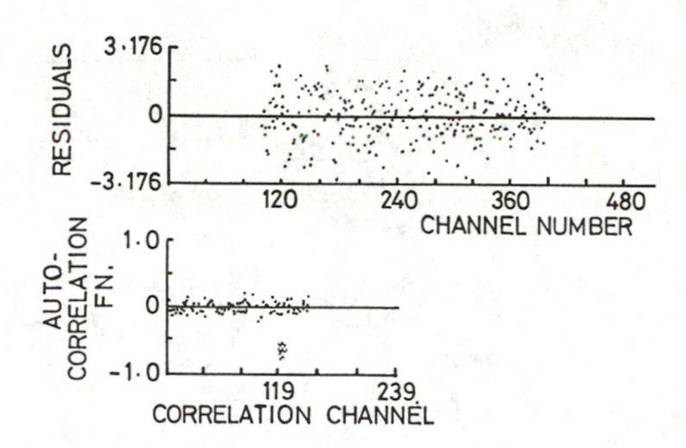


Figure 6. Residuals and autocorrelation for plots in Figure 5.

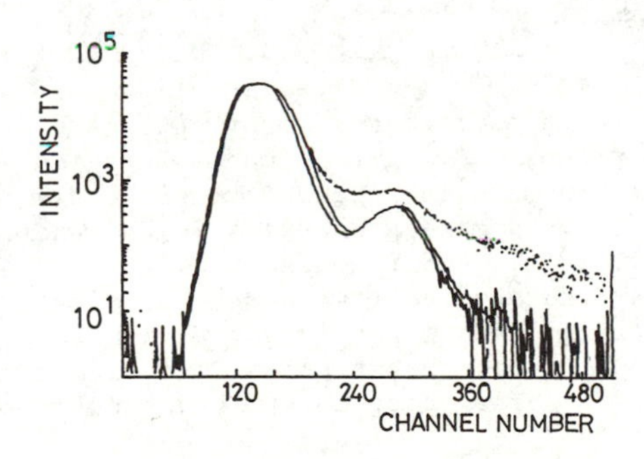


Figure 7. Fluorescence decay of polystyrene measured at 270 nm fitted to Equation 15.

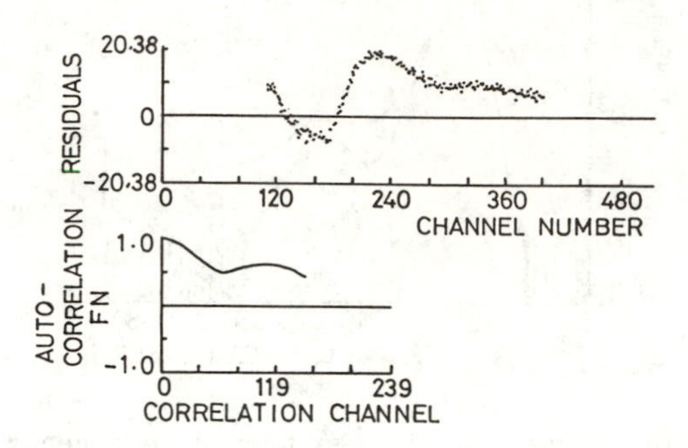


Figure 8. Residuals and autocorrelation for plots in Figure 7.

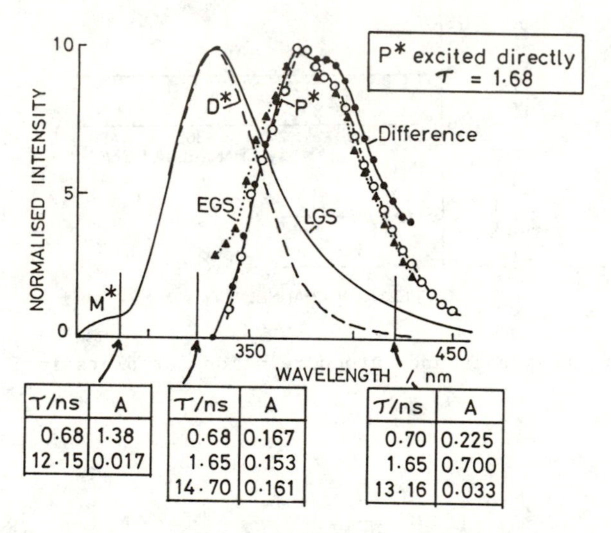


| τ/ns | A |
|---|---|
| 0.68 | 1.38 |
| 12.15 | 0.017 |

| τ/ns | A |
|---|---|
| 0.68 | 0.167 |
| 1.65 | 0.153 |
| 14.70 | 0.161 |

| τ/ns | A |
|---|---|
| 0.70 | 0.225 |
| 1.65 | 0.700 |
| 13.16 | 0.033 |

Figure 9. Fluorescence spectra and decay characteristics of POS containing polystyrene. M*, styrene monomer region, dual decay kinetics. D*, styrene excimer region, triple decay characteristics (double fit shown does not correlate with other wave lengths, thus meaningless). P* is POS fluorescence, triple decay characteristics when styrene excited (see box), but single, $\tau$ = 1.68 ns when excited directly. EGS is early-gated time-resolved spectrum which matches closely spectrum of P* excited directly, and difference between late-gated spectrum LGS and known spectrum of D*.

Table 1 - Summary of Lifetime Data Derived by Tail Fitting the Fluorescence Decays Recorded in the Excimer Spectral Region of Polymer Sample b †

| Emission Wavelength | Tail Fitting | | | | | |
|---|---|---|---|---|---|---|
| nm | 83-220 ns (67%-1%)* | 95-220 ns (33%-1%)* | 108-220 ns (30%-1%)* | 120-220 ns (20%-1%)* | 133-220 ns (13%-1%)* | 145-220 ns (10%-1%)* |
| 450 | 47.0 ± 0.4 | 47.6 ± 0.5 | 48.2 ± 0.5 | 47.7 ± 0.4 | 49.1 ± 0.4 | 49.7 ± 0.7 |
| 460 | 46.5 ± 0.9 | 46.9 ± 0.9 | 47.3 ± 0.9 | 47.4 ± 0.9 | 47 ± 1 | 47.8 ± 0.8 |
| 470 | 46.3 ± 0.4 | 46.8 ± 0.3 | 47.2 ± 0.3 | 47.7 ± 0.7 | 48.1 ± 0.6 | 48.3 ± 0.9 |
| 480 | 45.5 ± 0.4 | 46.0 ± 0.5 | 46.5 ± 0.6 | 46.8 ± 0.7 | 47.1 ± 0.8 | 47.7 ± 0.5 |
| 500 | 45.4 ± 0.8 | 45.9 ± 0.9 | 46.3 ± 0.9 | 47 ± 1 | 47 ± 1 | 47 ± 1 |

*Fitting range expressed as percentage of the number of counts in the channel containing the maximum number of counts.

†Details of sample as in Table 2.

Table 2 - Summary of the Lifetime Data Derived from Triple Exponential Analyses of the Fluorescence Decay Curves Recorded at 330 nm and 480 nm for Polymer Samples a, b, c, and d †

| Polymer Sample | Emission Wavelength (nm) | $A_1$ | $\tau_1$ (n sec) | $A_2$ | $\tau_2$ (n sec) | $A_3$ | $\tau_3$ (n sec) |
|---|---|---|---|---|---|---|---|
| a: $f_n$ = 0.27<br>$f_{nm}$ = 0.055<br>$\bar{\ell}_n$ = 1.32 | 330 | 0.18 ± 0.01 | 3.4 ± 0.4 | 0.76 ± 0.01 | 13.2 ± 0.2 | 0.06 ± 0.01 | 36 ± 2 |
| | 480 | −0.23 ± 0.09 | 4.0 ± 0.9 | −0.3 ± 0.2 | 4.7 ± 0.2 | 1.00 | 45 ± 1 |
| b: $f_n$ = 0.58<br>$f_{nm}$ = 0.319<br>$\bar{\ell}_n$ = 2.27 | 330 | 0.42 ± 0.01 | 2.3 ± 0.2 | 0.49 ± 0.01 | 9.1 ± 0.2 | 0.088 ± 0.009 | 39 ± 2 |
| | 480 | −0.21 ± 0.05 | 1.0 ± 0.3 | −0.46 ± 0.02 | 5.6 ± 0.1 | 1.00 | 41.6 ± 0.2 |
| c: $f_n$ = 0.75<br>$f_{nm}$ = 0.557<br>$\bar{\ell}_n$ = 3.95 | 330 | 0.58 ± 0.02 | 1.5 ± 0.3 | 0.32 ± 0.02 | 7.8 ± 0.8 | 0.09 ± 0.01 | 35 ± 2 |
| | 480 | −0.19 ± 0.02 | 1.4 ± 0.4 | −0.31 ± 0.01 | 9.3 ± 0.9 | 1.00 | 40 ± 2 |
| d: $f_n$ = 1.00<br>$f_{nm}$ = 1.00 | 330 | 0.83 ± 0.02 | 0.45 ± 0.06 | 0.12 ± 0.01 | 5.5 ± 0.5 | 0.051 ± 0.007 | 23 ± 2 |
| | 480 | | | | | 1.00 | 35 |

***evaluated by tail fits.**

† copolymers of vinyl naphthalene and methyl methacrylate with the composition given, where $f_n$ is mole fraction of naphthalene chromophores, $f_{mn}$ is the fraction of linkages between naphthalene species, and $\bar{l}_n$ the mean sequence length of aromatic species.

(3) Pre-exponential factors for excimer decay do not sum to zero. Analysis of 'synthetic' data recovers input parameters perfectly, thus the result must be taken as real.

In the light of these new experiments, the kinetic model based upon heterogeneity is no longer internally consistent, and future work on these polymers must consider other appropriate models, including those of Itagaki et al, (16) and Frederickson and Frank (22).

## CONCLUSIONS

We conclude that interpretation of even the most precise experiments on fluorescence decay in synthetic polymers will be difficult, given the complexities of such systems. However, with physically well-defined systems, adequate discrimination between models should be possible. In less well defined systems, such as co-polymers of vinyl aromatic monomers in fluid solution, discrimination between models may be more intractable, but some progress can be made if extensive, and complete, experiments are attempted.

## Acknowledgments.

It is a pleasure to acknowledge the contributions to this work of A.J.Roberts,G.Rumbles,J-L.Gardette,S.D.D.V.Rughooputh,R.Drake,I.Soutar and D.Bloor.Financial support form the Science and Engineering Research Council, UK and the United States Army European Research Office is gratefully acknowledged.

## Literature Cited.

1. 'Analysis of fluorescence decay data from synthetic polymers: heterogeneity, motion and migration.' Phillips, D.and Soutar, I. in NATO ASI 'Photophysical and Photochemical Tools in Polymer Science'; Winnik M., Ed.; D. Reidel Publishers; 1986.
2. Albery,W.J.; Bartlett,P.N.; Wilde, C.P.; Darwent,J.R. J.Amer.Chem.Soc. 1985, 107, 1854.
3. Roberts, A.J.; O'Connor, D.V.; Phillips, D. Ann.N.Y.Acad.Sci. 1981, 366, 109.
4. Roberts, A.J.; Soutar I.; Phillips, D. J.Polym.Sci., (Polymer Letters), 1980, 18, 123.
5. Phillips, D.; Roberts, A.J.; Soutar, I. J.Polym.Sci. Polymer Physics., 1980, 18, 2401.
6. Phillips, D.; Roberts A.J.; Soutar, I. Polymer, 1981, 22, 293.
7. Phillips, D.; Roberts A.J.; Soutar, I. Eur.Polym.J., 1981, 17, 101.
8. Phillips, D.; Roberts A.J.; Soutar, I.Polymer, 1981, 22, 427.
9. Phillips, D.; Roberts A.J.; Soutar, I.J.Polym.Sci., (Polymer Physics), 1982, 20, 411.
10. Phillips, D.; Roberts A.J.; Soutar, I. Macromolecules, 1983, 16, 1593.
11. Phillips, D.; Roberts A.J.; Rumbles, G.; Soutar, I. Macromolecules, 1983, 16, 1597.
12. Rumbles,G. Ph.D.Thesis, University of London, 1984.
13. Vandendriessche,J.; Goedeweeck,R.; Collart P.; de Schryver ,F.C.; in 'Photophysical and Photochemical Tools in Polymer Science.' Winnik, M.A. Ed.; D.Reidel Publishing Co., 1986; p.225.

14. Nemzek, T.L.; Ware, W.R.J.Chem.Phys., 1975, 62, 477.
15. Itagaki, H.; Horiè, K.; Mita, I.; Macromolecules, 1983, 16, 1395.
16. 'Exciton coherence.' Fayer ,M.D.in 'Spectroscopy and Excitation Dynamics of Condensed Molecular Systems.'Agranovich,V.M.; Hochstrasser, R.M. Eds. ; North-Holland: Amsterdam, 1983, p.185.
17. Hunt , I.G.; Bloor, D.; Movaghar, B. J.Phys.C.Solid State Phys., 1983, 16, L623.
18. Gochanour, C.R.; Anderson, H.C.; Fayer, M.D.J.Chem.Phys., 1979, 70, 4254.
19. Ediger, M.D.; Fayer, M.D. J.Chem.Phys., 1983, 78, 2578.
20. Dwayne Miller, R.J.; Pierre ,M.; Fayer, M.D. J.Chem.Phys., 1983, 78, 5138.
21. Frederickson, G.H.; Frank, C.W. Macromolecules, 1983, 16, 572.
22. Movaghar, B.; Sauer, G.; Wurtz, D. Solid State Communications, 1981, 39, 1179.
23. Movaghar, B.; Sauer , G.; Wurtz, D. J.Stat.Phys., 1982, 27, 473.
24. Movaghar, B.; Pohlmann , B.; Wurtz, D. Phys.Rev.A, 1984, 29, 1568.
25. Klaftov, J.; .Blumen, A .J.Chem.Phys., 1984, 80, 875.
26. Rughooputh, S.D.D.V.Ph.D.Thesis, University of London, 1985.
27. Webman, I. Phys.Rev.Lett., 1984, 52, 220.
28. Zumofen, G.; Blumen , A.;Klaftov, S. J.Physique Lett., 1984, 45, L49.
29. Rumbles, G.Ph.D.Thesis, University of London, 1984.
30. Drake,R.C. Ph.D.Thesis, University of London, 1986.

RECEIVED August 28, 1987

# Chapter 25

# Ensemble Average Conformation of Isolated Polymer Coils in Solid Blends Using Electronic Excitation Transport

K. A. Peterson, M. B. Zimmt, S. Linse, and M. D. Fayer

Department of Chemistry, Stanford University, Stanford, CA 94305

A method for calculating observables resulting from incoherent excitation transport among chromophores randomly tagged in low concentration on isolated, flexible polymer chains is described. The theory relates the ensemble average root-mean-square radius of gyration ($\langle R_g^2\rangle^{1/2}$) of a polymer coil to the rate of excitation transport among the chromophores. Analysis of experiments which monitor the rate of excitation transport among naphthyl chromophores in low concentration on isolated coils of poly-(2-vinylnaphthalene-co-methyl methacrylate) in poly-(methyl methacrylate) (PMMA) host by time-resolved fluorescence depolarization spectroscopy allows the quantitative determination of the copolymer $\langle R_g^2\rangle^{1/2}$. $\langle R_g^2\rangle^{1/2}$ is determined for 23,000 $M_w$ and 60,000 $M_w$ copolymers containing 9% 2-vinylnaphthalene (2-VN) monomers. The results are identical to determinations of $\langle(R_g^2\rangle^{1/2}$ for PMMA in $\theta$-solvents by other methods. The dependence of $\langle R_g^2\rangle^{1/2}$ on the fraction of 2-VN monomer in the copolymer is determined for three 23,000 $M_w$ copolymers containing 9%, 6% and 4% 2-VN. The measured $\langle R_g^2\rangle^{1/2}$ are independent of 2-VN concentration. The results demonstrate the quantitative utility of excitation transport techniques in the investigation of polymer conformation in the solid state.

There has been increasing interest in recent years in using incoherent electronic excitation transport as a probe of molecular interactions in solid state polymer systems. The macroscopic properties of such systems arise from the microscopic interaction of the individual polymer chains. The bulk properties of polymer blends are critically dependent on the mixing of blend components on a molecular level. Through the careful adjustment of the composition of blends technological advances in the engineering of polymer materials have been made. In order to understand these systems more fully, it is desirable to investigate the interactions

0097-6156/87/0358-0323$06.00/0

of individual polymer chains with the host environment. Excitation transport observables are sensitive to the spatial orientation and separation of chromophores and can provide a direct probe of these microscopic interactions.

The dependence of excitation transport on local chromophore concentration has been used to provide qualitative information on the characteristics of polymers in blends. Excimer fluorescence resulting from excitation transport has been employed to characterize polymer miscibility, phase separation and the kinetics of spinodal decomposition (1-3). Qualitative characterization of phase separation in blends (4,5) and the degree of chain entanglement as a function of sample preparation and history (6,7) has also been investigated through transport with trapping experiments. In these experiments one polymer in the blend contains donor chromophores and the second contains acceptors. Selective excitation of the former and detection of the latter provides a qualitative measure of interpenetration of the two components.

Due to the sensitivity of electronic excitation transport to the separation and orientation of chromophores, techniques which monitor the rate of excitation transport among chromophores on polymer chains are direct probes of the ensemble average conformation (8). It is straightforward to understand qualitatively the relationship between excitation transport dynamics and the size of an isolated polymer coil which is randomly tagged in low concentration with chromophores. An ensemble of tagged coils in a polymer blend will have some ensemble averaged root-mean-squared radius of gyration, $\langle R_g^2\rangle^{1/2}$. If in one host $\langle R_g^2\rangle^{1/2}$ for the guest coils is large because of favorable guest-host thermodynamic interactions, the average distance between chromophores will be large. Since the rate of excitation transport depends on $1/r^6$ where r is the chromophore separation, transport will be slow. If the same guest polymer is placed in a different host in which the guest-host thermodynamic interactions are less favorable, the coils will contract, and the average chromophore separation will decrease. This decrease will result in more rapid excitation transport. The $1/r^6$ distance dependence makes the excitation transport observables very sensitive to small changes in $\langle R_g^2\rangle^{1/2}$.

The two most common methods used to determine $\langle R_g^2\rangle^{1/2}$ of individual polymer chains are Rayleigh scattering and neutron scattering. Rayleigh, or light scattering has been extremely useful in studying polymers in solution (9,10). However, this technique is not easily applied to polymers in the amorphous solid state due to lack of contrast in refractive index of the blend components. Neutron scattering has been successfully applied to studying isolated polymer coils in solid blends (11-15). Although the neutron scattering technique has the advantage of providing a direct probe of chain structure, it involves several limitations which may be overcome by excitation transport techniques. First, in order to produce contrast, the polymer component being investigated, or the host polymer must be deuterated. Deuteration has been shown to affect the mechanical and thermodynamic properties of a number of polymers and polymer blends (16,17). Second, a monochromatic neutron source is required, and finally, it is difficult to investigate the behavior of isolated guest polymers in a solid at very low concentrations (< 1%) due to scattering from the host polymer. Many blends are already phase separated even at these low concentrations. For these types of blends in particular, excitation transport techniques will prove extremely useful. The excellent signal-to-noise ratio obtainable in excited state transport experiments allows measurements on blends with guest polymer concentrations that are orders of magnitude lower than the sensitivity limits of neutron scattering. In addition, for chains with multiple, randomly placed chromophores, the effect of the labeled monomer units on the coil conformation can be determined by performing a series of experiments on copolymers containing various mole fractions of label. This allows either the elimination of any such effects or an extrapolation to zero chromophore concentration.

Previous experiments measuring fluorescence depolarization arising from excitation transport among chromophores on isolated guest coils in solid polymer blends demonstrated the feasibility of determining the relative size of individual chains in various host environments (18). The ability of these experiments to provide a quantitative measure of $\langle R_g^2 \rangle^{1/2}$ was limited by the theories used to analyze the results, and not by the technique itself. Each of the two theories which were initially applied to the experiments includes some aspects of polymer chain structure and both provided qualitatively correct experimental predictions, but neither were quantitatively correct. Fredrickson, Andersen, and Frank (FAF) (19-22) developed a theory for energy transport (with or without traps) on isolated, Gaussian polymer coils. They used an extension of the diagrammatic self-consistent method developed by Gochanour, Andersen, and Fayer (23) for excitation transport in solution. The FAF theory used the infinite chain limit (Debye form) for the segment pair-correlation function of an ideal polymer coil. Ediger and Fayer (EF) (8) modeled chromophores randomly tagged on a polymer chain of a particular radius of gyration as chromophores randomly distributed in a sphere; relating the radius of gyration of the sphere to the radius of gyration of the polymer. Polymer statistics were included by averaging over the distribution of radii of gyration using the Flory-Fisk distribution function.

Although both theories correctly predicted the shape of the time-dependent anisotropy, the EF theory predicted values for theta-condition PMMA which were lower than expected (as compared to determinations by light scattering and neutron scattering) while the FAF theory resulted in values which were too high. The reason for the disparities in the excitation transport size determinations has been shown to be due to inadequate models of the spatial distribution of chromophores on a polymer chain (18).

The accurate description of excitation transport on isolated polymer coils is an interesting and difficult problem. Chromophores attached to a polymer present an inhomogeneous medium for excitation transport. Rather than being randomly distributed, as in a solution, the positions of the chromophores are correlated through the covalent bonds of the polymer. Also, the finite size of the polymer limits the number of sites the excitation can sample. This inhomogeneity in the chromophore distribution resulting from the requirements of polymer chain structure can lead to complicated expressions for the chromophore pair correlation function. Both the FAF and EF theories involve complicated series expansions in Laplace space, making inclusion of accurate forms for the chromophore distribution function computationally difficult.

Recently, we have developed a theory for excitation transport on isolated, finite size, flexible polymer coils which contains a more reasonable model of the pair correlation function, and which is amenable to the inclusion of any form of the pair correlation function (24). The theory is based on a method developed by Huber (25,26) for describing excitation transport in infinite, disordered systems. This method has the advantage of being a relatively straightforward time-domain calculation, making the inclusion of different chromophore distribution functions mathematically tractable. The method in general, can be applied to many systems where there are spatial restrictions on and/or correlations between chromophore sites (24,27). Here, we will discuss the application of this theory to excitation transport among chromophores randomly distributed in low concentration on isolated, flexible polymer chains. The theory utilizes a freely jointed chain pair correlation function within a truncated cumulant expansion expression for the excitation transport dynamics. Recent experimental results (28) are also described which demonstrate that excitation transport experiments, in conjunction with this theory, are capable of providing quantitative information on the ensemble average configuration of polymer coils in solid blends.

## Theory

In this section, we describe a theory for calculating observables resulting from incoherent excited state transport among chromophores randomly distributed in low concentration on isolated, flexible polymer chains. The pair correlation function used to describe the distribution of the chromophores is based on a Gaussian chain model. The method for calculating the excitation transfer dynamics is an extension to finite, inhomogeneous systems of a truncated cumulant expansion method developed by Huber for infinite, homogeneous systems (25,26).

Briefly, the truncated cumulant expansion method for describing energy transport (with or without traps) among chromophores randomly distributed in solution considers only interactions between pairs of chromophores. An exact configuration average leads to a power series in the trap concentration (for donor to trap transport) or in the donor concentration (for donor to donor transport) which is then truncated to first order. The result is identical to Förster-type equations for donor-trap transport in the limit of high trap concentration. It is an approximation for donor-donor transport since any paths involving more than two chromophores which leads to the return of the excitation to the originally excited molecule are excluded. Only the result for excitation transport among identical chromophores (no traps) will be given here. The extension to systems with traps is straightforward (27).

In systems involving donor-donor excited state transport, the fundamental quantity of theoretical and experimental interest is $G^s(t)$; the ensemble averaged probability that an originally excited chromophore is excited at time t (23). $G^s(t)$ contains contributions from excitations that never leave the originally excited chromophores, and from excitations that return to the initially excited chromophores after one or more transfer events. $G^s(t)$ does not contain loss of excitation due to lifetime (fluorescence) events.

First, the general method for calculating $G^s(t)$ for any system for which the possible sites of the chromophores are nonequivalent will be outlined. Then, the necessary equations for treating the problem of isolated polymer chains will be given.

General Method. In systems where the chromophore distribution is other than random in an infinite volume, two important points must be taken into account in the calculation of $G^s(t)$, regardless of which formalism is used to describe the transport dynamics. First, in performing the spatial average over all configurations of molecules surrounding the initially excited molecule, the appropriate radial distribution function must be employed. This is true for nonrandom distributions in infinite or finite systems. The second change applies only to finite systems. In an infinite system, the ensemble average over all configurations is translationally invariant. Therefore the average can be performed about any starting point. In a finite system the translational invariance is lost. The ensemble averaged observable must be calculated for a particular point of initial excitation and then the observable must be averaged over all possible starting locations.

Using the truncated cumulant treatment, first $G_1^s(t)$ is found for an initially excited chromophore (labeled 1) fixed in space, averaged over all possible positions of a second chromophore:

$$\ln G_1^s(t) = \frac{(N-1)}{2A} \int (e^{-2\omega_{12}t} - 1) P(r_{12}) d\vec{r}_{12} \quad (1)$$

and where for incoherent, dipole-dipole (Förster type) energy transport with which we are concerned with here,

$$\omega_{12} = \frac{1}{\tau} \left( \frac{R_0}{r_{12}} \right)^6 \quad (2)$$

$\tau$ is the fluorescence lifetime and $R_0$ the critical transfer radius for donor-donor transport (29). For experiments in high viscosity media, the chromophores are essentially static and the appropriate orientation dependent $R_0$ must be used (8,21). N is the number of chromophores in the system. $P(r_{12})d\vec{r}_{12}$ is the probability that there is a chromophore at a distance $r_{12}$ from chromophore 1. The integral is over the physical space and A is the appropriate normalization factor. Next, it is necessary to average over the possible positions of the initially excited chromophore

$$G^s(t) = \frac{1}{A} \int G_1^s(t) d\vec{r}_1 \qquad (3)$$

For an infinite, disordered system, $P(r_{12})d\vec{r}_{12}$ is $4\pi r_{12}{}^2 dr_{12}$ and Equations 1 and 3 become identical to the results of Huber. This cumulant expansion method has been shown to be in excellent agreement with the infinite order diagrammatic theory of Gochanour, Andersen, and Fayer (23) which has been demonstrated theoretically and experimentally to be an accurate description of energy transport in homogeneous solutions (30,31). This truncated cumulant expansion method has also been shown to provide a reasonable description of energy transport among chromophores randomly distributed in a spherical volume (24). The method has also been applied to transport in two-dimensions and transport between planes (27). In addition to describing excitation transport among chromophores randomly distributed along a polymer chain, transport between two chromophores, one attached to each end of a polymer chain has been treated (24). In this case, the cumulant expansion is not necessary as excitation transport between two chromophores can be treated exactly.

Other investigators have also applied the truncated cumulant expansion to energy transport in a sphere and to polymer systems (32). However, they chose to approximate $G^s(t)$ rather than $G_1^s(t)$ by the cumulant expansion. This approach approximates the average over the initial excitation location. Performing the expansion on $G_1^s(t)$ allows an exact average over the position of the initial excitation and results in a more accurate approximation to $G^s(t)$.

Excitation Transport on Isolated, Flexible Polymer Chains Randomly Tagged with Chromophores. In this section, the expressions to describe energy transport on isolated polymer chains with a small number of chromophores randomly distributed along the chain are given. The chromophore concentration is chosen low enough that there is a low probability of having more than one chromophore per statistical segment of the chain. This assures that there are no angular correlations among the chromophores' transition dipoles and that transport is three dimensional rather than quasi one dimensional (along the chain backbone). The overall concentration of tagged coils in the blend is taken to be low enough that there is no interchain transfer, i.e., all excitation transfer occurs among chromophores tagged on the same chain.

The forms of Equations 1 and 3 permit any model of the polymer pair correlation function to be used in the transport calculation. Here we are considering flexible chains and it is reasonable to use a freely jointed chain model, which obeys Gaussian statistics, in the calculations. The chain segment length is identified with the statistical polymer segment (Kuhn segment) length. In this model, the conformation averaged radial probability of finding any other polymer segment a distance r from a segment i is (33)

$$p_i(r)dr = \frac{4\pi}{(\bar{N}-1)} \sum_{\substack{j=1 \\ j\neq 1}}^{\bar{N}} \left(\frac{3}{2\pi a^2|i-j|}\right)^{3/2} \exp\left(\frac{-3r_{12}^2}{2a^2|i-j|}\right) r_{12}^2 dr_{12} \qquad (4)$$

where $\bar{N}$ is the number of statistical segments and a is the statistical segment length. Since the chromophore concentration on the chain is small we identify the probability of finding a second chromophore (labeled 2) at a distance $r_{12}$ from an initially excited chromophore (labeled 1) placed on segment i with the probability that there is a polymer segment at that distance. Replacing r with $r_{12}$ in Equation 4 (this will be denoted by $p_i(r_{12})$) and multiplying by N - 1 (N is the number of chromophores on the chain) yields the conformation averaged probability of finding another chromophore a distance $r_{12}$ from the initially excited chromophore, excluding the small correction due to additional, unexcited chromophores on segment i.

The correction due to unexcited chromophores on the segment containing the initially excited chromophore, i, can be included by calculating the average number (n) of unexcited chromophores on segment i and using some reasonable chromophore distribution function ($p'(r_{12})$) for this segment. Then the probability of an unexcited chromophore being a distance $r_{12}$ away from the initially excited chromophore on the ith segment is

$$P_i(r_{12})d\vec{r}_{12} = \left\{n\, p'(r_{12}) + (N - 1 - n)\, p_i(r_{12})\right\} d\vec{r}_{12} \qquad (5)$$

For low concentrations of chromophores, n is a small number and the choice of $p'(r_{12})$ is not critical. It is only necessary to concentrate the appropriate chromophore density in the region of space of the ith segment. A simple model for $p'(r_{12})$ is to assume that the initially excited chromophore is in the center of the segment and surrounded by a random distribution of the n chromophores, i.e.,

$$\begin{aligned} p'(r_{12})d\vec{r}_{12} &= 4\pi r_{12}^2 dr_{12} \qquad 0 \le r_{12} \le a/2 \\ &= 0 \qquad\qquad\qquad r_{12} \ge a/2 \end{aligned} \qquad (6)$$

If Equation 5 with Equation 6 is substituted for (N-1)-$P(r_{12})d\vec{r}_{12}$ in Equations 1-3 and the position of the initially excited chromophore is averaged over by summing the equally weighted contributions from each segment i, the expression for $G^s(t)$ is

$$G^s(t) = \frac{1}{\bar{N}} \sum_{i=1}^{\bar{N}} \left\{ \mathrm{Exp}\left[\frac{4\pi}{2} \int_0^{\infty} P_i(r_{12})(e^{-2\omega_{12}t} - 1) r_{12}^2 dr_{12}\right]\right\} \qquad (7)$$

The upper limit of integration in Equation 7 can be extended to infinity since contributions to the energy transport from distances greater than several $R_0$ are negligible and also since $P_i(r_{12})$ becomes very small for large values of $r_{12}$ relative to

$\langle R_g^2 \rangle^{1/2}$. For small coils with large values of $R_0$ a limit of integration appropriate to the finite coil size may be used.

Relation of Theory to Experimental Observables

In this section we discuss how $G^s(t)$ can be obtained experimentally from time-resolved fluorescence depolarization data. $G^s(t)$ is directly related to the time-dependent fluorescence anisotropy which can be measured through time-resolved fluorescence depolarization, polarized transient grating or polarized absorption techniques (30,31).

In time-resolved fluorescence depolarization experiments a sample of randomly oriented chromophores is excited by a short pulse of plane polarized light. The decay of the fluorescence intensities polarized parallel ($I_\parallel(t)$) and perpendicular ($I_\perp(t)$) to the exciting light can be written as

$$\begin{aligned} I_\parallel(t) &= e^{-t/\tau}(1 + 2\, r(t)) \\ I_\perp(t) &= e^{-t/\tau}(1 - r(t)) \end{aligned} \tag{8}$$

$\tau$ is the fluorescence life time and $r(t)$ is the time dependent fluorescence anisotropy. $r(t)$ contains information about all sources of depolarization and can be obtained from the individual parallel and perpendicular components of the fluorescence intensity. From Eqation 8

$$r(t) = \frac{I_\parallel(t) - I_\perp(t)}{I_\parallel(t) + 2\, I_\perp(t)} \tag{9}$$

If the transition dipoles of the chromophores in a solid polymer matrix are randomly oriented, the main source of depolarization in these experiments will be due to excitation transport. The initially excited ensemble is polarized along the direction of the excitation $\bar{E}$ field and gives rise to polarized fluorescence. Transport occurs into an ensemble of chromophores with randomly distributed dipole directions and the fluorescence becomes unpolarized. The random distribution is assured by the low concentration of the chromophores. To a slight extent, on the time scale of interest, depolarization also occurs as a result of chromophore motion. In this case the fluorescence anisotropy is approximately

$$r(t) = C\, \phi(t)\, G^s(t) \tag{10}$$

where $\phi(t)$ is the rotational correlation function which contains the effects due to motion of the chromophores. C is a time-independent constant that describes the degree of polarization of the excitation and emission transitions involved. There are two approximations in Equation 10. The first is that the rotational and energy transport contributions to depolarization are independent. This is an excellent approximation for the very slow and small extent of rotational depolarization in polymer blends on the time scale of interest. The second approximation is that

$G^s(t)$ decays to zero resulting in complete depolarization, i.e., the irreversible transfer of excitation from the initially excited donor into the ensemble of unexcited donors results in total loss of polarization. For coils with a low concentration of randomly placed chromophores this is approximately true. The residual polarization is only 4%, which results in an insignificant error (34).

In order to obtain $G^s(t)$ for a given polymer, experiments on two different samples must be performed. These samples differ only in that the guest copolymers have a different fraction of chromophore containing monomers. Copolymer A is the polymer of interest and has an appreciable number of chromophores, such that excitation transport will occur. Its fluorescence anisotropy, $r_A(t)$, is given by Equation 10. Copolymer B has such a small number of chromophores that excitation transport is negligible ($G^s(t) = 1$) and only chromophore motion contributes to the anisotropy,

$$r_B(t) = C\,\phi(t) \qquad (11)$$

$G^s(t)$ arising from the excitation transport on copolymer A can be calculated from the two experimental anisotropies:

$$G^s(t) = r_A(t)/r_B(t) \qquad (12)$$

This method of determining $G^s(t)$ has the advantage that detailed knowledge of the parameters C and $\phi(t)$ is unnecessary.

To obtain a value for $\langle R_g^2\rangle^{1/2}$ of the copolymer A, we compare the experimentally determined $G^s(t)$ to a theoretical calculation of $G^s(t)$ (from Equations 4-7) for the same copolymer. Once the molecular weight and number of chromophores are known (these can be independently determined), the theory has only one adjustable parameter; the statistical segment length. Since Gaussian chain statistics are assumed, this parameter is directly related to $\langle R_g^2\rangle^{1/2}$ by

$$\langle R_g^2\rangle = \frac{1}{6}(\bar{N}\,a^2) \qquad (13)$$

Thus, a fit of the experimentally determined $G^s(t)$ with a theoretically calculated $G^s(t)$ determined by adjusting the statistical segment length will give a measure of $\langle R_g^2\rangle^{1/2}$ for the copolymer.

## Experimental Results and Discussion

Earlier experiments have shown the utility of excitation transport measurements in providing relative information regarding coil size in polymer blends (18). Here, we will summarize the results of recent experiments (28) which demonstrate that monitoring excitation transport on isolated coils in solid blends through time-resolved fluorescence depolarization techniques provides a quantitative measure of $\langle R_g^2\rangle^{1/2}$ for the guest polymer. Experiments on different molecular weight guest copolymers are necessary to show the general applicability of the theory relating

$G^s(t)$ to $\langle R_g^2\rangle^{1/2}$ and to confirm the utility of assuming a Gaussian segment distribution for PMMA. Since the presence of the naphthalene groups could perturb the average chain conformation, $\langle R_g^2\rangle^{1/2}$ was determined for a series of copolymers of essentially the same molecular weight but differing in the average number of naphthalenes on the chains. The results demonstrate that the copolymer size is independent of the number of naphthalenes per coil for relatively low naphthalene concentrations.

Materials and Technique. Five copolymers of methylmethacrylate (MMA) and 2-vinylnaphthalene (2-VN) were used in the fluorescence depolarization experiments. Copolymer (I) contains so few chromophores that depolarization occurs only due to molecular motion. Three of the copolymers (4-23, 6-22 and 9-23) were of the same weight average molecular weight (~ 23,000) but varied in the percentage of chromophore containing monomers. Copolymer (9-60) was of a higher molecular weight (~ 60,000). The details of the copolymer synthesis, characterization and solid blend preparation are to be found in reference (28). Table I gives a summary of the characteristics of the five copolymers. The host polymer for all blends was PMMA, number average molecular weight ~ 120,000.

**Table I. Copolymer Characteristics**

| Copolymer | Mole fraction 2-VN | Molecular weight | $M_w/M_n$ | Average number Chromophores/chain |
|---|---|---|---|---|
| I | 0.0015 | ~50,000 | ~2.5 | < 1 |
| 4-23 | 0.040 | 22,700 | 1.4 | 9 ± 1 |
| 6-22 | 0.059 | 22,400 | 1.5 | 13 ± 2 |
| 9-23 | 0.087 | 23,400 | 1.5 | 20 ± 1 |
| 9-60 | 0.087 | 59,800 | 1.3 | 49 ± 2 |

The experimental apparatus and methods of data acquisition and analysis are also fully described in reference (28). Briefly, the samples were excited 25 psec laser pulses at a wavelength of 320 nm (320 nm is the peak of the absorption origin). The spot size of the excitation beam at the sample was 1 mm and the energy per pulse was ≤ 0.5 μJoule. The fluorescence was detected at 337 nm. Great care was taken to ensure that there were no significant effects due to impurity fluorescence, readsorption, systematic polarization biases, or birefringence in the samples. Concentrations of the copolymers in the blends such that the copolymers chains were isolated were determined through fluorescence depolarization concentration studies (18,28). In addition, emission spectra at these low concentrations showed no significant naphthalene excimer emission, indicating that there is no aggregation of naphthalene chromophores on different chains or within the same chain.

Determination of $\langle R_g^2\rangle^{1/2}$ for Copolymers of Different Molecular Weights. Time-resolved fluorescence depolarization experiments were performed on samples made from copolymer 9-23 (0.087 mole fraction 2-VN, 23,400 $M_w$) and from copolymer 9-60 (0.087 mole fraction 2-VN, 59,800 $M_w$) in 120,000 $M_w$ PMMA. In both cases, the amount of copolymer in the host polymer was 3/8 weight percent. Figure 1a shows the polarized fluorescence decays ($I_\parallel(t)$ and $I_\perp(t)$) for the copolymer 9-23/PMMA blend. Figure 1b (lower curve) shows the fluorescence anisotropy, r(t), calculated from the data, using Equation 9. Also shown (upper curve) is r(t) obtained in the same manner for the 0.0015 mole fraction 2-VN copolymer I. This sample contains so few chromophores that depolarization occurs solely as a result of chromophore motion. There is no excitation transport.

As discussed above, to obtain the experimental $G^s(t)$ for copolymer 9-23, it is necessary to take the ratio of the anisotropies of copolymer 9-23 to copolymer I. The resulting experimental $G^s(t)$ curve is shown in Figure 2. Figure 3 shows the experimental $G^s(t)$ obtained in the same manner for copolymer 9-60. In both, we also show the best fits (smooth curves) obtained from the theory described above. Variation of the single adjustable parameter, the statistical segment length a, results in $\langle R_g^2\rangle^{1/2}$ of 37 ± 3 Å and 61 ± 3 Å for the 23,400 $M_w$ and 59,800 $M_w$ copolymers, respectively. The sensitivity of the technique is demonstrated by theoretical curves at ±2 Å from the best fits.

The measured $\langle R_g^2\rangle^{1/2}$ of these two polymers varies with the square root of the chain length. This is as expected for flexible coils (35) and indicates that the Gaussian segment distribution function utilized in the analysis of the data is applicable to these copolymers. Neutron scattering experiments on isolated PMMA coils in a deuterated PMMA host show this same scaling of $\langle R_g^2\rangle^{1/2}$ with molecular weight (15).

Since these blends are essentially PMMA in PMMA, the copolymers are approximately in $\theta$-conditions (36). (The effect of the naphthalene groups is discussed below.) True $\theta$-conditions for a solid blend would be PMMA in a PMMA host of the same molecular weight. We do not expect that a host $M_w$ of 120,000 results in a significant deviation from $\theta$-conditions since, in a previous fluorescence depolarization experiment, nearly identical r(t) curves were obtained for a 20,000 $M_w$ 2-VN/MMA copolymer in both 20,000 $M_w$ and 120,000 $M_w$ PMMA hosts (18). Assuming that these copolymers are essentially PMMA at $\theta$-conditions, it is possible to compare these results with light scattering measurements of $\langle R_g^2\rangle^{1/2}$ for equivalent molecular weight PMMA in $\theta$-solvents. The values can be easily caculated from tabulated data (37). For 23,000 $M_w$ and 60,000 $M_w$ PMMA at $\theta$-condition, $\langle R_g^2\rangle^{1/2}$ should be 39 ± 4 Å and 64 ± 7 Å, respectively. The results from the excitation transport experiments are in excellent agreement with these values.

Effect of the Presence of Naphthalene Groups. Although the results for the two copolymers with 0.087 mole fraction of naphthalene containing monomers agree very well with other determinations of $\langle R_g^2\rangle^{1/2}$ for $\theta$-condition PMMA, it is still necessary to ascertain that the presence of the chromophores does not significantly perturb the average chain conformation. This was accomplished by obtaining experimental $G^s(t)$ curves for three copolymers of essentially the same molecular weight, but varying in the amount of

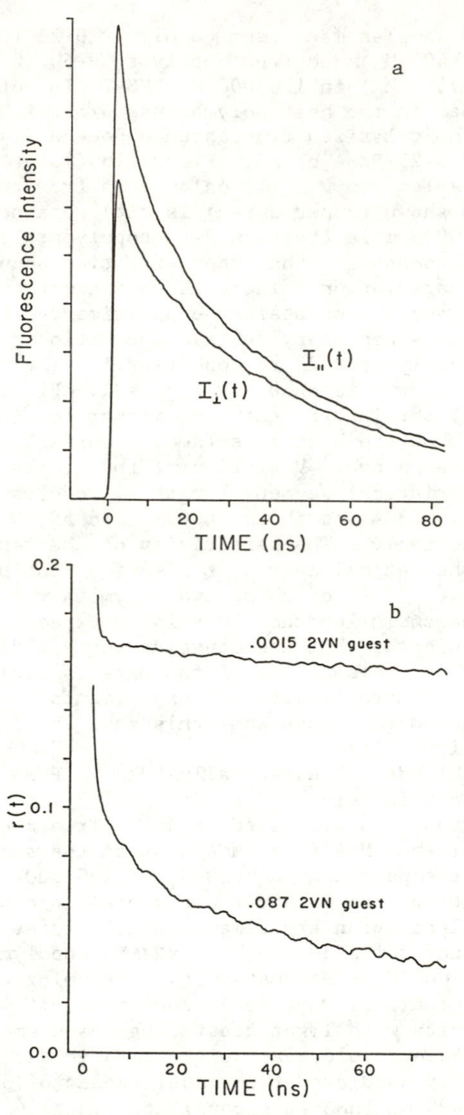


Figure 1. (a) Polarized fluorescence decays for detection parallel ($I_{\parallel}(t)$) and perpendicular ($I_{\perp}(t)$) to the excitation polarization. The sample is copolymer 9-23 (23,400 $M_w$, 0.087 mole fraction 2-VN) in a 120,000 $M_w$ PMMA host. Copolymer concentration is 0.38 wt. percent. (b) Fluorescence anisotropies, r(t), calculated from experimental $I_{\parallel}(t)$ and $I_{\perp}(t)$ (see Equation 9). The upper curve is a copolymer with a very small mole fraction (0.0015) of 2-vinylnaphthalene (copolymer I). Excitation transport does not occur and the time dependence is due to very slow rotational depolarization. The lower curve is from the data shown in Figure 1a. Here, the time dependent depolarization is due to both excitation transport and the small amount of rotational depolarization.

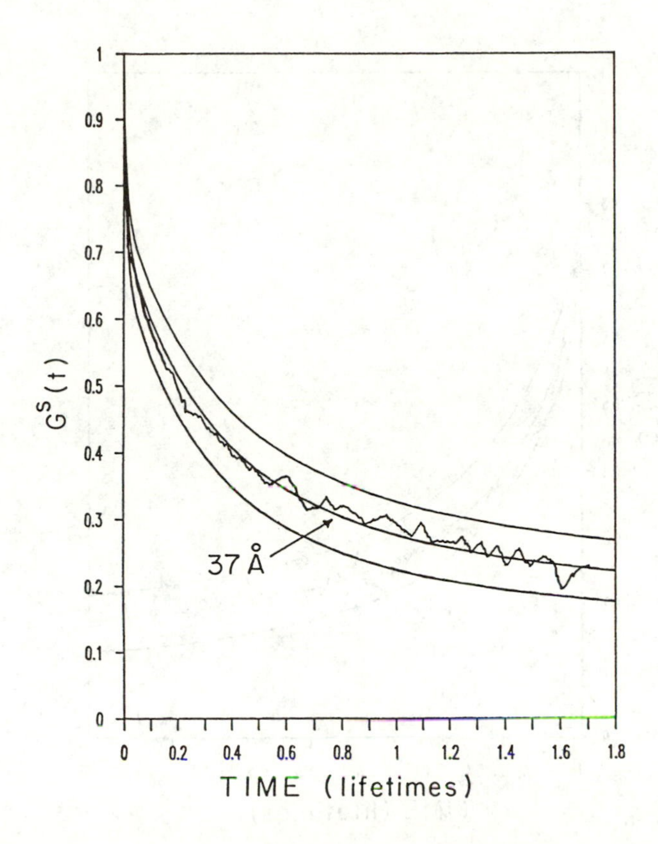


Figure 2. The experimental $G^S(t)$ for copolymer 9-23 (23,400 $M_w$, 0.087 mole function 2-VN) in a 120,000 $M_w$ PMMA host. Also shown are theoretical $G^S(t)$ curves calculated using Equation 7 for a 23,400 $M_w$ PMMA chain with 20 naphthalenes. The best fit results in an $\langle R_g^2 \rangle^{1/2}$ of 37 Å. This result is in close agreement with light scattering determinations of $\langle R_g^2 \rangle^{1/2}$ for a PMMA coil in a $\theta$-condition media. The upper and lower theoretical curves are for 39 Å and 35 Å, respectively.

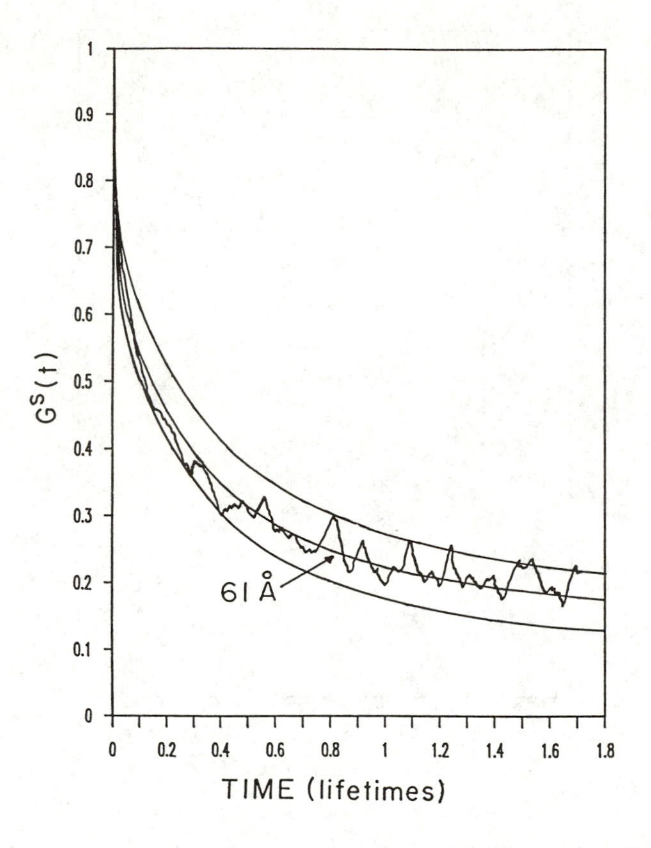


Figure 3. The experimental $G^s(t)$ is for copolymer 9-60 (59,800 $M_w$, 0.087 mole fraction 2-VN) in a 120,000 $M_w$ PMMA host. Also shown are theoretical $G^s(t)$ curves for a 59,800 $M_w$ PMMA chain with 49 naphthalenes. The best fit yields an $\langle R_g^2 \rangle^{1/2}$ of 61 Å. The upper and lower theoretical curves are for 63 Å and 59 Å, respectively.

naphthalene in the chains. These three copolymers are designated 9-23, 6-22, and 4-23 in Table I and contain 8.7, 5.9, and 4.0 mole percent of naphthalene containing monomers, respectively. The samples consisted of 3/8 weight percent of the desired copolymer in the 120,000 $M_w$ host PMMA. Figure 4 shows the experimental $G^s(t)$ curves obtained for these copolymers along with the theoretical best fits. The resulting values for $\langle R_g^2 \rangle^{1/2}$ are 37 ± 3 Å, 39 ± 3 Å, and 38 ± 3 Å for the three copolymers in order from the highest to the lowest naphthalene content. These results are in quantitative agreement with solution $\langle R_g^2 \rangle^{1/2}$ measurements under $\theta$-conditions. In addition, these results indicate that the presence of the naphthalene containing monomers does not significantly perturb the average coil dimensions, at least up to naphthalene concentrations of nine mole percent.

It is also necessary to consider the possibility that there are angular correlations between naphthalene chromophores, as they could effect the depolarization observables. Angular correlations between chromophores could result from local chain structure. The theory assumes there are no such correlations. The chain is treated as a random flight, i.e. no angular correlations between different statistical segments. The requirement then is that the chromophores are separated by a distance at least as great as the statistical segment length. Or equivalently, for randomly distributed chromophores, the number of chromophores should be considerably less than the number of statistical segments.

The 2-VN monomers are randomly incorporated into the copolymers used in this study. The reactivity ratios for copolymerization of 2-VN (monomer 1) with MMA (monomer 2) are $r_1$ = 1.0 and $r_2$ = 0.4 at 60 °C (38). The copolymerizations were carried to only 15% conversion, ensuring there was no significant depletion of either monomer species during the polymerization. The 2-VN is present as the minor component of the copolymer, in all cases the mole percent of 2-VN is less than 10. These conditions are sufficient to guarantee a random distribution of 2-VN in the copolymers.

Calculating from the $\theta$-condition $\langle R_g^2 \rangle^{1/2}$ for a 23,000 $M_w$ PMMA coil, the number of statistical segments is 37 and the statistical segment length is 15.7 Å with 6.2 monomer units per segment. Since neither of the monomer species are depleted during polymerization, the polymerization process can be treated statistically as random sampling with replacement. Then, for a copolymer containing 9 mole percent 2-VN, the probability that any segment of 6 monomer units contains 2 or more naphthalene chromophores is 9.5%. For a 4% 2-VN copolymer of the same length this probability is only 2.2%. Therefore, the likelihood that there is some angular correlation between chromophores is small.

Although the probability of having two chromophores within the same statistical segment is small for the copolymers used in this study, it is still important to consider what effect two chromophores in the same statistical segment will have on the depolarization observables. The greatest correlation would occur between 2-VN monomer units which are nearest neighbors, as in the case of poly(2-vinyl naphthalene) (P2VN). Ediger and Fayer (8) have calculated the expected emission anisotropy for a naphthalene which is a nearest neighbor of the originally excited chromophore.

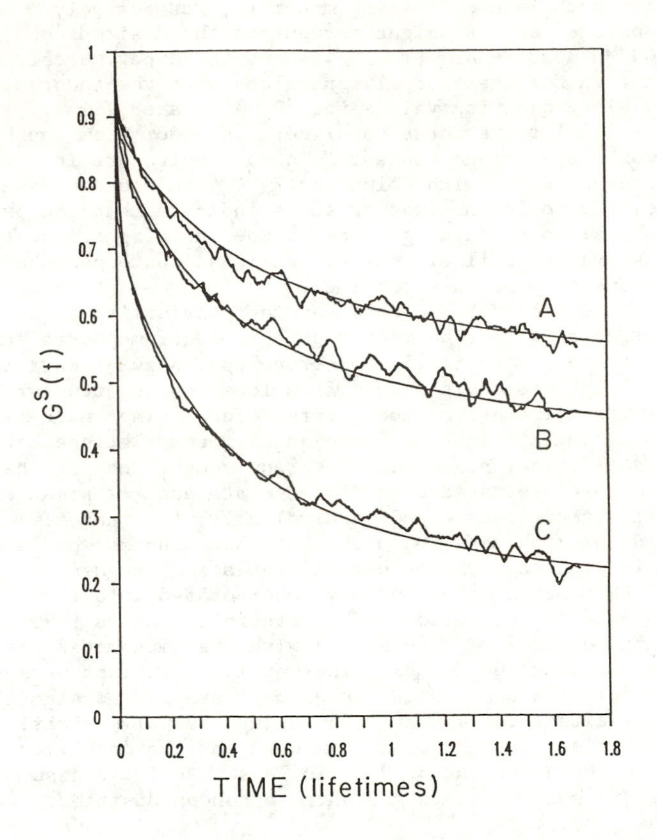


Figure 4. Data is shown for three copolymers of essentially the same molecular weight but differing in mole fraction of 2-VN. Curve A is copolymer 4-23 (22,700 $M_w$, 0.04 mole fraction 2-VN), curve B is copolymer 6-22 (22,400 $M_w$, 0.087 mole fraction 2-VN), and curve C is copolymer 9-23 (23,000 $M_w$, 0.087 mole fraction 2-VN). Also shown are the best fit theoretical $G^s(t)$ curves for equivalent molecular weight PMMA chains with the appropriate number of naphthalenes. The resulting values for $\langle R_g^2 \rangle^{1/2}$ are 38 Å, 39 Å, and 37 Å in order from A to C.

They used the most stable configurations of 1,3-di(2-naphthyl) propane and their relative weights determined by Ito *et al.* (39) as an approximation of the chromophore configurations in P2VN. The ratio of the anisotropy of the acceptor to that of the donor for this worst case is only ~ 2.5 times greater than for a randomly distributed solution. For the 2-VN/MMA copolymers used in this study the occurrence of 2-VN diads is very small and the influence on the time dependent polarization is certainly negligible. The equivalence of the $\langle R_g^2 \rangle^{1/2}$ determinations for copolymers 9-23, 6-23 and 4-23 supports the conclusion that there is no significant effect on the observables due to angular correlations between chromophores.

Table II summarizes the results for the various copolymers in this study. A comment on the error bars associated with the measurements of $\langle R_g^2 \rangle^{1/2}$ in this study is necessary. The theoretical $G^s(t)$ curves at ±2 Å shown in Figures 2 and 3 are to show the sensitivity of $G^s(t)$ to small changes in $\langle R_g^2 \rangle^{1/2}$. The reported errors in Table II are estimated from errors in the measurement of the average number of naphthalenes per chain and the weight average molecular weight. The fact that there is really a distribution of the number of chromophores and the molecular weight does not significantly affect the result. The molecular weight distribution can be controlled and tested during sample preparation. The samples used in this study have sufficiently narrow molecular weight distributions to avoid problems. The effect of the distribution in the number of chromophores per chain about the average number has been tested by theoretical calculations and shown to be insignificant (24). However, error in the determination of their average values, especially in the number of chromophores per chain for the smaller copolymers can lead to errors of a few angstroms in the theoretical fit.

TABLE II. Summary of Results

| Copolymer | $\langle R_g^2 \rangle^{1/2}$ Determined by Excitation Transport | $\langle R_g^2 \rangle^{1/2}$ for equivalent $M_w$ $\theta$-condition PMMA |
|---|---|---|
| 4-23 | 38 ± 3 | 39 ± 4 |
| 6-22 | 39 ± 3 | 39 ± 4 |
| 9-23 | 37 ± 3 | 39 ± 4 |
| 9-60 | 61 ± 3 | 64 ± 7 |

## Concluding Remarks

These experiments, in conjunction with the theory described here, demonstrate the utility of excitation transport induced fluorescence depolarization in the study of polymer blends. The technique allows quantitative determination of $\langle R_g^2 \rangle^{1/2}$ for isolated guest coils in a polymer matrix. The accuracy of the measurements is comparable to determinations performed with neutron scattering methods. However, we believe that the excitation transfer technique offers greater flexibility in two ways. First, the signal-to-noise ratio achievable in excitation transport experiments allows measurements on blends with guest polymer concentrations that are one or more orders of magnitude lower than the sensitivity limits of neutron scattering. Second, and most importantly, the effect of the labeled monomer on the copolymer conformation can be ascertained by performing a series of experiments on copolymers containing various mole fractions of label. Despite the insensitivity of the present results to the mole fraction of 2-VN, label concentration dependence studies must be performed on every blend system in order to quantify the effect of label induced perturbation.

The accuracy of the analysis presented in this paper is determined by the validity of two key approximations: (1) the description of the energy transfer dynamics by the first order cumulant expansion method, (2) the use of a Gaussian chromophore pair distribution function. Although originally developed for, and successfully applied to, the problem of energy transfer in disordered infinite volume systems, the cumulant method can be modified to provide a highly accurate description of energy transfer in finite volume systems such as polymer coils (24).

Gaussian pair distribution functions are commonly employed in calculations involving freely jointed chains. Except in the limit of infinitely long polymer molecules, this description is approximate. The quantitative agreement between the $\theta$-condition $\langle R_g^2 \rangle^{1/2}$ values determined for the 23,000 $M_w$ and 60,000 $M_w$ polymer coils in solid and solution states shows that the Gaussian pair distribution function is adequate to describe these chains.

In the experiments discussed here, we have determined $\langle R_g^2 \rangle^{1/2}$ for PMMA in the solid state under $\theta$-conditions. The techniques and theory employed are easily extended to other solid state polymer systems. For example, changes in average coil conformation with various host polymers in both miscible and phase separated blends can be interrogated. Similarly, the effects of residual solvent in solvent cast systems and of sample processing and history can be investigated.

In addition, the theoretical analysis used in this paper to derive $\langle R_g^2 \rangle^{1/2}$ from the data is sufficiently powerful and flexible that transport observables for any pair distribution function can be calculated simply by replacing the Gaussian pair distribution function in Equation 7. In fact, use of the theory in conjunction with a series of depolarization experiments covering a wide range of polymer types and molecular weights could serve to determine the applicability of the segmental distributions functions proposed for various types of polymers. However, it is necessary to be careful in drawing conclusions about model pair-correlation functions based on the qualitative shape of the theoretical $G^s(t)$ curves alone (28). When possible, it is useful to check the validity of the chromophore distribution model employed by making quantitative comparisons of excitation transfer results in $\theta$-condition media to structural determinations by other methods such as light scattering.

Excitation transport experiments have already provided useful qualitative information on polymer systems. However, the additional ability to obtain quantitative information on polymer-environment interactions through changes in the coil size makes energy transport techniques even more powerful. The sensitivity of excitation transport observables to small changes in chromophore distribution coupled with the relative simplicity of time resolved fluorescence experiments and the flexibility of the transport theory makes excitation transport induced fluorescence depolarization spectroscopy a valuable and unique tool for the study of solid state polymer systems.

## Acknowledgments

This work was made possible by a grant from the Department of Energy, Office of Basic Energy Sciences (DE-FG03-84ER13251). KAP would like to thank the IBM Corporation for a Predoctoral Fellowship. We would also like to thank the Stanford Center for Materials Research and the members of the Polymer Thrust Program for invaluable contributions to this work. In particular we would like to acknowledge Dr. Jürg Baumann for his help in applying the cumulant treatment to the polymer problem, and Professor C. W. Frank for many stimulating conversations in connection with this work.

## Literature Cited

1. Fitzgibbon, P. D.; Frank, C. W. Macromolecules 1982, 15, 733.
2. Gelles, R.; Frank, C. W. Macromolecules 1983, 16, 1448.
3. Semerak, S. N.; Frank, C. W. Can. J. Chem. 1983, 63, 1328.
4. Morawetz, H.; Amrani, F. Macromolecules 1978, 11, 281.
5. Morawetz, H. Pure and Appl. Chem. 1980 52, 277.
6. Shiah, T. Y.-J.; Morawetz, H. Macromolecules 1984, 17, 792.
7. Morawetz, H. Polymer Preprints 1986, 27, 62.
8. Ediger, M. D.; Fayer, M. D. Macromolecules 1983, 16, 1839.
9. "Light Scattering from Polymer Solutions," M. B. Huglin, Editor, Academic Press, New York, 1972.
10. Berne, B. J. and Pecora, R., Dynamic Light Scattering, Wiley Interscience: New York, 1976.
11. Kruse, W. A.; Kirste, R. G.; Haas, J.; Schmitt, B. J.; Stein, D. J. Makromol. Chem. 1976, 177, 1145.
12. Jelenic, J.; Kirste, R. G.; Oberthür, R. C.; Schmitt-Strecker, S.; Schmitt, B. J. Makromol. Chem. 1984, 185, 129.
13. Dettenmaier, M.; Maconnachie, A.; Higgins, J. S.; Kausch, H. H.; Nguyen, T. Q. Macromolecules 1986, 19, 773.
14. Schmitt, B. J.; Kirste, R. G.; Jelenic, J. Makromol. Chem. 1980, 181, 1655.
15. Kirste, R. G.; Kruse, W. A.; Ibel, K. Polymer 1975, 16, 120.
16. Ben Cheikh Larbi, F.; Leloup, S.; Halary, J. L.; Monnerie, L. Polym. Comm. 1986, 27, 23.
17. Bates, F. S.; Wignall, G. D. Macromolecules 1986, 19, 932.
18. Ediger, M. D.; Domingue, R. P.; Peterson,K. A.; Fayer, M. D. Macromolecules 1985, 18, 1182.
19. Fredrickson, G. H.; Andersen, H. C.; Frank, C. W. Macromolecules 1983, 16, 1456.

20. Fredrickson, G. H.; Andersen, H. C.; Frank, C. W. Macromolecules 1983, 79, 3572.
21. Fredrickson, G. H.; Andersen, H. C.; Frank, C. W. Macromolecules 1984, 17, 54.
22. Fredrickson, G. H., Andersen, H. C.; Frank, C. W. Macromolecules 1984, 17, 1496.
23. Gochanour, C. R.; Andersen, H. C.; Fayer, M. D. J. Chem. Phys. 1979,70, 4254.
24. Peterson, K. A.; Fayer, M. D. J. Chem. Phys. 1986, 85, 4702.
25. Huber, D. L. Phys. Rev. B 1979, 20, 2307.
26. Huber, D. L. Phys. Rev. B 1979, 20, 5333.
27. Baumann, J; Fayer, M. D. J. Chem. Phys. 1986, 85, 4087.
28. Peterson, K. A.; Zimmt, M. B.; Linse, S.; Domingue, R. P.; Fayer, M. D. Macromolecules 1987, 20, 168.
29. Förster, Th. Ann. Phys. 1948, 2, 55.
30. Gochanour, C. R.; Fayer, M. D. J. Phys. Chem. 1981, 85, 1989.
31. Miller, R. J. D.; Pierre, M.; Fayer, M. D. J. Chem. Phys. 1983, 78, 5138.
32. Fredrickson, G. H.; Andersen, H. C.; Frank, C. W. J. Polym. Sci. Polym. Phys. Ed. 1985, 23, 591.
33. Yamakawa, H., Modern Theory of Polymer Solutions, Harper and Row: New York, 1971; Chapt. 2.
34. Galanin, M. D. Tr. Fiz. Inst. I. P. Pavolova, 1950, 5, 341.
35. Flory, P. J. Statistical Mechanics of Chain Molecules; Interscience Publishers: New York, 1969.
36. Hayashi, H.; Flory, P. J.; Wignall, G. D. Macromolecules 1983, 16, 1328.
37. Brandrup, J.; Immergut, E. H. Eds. Polymer Handbook; John Wiley and Sons: New York, 1975; Section IV.
38. ibid., Section II.
39. Ito, S.; Yamamoto, M.; Nishijima, Y. Bull. Chem. Soc. Jpn. 1986, 55, 363.

RECEIVED April 27, 1987

# Chapter 26

# Chromophoric Polypeptides

## One-Dimensional Chromophoric Assembly in Solution

**Masahiko Sisido**

**Research Center for Medical Polymers and Biomaterials, Kyoto University, 53 Kawahara-cho, Shogoin, Sakyo, Kyoto 606, Japan**

A new class of polypeptides carrying aromatic chromophores, such as naphthyl and pyrenyl groups were synthesized in the form of homopolypeptides of aromatic amino acids or sequential polypeptides with nonchromophoric amino acids, which favor α-helical main-chain conformation. Ground-state conformations of these polypeptides in solution were determined by the comparison of the experimental circular dichroic (CD) spectrum with the theoretical one, calculated for some minimum-energy conformations predicted by the ECEPP conformational analysis. In most cases, these polypeptides were found to take an α-helical main-chain conformation in solution and the aromatic groups were arranged regularly along the helix. Interchromophoric interactions in the excited state was studied by static and dynamic fluorescence spectroscopy including circularly polarized fluorescence (CPF) spectroscopy. It was found that the mode and extent of interchromophoric interactions depend strongly on spatial arrangement of the chromophores along the helical main chain.

Polypeptides carrying aromatic chromophores on their side chains are a new class of chromophoric polymers which may show the following characteristic features (1). (1) In principle, one can prepare any peptide or polypeptide which has a particular sequence of different types of chromophores. (2) Polypeptides can take regular and rigid conformation even in solution. Therefore, one can fix the side-chain chromophores in a one-dimensional arrangement. (3) Chirality of polypeptides allows one to use chiroptical spectroscopy, such as circular dichroism (CD), fluorescence-detected circular dichroism (FDCD), and circularly polarized fluorescence (CPF) (2), besides the conventional static and dynamic spectroscopy. The use of chiroptical spectroscopy sometimes increases greatly the information on the structure of the side-chain arrangement and on the

0097–6156/87/0358–0343$06.00/0

interactions among the side-chain chromophores. (4) Chromophoric peptides and polypeptides, being biologically relevant substances, can be used as probes for biological studies and be considered as potential candidates for "bioelectronic devices" (3).

This article reviews recent progress in the preparative, structural, and the spectroscopic study of homo and sequential polypeptides containing optically active aromatic amino acids.

## Synthesis of Optically Active Aromatic Amino Acids and Polypeptides

Since the primary purpose of this study is to obtain stereoregular aromatic polypeptides which take a regular conformation, it is essential to synthesize optically pure aromatic amino acids. Although many aromatic amino acids which do not occur in nature have been synthesized in the past, few have been optically resolved. We have undertaken the synthesis and optical resolution of a variety of amino acids carrying aromatic substituents. These include an electron-donating amino acid (p-dimethylaminophenylalanine), an electron-accepting one (p-nitrophenylalanine), and those with large π-electron conjugation (1- and 2-naphthylalanines, anthrylalanine, and pyrenylalanine). Other aromatic amino acids, such as carbazolylalanine and ferrocenylalanine, have been synthesized, but their optical resolution has not been successfully achieved yet.

There are several ways to prepare peptides, polypeptides, and artificial proteins carrying these aromatic amino acids as their constituents. Homopolypeptides can be prepared by the polymerization of amino acid N-carboxyanhydrides (NCAs) derived from the corresponding amino acids (4). Sequential polypeptides of the form $(ABC...W)_n$ are prepared by the polymerization of the corresponding oligopeptide active esters ABC...W-X. Small peptides of any sequence can be synthesized by a step-by-step procedure (5). Finally, cultivation of some bacteria in the presence of artificial amino acids will possibly incorporate them into the proteins produced by the bacteria. In the following, attention will be focused on homopolypeptides and sequential copolypeptides carrying one type of chromophore on a chain.

## Poly(L-Arylalanines)(6-9)

Synthesis and Ground-State Conformation  Homopolypeptides carrying 1- and 2-naphthyl groups, poly(L-1- and 2-naphthylalanine) = poly(1- and 2-napAla) (6,7), and 1-pyrenyl group, poly(L-1-pyrenylalanine) = poly(1- pyrAla) (8,9), as the aromatic group have been synthesized in a form of block copolymers with poly(γ-benzyl DL-glutamate) unit which functions as a solubilizing agent (I).

$$\left(\mathrm{NH\text{-}CH(CH_2Ar)\text{-}CO}\right)_n\!\!-\!\left(\mathrm{NH\text{-}CH((CH_2)_2CO\text{-}O\text{-}CH_2\text{-}C_6H_5)\text{-}CO}\right)_{150} \qquad \text{(I)}$$

CD spectra of poly(1- and 2-napAla)s in trimethyl phosphate (TMP) solution showed an intense exciton splitting at the $^{1}Bb$ absorption band of naphthyl chromophore around 230 nm. The molar ellipticity reached a maximum of $6x10^{5}$ (deg $cm^{2}$ $dmol^{-1}$) with respect to the naphthyl chromophore. The observation of strong exciton splitting indicates a regular helical arrangement of the naphthyl groups along a helical polypeptide main chain. A theoretical calculation of the CD spectrum based on exciton theory (10) was performed for possible main-chain and side-chain conformations predicted from the ECEPP (11) conformation energy calculation. It was concluded that poly(1-napAla) has a right-handed α-helical main-chain, whereas poly(2-napAla) is in a left-handed helix.

Exciton couplet was also observed in the $^{1}Ba$ absorption band (around 240 nm) of poly(pyrAla) in TMP. Conformational analysis based on the ECEPP potential functions and the theoretical CD computation suggested a left-handed α-helical main chain along which the side-chain pyrenyl groups are regularly arranged.

Interactions in the Excited-State of Poly(arylalanines) The interchromophore distances between the nearest pair of naphthyl groups in the most probable ground-state conformation is estimated to be 7.1 A (1-5 pair, the numbers indicate the amino acid units carrying the nearest pair of naphthyl groups) for poly(L-1-naphthylalanine) and 6.0 A (1-5 pair) for poly(L-2-naphthylalanine). The interchromophore distances are too long to form excimers in the excited state. Actually no excimer was found in poly(1-napAla) when its chain length was long enough (degree of polymerization > about 40) to hold stable helix conformation. However, a small but definite amount of excimer was found in poly(2-napAla) in TMP.

CPF spectroscopy measures a difference in intensities of left- and right-circularly polarized components of a fluorescence emitted from a chiral fluorophore and reflects the chirality of conformations and interactions in the excited state (2). Figure 1 shows CPF spectra of poly(1- and 2-napAla)s measured in DMF solution (12). The ordinate of the CPF spectra is the Kuhn's emission dissymmetry factor $g_{em}$,

$$g_{em} = 2(I_L - I_R)/(I_L + I_R),$$

where $I_L$ and $I_R$ indicate intensities of left- and right-circularly polarized components of fluorescence.

The excimer of poly(2-napAla) showed an intense circular polarization, indicating that the excimer has a skewed chiral configuration. Therefore, the excimer in poly(2-napAla) may be formed not in the randomly coiled part (e.g., at the termini) ,where no particular excimer configuration is expected, but in the helix part of the polypeptide. Since the interchromophore distance in the helix part (6.0 A) may be too far to form an excimer, a small distortion of the side-chain and/or main-chain will be necessary to form the excimer.

It was also found in the CPF spectra that no circular polarization was detected in the monomer fluorescence of poly(1- and 2-napAla)s. This indicates that the lowest-energy excited state of the naphthyl polymers is virtually localized on a monomeric naphthyl group and not delocalized over neighboring chromophores to form an exciton state.

Although the center-to-center interchromophore distance in the most probable ground-state conformation of poly(pyrAla) has been estimated to be about 7.4 A (9), a moderately strong excimer was observed. The intensity of the excimer emission is however smaller than that of a random DL-copolypeptide of pyrenylalanine, which does not form any regular conformation. Therefore the effect of regular conformation to supress excimer formation is again exemplified in this polypeptide. CPF spectroscopy was again used to study the nature of the excimer in poly(pyrAla) (8). Figure 2 shows CPF spectra of poly(pyrAla) (average degree of polymerization = 20) in TMP. The spectra clearly indicate the presence of two types of excimers: one with a positive CPF signal (Left-handed circularly polarized light is more intense than the right-handed one) centered at longer wavelength than about 550 nm, the other with a negative CPF signal at around 460 nm. The contribution of the positive component is greater at low temperatures and that of the negative one increases at higher temperatures, suggesting that the formation of the latter excimer requires large molecular motions, whereas the excimer with positive circular polarization forms without large conformational relaxation. This multiplicity in excimer emission may be interpreted by the multiplicity in the mode of side-chain interactions, e.g., excimer can be formed between both 1-4 and 1-5 pairs of the pyrenyl groups, alternatively, it can be formed in the helical part as well as in the randomly coiled part of the polypeptide.

Rate of Energy Migration along Poly(L-1-naphthylalanine) Chain A preliminary experiment on the rate of energy migration along poly(1-napAla) has been carried out (13). Polymer samples containing a diethylanilino (DEA) group covalently attached at the terminal (II) were prepared for this purpose. In structure II the poly(γ-benzyl DL-glutamate) unit was attached as a solubilizing agent, as before.

$$(C_2H_5)_2N\text{-}C_6H_4\text{-}(CH_2)_3CO\text{-}(NH\text{-}\underset{\underset{C_{10}H_7}{|}}{\underset{|}{CH}}\!\!\underset{}{}CH_2\text{-}CO)_n\text{-}(NH\text{-}\underset{(CH_2)_2COOCH_2C_6H_5}{CH}\text{-}CO)_{120}\text{-}NH(CH_2)_5CH_3 \quad \text{(II-n)}$$

The number-average degree of polymerization of the poly(1-napAla) unit n, was 10 and 30, which was adjusted by the polymerization conditions. As a reference sample, compound III, which has the same structure as the terminal unit of polymer II, was also synthesized.

$$(C_2H_5)_2N\text{-}C_6H_4\text{-}(CH_2)_3CO\text{-}NH\text{-}\underset{CH_2C_{10}H_7}{CH}\text{-}CO\text{-}O\text{-}CH_3 \quad \text{(III)}$$

Fluorescence spectra of the polypeptides and the model compound showed an exciplex emission centered at 500 nm in dichloroethane (95%)/dimethylformamide (5%) mixed solvent under nitrogen atmosphere at room temperature. Since the attachment of the DEA group to the chain end was not quantitative, the dynamics of energy migration was

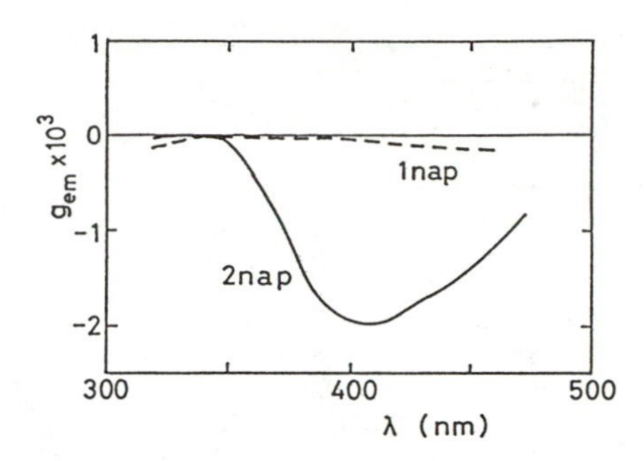


Figure 1. CPF spectra of poly(1-napAla) (----) and poly(2-napAla) (——) in $N_2$-bubbled DMF, [Nap] = $3.0 \times 10^{-5}$ M.

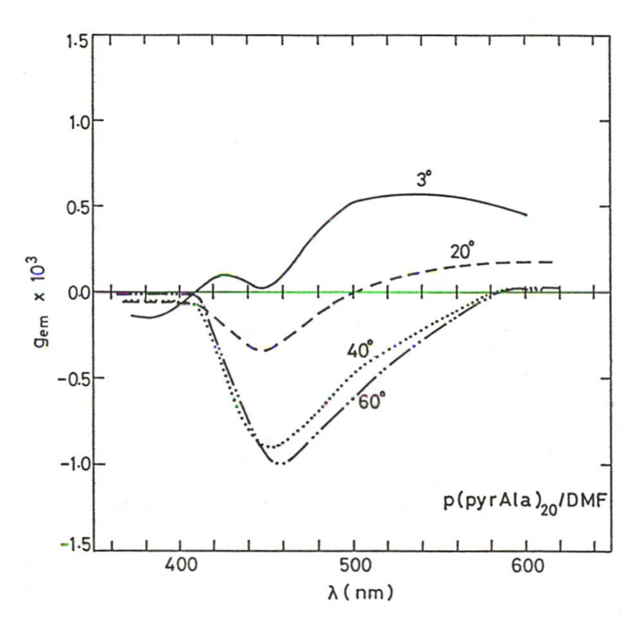


Figure 2. CPF spectra of poly(pyrAla) (number average degree of polymerization = 20) in $N_2$-bubbled DMF at different temperatures. [Pyr] = $6 \times 10^{-5}$ M.

discussed from the rise and decay curves of the exciplex emission. The results are listed in Table I.

Table I. Results of rise and decay curve analysis of the exciplex emission from polymer II and compound III

| Sample | $\tau_1$ (nsec) | $\tau_2$ (nsec) |
|---|---|---|
| II-10 | 2.5 | 45 |
| II-30 | 4.5 | 55 |
| III | -- | 46 |

The rise and decay curves were fitted to the followeing equation: $I(t) = a[\exp(-t/\tau_2) - \exp(-t/\tau_1)]$.

The rise time for the model compound could not be measured by the instrument employed (time resolution may be about one nanosecond). On the other hand, a delay of the order of a few nanosecond was detected in the rise of the exciplex emission in the polypeptide systems and the delay is longer for the longer polymer. If one assumes random hopping processes for the energy migration and a uniform probability for the excitation of one of the naphthyl group, the time required for an elementary hopping process may be calculated to be about a few hundred picosecond or less. For a more detailed study, however, polymer samples carrying a definite number of naphthyl chromophores will be needed.

## Sequential Polypeptides Carrying Naphthyl or Pyrenyl Chromophores (14,15)

As described above, there is a multiplicity in the interactions among chromophores in the aromatic homopolypeptides. One of the reasons for the multiplicity is that 1-4, 1-5, and other interactions are possible in the homopolypeptides. In this sense, the homopolypeptides are not true one-dimensional chromophoric systems. To attain one-dimensional systems, sequential polypeptides carrying one chromophore in each helix turn were prepared.

Sequential Polypeptide Carrying 1-Naphthyl Groups, Ground-State Conformations Polypeptides having a regular array of chromophores with varying spacings can be prepared by polymerization of oligopeptide active esters of the form, $H\text{-}A_m\text{-}B\text{-}X$, where A is a spacer amino acid having no chromophore and B is an aromatic amino acid such as naphthylalanines and pyrenylalanine, and X is an active ester group. In particular, if one selects an amino acid which favors α-helical conformation as A, the chromophore attached to B will arrange regularly along a rigid framework of an α-helix. This expectation was actually realized in the case of poly[Lys(Z)$_m$-1-napAla] (m = 1 - 4, all amino acids are L-form) (IV), where Lys(Z) unit functions as an α-helix-forming spacer (14).

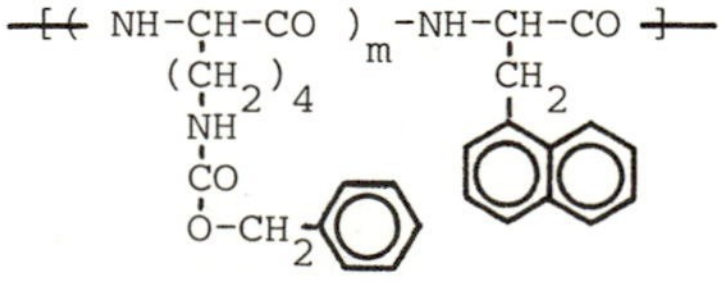

$$-[(\text{NH-CH-CO})_m\text{-NH-CH-CO}]- \quad (m=1 - 4) \quad (IV)$$

(side chains: $(CH_2)_4$-NH-CO-O-$CH_2$-phenyl; $CH_2$-1-naphthyl)

Figure 3 shows experimental CD spectra of the sequential polypeptides in trimethyl phosphate (TMP) solution. Strong exciton couplets in the $^1$Bb absorption band of naphthyl group, particularly in the case of m=2, indicate regular and helical arrangement of the naphthyl groups. Theoretical CD spectra were calculated assuming α-helical main chain and the minimum-energy side-chain conformation predicted from the ECEPP conformational calculations. The results are shown in Figure 4. The theoretical CD curves of poly[$Ala_m$-1-napAla] qualitatively reproduce the features of the experimental spectra for the four sequential polypeptides. Figure 5 illustrates the most probable α-helical conformations of the sequential polypeptides drawn by using the NAMOD (16) molecular display program. The interchromophore distances between the centers of the nearest pair of the naphthyl groups in the conformations shown in Figure 5 are: 7.1 (m=1, 1-5); 6.7 (m=2, 1-4); 6.7 (m=3,1-5); and 13.4 (m=4, 1-6) A.

Energy Migration along Helix Fluorescence spectra of the sequential polypeptides showed virtually no excimer emission. This is in accordance with the absence of excimer emission in poly(1-napAla), and again suggests the regular and rigid arrangement of naphthyl groups along the α-helical main chain.

The extent of the intramolecular excitation energy migration was evaluated from the efficiency of the fluorescence quenching of the polypeptides by using diacetyl as an external quencher (17). Figure 6 shows Stern-Volmer plots for the sequential polypeptides and the homopolypeptides of L-1-napAla, poly(1-napAla), and DL-1-napAla, poly(DL-1-napAla). The energy migration is most efficient in the homopolypeptide of L-napAla. The efficiency of poly(DL-1-napAla) is a little lower than the L-polymer, but substantially higher than the four sequential polypeptides. The high transfer efficiency of L- and DL-homopolypeptides can be interpreted by the multiplicity in the paths for energy migration along the polymer chain. That is, in the homopolypeptides energy migration between both 1-4 and 1-5 pairs of naphthyl groups may be frequent, whereas in the sequential polypeptides only 1-5 (m=1 and 3), 1-4 (m=2), or 1-6 (m=4) energy migration may be allowed (as above, the numbers indicate amino acid units).

It is interesting that the sequential polypeptide with two Lys(Z) units as a spacer (m=2) is more efficient than that with one Lys(Z) unit as a spacer (m=1), although the density of naphthyl chromophore is lower in the former polypeptide. The efficiency for m=1 is similar to that of m=3. These findings can be interpreted again by the differences in the positions of the nearest pair of naphthyl groups: 1-5 for m=1 and 3, 1-4 for m=2, and 1-6 for m=4. The similar energy migration efficiencies of m=1 and m=3 indicate a negligible energy migration between 1-3 pair of m=1.

It is also interesting that the polypeptide of m=4 showed moderately efficient energy migration when it is compared with the efficiency in a polypeptide (m=49r) containing 2% of napAla unit, in which an energy migration is virtually impossible. The center-to-center distance between the nearest naphthyl groups in m=4 is 13.4 A, which is a little longer than the critical transfer distance $R_0$ for a naphthyl-naphthyl pair (12 A).

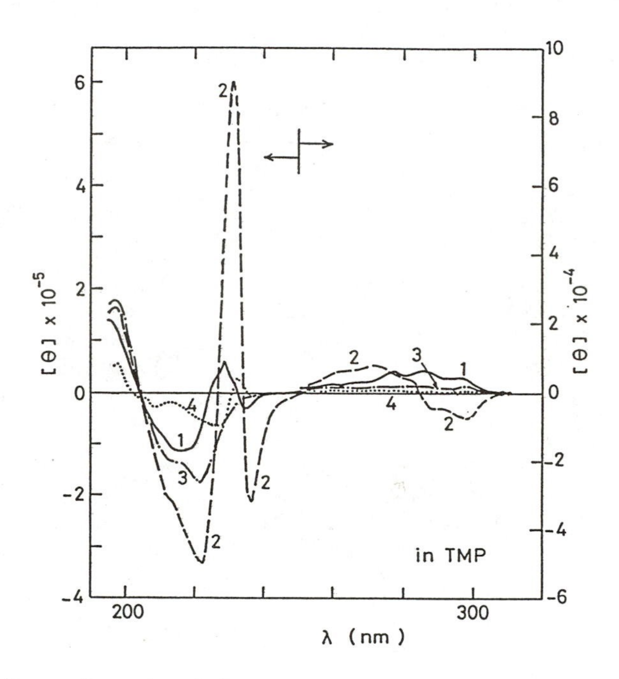


Figure 3. Experimental CD spectra of poly[Lys(Z)$_m$-1-napAla] in trimethyl phosphate at room temperature, [Nap] = 2.0x10$^{-4}$ M. Numbers in the figure indicate the number of Lys(Z) units, m.

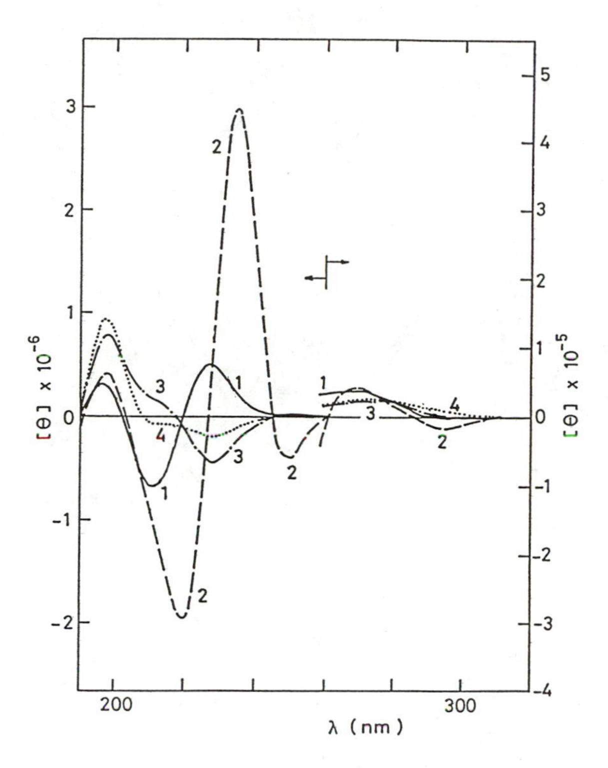


Figure 4. Theoretical CD spectra of poly[$Ala_m$-1-napAla] in α-helical conformation with the least-energy side-chain conformation predicted from the ECEPP energy calculation. Numbers in the figure indicate m.

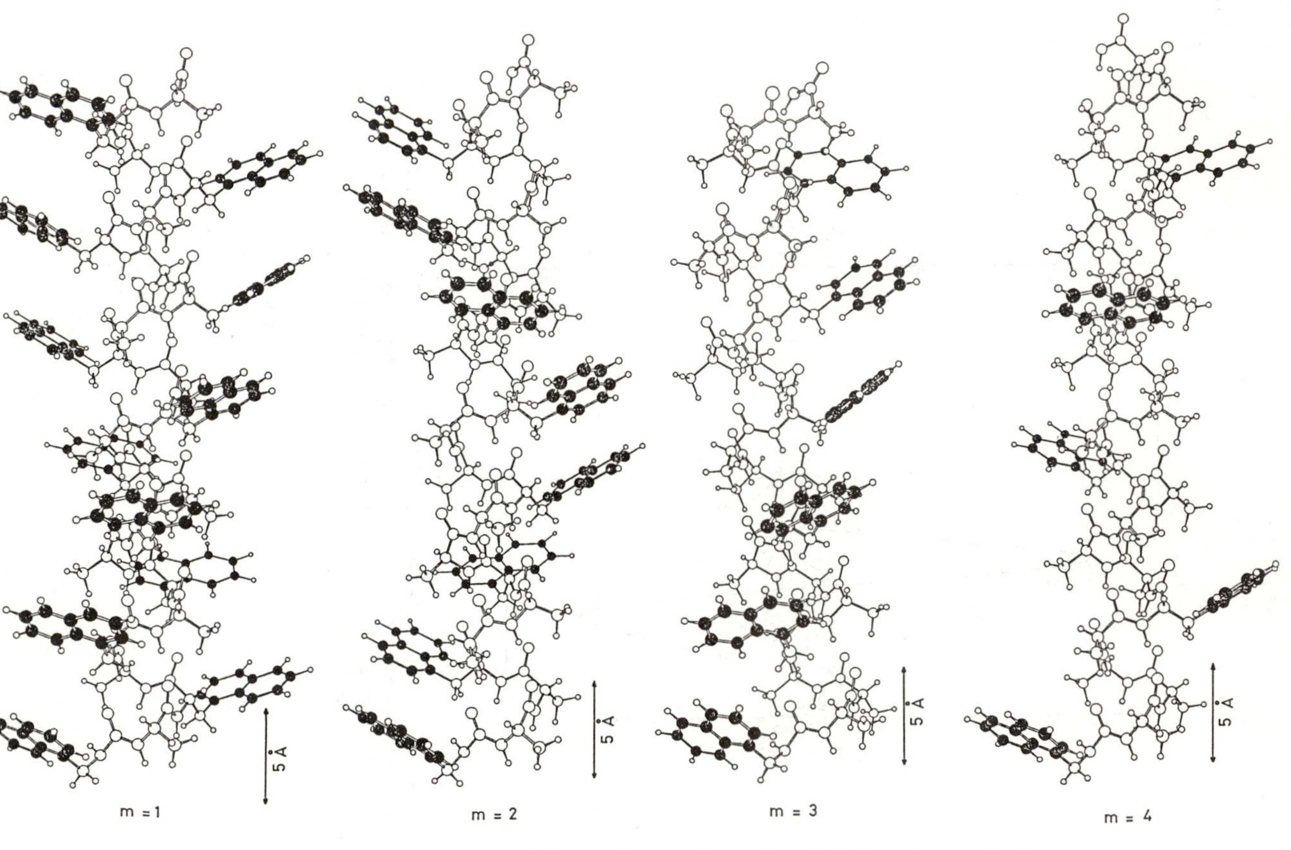


Figure 5. Most probable conformation of poly[Ala$_m$-1-napAla] predicted from the ECEPP energy calculation. NAMOD (version 3) program (ref. 16) was used to draw the molecular models.

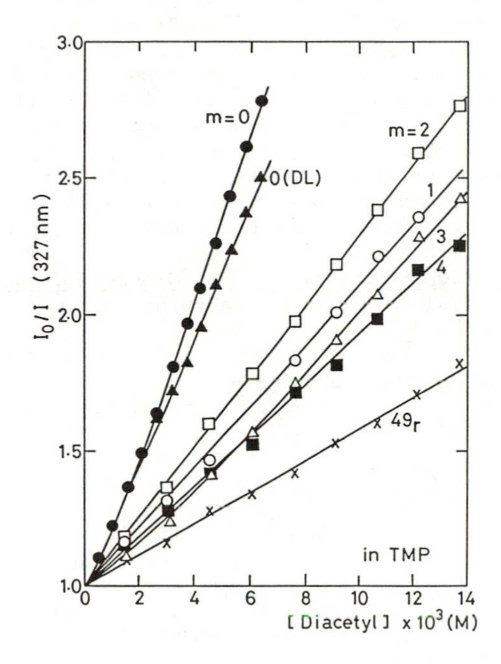


Figure 6. Stern-Volmer plots for the quenching of the monomer emission of poly[Lys(Z)$_m$-1-napAla] in $N_2$-bubbled trimethyl phosphate by diacetyl as a quencher. [Nap] = $2.0 \times 10^{-5}$ M, room temperature.

Fluorescence Decay Analysis Fluorescence decay curves were measured for the monomer emission of the sequential polypeptide and the random copolypeptide having a very small amount of naphthyl groups (17). Although no excimer was observed in the stationary fluorescence spectra, the decay curves did not fit single exponential functions, but fitted dual-exponential curves. The lifetime of the fast-decaying component is about 12 ns, whereas that of the slow component is between 50 - 60 ns irrespective of the length of the spacer. Since the polypeptide with a very small content of the naphthyl group (m=49r) showed a single exponential decay with a lifetime of 55 ns, the presence of the fast-decaying component cannot be explained by any interaction of excited naphthyl chromophore with amide or other chromophores of the polypeptide chain. A possible interpretation for the fast component is that a special pair of naphthyl groups deactivates very rapidly the excited state. However, the nature of the interaction is not clear at present.

Sequential Polypeptides Carrying Pyrenyl Groups, Ground-State Conformations Sequential polypeptides carrying pyrenyl chromophores (V) were also prepared (15).

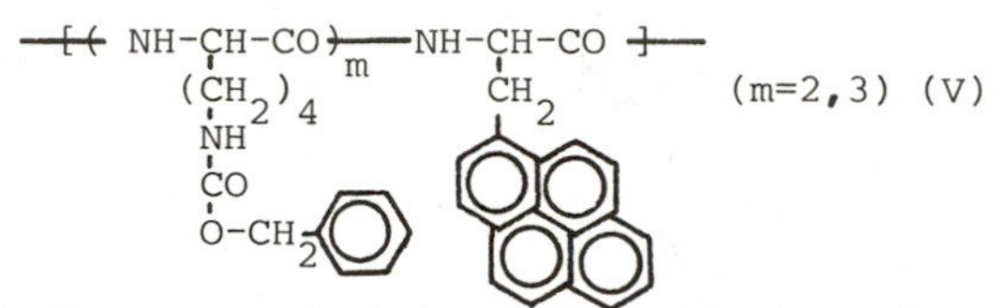


No particular ground-state interaction was detected in the absorption spectra of the two polypeptides. Circular dichroic spectra of the polypeptides are very similar to each other and showed an exciton splittings at the $^1$La and the $^1$Bb absorption bands: $[\theta]353$ = 17300, $[\theta]348$ = -36500, $[\theta]283$ = -30200, $[\theta]279$ = 21100 (m=2); $[\theta]355$ = 10200, $[\theta]349$ = -30100, $[\theta]283$ = -21000, $[\theta]280$ = 18400 (m=3). However only a small negative peak was observed at the $^1$Ba band: $[\theta]241$ = -26800 (m=2); $[\theta]247$ = -26300 (m=3). The shape of CD spectra at the amide transition band around 200-220 nm suggested right-handed $\alpha$-helical main-chain conformations for the two polypeptides. These features indicate that the main chain takes $\alpha$-helical conformation but the orientation of the side-chain pyrenyl groups are more fluctuated than in the case of naphthyl groups.

An empirical conformation energy calculation based on the ECEPP functions indicated two stable side-chain orientations (A and B) are possible for $\alpha$-helical main chain. A theoretical CD computation, however, suggested that orientation B ($\chi_1$ = 290°, $\chi_2$ = 285°) is more populated than A.

In the case of m=2 in dimethylsulfoxide solution, the right-handed $\alpha$-helical main chain conformation has been confirmed by the circular dichroism in the vibrational region (VCD). (A preliminary result of joint research with Professor T. Keiderling, University of Illinois at Chicago). By the VCD spectroscopy one can obtain information on the main chain conformation of polypeptides without any interference by the side-chain chromophores (18). The VCD couplet at the amide I region of poly[Lys(Z)$_2$-1-pyrAla] showed a typical pattern for right-handed α-helical conformation.

<u>Excimer Emission and CPF Spectra</u> Fluorescence spectra of the two polypeptides in TMP solution are shown in Figure 7 (lower curves). Small excimer emissions are observed in the two polymers. The monomer/excimer intensity ratio was independent of the polymer concentration at least down to [pyr] = $1 \times 10^{-7}$ mol $L^{-1}$, suggesting an intramolecular character of the excimer. Since the interchromophore distances in the most probable conformations predicted from the conformational energy calculation are much longer than the excimer-forming distance, the excimers should be formed at the point where conformations of the main chain and/or the side chain are largely distorted.

The upper curves of Figure 7 shows CPF spectra of the two polypeptides. A large positive CPF signal is seen at the excimer region, indicating that the excimers in the polypeptides have skewed configurations. The CPF spectra for m=2 and 3 are virtually the same, suggesting that the geometrical configurations of the two excimers are very similar to each other. The dissymmetry factor is virtually constant over the excimer region, if the contribution of monomer emission which showed no circular polarization was subtracted. The spectra in Figure 7 contrasts that of poly(L-1-pyrAla) shown in Figure 2, where negative and positive signals were observed. The CPF spectra of the sequential polypeptides were virtually temperature independent over the temperature range of 5 - 60°C. This again contrasts the case of poly(L-1-pyrAla), for which a marked temperature dependence was observed. These findings suggest that, at least in the stationary spectrum, contribution of one type of excimer is predominant in the sequential polypeptide. This may reflect the fact that only one type of side chain interaction (1-4 for m=2, 1-5 for m=3) is possible in the latter polymers.

<u>Time-Resolved Fluorescence Spectra---Presence of Fast-Decaying Excimer</u> Time-resolved fluorescence spectra were measured on a time-correlated single-photon counting apparatus with a mode-locked synchronously pumped dye laser as a light source. The result is shown in Figure 8 (a preliminary result of joint research with Dr. Y. Taniguchi, Advanced Research Lab., Hitachi, Ltd, Tokyo, Japan). The time-resolution of the system is a few hundred picosecond, which is determined by the time constant of the detector. In Figure 8 an excimer emission centered around 430 nm appears shortly after excitation (risetime = 0.3 ns), which decays with a lifetime of 5.0 ns. After 40 ns or more, the fluorescence spectrum becomes similar to the stationary spectrum (Figure 7, bottom). The lifetime of the slow-decaying component is 100-120 ns. The relative populations (preexponential factors) of the fast component measured at 440 and 480 nm are about the same (0.69 at 440 nm and 0.76 at 480 nm, for fast component) indicating that the emission bands of the two excimers are largely overlapped. At present we have no clear-cut interpretation for the fast-decaying excimer.

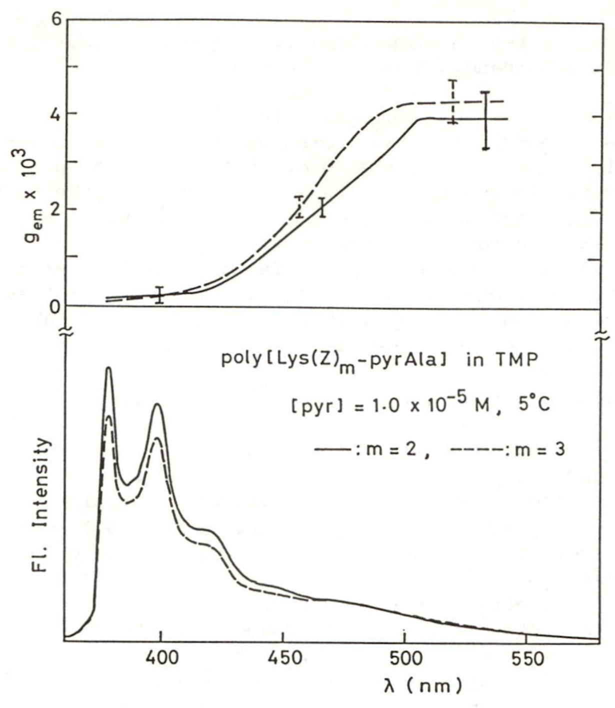


Figure 7. Fluorescence (lower curves) and CPF (upper curves) spectra of poly[Lys(Z)$_m$-pyrAla] in $N_2$-bubbled trimethyl phosphate, 5°C. [Pyr] = 1.0x$10^{-5}$ M. (——): m=2, (----): m=3.

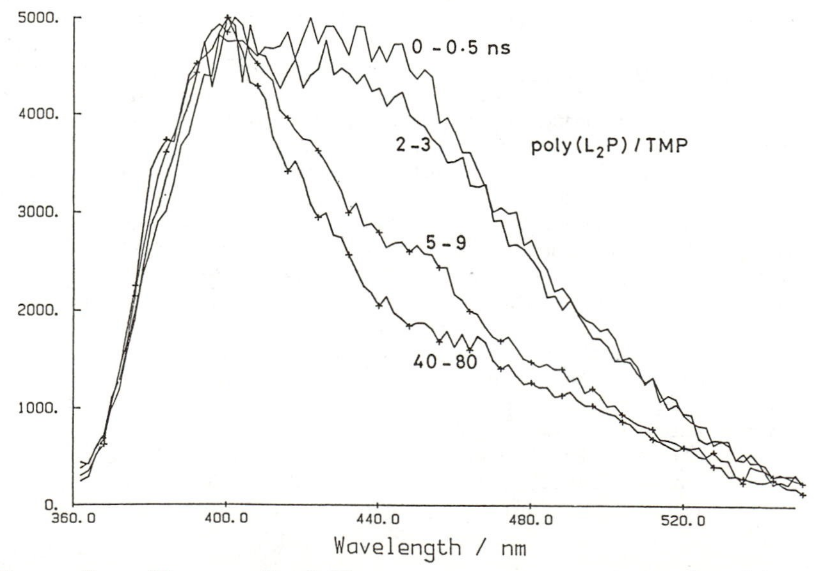


Figure 8. Time-resolved fluorescence spectra of poly[Lys(Z)$_2$-pyrAla] in $N_2$-bubbled trimethyl phosphate at room temperature. [Pyr]=2x$10^{-5}$M.

## Concluding Remarks

Most of the aromatic polypeptides presented in this review were found to exist in α-helical conformations in solution and the side-chain chromophores were shown to be arranged along the helix. Furthermore, in some sequential polypeptides, only one type of side-chain interaction was thought to be allowed. However, CPF spectra and time-resolved spectra showed some multiplicity in the side-chain interactions. In particular, fast-decaying components were detected both in the monomer emission of the sequential polypeptides carrying naphthyl groups and in the excimer emission of those carrying pyrenyl groups. It is a future problem whether any new and special mechanism is required to interpret these apparently complex results.

Acknowledgments The author wishes to thank Professor T. Keiderling, University of Illinois at Chicago and Dr. Y. Taniguchi, Adv. Res. Lab., Hitachi Ltd. for allowing the use of the preliminary results. Acknowledgement is also made to the Donors of the Petroleum Research Fund, administered by the Americal Chemical Society, for partial support of this activity.

## Literature Cited

1. Sisido, M, Makromol. Chem. 1985, 14, 131.
2. Riehl, J.P.; Richardson, F.S. Chem.Rev. 1986, 86, 1.
3. Molecular Electronic Devices; Carter, F.L. ed.; Marcel Dekker, Inc.: New York, 1982.
4. Bamford, C.H.; Elliott, A.; Hanby, W.E.; Synthetic Polypeptides; Academic Press: New York, 1956.
5. Izumiya, N.; Kato, T.; Aoyagi, H; Waki, M. Fundamentals and Practices in Peptide Synthesis (in Japanese); Maruzen, Tokyo, 1985.
6. Sisido, M; Egusa, S.; Imanishi, Y. J. Am. Chem. Soc. 1983, 105, 1041.
7. Sidido, M; Egusa, S.; Imanishi, Y. J. Am. Chem. Soc. 1983, 105, 4077.
8. Egusa, S.; Sisido, M; Imanishi, Y. Macromolecules 1985, 18, 882.
9. Sisido, M; Egusa, S.; Imanishi, Y. Macromolecules 1985, 18, 890.
10. Woody, R.W. J. Polym. Sci., Macromol. Rev. 1977, 12, 181.
11. Momany, R.A.; McGuire, R.F.; Burgess, A.W.; Scheraga, H.A. J. Phys. Chem. 1975, 79, 2361.
12. Sisido, M; Egusa, S.; Okamoto, A.; Imanishi, Y. J. Am. Chem. Soc. 1983, 105, 3351.
13. Egusa, S.; Sisido, M.; Imanishi, Y. unpublished work.
14. Sisido, M.; Imanishi, Y. Macromolecules 1986, 19, 2187.
15. Sisido, M.; Imanishi, Y. to be published.
16. Beppu, Y. Quantum Chemistry Program Exchange No. 370, 1979.
17. Sisido, M.; Imanishi, Y. unpublished work.
18. Yasui, S.C.; Keiderling, T.A. Biopolymers 1986, 25,5.

RECEIVED March 13, 1987

# Chapter 27

# Exciton Migration in Copolymers of Acenaphthylene

**W. R. Cabaness, S. A. Zamzam, and C. T. Chen**

**Department of Chemistry, The University of Texas at El Paso, El Paso, TX 79968–0513**

Excited states and energy migration by way of excitons have been studied in random copolymers of acenaphthylene (ACN) with acrylonitrile (AN), methacrylonitrile (MAN) and 2-vinyl napthalene (2VN). Also ACN was copolymerized with vinyl acetate and hydrolyzed to give free hydroxyl groups. The latter were reacted with acyl chlorides: 1-naphthoyl, 2-naphthoyl, benzoyl and cinnamoyl. Exciton migration lengths, L, quenching rate constants, $k_{qe}$, and efficiencies of energy transfer were calculated based on steady state fluorescence spectra using oxygen or dimethylaniline as quenchers. The average exciton migration lengths are: 92.5 poly(ACN-co-AN), 130 poly(ACN-co-MAN), 100.5 poly(ACN-co-2VN) and 79.4 Å for polymers containing ester groups.

The absorption of UV energy by fluorophores attached to a polymer chain and the subsequent energy migration by way of excitons are subjects of theoretical interest (1-3). Other workers (4) have described polymers containing pendant chromophores as "antenna polymers" which may function as energy gathering devices or allow singlet energy to proceed to a trap. More fundamental questions as the exact mechanism of exciton migration between fluorophores, the required orientation and maximum distance remain unanswered for many systems.

We have studied excited states in polyacenaphthylene (PACN) and in random copolymers of acenaphthylene with the following comonomers: 2-vinylnaphthalene (2VN), acrylonitrile (AN), methacrylonitrile (MAN) and N-vinylcarbazole (VCz) (5.6). Also ACN was copolymerized with vinyl acetate, and the acetate groups were removed by hydrolysis. The hydroxylated copolymers, poly(acenaphthylene-co-vinyl alcohol), were reacted with the following acyl chlorides: 1-naphthoyl, 2-naphthoyl, benzoyl and cinnamoyl. The latter polymers contain an ester group as a spacer between the main chain and one of the fluorophores. The choice of an ester group as a connecting unit between the polymer chain and fluorophore is probably a poor one since it is known that carbonyl groups easily form triplets and photo-Fries rearrangements can occur (7).

A copolymer containing a dimethylsiloxy spacer was prepared by reacting poly(acenaphthylene-co-vinyl alcohol) with dimethyldichlorosilane followed by reaction with 2-naphthol. (See Figure 1).

0097–6156/87/0358–0358$06.00/0

Two types of copolymers of ACN were investigated in this work. The first type contains ACN units separated by photo-inactive groups in the main chain such as AN or MAN. This permits a comparison of the homopolymer, PACN, to polymers containing smaller numbers of chromophores and the latter being randomly spaced along the polymer chain. The second type is a copolymer of ACN and another monomer containing different chromophore. In the first type only excited singlet, $A^*$, and excimers, $(AA)^*$ are important states while in the second type two excimers, $(AA)^*$ and $(BB)^*$, and the exciplex $(AB)^*$ are possible.

Exciton hopping to the same chromophore and using equi-energetic states depend on dipole-dipole orientation and the interaction distance, $R_0$.

$$A_i^* + A_j^o \longrightarrow A_i^o + A_j^*$$

In copolymers the exciton migration length is limited by the polymer structure available, and most importantly, by polymer conformation. When exciton migration involves excimers or exciplexes, long sequences of excimer-forming-sites (EFS) must be correctly orientated for maximum distance.

Poly(ACN) has a rigid chain structure yet can form excimers with alternate units along the chain (8), or by stacking in a helical conformation. Excimer formation has been reported for alternate copolymers of ACN with styrene (9) and for ACN with maleic anhydride (10). The situation is different for 2-vinylnaphthalene since alternating copolymers of 2VN with methyl methacrylate or methacrylic acid did not form excimers, yet random copolymers of the same systems showed excimer fluorescence (11). Only random copolymers of ACN were prepared in this work.

Quenching was carried out using oxygen or N,N-dimethylaniline to obtain steady state fluorescence spectra. Then exciton migration lengths were calculated by three methods described below with the assumption that the fluorophores show exponential decay. An arbitrary comparison between the migration length and the polymer's end-to-end distance was made. Since the migration length is based on a one-dimensional random walk, a more meaningful comparison is between migration lengths in copolymers and homopolymers.

In equation 1 the singlet exciton migration length, L, is related to the Stern-Volmer constant, $K_{sv}$, and the Förster encounter radius, $R_0$, and $N_0$ is Avogadro's number (12).

$$L = [K_{sv}/2\pi R_0 N_0]^{\frac{1}{2}} \quad (1)$$

Webber and co-workers have proposed two equations which allow the calculation of $\Lambda/D$ where $\Lambda$ is the singlet energy migration rate and D is the diffusion constant for the quencher (13). In equation 2, $k_{qe}^p$ is the bimolecular rate constant for quenching of the polymer while $k_{qe}^m$ refers to a model compound of low molecular weight such as ethylnaphthalene.

$$\Lambda/D = [k_{qe}^p - (1/2)k_{qe}^m]/(1/2)k_{qe}^m \quad (2)$$

A similar equation 3 gives the $\Lambda/D$ ratio when the model system is a polymer containing only a small percentage of chromophore (11).

$$\Lambda/D = [k_{qe}^{p} - k_{qe}^{m}]/k_{qe}^{m} \quad (3)$$

Then the exciton migration length, L, assuming a one-dimensional random walk, is given by equation 4:

$$L = [2\Lambda\tau^{p}]^{\frac{1}{2}} \quad (4)$$

where $\tau^{p}$ is the lifetime of the excited state in the polymeric system.

## Experimental

Copolymers of ACN and AN were prepared with varied compositions by emulsion polymerization in water-ethanol (4:1 by vol) degassed with nitrogen. A surfactant was added and potassium persulfate was the initiator. Polymerizations were carried out at ca. 90°C for 24 h. Copolymers were precipitated into methanol and purified by subsequent solution in benzene and precipitation into methanol. The preparation of poly(ACN-co-2VN) was reported previously (6).

All samples of ACN used in polymerizations were recrystallized twice from methanol-pentane (3:1). AN and MAN were distilled prior to use. Copolymers of ACN and MAN were prepared by solution polymerization in benzene at 60°C for 36 h. AIBN was the initiator, and the polymers were isolated by precipitation into 600 mL of methanol followed by two subsequent precipitations from benzene into methanol.

Vinyl acetate and ACN were copolymerized in dry benzene at 60°C for 36 h using AIBN as initiator. The molar ratio of monomers, vinyl acetate to ACN was 20 to 1. Polymers were isolated by precipitation into methanol. followed by two subsequent precipitations from benzene into methanol. Poly(ACN-co-vinyl acetate) (0.5 g) was dissolved in 100 mL of THF and mixed with 25 mL of 1 N methanolic NaOH. After heating to 60°C for 10 h, the mixture was precipitated into 600 mL of methanol. The polymer was purified by dissolving in THF and precipitation in methanol. Poly(ACN-co-vinyl alcohol) (0.4 g) was dissolved in 50 mL of pyridine and treated with benzoyl chloride in benzene. The mixture was stirred at room temperature for 10 h and precipitated into 600mL of methanol. The copolymer was purified by two precipitations from THF into methanol. A similar procedure was used for reacting other acyl chlorides with poly(ACN-co-vinyl alcohol).

Poly[acenaphthyylene-co-dimethyl-(2-naphthoxy)-vinyloxysilane] was prepared by treating poly(ACN-co-vinyl alcohol)(0.4 g) in 50 mL of THF with dichlorodimethylsilane (0.8 mL) and triethylamine (0.5 mL) at 60°C for 10 h. Then, 2-naphthol (1.0 g) in 20 mL of THF was added and heating was continued for another 10 h at 60°C. The mixture was cooled and poured into 600 mL of diethyl ether. The polymer was purified by two precipitations from THF into ether.

Copolymer compositions of poly(ACN-co-AN) and poly(ACN-co-MAN) were determined by elemental nitrogen analyses. The percentages of ACN in hydrocarbon copolymers were determined by UV spectroscopy, using the measured optical density for ACN in PACN at 310 nm in 1,2-dichloroethane as solvent. For poly(ACN-co-2VN) the compositions were checked by NMR spectroscopy (300 MHz). In all cases, conversions were held to ten percent or less in order to avoid drift in copolymer composition.

Steady state fluorescence spectra were obtained on an Amino-Bowman J4 instrument using a Hg-Xe light source. Samples were prepared immediately before use with 1,2-dichloroethane. Several excitation wavelengths 290, 325 and 353 nm were tried with each series of ACN copolymers to obtain emission bands of highest intensity. For quenching experiments using oxygen, samples were degassed with nitrogen, air and oxygen in sequence (14). The concentrations of N,N-dimethylaniline when used as a quencher were zero, 2.3 and 12.9 x $10^{-3}$ M. The diffusion coefficient for oxygen in 1,2-dichloroethane at 298°K was calculated to be 8.79 x $10^{-6}$ $cm^2/sec$ by the Stokes equation (15). The following values were used in calculations: $\tau^p$, excimer in P2VN, is 45 ns (17); $k^m_{qe}$, model polymer, is 1.52 x $10^9$ l $m^{-1}s^{-1}$ (11), $k^m_{qe}$, 2-ethylnaphthalene, is 4.96 x $10^9$ l $m^{-1}s^{-1}$ (11).

## Results and Discussion

Stern-Volmer plots were made from fluorescence spectral data and $K_{SV}$ values calculated. Exciton migration lengths were calculated using equation (1) and equations (2), (3) and subsequently equation (4). This procedure gave three values of the migration length for comparison.

Migration lengths in random copolymers of ACN with AN are not greatly affected by composition. Copolymers containing 47 and 80% ACN had the longest migration length of 70Å compared to 63 Å for PACN using equation (1). The above percentages correspond to an overall monomer composition of 1:1 and 4:1. Random poly(ACN-co-AN) showed monomeric fluorescence at 355 nm and excimer fluorescence at 405 nm. The $I_D/I_M$ ratios were 0.88 and 0.61 for the 47 and 80% ACN copolymers, respectively, and the same value for PACN was 0.52. The incorporation of photo-inactive units such as AN in the copolymer increase chain flexibility and allows more excimer formation. Exciton migration lengths are listed in Table I.

For a series of poly(ACN-co-MAN) copolymers the amount of chromophore, ACN, was limited to five, seven and 14 percent to determine the effect on migration length and excimer formation. The results are given in Table II including the end-to-end distances for comparison. A copolymer containing five percent ACN had the longest exciton migration distance of 149 Å which exceeds its end-to-end distance of 31 Å. The migration lengths decreased as the ACN content increased from five to 14 percent but in all cases were longer than the end-to-end distances.

The fluorescence spectra of poly(ACN-co-MAN) showed monomer emission at 346 nm and excimer emission at 405 nm. The intensity of the excimer emission band decreased as the percentage of ACN decreased from 14 to 5% and shifted to a longer wavelength at 412 nm. Corresponding $I_D/I_M$ ratios decreased from 0.33 ro 0.13, respectively.

The second type of copolymer studied in this work contained ACN and another monomer which was also a fluorophore. A series of copolymers of ACN and 2VN were prepared which varied in composition from 5.5 to 94.5% ACN. Calculated exciton migration lengths and $K_{SV}$ values are given in Table III. A copolymer containing 53.5% ACN or ca. 1:1

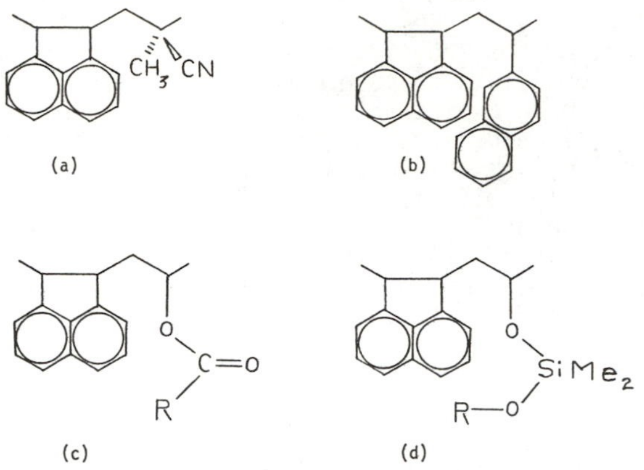

(a) (b)

(c) (d)

Figure 1. Copolymers of acenaphthylene: (a) poly(ACN-co-MAN); (b) poly(ACN-co-2VN); (c) poly(ACN-co-vinyl ester) where acyl is benzoyl, cinnamoyl, 1-naphthoyl and 2-naphthoyl; (d) poly [ACN-co-dimethyl-(2-naphthoxy)-vinyloxysilane].

Table I
Compositions, Quenching Constants and Migration Lengths in Poly(ACN-co-AN)

| mol % ACN in copolymer | $K_{SV}, M^{-1}$ [a] | L, Å [b] | L, Å [c] | L, Å [d] |
|---|---|---|---|---|
| 36 | 157 | 68 | 62 | 101 |
| 47 | 168 | 70 | 68 | 107 |
| 66 | 134 | 63 | 46 | 87 |
| 76 | 113 | 58 | 24 | 72 |
| 78 | 139 | 64 | 50 | 90 |
| 80 | 169 | 70 | 69 | 108 |
| 90 | 131 | 62 | 44 | 85 |
| 95 | 138 | 64 | 50 | 90 |
| 100 | 137 | 63 | 49 | 89 |

[a] $K_{SV}$ for excimer emission at 405 nm.

[b] $L = [K_{SV}/2\pi R_0 N_0]^{1/2}$.

[c] $\Lambda/D = [k_{qe}^{p} - (\frac{1}{2})k_{qe}^{m}]/(\frac{1}{2})k_{qe}^{m}$.

[d] $\Lambda/D = [k_{qe}^{p} - k_{qe}^{m}]/k_{qe}^{m}$ and for column b and c, $L = [2\Lambda\tau^{p}]^{\frac{1}{2}}$. The diffusion coefficient, D, is $8.79 \times 10^{-6} cm^2/s$ for oxygen in 1,2-dichloroethane at 298 K. Oxygen was used as quencher.

Table II
Compositions, End-to-End Distances and Migration Lengths in Poly(ACN-co-MAN)

| mol % ACN in copolymer | $\langle \bar{r}^2 \rangle^{1/2}$, Å [a] | L, Å [b] | L, Å [c] | L, Å [d] |
|---|---|---|---|---|
| 5 | 31 | 88 | 107 | 149 |
| 7 | 70 | 78 | 86 | 125 |
| 14 | 62 | 69 | 77 | 116 |

[a] Calculated using Flory's equation: $[\eta] = \Phi \langle \bar{r}^2 \rangle^{3/2} M^{-1}$
[b] Calculated using equation (1).
[c] Calculated using equations (2) and (4). [d] Calculated using equations (3) and (4).

Table III
Compositions, Quenching Constants and Migration Lengths in Poly(ACN-co-2VN)

| mol% ACN in copolymer | $K_{SV}$, $M^{-1}$ [a] | L, Å [b] | L, Å [c] | L, Å [d] |
|---|---|---|---|---|
| 5.5 | 126 | 61 | 39 | 82 |
| 17.3 | 148 | 66 | 56 | 96 |
| 41.7 | 130 | 62 | 43 | 85 |
| 53.5 | 291 | 93 | 118 | 160 |
| 59.8 | 139 | 64 | 50 | 90 |
| 80.3 | 165 | 70 | 67 | 106 |
| 94.5 | 130 | 62 | 43 | 85 |

[a] $K_{SV}$ for exciplex emission at 400 nm.

[b-d] Same as Table I. Oxygen was used as quencher.

composition had the longest L value of 160 Å. This ratio favors exciplex formation and may indicate that exciplexes provide and adventitous pathway for exciton migration. Copolymers of other compositions, both greater than and less than 50% ACN, gave approximately 60Å for migration lengths as calculated by equation (1). A regrouping of fluorophores may occur during exciton migration in the 1:1 copolymer (19). While such an effect might be expected to be temperature dependent, no studies in this area were attempted.

$$B\,(A\,B)_i^{*}\,A\,B \xrightarrow{\text{migration}} B\;A\,(B\,A)_j^{*}\,B$$

Table VI gives the exciton migration lengths for five copolymers of ACN obtained by reacting poly(ACN-co-vinyl alcohol) with different acyl chlorides. The migration length as a percentage of the end-to-end distance is 167% for poly(ACN-co-1-vinyl naphthoate). For fluorophores 2-vinyl naphthoate, vinyl benzoate and vinyl cinnamate the average is 116% and for dimethyl-(2-naphthoxy)vinyloxysilane, 146%.

Table IV
Compositions, End-to-End Distances and Migration Lengths in Copolymers of ACN

| copolymers[a] | ACN in copolymer, % | calc, $\langle \overline{r}^2 \rangle^{\frac{1}{2}}$, Å | L, Å[b,c] |
|---|---|---|---|
| poly(ACN-co-1VNo) | 78.0 | 56 | 94 |
| poly(ACN-co-2VNo) | 85.2 | 71 | 80 |
| poly(ACN-co-VB) | 87.4 | 74 | 87 |
| poly(ACN-co-VC) | 84.5 | 72 | 85 |
| poly(ACN-co-DMVS) | 81.5 | 35 | 51 |

[a] leg. 1-vinyl naphthoate, 1VNo; 2-vinyl naphthoate, 2VNo; vinyl benzoate, VB; vinyl cinnamate, VC; dimethyl-(2-naphthoxy) vinyloxysilane, DMVS.
[b] Calculated using equation (1).
[c] N,N-Dimethylaniline was used as quencher. $K_{SV}$ for exciplex emission was used. Exciplex emission occurred in the 420 to 440 nm region.

Poly(ACN-co-vinyl cinnamate) showed an increase in intensity of the exciplex peak (433 nm) on increasing the concentration of oxygen. This behavior is similar to the formation of a CT complex (16). When dimethylaniline was used as the quencher, normal quenching curves were observed.

The efficiencies of energy transfer were calculated using equation (5) and listed in Table V.

$$\chi = K_{SV}(Q)/[1 + K_{SV})(Q)] \tag{5}$$

Oxygen was used as quencher at a concentration of 12.9 x $10^{-3}$M. Solvent in all cases was 1,2-dichloroethane. The intensity of the excimeric (or exciplex) emission band was selected to calculate $K_{SV}$ values.

## Summary

In random copolymers of ACN and AN, compositions of 47 and 80% ACN gave the longest exciton migration distances. In poly(ACN-co-MAN) a copolymer containing only five percent ACN had a migration distance of 149Å. This indicates chromophores remotely situated on a polymer chain can interact with energy transfer (18).

Table V
Efficiencies of Energy Transfer $\chi$ in Copolymers of ACN

| copolymers | ACN in copolymers, % | $\chi$ excimeric |
|---|---|---|
| poly(ACN-co-AN) | 47.0 | 0.66 |
| poly(ACN-co-AN) | 80.0 | 0.59 |
| poly(ACN-co-MAN) | 5.0 | 0.63 |
| poly(ACN-co-2VN) | 53.5 | 0.65 |
| poly(ACN-co-1VNo) | 78.0 | 0.67 |
| poly(ACN-co-2VNo) | 85.2 | 0.76 |
| poly(ACN-co-VB) | 87.4 | 0.78 |
| poly(ACN-co-VC) | 84.5 | 0.76 |
| poly(ACN-co-DMVS) | 81.5 | 0.50 |

In a series of copolymers of ACN and 2VN in which the compositions varied from 5.5 to 94.5% ACN, a copolymer containing 53.5% ACN had the longest migration distance of 160 Å. Energy from an exciplex may transfer to an exciplex-forming site.

Copolymers of ACN with other monomers containing fluorophores and attached to the polymer chain by ester groups had exciton migration distances which are greater than their calculated end-to-end distances, (Table IV). It is noted that ester groups may be photolabile.

Efficiency of energy transfer, $\chi$, is approximately 0.63 for copolymers of ACN with AN, MAN or 2VN. The efficiency is higher for copolymers with ester groups, ca. 0.74 and is lowest for the polymer containing silicon, 0.50, (Table V).

## Literature Cited

1. Guillet, J.E.; Rendall, W.A. Macromolecules, 1986, 19, 224.
2. Burkhart, R.D.; Dawood, I. Macromolecules, 1986, 19, 447.
3. Winnik, M. A.; Slomkowski, S. Macromolecules, 1986, 19, 500.
4. Ren, X.; Guillet, J.E. Macromolecules, 1985, 18, 2012.
5. Cabaness, W.R.; Lin, C.L. J. Polym. Sci. Polym. Chem. Ed. 1984, 22, 857.
6. Cabaness, W.R.; Zamzam, S.A. Polymer Preprints, 1985, 26(1), 102.
7. Holden, D.A.; Shephard, S.E.; Guillet, J.E. Macromolecules, 1982, 15, 1481.

8. David, C. Lempereur, M.; Gauskens, G.; Eur. Polym. J., 1972, 8, 417.
9. Wang, Y.C.; Morawetz, H. Makromol, Chem., Suppl. 1975, 1, 283.
10. Reid, R.R.; Soutar, I. J. Polym. Sci. Letters Ed., 1980, 18, 123.
11. Bai, F.; Chang, C.H.; Webber, S.E. Macromolecules, 1986, 19, 588.
12. Johnson, G.E. Macromolecules, 1980, 13, 145.
13. Webber, S.E.; Avots-Avotins, P.E.; Deumie, M. Macromolecules, 1981, 14, 105.
14. Ishii, T.; Handa, T.; Matsunaga, S. Macomolecules, 1978, 11, 40.
15. Birks, J.B. Organic Molecular Photophysics; John Wiley and Sons: New York, 1973; Vol 1. 1, p 454.
16. Webber, S.E.; Kim, N. Macromolecues, 1985, 18, 741.
17. Pratte, J.F.; Webber, S.E. Macromolecules, 1984, 17, 2116.
18. Zachriasse, K.A.; Kuhnle, W.; Weller, A. Chem. Phys. Lett. 1978, 59, 375.
19. Birks, J.B. Organic Molecular Photophysics; John Wiley and Sons: New York, 1975; Vol. 2, p. 573.

RECEIVED March 13, 1987

# PHOTOPHYSICS OF POLYELECTROLYTES

Chapter 28

# Kinetic Spectroscopy of Relaxation and Mobility in Synthetic Polymers

**K. P. Ghiggino, S. W. Bigger, T. A. Smith, P. F. Skilton, and K. L. Tan**

**Department of Physical Chemistry, University of Melbourne, Parkville, Victoria, Australia 3052**

Time-resolved fluorescence spectroscopy and fluorescence anisotropy measurements have been applied to study (i) excimer formation and energy transfer in solutions of poly(acenaphthalene) (PACE) and poly(2-naphthyl methacrylate) (P2NMA) and (ii) the conformational dynamics of poly(methacrylic acid) (PMA) and poly (acrylic acid) as a function of solution pH. For PACE and P2NMA, analysis of projections in which the spectral, temporal and intensity information are simultaneously displayed have been used to re-examine kinetic models proposed to account for the complex fluorescence decay behaviour that is observed. Time-resolved fluorescence anisotropy measurements of fluorescent probes incorporated in PMA have led to the proposal of a "connected cluster" model for the hypercoiled conformation of this polymer existing at low pH.

For many years, fluorescence techniques have proved to be extremely powerful and sensitive tools for studying the mechanisms of excitation energy dissipation in polymers and for probing macromolecular structure and dynamics (1-4). However, recent developments in time-resolved fluorescence instrumentation have led to dramatic improvements in the quality of fluorescence decay data obtainable and, when combined with computer-aided data analysis procedures, give rise to new possibilities for resolving the heterogeneous emission and complex fluorescence decay behaviour observed for many polymer systems. In this paper the application of such time-resolved fluorescence techniques is described for investigating (i) excimer formation and energy transfer in polymers containing naphthyl and anthryl chromophores, and (ii) cluster size and chain mobility in poly(methacrylic acid) (PMA) and poly(acrylic acid) (PAA) as functions of solution pH.

It is now well established that the temporal decay of fluorescence from solutions of polymers containing pendant aromatic

0097-6156/87/0358-0368$06.00/0

chromophores rarely follows the conventional kinetic schemes that are applicable to simple molecules (2,3). The multi-exponential (or non-exponential) fluorescence decay kinetics observed in the monomer and excimer regions of the emission spectrum have been variously attributed to more than one kinetically distinct monomer and/or excimer species (3,5), spectrally and temporally distinct emitting sites (6,7), configurational and conformational influences upon excimer formation (8,9), non-equilibrium diffusion-controlled excimer sampling (10) and time-dependent energy trapping (11). The mechanisms and rates of the transfer of energy to incorporated chromophores which act as trap sites also have attracted considerable attention due to their relevance to photodegradation and light-harvesting phenomena. Guillet and co-workers (12,13) have demonstrated clearly that transfer within aromatic polymer chains to chemically bound acceptor chromophores can occur with high efficiency and is significantly influenced by polymer structure and conformation.

The kinetics of these excited-state processes in polymer systems are undoubtedly complex. However, it is possible to obtain an overview of the relaxation behaviour following light absorption through the recording of time/fluorescence intensity/wavelength hypersurfaces. These surfaces, when combined with fluorescence decay analyses, can provide additional insight into the number and identity of the emitting species present as well as the rates of energy relocation in the polymer chain. In the present work, time-resolved fluorescence analyses on poly(acenaphthalene) (PACE) with and without anthryl end-groups have been undertaken and compared with previous studies on poly(2-naphthyl methacrylate) (P2NMA).

A further application of time-resolved fluorescence measurements is in the study of conformational dynamics of polymer chains in solution. Fluorescence anisotropy measurements of macromolecules incorporating suitable fluorescent probes can give details of chain mobility and polymer conformation (2,14). A particular example studied in this laboratory is the conformational changes which occur in aqueous solutions of polyelectrolytes as the solution pH is varied (15,16). Poly(methacrylic acid) (PMA) is known to exist in a compact hypercoiled conformation at low pH but undergoes a transition to a more extended conformation at a degree of neutralization ($\alpha$) of 0.2 to 0.3 (16). Similar conformational transitions are known to occur in biopolymer systems and consequently there is considerable interest in understanding the nature of the structures present in model synthetic polyelectrolyte solutions.

## Experimental

Materials. PACE and P2NMA (see Figure 1) were prepared by free radical polymerization of the purified monomers in degassed benzene using azo bis-isobutyronitrile (AIBN) initiator at 75°C. Anthracene end-groups were incorporated in the polymers by polymerization in the presence of 9-chloromethylanthracene (6% by weight) which acts as a chain transfer agent. The polymers were purified by multiple reprecipitations using benzene/methanol as the solvent/non-solvent until no further changes in the fluorescence spectra were observed. Polymerization yields of approximately 15% were obtained and the weight average molecular weights (in polystyrene equivalents) of the terminated polymers as determined by gel permeation chromatography were 24000 (PACE) and 23300 (P2NMA). Solution absorbances of less than 0.5 at the wavelength of maximum absorbance were used in order to avoid interchain interactions during the excited-state lifetime.

Poly(methacrylic acid) and poly(acrylic acid) (PAA) were prepared by AIBN initiated polymerization of the freshly distilled monomer in deoxygenated methyl ethyl ketone at 60°C. The incorporation of 9,10-dimethylanthracene (9,10-DMA) end-groups in the polymer was achieved by the addition of the chain transfer agent (1% by weight) to the polymerization mixture. Unreacted 9,10-DMA was separated from the polymer by gel permeation chromatography using a column packed with Sephadex LH-20 and methanol as the eluent. Analysis of the PMA sample by NMR indicates that the polymer produced under these conditions consists of 57% syndiotactic, 33% heterotactic and 10% isotactic triads (15). Solution concentrations were 0.02 M in repeating units of the polymer.

All solvents were freshly distilled before use and solutions in organic solvents were degassed by multiple freeze-pump-thaw cycles.

Instrumentation. A schematic representation of the time-resolved fluorescence spectrophotometer developed in this laboratory is given in Figure 2. The excitation source is a synchronously mode-locked and cavity dumped dye laser (Spectra Physics 171 mode-locked argon ion laser/377 dye laser/344 cavity dumper). The laser provides tunable light pulses of approximately 10 ps duration at repetition rates up to 82 MHz in the spectral region 565 to 630 nm (using Rhodamine 6G dye). Laser alignment and pulse quality are monitored with a Spectra Physics 409 autocorrelator and 403B high speed photodiode. Studies on naphthalene-containing polymers were performed using ultraviolet radiation at 295 nm which is produced by means of a temperature-tuned second harmonic generation (SHG) ADA crystal (J. K. Lasers). Pulse repetition rates of 4 MHz or 800 kHz were employed. Fluorescence from the sample is passed through a Jobin-Yvon H-20 holographic grating scanning monochromator before detection by a Philips XP2020Q photomultiplier tube.

The time-correlated single photon counting electronics incorporate ORTEC modules including a 454 timing filter amplifier, 583 constant fraction discriminator, 457 time-to-amplitude converter (TAC) and a Tracor-Northern NS-710A multichannel analyzer (MCA). The optimum response function of the photomultiplier tube and electronics is 400 ps FWHM. Data from the MCA are stored using a dedicated 6802 microcomputer and are then transferred to a VAX 11/780 system for all subsequent manipulation and analyses using Fortran 77 programs that were developed in this laboratory.

Fluorescence decay profiles are analysed as a sum of up to three exponentials by an iterative reconvolution procedure which has been described elsewhere (17-19) and is based on the Marquardt algorithm. Goodness of fit is judged by inspection of the weighted residuals, autocorrelation function of the weighted residuals, reduced chi-square value and the Durbin-Watson parameter.

Time-resolved emission (TRE) spectra are generated by setting voltage discriminators on the output of the TAC to select a "time window" in the decay profile. The monochromator is then

PACE P2NMA

Figure 1. Structures of poly(acenaphthalene) (PACE) and poly(2-naphthyl methacrylate) (P2NMA).

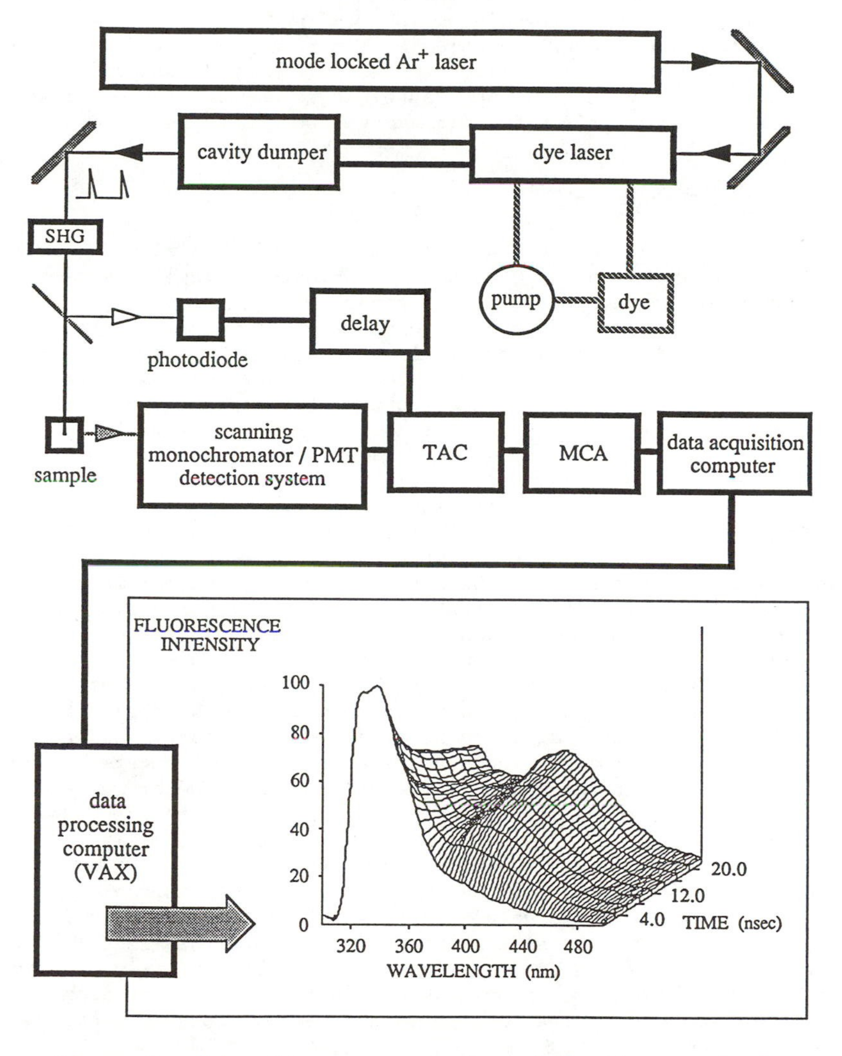


Figure 2. Schematic representation of the time-resolved fluorescence spectrometer. Inset diagram: intensity/time/wavelength hypersurface for PACE in benzene at 25°C.

synchronously scanned over the desired wavelength range with the MCA operating in multichannel scaling mode. Multiple TRE spectra can be displayed as either a two or three dimensional intensity/wavelength/time representation. Three dimensional hypersurface representations are generated by non-linearly interpolating between, and connecting intensity values of the gated spectra for a given wavelength. Intensity/wavelength/time cross-sectional diagrams (or time-resolved fluorescence "contour" diagrams) are generated using a weighted non-linear least squares polynomial surface procedure (20). Area-normalized TRE spectra can be used for convenient pictorial representation, since the absolute emission intensity of individual time-resolved spectra vary substantially with time after excitation.

An Applied Photophysics Model SP 2X nanosecond spectrometer incorporating an alternating polarization rotation unit (15) was used for the time-resolved fluorescence anisotropy measurements. An excitation wavelength of 365 nm was employed for excitation of the anthracene end-groups and emission above 400 nm was isolated with a Schott GG 400 filter.

## Results and Discussion

TRE Studies of PACE and P2NMA. An intensity/time/wavelength surface for PACE in benzene at 25°C is shown in Figure 2 (inset diagram) and the corresponding contour diagram is presented in Figure 3. Fluorescence arising from monomeric sites with a maximum near 340 nm and an excimer emission with a maximum at 400 nm are clearly evident. It should be noted that these surfaces are generated using area-normalized time-resolved spectra and thus the apparent rates of monomer fluorescence decay and "grow-in" of excimer emission depicted in Figures 2 and 3 are distorted. Analysis of these diagrams indicates that there are only two spectrally distinct species present. Support for this conclusion is provided by the presence of one isoemissive line in each surface at 373 nm which would be most unlikely to occur if more than two different species contribute to fluorescence. In addition, the application of Principal Component Analysis (21) to the TRE spectra is able to clearly resolve two spectral components with a high degree of confidence.

Analysis of the fluorescence decay profiles indicates that a two exponential fit, which would be expected for conventional monomer/excimer kinetics (2-4), is inadequate to describe the decay at any wavelength of observation throughout the emission band. A fit to a sum of three exponentials leads to significant improvements in the fitting criteria and examples of the results obtained at wavelengths in the monomer and excimer regions of the spectrum are presented in Table I.

Lifetimes obtained from triple exponential fits at 25 nm wavelength intervals across the emission band are reasonably consistent and give mean values of $\tau_1$ = 0.80 ± 0.14 ns, $\tau_2$ = 6.8 ± 0.8 ns and $\tau_3$ = 33 ± 4 ns (1 standard deviation error limits). However, the pre-exponential factor for $\tau_1$ becomes negative in the excimer region while the magnitude of the contribution of each lifetime component to the total emission remains constant throughout the excimer band.

Table I. Fluorescence Decay Data for PACE and P2NMA Homopolymers in Benzene at 25°C

Data are fitted to functions of the form $I(t) = \Sigma A_i \exp(-t/\tau_i)$

| POLYMER | $\lambda_{obs}$(nm) | $\tau_1$(ns)/$A_1$ | $\tau_2$(ns)/$A_2$ | $\tau_3$(ns)/$A_3$ |
|---|---|---|---|---|
| PACE | 325 | 0.97/42359 | 6.1/29170 | 38.7/5126 |
| | 450 | 0.74/-12474 | 7.7/31535 | 33.1/27775 |
| P2NMA | 340 | 3.1/37679 | 12.0/18930 | 34.1/3369 |
| | 460 | 3.1/-9717 | - | 34.1/25445 |

For P2NMA in benzene at 25°C, analysis of the time-resolved fluorescence surface (not shown) also indicates that only two spectrally distinct species contribute to emission - a monomer with a fluorescence maximum at 335 nm and an excimer with an emission maximum at 395 nm. A minimum of a sum of three exponentials is required to fit the fluorescence decay profiles in the monomer spectral region whilst in the excimer emission band a difference of two exponentials proves to be adequate (see Table I). The lifetimes extracted are consistent throughout the spectrum resulting in the mean values of $\tau_1 = 3.00 \pm 0.14$ ns, $\tau_2 = 12.6 \pm 1.5$ ns and $\tau_3 = 33.88 \pm 0.74$ ns (1 standard deviation error limits). In this case the $\tau_2$ component is significant only in the monomer spectral region and its contribution to the total emission follows the same trend as the monomer emission band (Figure 4). The pre-exponential factor associated with the $\tau_1$ component becomes negative in the excimer band although it should be noted that the absolute values of the pre-exponential factors associated with $\tau_1$ and $\tau_3$ do not become equal in this spectral region, contrary to the situation predicted by conventional excimer kinetic schemes (2-4).

The results presented above may be compared with previous studies on these polymers. It has been suggested that more than one spectrally distinct excimer species may exist in some naphthalene-containing polymers (6,22). Analyses of the spectral surfaces for the polymers studied in this work clearly indicate that only one spectrally distinct monomer and one excimer species contribute to emission. Phillips and co-workers (3,23) also found that the monomer fluorescence decay from PACE requires a minimum of three exponential terms to provide adequate fitting. They explained their results on the basis of a kinetic model involving two temporally distinct excited monomers ($M_1$*,$M_2$*) which can form the excimer (D*) as outlined in Scheme 1. Reverse dissociation of the excimer was also considered important.

$$M_2^* \rightarrow M_1^* \rightleftharpoons D^* \qquad \text{Scheme 1}$$

This scheme leads to a sum of three exponential terms being required to fit the monomer and excimer fluorescence decay curves (2,23). The results obtained in the present study are consistent with the earlier conclusions of Phillips in so far as the presence of monomer emission at all decay times after excitation confirms

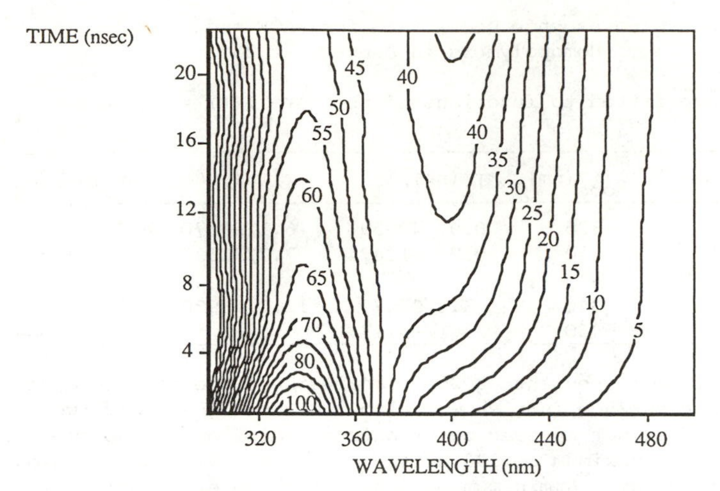


Figure 3. Time-resolved fluorescence contour diagram for PACE in benzene at 25°C.

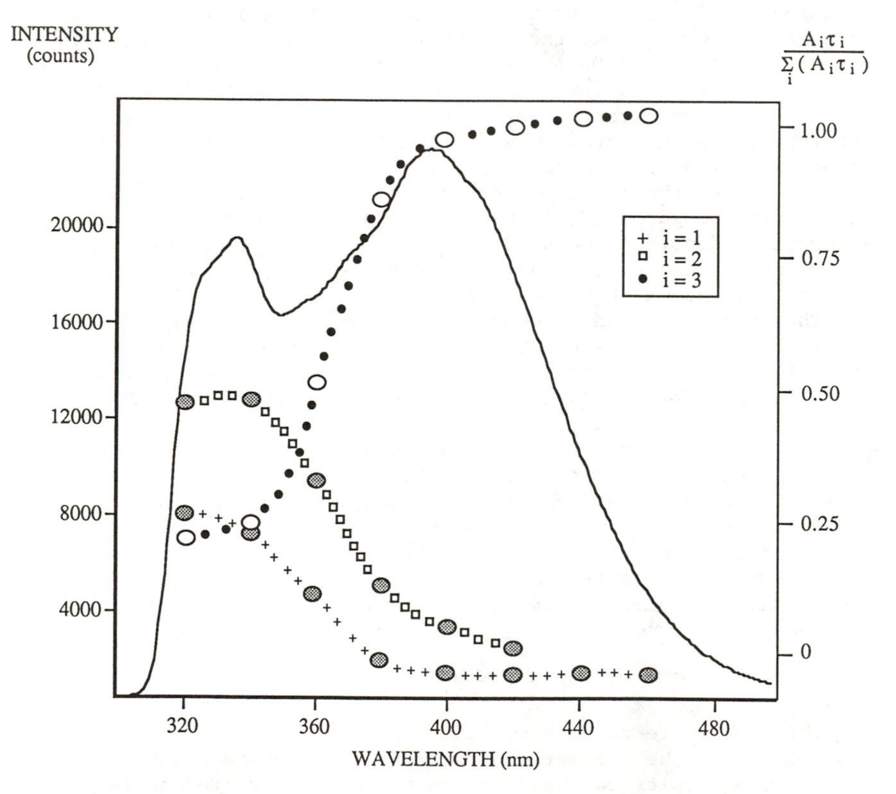


Figure 4. Total fluorescence spectrum for P2NMA in benzene at 25°C. Overlaid data (large circles) indicate the contribution of the components with lifetimes $\tau_1$, $\tau_2$ and $\tau_3$ to the total emission.

that excimer dissociation occurs. Three decay times would also arise from a kinetic scheme whereby two distinct subgroups of monomer chromophores are able to form the excimer with different average rate constants (2,24)(c.f. Scheme 2). It should be noted, however, that the relative magnitudes of the pre-exponentials obtained for PACE are not as expected from analysis of Schemes 1 and 2 (24).

$$M_2^* \rightleftharpoons D^* \leftarrow M_1^* \qquad \text{Scheme 2}$$

The results obtained with P2NMA also indicate the presence of only two spectrally distinct emitting sites. The $\tau_2$ component present in the short wavelength region, whose spectral characteristics correspond to monomer fluorescence, may be attributed to chromophores which do not participate in excimer formation since no corresponding decay component is observed at wavelengths above 400 nm. The $\tau_3$ component observed in the monomer fluorescence decay curves must arise from excimer dissociation. The identification of isolated monomer chromophores which do not form excimers is in agreement with the scheme proposed by Guillet and co-workers (25) for a range of naphthyl alkyl methacrylate polymers. However, the fact that the pre-exponential terms for the $\tau_1$ and $\tau_3$ components are not equal and opposite in the excimer region suggests that the conventional monomer/excimer kinetic scheme is still not applicable to this polymer.

Recent theoretical approaches indicate that transient diffusional effects on excimer sampling (10) and time-dependent energy trapping phenomena (11) may lead to functional forms for fluorescence decay in polymers that are different to the sum of exponential terms implemented here. The treatment of one-dimensional electronic energy transport as applied to vinyl aromatic polymers (11), indicates that the sum of exponential terms in the fluorescence decay that is predicted from Schemes 1 and 2 might be applicable to the polymers under investigation if the rate of energy migration is low. Alternatively, if efficient energy migration is present then non-exponential fluorescence decay kinetics might be expected. In certain cases, it has been shown that data described by a sum of exponentials can also be simulated by non-exponential decay functions (10). In the present work functions comprising the sums of two or three exponential terms could be used to adequately describe the data. However, the extracted pre-exponential parameters are inconsistent with monomer/excimer kinetic schemes that are based on distinct excited-state species. The unambiguous assignment of the appropriate kinetics that describe the excited-state dynamics in these vinyl aromatic polymers awaits further improvement in the quality and time resolution of fluorescence decay data.

A further illustration of the application of time-resolved fluorescence surfaces in studying energy distribution in polymers is provided by the intensity/time/wavelength surface for PACE with a 9-methylanthryl end-group (see Figure 5). Although nearly all the excitation radiation at 295 nm is absorbed by the acenaphthyl chromophores, there is considerable emission that is characteristic of the anthryl groups present in the total fluorescence spectrum indicating that energy transfer to the end-group occurs. The anthracene emission very rapidly "grows in" (<2 ns) and then the area-normalized spectra remain reasonably constant up to approximately 20 ns after excitation. This behaviour suggests that energy transfer to the anthryl groups fron non-adjacent acenaphthyl chromophores occurs during the excited state lifetime of the latter. At longer time periods after excitation, weak monomer/excimer emission remains indicating that not all excited monomer sites participate in the energy transfer process. The efficiency of energy transfer (determined as described by Guillet (12)) increases with the addition of the non-solvent ether suggesting that long-range dipole-dipole interaction is the most likely mechanism for excitation energy relocation in this polymer (c.f. Figure 6).

Although the emission from polymers with pendant aromatic groups is complex and there remains controversy concerning the specific kinetic models required to describe their fluorescence decay behaviour, the results presented above illustrate the usefulness of time-resolved fluorescence surfaces for obtaining an overview of the energy relaxation process in polymeric systems.

Fluorescence Polarization Studies of PMA and PAA. Time-resolved fluorescence polarization measurements are potentially a powerful means for studying molecular mobility. The fluorescence anisotropy function r(t) may be generated by monitoring the decay of vertically ($I_V(t)$) and horizontally ($I_H(t)$) polarized components of emission following excitation by vertically polarized light pulses (Equation 1).

$$r(t) = [I_V(t) - I_H(t)]/[I_V(t) + 2I_H(t)] = d(t)/s(t) \qquad (1)$$

The orientation autocorrelation function $P_2[\cos\theta(t)]$ is given by r(t) and reflects the motion undergone by the fluorescent chromophore (2,14). A number of models for Brownian motion have been proposed (14) but in the simple case of a rigid sphere, r(t) is described by a single exponential decay where $\tau_c$, the rotational correlation time is related to the hydrodynamic volume of the sphere and the viscosity of the medium through the Stokes-Einstein relation (14,16). More complex motions of fluorophores necessitate the development of models which fit the functional form of r(t) experimentally obtained (14).

For the case of a fluorophore incorporated in two spherical rotating bodies of different size, the difference decay function, $d(t) = I_V(t) - I_H(t)$ would be expected to be a double exponential leading to two rotational correlation times $\tau_{c1}$ and $\tau_{c2}$ (16). The motion of fluorescent anthracene probes solubilized in acid solutions of the polyelectrolyte PMA was recently described using such a model (16). In this earlier work, the size of the rotating units was shown to be much smaller than that expected for whole polymer rotation. This leads to the conclusion that the hypercoiled conformation of PMA that exists in aqueous solutions at low pH consists of a number of smaller compact structures. Further investigations of the mobility of 9,10-DMA end-groups incorporated in PAA and PMA, and studied over the whole neutralization range of the carboxylic acid groups of the polymers, are reported here.

The steady-state fluorescence polarization measured for 9,10-DMA attached to PAA and PMA at degrees of neutralization (α) in the range

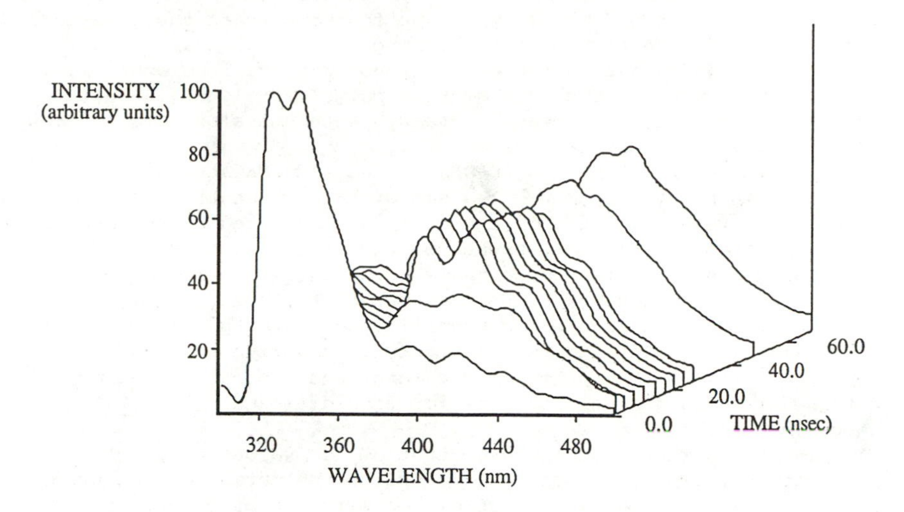


Figure 5. Time-resolved fluorescence spectra for PACE with a 9-methylanthryl end-group in benzene at 25°C. Excitation wavelength 295 nm.

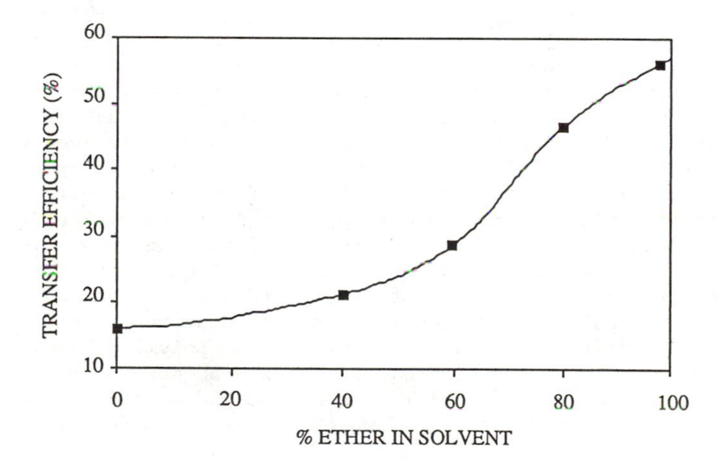


Figure 6. Efficiency of energy transfer from acenaphthyl chromophores to the 9-methylanthryl end-group in benzene/ether solvent mixtures.

0 to 1 is given in Figure 7. There is a dramatic decrease in the emission polarization of PMA at an $\alpha$ value between 0.2 and 0.3 which reflects the increased mobility of the fluorophore as the hypercoiled conformation of PMA collapses. The absence of this transition region in the case of PAA has been previously noted in various other measurements made on solutions of this polyelectrolyte (16,26). There is a small increase in the steady-state emission polarization of both polymers at higher values of $\alpha$.

The decay of d(t) for 9,10-DMA in PAA and PMA at $\alpha = 0$ is presented in Figure 8. For PMA a sum of two exponentials is required to obtain a reasonable fit of the data (as indicated by the random distribution of residuals for the two exponential fit) while for PAA a single exponential function is adequate. The fluorescence decay (s(t)) behaviour of 9,10-DMA attached to these polymers is well described by a single exponential function with a lifetime of approximately 13.5 ns and is relatively insensitive to the degree of neutralization of the polymer. The rotational correlation times obtained from an analysis of r(t) (or from d(t) and s(t) (16)) are given in Table II.

It is apparent from Table II that at low values of $\alpha$ two correlation times are required to describe the motion of the probe in PMA whilst only one $\tau_c$ is associated with the probe attached to PAA. The rotational correlation time that is expected for whole polymer rotation can be calculated from Equation 2 (26).

$$\tau_c = \gamma M \eta [\eta]/3RT \tag{2}$$

where $\eta$ is the viscosity of the solvent, $[\eta]$ is the intrinsic viscosity of the polymer, M the polymer molecular weight and the coefficient $\gamma$ depends on the selected chain model ($\gamma = 1.2$ for a spherical impermeable particle and $\gamma = 2$ for a Gaussian coil (15)). Values of $\tau_c$ equal to 346 ns (spherical particle) or 576 ns (Gaussian coil) are obtained for the PMA sample with a viscosity average molecular weight of 120000 and $[\eta] = 0.228$ dl/g. It is apparent that the rotational correlation times obtained from the analysis of the time-resolved fluorescence anisotropy data are much smaller than the values expected for whole polymer rotation, although they are considerably larger than that of the free 9,10-DMA probe.

One explanation that can be offered to explain the two $\tau_c$ values obtained for PMA at low values of $\alpha$ is that they represent the independent rotation of small clusters of the polymer chain. The larger value of $\tau_c$ (approximately 50 ns) can be associated with a rotating spherical cluster of radius 3.8 nm and of polymer molecular weight equal to 19000. Rotating units of similar size have been observed when the probes 9-methylanthracene and 9,10-DMA are solubilized in the PMA hypercoil structure (15,16) and when the more polar fluorescent probes Rhodamine B (27) and 1,8-anilinonaphthalene sulfonic acid (1,8-ANS) (28) are bound to PMA for a value of $\alpha$ equal to 0. The smaller rotating unit present in PMA and PAA whose value of $\tau_c$ is approximately equal to 5 ns (which corresponds to particles whose radii are approximately equal to 2 nm) may arise from the rotation of a small section of the chain which is just sufficient to surround the 9,10-DMA probe and protect it from unfavourable entropic interactions with water. This shorter $\tau_c$ was

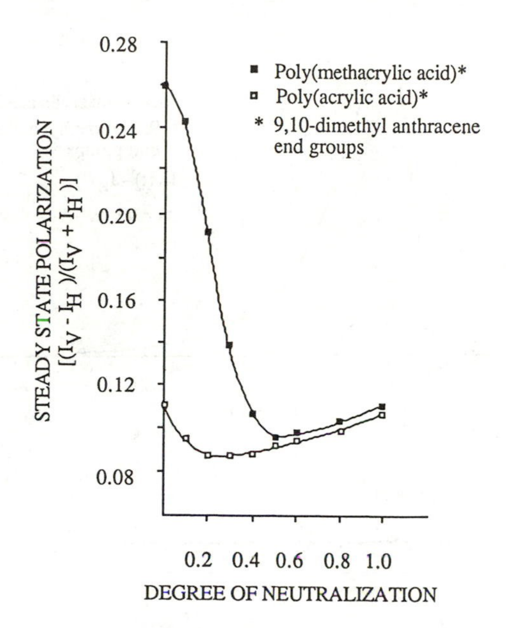


Figure 7. Steady-state fluorescence polarization for PMA and PAA with 9,10-DMA end-groups as a function of degree of neutralization (α) of the carboxylic acid groups.

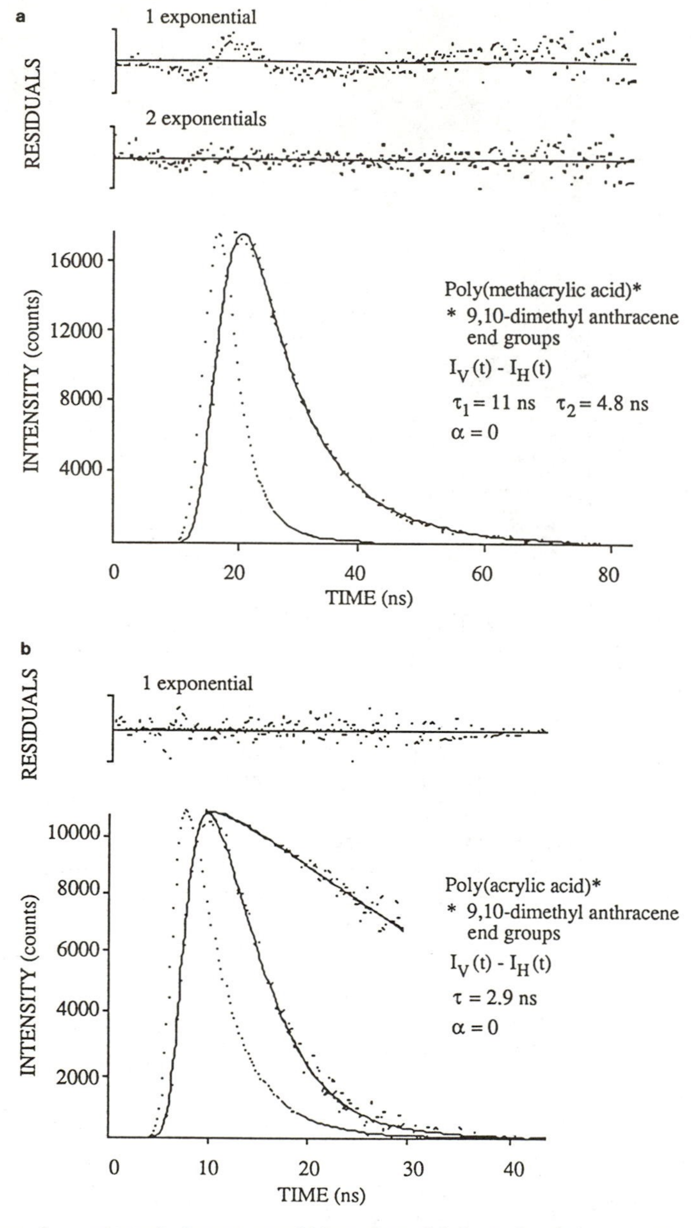


Figure 8. Fitted decay profiles for $d(t) = (I_V(t) - I_H(t))$ for 9,10-DMA end-groups attached to (a) PMA at $\alpha = 0$, (b) PAA at $\alpha = 0$. Points: experimental data; solid line: fitted curve.

not observed when the more polar Rhodamine B probe, bound to PMA at $\alpha = 0$, was studied (27). In addition, the value of $\tau_c$ for the 9,10-DMA probe attached to the PMA polymer in methanol (MeOH) is less than 1 ns suggesting that the larger structures only arise in aqueous media.

Table II. Steady-State Fluorescence Anisotropy $\bar{r}$ and Rotational Correlation Times ($\tau_c$) from Time-Resolved Measurements for PMA and PAA with 9,10-DMA End-Groups as a Function of $\alpha$ (Polymer Concentration 0.02 M in Repeat Units)

| $\alpha$ | Polymer | $\bar{r}$ | $\tau_{c1}$(ns) | $\tau_{c2}$(ns) |
|---|---|---|---|---|
| 0 | PAA | 0.076 | - | 4.0 |
| | PMA | 0.191 | 56.8 | 9.1 |
| 0.1 | PAA | 0.064 | - | 4.4 |
| | PMA | 0.176 | 36.3 | 4.6 |
| 0.2 | PAA | 0.059 | - | 4.5 |
| | PMA | 0.136 | 13.1 | 4.3 |
| 0.4 | PAA | 0.060 | - | 6.3 |
| | PMA | 0.073 | - | 4.2 |
| 0.8 | PAA | 0.067 | - | 8.7* |
| | PMA | 0.071 | 11.6 | 1.6 |
| 1.0 | PAA | 0.073 | - | 7.8* |
| | PMA | 0.075 | 9.1 | 0.8 |
| MeOH | PMA | - | - | 0.8 |

* poor fit to single exponential

The data in Table II indicate that as the carboxylic acid groups of the PMA polymer are ionized the larger rotating clusters collapse, leaving the smaller rotating units that are associated with the 9,10-DMA probe as well as a short length of polymer chain. The mobilities of the probe attached to the PMA and PAA polymers are similar in this intermediate region. As $\alpha$ approaches unity the decay of d(t) once again deviates from single exponential behaviour for both PMA and PAA. It is known that electrostatic repulsions between the charged carboxylic acid groups at high values of $\alpha$ encourage the polyelectrolyte chains to assume an extended rod-like conformation (15). The motions of such non-spherical structures are likely to be reflected in complex fluorescence anisotropy decay behaviour from the 9,10-DMA probe. Due to the quality of d(t) data presently available, a more comprehensive analysis at this stage does not seem warranted.

It should be noted that the short excited-state lifetime of the 9,10-DMA probe makes the accurate determination of slower polymer rotations difficult. Preliminary studies using pyrene (a longer-lived fluorescent probe) solubilized in PMA at $\alpha = 0$, have indicated an additional larger value of $\tau_c$ than is observed using the 9,10-DMA end-group probe (Ghiggino and Tan unpublished observations), suggesting that the polymer chain conformation may consist of clusters of various sizes. These results demonstrate the usefulness of time-resolved fluorescence anisotropy measurements for investigating the dynamics of polyelectrolyte chains in solution. The information on anisotropic particle motions and heterogeneity in probe siting and mobility which can be obtained is not readily available from steady-state emission experiments or other experimental techniques.

## Conclusion

The application of time-resolved fluorescence spectroscopy to studies of excimer formation and energy transfer in PACE and P2NMA provides an overview of the emitting species present and the dynamics of energy relaxation in these polymers. The results of fluorescence decay analyses suggest that kinetic models which have been proposed to explain monomer/excimer kinetics may require further refinement. In particular, the application of multi-exponential decay kinetics anticipated from models that assume distinct photophysical species within polymer chains may be inappropriate in some cases. The possibility of non-exponential fluorescence decay behaviour arising from energy migration and trapping (11) should also be considered. Additional studies of the mobilities of fluorescent probes incorporated in PMA using time-resolved fluorescence anisotropy measurements provide further support for a "connected cluster" model to describe the conformation of this polyelectrolyte in aqueous solution at low pH.

## Acknowledgments

K.P.G. is grateful for the continuing support of this research by the Australian Research Grants Scheme. Acknowledgment is made to the Donors of the Petroleum Research Fund, administered by the American Chemical Society, for partial support of this activity.

## Literature Cited

1. Beavan, S.W.; Hargreaves, J.S.; Phillips, D. Advances in Photochemistry 1979, 11, 207-303.
2. Ghiggino, K.P.; Roberts, A.J.; Phillips, D. Advances in Polymer Science 1981, 40, 69-167.
3. Roberts, A.J.; Soutar, I. In Polymer Photophysics; Phillips, D., Ed.; Chapman and Hall: London, 1985; Chapter 5.
4. Frank, C.W.; Semerak, S.N. Advances in Polymer Science 1984, 54, 31-85.
5. Phillips, D.; Roberts, A.J.; Soutar, I. J. Polym. Sci. Polym. Phys. Ed. 1982, 20, 411-421.
6. Itagaki, H.; Okamoto, A.; Horie, K.; Mita, I. Eur. Polym. J. 1982, 18, 885-891.
7. Roberts, A.J.; Cureton, C.G.; Phillips, D. Chem. Phys. Lett. 1980, 72, 554-556.
8. De Schryver, F.C.; Demeyer, K; van der Auweraer, M.; Quanten, E. Ann. N.Y. Acad. Sci. 1981, 366, 93-108.
9. De Schryver, F.C., Vandendriessche, J.; Toppet, S.; Demeyer, K.; Boens, N. Macromolecules 1982, 15, 406-408.

10. Itagaki, H.; Horie, K.; Mita, I. Macromolecules 1983, 16, 1395-1397.
11. Fredrickson, G.H.; Frank, C.W. Macromolecules 1983, 16, 572-577.
12. Guillet, J.E. In New Trends in the Photochemistry of Polymers; Allen, N.S.; Rabek, J.K., Eds.; Elsevier: London, 1985; Chapter 2.
13. Guillet, J.E.; Randall, W.A. Macromolecules 1986, 19, 224-230.
14. Monnerie, L. In Polymer Photophysics; Phillips, D., Ed.; Chapman and Hall: London, 1985; Chapter 6.
15. Tan, K.L. Ph.D. Thesis, University of Melbourne, Melbourne, 1985.
16. Ghiggino, K.P.; Tan, K.L. In Polymer Photophysics; Phillips, D., Ed.; Chapman and Hall: London, 1985; Chapter 7.
17. Phillips, D.; O'Connor, D.V. Time-Correlated Single Photon Counting; Academic Press: New York, 1983.
18. Demas, J.N. Excited State Lifetime Measurements; Academic Press: New York, 1983.
19. Bevington, P.R. Data Reduction and Error Analysis for the Physical Sciences; McGraw-Hill: New York, 1969.
20. McLain, D.H. The Computer J. 1974, 17, 318-324
21. Lawton, W.H.; Sylvestre, E.A. Technometrics 1971, 13, 617-633.
22. Gupta, A.; Liang, R.; Moacanin, J.; Kliger, D.; Goldbeck, R.; Horwitz, J.; Miskowski, V.M. Eur. Polym. J. 1981, 17, 485-490.
23. Phillips, D.; Roberts, A.J.; Soutar, I. Eur. Polym. J. 1981, 17, 101-106.
24. Snare, M.J.; Thistlethwaite, P.J.; Ghiggino, K.P. J. Am. Chem. Soc. 1983, 105, 3328-3332.
25. Holden, D.A.; Wang, P.Y.; Guillet, J.E. Macromolecules 1980, 13, 295-302.
26. Anufrieva, E.V.; Gotlib, Yu. Ya. Advances in Polymer Science 1981, 40, 1-68.
27. Snare, M.J.; Tan, K.L.; Treloar, F.E. J. Macromol. Sci.-Chem. 1982, A17, 189-201.
28. Treloar, F.E. Chemica Scripta 1976, 10, 219-224.

RECEIVED March 13, 1987

Chapter 29

# Alternating Copolymers of 2-Vinylnaphthalene and Methacrylic Acid in Aqueous Solution

**Fenglian Bai[1], Chia-Hu Chang, and S. E. Webber**

**Department of Chemistry and Center for Polymer Research, University of Texas at Austin, Austin, TX 78712**

Alternating copolymers of 2-vinylnaphthalene and methacrylic acid (P(2VN-alt-MA)) have been prepared and found to be soluble in water for pH's above ca. 6.5. There is a relatively small component of excimer fluorescence which is dependent on pH. An average hydrodynamic diameter of this polymer has been measured for $H_2O$, THF and methanol (plus sodium methoxide). It has been found that the intensity of the excimer fluorescence is not a simple function of coil density. The naphthalene fluorescence was found to be quenched very efficiently by $Cu^{2+}$ with an apparent second order rate constant of ca. $4x10^{14}M^{-1}s^{-1}$. This quenching is dynamic (i.e. intensity and lifetime quenched at equivalent rates) which is interpreted as the joint result of rapid down-chain energy migration and electrostatic binding of $Cu^{2+}$ to the polymer. Energy trapping by adsorbed anthracene and 9-anthracene methanol is compared to 9-anthracene methanol covalently bound to P(2VN-alt-MA). The quantum efficiency of trapping is quite high for all cases, approaching 100% for the latter case. It is argued that the photophysics of this polymer system cannot be treated as a simple 1-D lattice, presumably because of a tightly coiled configuration in solution.

In an earlier paper we have reported on the photophysical behavior of alternating 2-vinylnaphthalene - methacrylic acid or methylmethacrylate copolymers (P(2VN-alt-MA) or P(2VN-alt-MMA))[1]. In agreement with earlier reports[2], the alternating copolymers

[1]Current address: Institute of Chemistry, Academia Sinica, Beijing 100 080, People's Republic of China

0097-6156/87/0358-0384$08.25/0

displayed greatly diminished excimer fluorescence relative to random copolymers of similar composition. Singlet energy migration constants were estimated by the "comparative quenching technique" to be on the order of $10^{-5} cm^2 s^{-1}$. Covalently bound anthracene was found to be an excellent energy trap for the naphthalene singlet state in P2VN-alt-MA-co-Anth with a quantum yield of anthryl sensitization ($\chi$) on the order of 0.6 for an anthryl loading of ca. 1.2 mol%[1].

The present paper presents the results of a study of P(2VN-alt-MA) and P(2VN-alt-MA-co-Anth) in aqueous solutions for pH's higher than 6.5. At lower pH's polymer dissolution is not possible and even in the basic pH range concentrations higher than $10^{-3}$M in naphthalene groups will become turbid after a period of time, presumably because of aggregation induced by hydrophobic interactions between the naphthalene groups. Our initial aim was to determine if the polymer configuration in aqueous solution permitted energy migration between naphthalene groups or if excimer traps would be produced. Results relating excimer formation to solvent type are presented. We find that a minor amount of excimer fluorescence is present in aqueous solvent but photon harvesting by a covalently bound anthryl species is even more efficient than in organic solvents.

The photophysics of this system was further characterized by 1) solubilizing anthracene (Anth) or 9-anthryl methanol (AnOH) in the polymer and determining the quantum efficiency of anthracene sensitization, and 2) studying $Cu^{2+}$ quenching of the naphthalene fluorescence. In the case of Anth or $Cu^{2+}$ an apparent second order rate constant for naphthalene fluorescence quenching in excess of $10^{14} M^{-1} s^{-1}$ was obtained. A rate constant 3-4 orders of magnitude larger than diffusion controlled is almost certainly the result of strong binding of the quenching species to the polymer coil, producing a high local concentration. In previous studies in which a small mole fraction of chromophore was covalently attached to a polyelectrolyte, this enhancement factor was on the order of 1-2 orders of magnitude[3]. We believe the additional enhancement we observe is the result of energy migration among the naphthalene groups. This is especially clear in the case of $Cu^{2+}$ quenching because the quenching mechanism appears to be a "dynamic" one, i.e. the fluorescence intensity and lifetime are quenched at essentially the same rates. In our previous experience with $Cu^{2+}$ primarily static quenching was observed in systems in which energy migration was impossible.

The present work illustrates the wide range of phenomena that are exhibited by amphiphilic alternating copolymers. It remains to be seen how large a range of chromophores can be polymerized in this manner and what kind of solution properties - photophysical behavior relations will be found.

## Experimental

### 1. Polymers, Solution Preparations

The P(2VN-alt-MA) (1) and P(2VN-alt-MA-co-Anth) (1.2 mol%) polymers are those reported earlier[1].

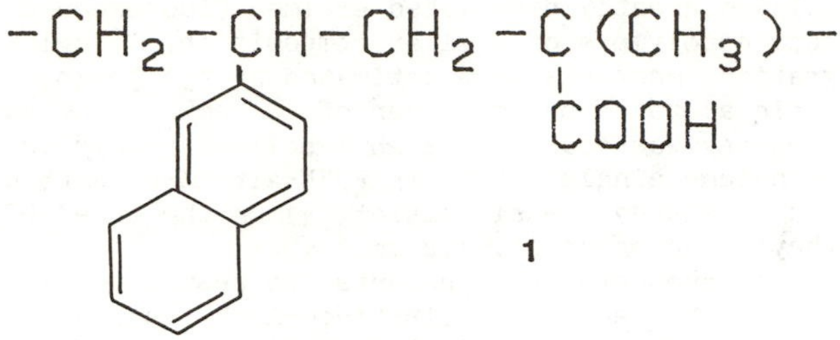

$$-CH_2-CH(C_{10}H_7)-CH_2-C(CH_3)(COOH)-$$

1

For the latter polymer the term "co-" denotes random placement of the anthracenes. The degree of polymerization (DP) of P(2VN-alt-MA) was previously estimated to be ca. 3500 based on GPC (using polystyrene calibration curves). P(2VN-alt-MA-co-Anth) was prepared by direct esterification of the former polymer with 9-anthryl methanol. Thus the degree of polymerization is the same for these two polymers.

The method of preparing aqueous or methanol solutions of these polymers is important because the degree of dispersion of the polymer is dependent on the rate of dissolution. Our method is as follows: A stock solution of the polymer (1.5mg/mL) in freshly distilled DMF is added to pH 11 water (NaOH) dropwise with vigorous stirring on a Vortex-Genie mixer. In the solubilization studies Anth or AnOH are dissolved in the same DMF solution that is to be added to the water. The Anth or AnOH and polymer DMF solution are stirred 30 minutes before adding to water. After addition of the polymer is complete the pH is adjusted downward by adding HCl as needed. The DMF is never more than 3.7 vol% of the final solution. Addition of DMF up to ca. 10 vol% is possible without modifying the fluorescence spectra. All aqueous and methanol solutions were placed in a cleaning bath sonicator for 2-3 minutes and for light scattering studies all solutions were centrifuged for ca. 30 minutes in a bench-top centrifuge to remove any large dust particles from the light path.

Methanol / sodium methoxide solutions were prepared similarly except that no sonication and centrifugation step was carried out. These solutions seem to be very unstable and polymer precipitation can occur under mechanical stress.

All solvents to be used in light scattering experiments were filtered through millipore 0.2µm PTFE or cellulose nitrate filters. Aqueous solutions were filtered several times through the latter to remove persistent dust particles. There does not seem to be any problem with the polymer adhering to the filter material based on UV absorption spectra taken before and after filtration.

2. Light Scattering Studies

All light scattering studies were carried out on a Brookhaven model BI-200 using a polarized HeNe laser (632.8nm). The samples were in a thermostatted index matching bath (xylene) and the results we report are based on Photon Correlation Light Scattering, using the Brookhaven model BI-2020 correlator and the supplied software. All data presented are for 90° scattering angle at 20°C.

The polymer samples we are using are polydisperse such that

the correlation function of the scattered light is not expected to decay as a single exponential. Hence the correlation function is assumed to be of the form

$$C(t) = \left|1 + bg(t)\right|^2 \quad (1)$$

with

$$\ln(g(t)) = \bar{\Gamma}t + \mu_2 t^2 + \cdots \quad (2)$$

where the average gamma value above is related to an "effective" hydrodynamic diameter ($\bar{d}_H$) by

$$\bar{\Gamma} = \bar{D}(4\pi n \sin(\theta_s/2)/\lambda_o)^2 \quad (3)$$

$$\bar{D} = kT/(3\pi\eta\bar{d}_H) \quad (4)$$

where in eqn. (3) $\lambda_o$ = 632.8nm, n = solvent refractive index and, $\theta_s$ = scattering angle. In eqn. (4) $\eta$ is the viscosity, which is assumed to be equal to that of the pure solvent (the concentration of the polymer is quite low, on the order of $10^{-1}$ - $10^{-2}$mg/mL). In the fit to the data the measured value of C(t) at long times was used to establish the relative magnitude of b in eqn. (1).

The value of $\bar{d}_H$ for various situations is presented in Table 1. We wish to emphasize sets of experiments in which changes in $\bar{d}_H$ are noted upon addition of a reagent or carrying out some other operation on the solution. (eg. group (b) of Table 1). Significant variations can occur in the value of $\bar{d}_H$ in nominally identical experiments. This variation seems to be a function of the age of the solution, rate of addition of the DMF stock solution to the solvent, etc., as described above. The data in Table 1 represent the values obtained after the solution preparation method had been refined to a consistent methodology. It is problematical if all the $\bar{d}_H$ values presented in Table 1 represent isolated or partially aggregated polymer coils for the case of $CH_3OH$ or $CH_3OH/CH_3ONa$ given the increase of $\bar{d}_H$ upon sonication for the former and the mechanical instability of the latter (see subsection 1 of this section). If these methanolic solutions are composed of aggregates the naphthalene - naphthalene interactions must be minimal based on the absense of excimer fluorescence (see next section).

3. Steady - State and Time - Dependent Fluorescence

All steady state fluorescence spectra were obtained using a Spex Fluorolog 2 with a 450W Xe lamp and a Hammamatsu R508 photomultiplier. Spectra were uncorrected for photomultiplier response except for quantum yield measurements. The latter were obtained by comparison to compounds of known fluorescence quantum yield. Unless noted otherwise solutions were not deaerated because of the relatively low solubility of $O_2$ in water.

All solutions were relatively dilute such that the OD at 290nm was << 0.1. The fluorescence intensity was sufficiently low that all decay curves were obtained at the Center for Fast Kinetics Research at the University of Texas. This system uses standard photon-counting electronics but the excitation source is a synch-pumped cavity-dumped dye laser pumped by a Nd:YAG laser

Table 1. Effective Hydrodynamic Diameters

| Conditions | $\bar{d}_H$(nm) |
|---|---|
| a) Effect of pH | |
| pH = 8.3 | 54.5(±1.8)[a] |
| pH = 10 | 66 |
| b) Effect of Added Salt (pH ~ 8-8.5) | |
| $[Cu^{2+}] = 5x10^{-7}M$ | 53 |
| $[KCl] = 3x10^{-5}M$ | 57 |
| $[KCl] = 1x10^{-2}M$ | 59 |
| c) Effect of Added Methanol (pH 8.5) | |
| V % MeOH = 12% | 62 |
| d) Addition of Anth or AnOH | |
| $[Anth] = 4x10^{-7}M$ | 87 |
| $[AnOH] = 4x10^{-7}M$ | 51 |
| e) Organic Solvents | |
| THF | 37 |
| DMF | 83 |
| $CH_3OH$ | 66 |
| $CH_3OH$, sonicate at 50°C | 105 |
| $CH_3OH/CH_3ONa$ | 261 |

a. Standard deviation based on 4 replica experiments.

mode-locked at 81.66 MHz. The dye-laser is cavity-dumped at 800kHz allowing a very rapid build-up of a satisfactory decay curve even for weakly emitting samples. The instrument response function for this system is ca. 400ps. Multiple exponential (eqn. (5)) fits were performed without reconvolution with the instrument response function. It is our experience with this system that for components longer than ca. 1 ns reconvolution is not essential:

$$I_{fl}(t) = \Sigma a_i e^{-t/\tau_i} \quad (5)$$

Emission wavelength selection is by means of interference or cut-off filters.

## Results

### A. Steady State Fluorescence: Excimer Formation

In our earlier report on the photophysics of 2-vinylnaphthalene - methacrylic acid alternating polymers (P(2VN-alt-MA)) the weakness of the excimer fluorescence was noted[1]. This observation was not unexpected since excimer formation in polymers is thought to occur primarily between nearest neighbor chromophores. Based on the work of Guillet et.al. it is reasonable to expect that in a poor solvent the relative intensity of the excimer fluorescence will increase as the coil diameter decreases[4]. For P(2VN-alt-MA) in water we observe an excimer emission which becomes more intense as the pH is decreased (Fig. 1). However, the intensity of this excimer peak is much smaller than for a random copolymer of 2-vinylnaphthalene and methacrylic acid (or methylmethacrylate) in a good solvent. Furthermore, when the spectrum in Fig. 1 is deconvoluted into monomer and excimer components, the excimer is blue-shifted from ca. 400-410nm for P(2VN-co-MA) in good organic solvents to 385-390nm for P(2VN-alt-MA) in $H_2O$. We presume this is the result of the inability of the naphthalene rings to achieve the optimal separation for excimer stabilization. However, one cannot simply ascribe the extent of excimer formation to the "coil diameter". From Table 1 it is observed that the hydrodynamic diameter for P(2VN-alt-MA) in THF is smaller than $H_2O$ (pH = 8.3 or 10.0). Since THF is a poor solvent for polymethacrylic acid we speculate that the -COOH groups are extensively hydrogen bonded to each other and this forces the naphthalene groups into a configuration that is not favorable for excimer formation. Additionally THF is a good solvent for the homopolymer P2VN such that there should be little tendency for naphthalene - naphthalene aggregation.

To test this idea P(2VN-alt-MA) was dissolved in methanol containing methoxy ion ($MeOH/MeO^-$) which we presume to completely deprotonate the polymer. The hydrodynamic diameter for P(2VN-alt-MA) in this solvent system was extremely large (see Table 1) which we assume must correspond to a nearly completely extended coil because of extensive electrostatic repulsion. The fluorescence spectrum for this solution in Fig. 1 appears to be completely devoid of excimer fluorescence and a close comparison of the THF and $MeOH/MeO^-$ spectra suggests that the former does

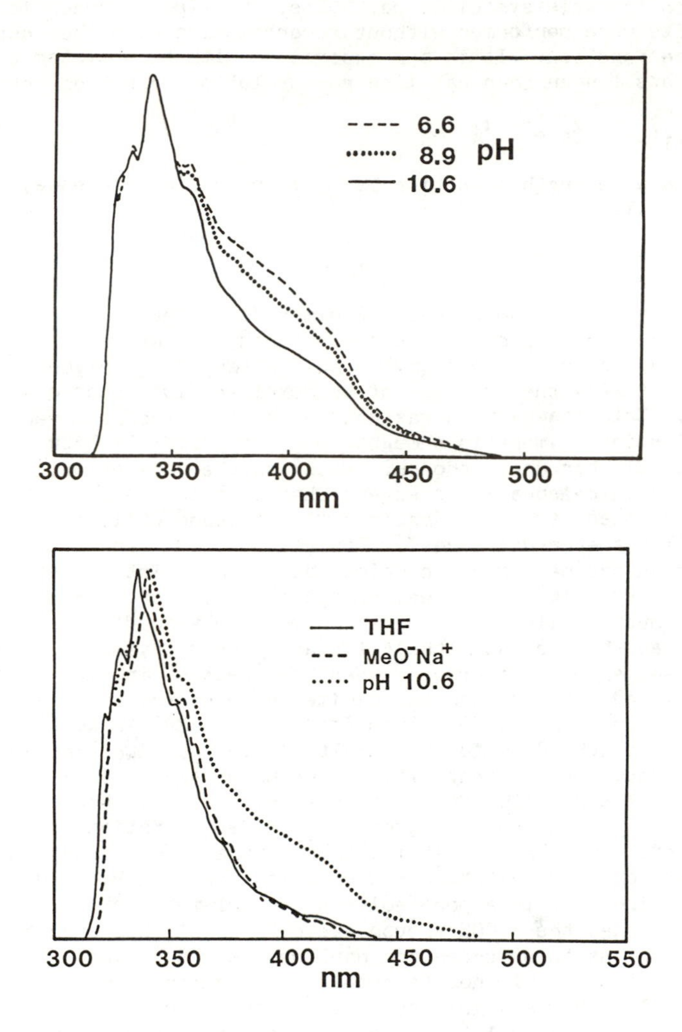


Fig. 1 Fluorescence spectra of a) P(2VN-alt-MA) in $H_2O$ (pH indicated), b) THF, $CH_3OH/CH_3O^-Na^+$, $H_2O$ (pH 10.6).

have a tail extending to long wavelengths that may be a vestigial excimer band.

Thus we see that for this class of polymer the relationship between hydrodynamic diameter and excimer/monomer ratio ($I_D/I_M$) is indirect. The ratio $I_D/I_M$ is low for the highest and the lowest $\overline{d}_H$ values respectively while for water $\overline{d}_H$ has an intermediate value and $I_D/I_M$ is a maximum. Clearly $I_D/I_M$ is primarily a function of the short range structure of the polymer. Thus we interpret our findings as follows:

1) In THF the solvation of the -COOH groups is very poor such that intracoil hydrogen bonding is encouraged. This produces a local polymer structure that disfavors naphthalene excimer formation.
2) In $H_2O$ the solvation of the naphthalene groups is very poor but the alternating structure makes intracoil naphthalene aggregation difficult. More complete deprotonation of the acid groups further discourages excimer formation.
3) In MeOH/MeO$^-$ solvation is excellent for the methacrylic acid moieties and fair for the naphthalene groups. The expansion of the coil further decreases excimer formation.

These speculations are illustrated in Fig. 2.

As we will discuss in the next section, energy transfer between the naphthalene groups is facile, thus we suppose that most excimer formation occurs via energy migration to excimer forming sites. Thus in the cases discussed above we associate the degree of excimer fluorescence with the mole fraction of excimer forming sites ($X_{efs}$). Thus we write: $X_{efs}$(MeOH/MeO$^-$) < $X_{efs}$(THF) << $X_{efs}$($H_2O$).

As is typical for polymeric systems, the fluorescence decay of the monomer and excimer is non-exponential. The fluorescence decay was fit satisfactorily to a biexponential function (see eqn. (5)) and the parameters for the monomer (observed at 340nm) and the excimer (observed at wavelengths longer than 400nm) are presented in Table 2. Also presented are the "quantum yield weighted average lifetimes" (eqn. (6)) and the average lifetime (eqn. (7)), defined by

$$\langle\tau\rangle = \Sigma a_i \tau_i^2 / \Sigma a_i \tau_i \qquad (6)$$

and

$$\overline{\tau} = \Sigma a_i \tau_i / \Sigma a_i \qquad (7)$$

The definition of the average decay in eqn. (6) uses as a weighting factor the relative fluorescence yield of the ith component:

$$\langle\tau\rangle = \Sigma a_i \phi_{fl}^i \tau_i \qquad (8)$$

where

$$\phi_{fl}^i = k_r \tau_i / k_r \Sigma a_i \tau_i \qquad (9)$$

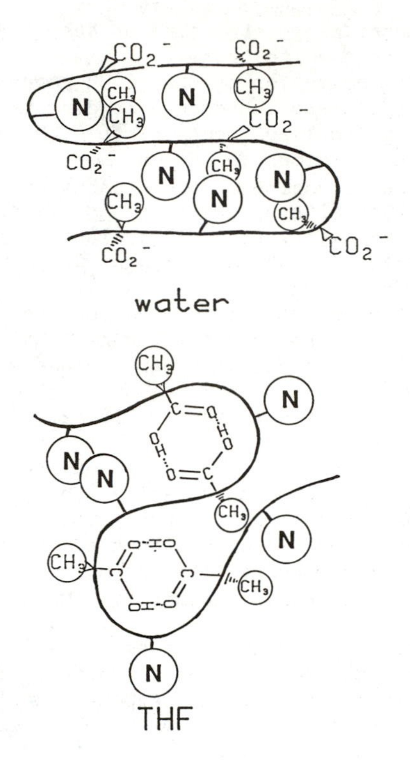


Fig. 2 Illustration of possible polymer structures in $H_2O$ and THF (speculative).

Table 2. Fluorescence Decay Parameters for Alternating Copolymer in Different pH Solution

| pH | 340 nm | | | | > 400 nm | |
|---|---|---|---|---|---|---|
| | $\tau_1/A_1$ | $\tau_2/A_2$ | $\langle\tau\rangle^a$ | $\bar{\tau}^b$ | $\tau_3/A_3$ | $\tau_4/A_4$ |
| 10.5 | 6.41/0.466 | 37.80/0.534 | 33.75 | 23.2 | 4.78/0.700 | 53.28/0.300 |
| 9.8 | 3.40/0.593 | 31.43/0.407 | 27.61 | 14.8 | 6.64/0.789 | 46.08/0.211 |
| 9.0 | 3.22/0.604 | 28.78/0.396 | 25.05 | 13.3 | 3.73/0.776 | 40.37/0.224 |
| 8.5 | 3.61/0.630 | 28.15/0.370 | 23.75 | 12.7 | 3.98/0.795 | 44.48/0.205 |
| 7.9 | 3.98/0.685 | 26.84/0.315 | 21.27 | 11.2 | 3.29/0.803 | 36.66/0.197 |
| 7.1 | 3.61/0.712 | 24.63/0.288 | 19.04 | 9.7 | 5.27/0.791 | 44.46/0.209 |
| 6.3 | 5.83/0.706 | 26.85/0.294 | 19.65 | 12.0 | 5.78/0.772 | 46.53/0.228 |

a. See eqn. (6) of text

b. See eqn. (7) of text

and $k_r$ is the radiative rate of the chromophore. Eqn. (8) and (9) are appropriate for chromophores in different environments that may modify the total decay rate but not the radiative rate constant. The quantity $\bar{\tau}$ will be discussed later (eqn. (16), (17)). It will be noted that $\langle\tau_M\rangle$ (monomer average lifetime measured at 340 nm) tends to decrease with decreasing pH, presumably as a result of a higher density of excimer forming sites (efs). The values of $\langle\tau_D\rangle$ (excimer average lifetime measured at wavelengths longer than 400 nm) do not vary systematically with pH. No rise time is observed for the excimer fluorescence decay, which may be the result of very fast transfer to the efs or the effect of overlapping monomer and excimer fluorescence. We have argued previously that the absence of a rise time for a sensitized trap emission can imply a common precursor state for both the monomer and trap excited states, in addition to the expected monomer to trap energy transfer[5].

B. Fluorescence Quenching by $Cu^{2+}$

We have previously used the $Cu^{2+}$ ion as a convenient cationic quencher for polyanion - bound chromophores[3]. In these previous papers it was found that the apparent second order rate constant was extremely high, presumably because of the electrostatic attraction between $Cu^{2+}$ and the polyelectrolyte. The quenching was found previously to be largely static in that the total fluorescence intensity was quenched very efficiently but there was little or no lifetime quenching.

The present polymer system exhibits unusual behavior on two counts: 1) The Stern-Volmer constant for intensity quenching is extremely large, leading to an apparent second order rate constant for quenching in excess of $10^{14}M^{-1}s^{-1}$; 2) The lifetime quenching constant is nearly as large as the intensity quenching constant, implying a "dynamic" quenching mechanism. Normally this mechanism is observed when both the chromophore and quencher interact through collisions rather than the formation of stable contact complexes.

The relevent equations in our case are:

$$I_0/I = 1 + K_{SV}^{(1)}[Cu^{2+}] \quad (10)$$

$$= 1 + k_q^{(1)}\langle\tau\rangle_0[Cu^{2+}] \quad (11)$$

$$\langle\tau\rangle_0/\langle\tau\rangle = 1 + K_{SV}^{(2)}[Cu^{2+}] = 1 + k_q^{(2)}\langle\tau\rangle_0[Cu^{2+}] \quad (12)$$

$$\bar{\tau}_0/\bar{\tau} = 1 + K_{SV}^{(3)}[Cu^{2+}] = 1 + k_q^{(3)}\bar{\tau}_0[Cu^{2+}] \quad (13)$$

In eqns. (10) through (13) the superscript 1, 2, and 3 refer to the steady state and lifetime measurements respectively, while $\langle\tau\rangle$ and $\bar{\tau}$ refer to the average lifetimes of the naphthalene monomer fluorescence defined in eqn. (6) and (7). The subscript 0 refers to the value in the absense of $Cu^{2+}$. The fluorescence decay parameters are given in Table 3. The values of $K_{SV}$ and $k_q$ are given in Table 4 and the quenching curves are displayed in Fig. 3.

The fact that the $Cu^{2+}$ quenching is so efficient implies that the $Cu^{2+}$ is bound to the polymer (presumably by electrostatic attraction), resulting in a high local concentration. This

Table 3. Lifetime of P2VN-alt-MA With Addition of $Cu^{2+a}$ at pH 8.0

a. $I_{emi}$ = 340 nm

| $[Cu^{+2}]$x$10^8$ | $A_1$ | $A_2$ | $\tau_1$ | $\tau_2$ | $\langle\tau\rangle_{ave}$ [b] | $\bar{\tau}$ [c] |
|---|---|---|---|---|---|---|
| 0 | 0.578 | 0.422 | 5.372 | 28.82 | 24.05 | 15.27 |
| 6.9 | 0.683 | 0.317 | 4.675 | 25.07 | 19.22 | 11.14 |
| 13.8 | 0.724 | 0.276 | 2.797 | 17.28 | 12.96 | 6.79 |
| 20.6 | 0.724 | 0.276 | 2.755 | 14.92 | 10.05 | 6.11 |
| 27.5 | 0.742 | 0.258 | 2.488 | 12.59 | 8.93 | 5.09 |
| 34.4 | 0.724 | 0.276 | 1.664 | 9.038 | 6.64 | 3.70 |

b. $I_{emi}$ > 400 nm

| $[Cu^{+2}]$x$10^8$ | $A_1$ | $A_2$ | $\tau_1$ | $\tau_2$ | $\langle\tau\rangle$ [b] | $\bar{\tau}$ [c] |
|---|---|---|---|---|---|---|
| 0 | 0.715 | 0.285 | 5.743 | 51.23 | 41.24 | 18.71 |
| 6.9 | 0.798 | 0.202 | 4.681 | 41.38 | 30.06 | 12.09 |
| 13.8 | 0.853 | 0.147 | 4.816 | 38.44 | 24.28 | 9.76 |
| 20.6 | 0.821 | 0.179 | 3.525 | 27.31 | 18.47 | 7.78 |
| 27.5 | 0.820 | 0.180 | 5.918 | 31.91 | 20.00 | 10.60 |
| 34.4 | 0.841 | 0.159 | 3.272 | 22.61 | 14.22 | 6.35 |

a. [Naph] = 1x$10^{-4}$M

b. $\langle\tau\rangle$ defined by eqn. (6)

c. $\bar{\tau}$ defined by eqn. (7)

Table 4. $K_{SV}$ and $k_q$ Value for $Cu^{2+}$ Quenching at pH 8.0

| Based on | $K_{SV}(M^{-1})$ [a] | $k_q(M^{-1}s^{-1})$ [b] |
|---|---|---|
| $I_o/I$ | 10.5x$10^6$ | 4.4x$10^{14}$ |
| $\bar{\tau}_o/\bar{\tau}$ | 8.4x$10^6$ | 5.5x$10^{14}$ |
| $\langle\tau\rangle_o/\langle\tau\rangle$ | 7.5x$10^6$ | 3.1x$10^{14}$ |
| | 8.8(±1.5)x$10^6$ | 4.3(±1.2)x$10^{14}$ |

a. From least square fit of data

b. From eqn. (10) to (13) respectively

"concentrating effect" has been observed many times for polyelectrolytes[3a,6]. For the very low concentrations of $Cu^{2+}$ used there is no evidence that the polymer coil density is perturbed, as might be expected if ionic cross-linking occurred (see Table 1). Since dynamic quenching is observed either (1) the singlet energy migrates rapidly along the coil until a bound $Cu^{2+}$ ion is encountered, or (2) the $Cu^{2+}$ ion can migrate until an excited naphthalene is encountered, or both. In any case essentially all excited naphthalenes are accessible to the quencher.

In order to obtain an estimate of the intracoil energy migration rate from the present results one must evoke some model for the system. This will be taken up in the Discussion section. Such an interpretation should not distract from the primary experimental observation presented in this subsection: excited state quenching by $Cu^{2+}$ is extremely efficient in this polymer system with an apparent second order rate constant higher than any we have encountered previously.

C. Energy Transfer to Solubilized and Covalently Bound Anthryl Groups:Steady State Results

We have previously reported energy trapping by anthryl groups covalently bound to P(2VN-alt-MA) in THF solutions. The quantum efficiency of energy trapping is given by $\chi$. $\chi$ is obtained from the expression

$$(I_A/I_N)(\phi_{fl}^N/\phi_{fl}^A) = \chi/(1-\chi) \qquad (14)$$

where $I_A$, $I_N$ are the intensities of corrected anthracene and naphthalene fluorescence respectively, $\phi_{fl}^A$ and $\phi_{fl}^N$ are the corresponding quantum yields of fluorescence and $\chi$ represents the ratio of photons that are initially absorbed by the sensitizer (naphthalene) that are transferred to the acceptor (anthracene) to total photons absorbed by the donor. (See Fig. 4 for spectra of P(2VN-alt-MA-co-Anth)) in THF and $H_2O$ (pH 10). $\chi$ can also be obtained from a comparison of the absorption spectrum and the acceptor excitation spectrum. The value of $\chi$ for P(2VN-alt-MA-co-Anth) (1.2 mol%) in THF solutions is approximately 0.65[1] while for $H_2O$ $\chi$ ranges from ca. .90 to 1.0 over the pH range 10.4 to 7.7 (Table 5). This increase in $\chi$ is consistent with a solution structure in which the average separation of naphthalene groups is diminished, thus enhancing down-chain energy migration as well as decreasing the average naphthalene - anthracene separation. Anthracene sensitization is also very efficient in the $MeOH/MeO^-$ system for which we postulate a very open structure for the polymer chain (see section (A)). We ascribe the high efficiency of the sensitization in this case to the absense of excimer traps. Thus we may write for intracoil coil sensitization: $\chi(THF) < \chi(MeOH/MeO^-) < \chi(H_2O)$. This is not the same ordering as found for the excimer fluorescence intensity (see previous section). This illustrates that the efficiency of photon-harvesting by polymers cannot be optimized exclusively by reducing the extent of excimer formation. However, the present system may not be typical for polymers in which exclusively down-chain energy transfer occurs because of the large Förster radius for naphthalene to anthracene dipole-dipole transfer (ca.

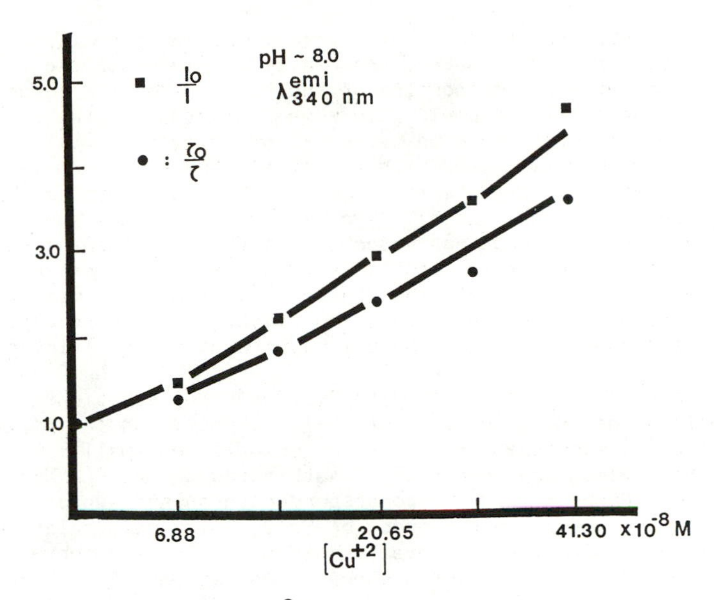


Fig. 3 $I_o/I$, $\langle\tau_o\rangle/\langle\tau\rangle$ $Cu^{2+}$ quenching curve for P(2VN-alt-MA).

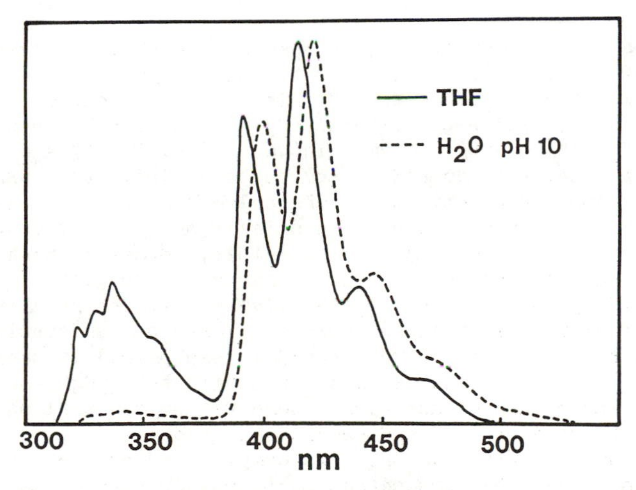


Fig. 4 Fluorescence spectra of P(2VN-alt-MA-co-Anth) in different solvents.

25.5Å[7]). Thus sensitization of the anthracene is most likely a combination of down-chain energy migration and a single-step Förster transfer.

It is well-known that polyelectrolytes can solubilize hydrophobic species in water, presumably by a hydrophobic attraction between some portion of the polymer and the adsorbate. It was thought that we could gain some insight as to the structure of P(2VN-alt-MA) polymers in water by comparing the $\chi$ values for solubilized anthryl groups with the covalently bound sample discussed above. As probe molecules we chose 9-anthryl methanol (AnOH) and anthracene (Anth). The method for preparing solutions containing these probes was described in the Experimental section. The results we report are sensitive to the solution preparation methodology which suggests that these "solutions" are metastable phases and not a true equilibrium solution. To a lesser extent the same may be said of the solutions without the addition of anthryl probes.

Addition of Anth and AnOH to P(2VN-alt-MA) solutions further emphasizes the subtle effects of small molecules on the polymer coil and the microenvironment of the solubilized small molecule.

The maximum concentration of Anth added to the P(2VN-alt-MA) solution ([Naph] = $1 \times 10^{-4}$M) corresponded to an average concentration of An on the order of $5 \times 10^{-7}$M (.5 mol% relative to naphthalene). Above this level turbidity could be observed in the solution (approximately 10x higher concentrations of AnOH are possible - see below). The hydrodynamic radius increased significantly upon the addition of Anth (see Table 1) which suggests the onset of aggregation since it does not seem reasonable that the Anth moieties could act as a better "solvent" for the polymer.

Sensitized anthracene fluorescence is easily observed for this system (Fig. 5), and $\chi$ tends to increase with increased anthryl loading and decreased pH (Fig. 6), as expected. The $\chi$ values compare well with P(2VN-alt-MA-co-Anth) in THF but are not high as for this latter polymer in water (cf. Fig. 4 and 5). Thus the solubilized Anth does not seem to be as intimately associated with the naphthalene regions of the polymer as was the case of P(2VN-alt-MA-co-Anth). Since essentially no dissolution of Anth in the absense of the polymer is possible, we assume that all Anth is adsorbed onto the chain. Thus our results suggest that there exists hydrophobic regions of the polymer that are relatively distant from the naphthalene regions. Based on our results with a random copolymer P(2VN(10 mol%)-co-MA) (see below) it seems likely that these regions might be associated with the $-CH_3$ group of MA. For later reference we note that there was no spectral shift between directly excited and sensitized Anth.

The use of AnOH as a probe is complicated by the fact that a much higher (ca. 10x) concentration is possible before solution turbity is noted. Thus it seems likely that a certain amount of AnOH is present in bulk solution. We note from Table 1 that AnOH does not significantly perturb $\overline{d}_H$. This is supported by the observed $\chi$ values, which are comparable to those of Anth for a much lower mole fraction of adsorbate. Likewise there is a distinct blue-shift in the fluorescence of directly excited AnOH relative to sensitized AnOH (see Fig. 5). Thus there are at least

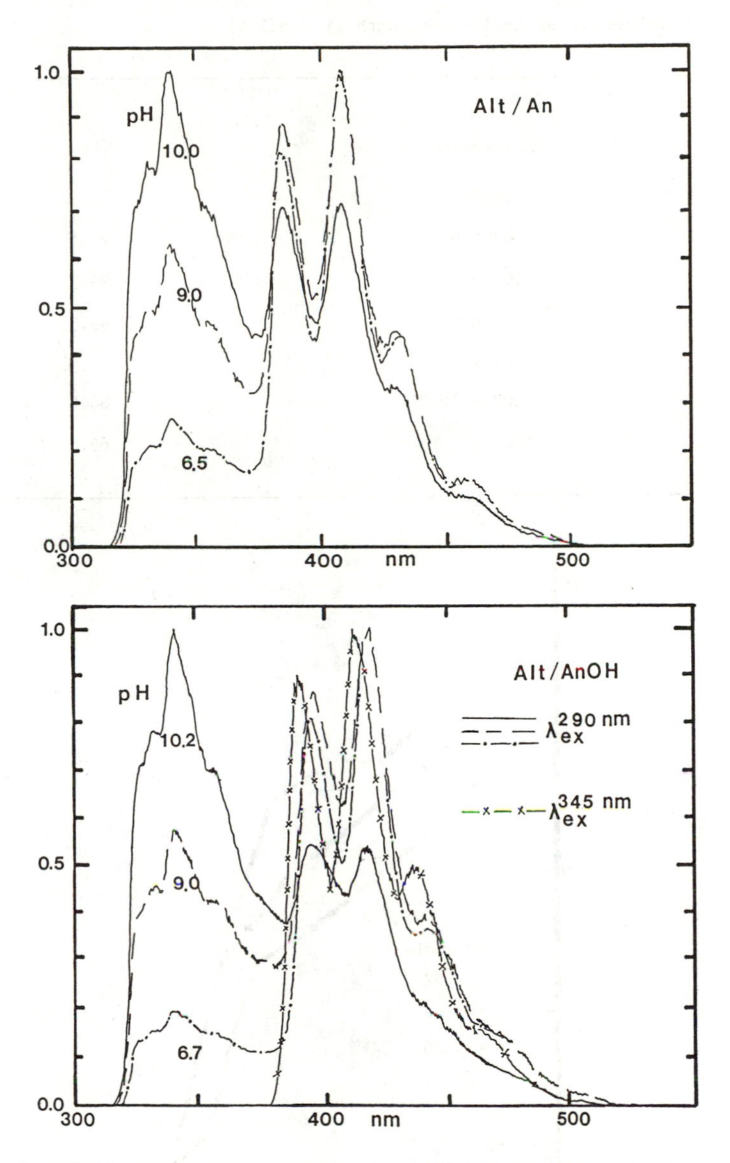


Fig. 5 a) Fluorescence spectra for P2VN-alt-MA/Anth(ads) at indicated pH; b) Fluorescence spectra for P2VN-alt-MA/AnOH(ads) at indicated pH. All spectra excited at 290 nm except x-x which was excited at 345 nm.

Table 5. χ Values for P2VN-alt-MA-co-Anth (1.2 mol %)

| Solvent/pH | χ(%) |
|---|---|
| $H_2O$/10.4 | 89.9 |
| $H_2O$/10.0 | 93.5 |
| $H_2O$/ 9.1 | 99.3 |
| $H_2O$/ 8.6 | 99.3 |
| $H_2O$/ 7.7 | 99.5 |
| MeOH, $Na^+MeO^-$ | 88.8 |
| THF | 65.0 |

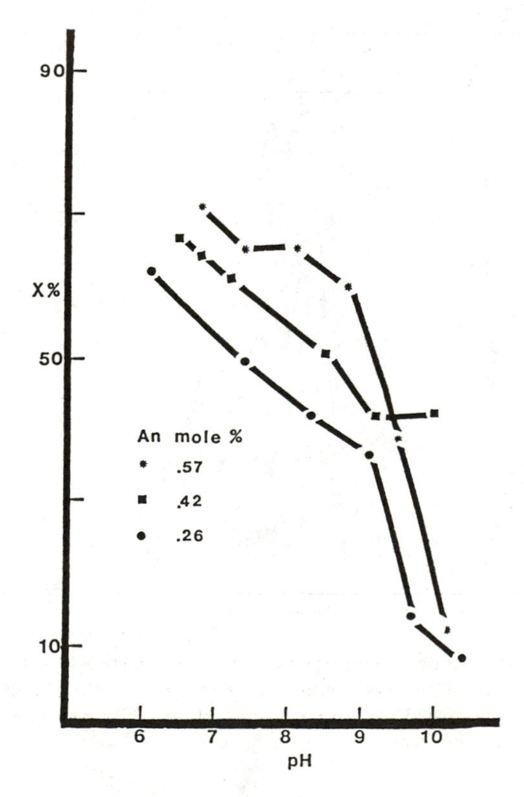


Fig. 6 Plot of $\chi_{SS}$ vs. pH for indicated loadings of anthracene.

two distinct environments for the AnOH, which we speculate to be bound and free AnOH.

For comparison with the alternating copolymer we synthesized a random polymer of 2-VN and MA containing ca. 10 mol% of 2-VN (higher naphthyl contents did not dissolve in basic $H_2O$). Solubilization by this copolymer demonstrates: 1) a much lower sensitization efficiency than the alternating copolymer, 2) no spectral shift between directly excited and sensitized AnOH, and 3) in the absense of AnOH a clearly discernible excimer component in the fluorescence (see Fig. 7). While comparison of the alternating and random copolymer is dangerous because of the different naphthyl contents, these results do imply that 1) the AnOH is not preferentially associated with the naphthyl moieties in the random copolymer, and 2) the red-shift observed in the sensitized AnOH fluorescence for P(2VN-alt-MA) is the result of naphthalene - anthryl interactions.

Thus we have learned the following from these sensitization experiments:

1) The local structure of the P(2VN-alt-MA) in water brings the naphthalene and covalently bound anthracene groups into closer proximity than for better polymer solvents.
2) Based on the $\chi$ values solubilized Anth or AnOH is apparently not as intimately associated with the naphthalene groups as covalently bound anthracene.
3) Even at low mole %'s a highly hydrophobic probe can stimulate polymer aggregation.
4) The local environment of P(2VN-alt-MA)-solubilized AnOH is different than free AnOH, but this is not the case for AnOH solubilized by P(2VN(10 mol%)-co-MA). Solubilized Anth or covalently bound Anth does not exhibit this phenomenon.
5) Based on the random copolymer results, there does not seem to be any preferential association of AnOH with the naphthalene groups.

The naphthalene monomer fluorescence lifetime is shortened by the solubilized Anth to an extent in reasonable agreement with the observed $\chi$ values. Following the general analysis of Fredrickson and Frank[8], we have argued that $\chi$ can be estimated from the fluorescence decay from the expression

$$\chi_t = 1-\bar{\tau}_x/\bar{\tau}_o \qquad (15)$$

where

$$\bar{\tau}_x = \int_o^\infty e^{-k_N t} G_N^x(t)dt \qquad (16)$$

The product $e^{-k_N t} G_N^x(t)$ is the decay function for the monomer naphthalene singlet state where $k_N$ is the unperturbed naphthalene lifetime and $G_N^x(t)$ is a general, non-exponential function that describes the naphthalene fluorescence decay in the presence of X mole fraction of energy accepter. Using our multiexponential fitting function $\bar{\tau}_x$ is given by:

$$\bar{\tau}_x = \Sigma a_i^x \tau_i^x / \Sigma a_i^x \qquad (17)$$

For the case of covalently bound anthracene (i.e. P(2VN-alt-MA-co-Anth) in an organic solvent (THF) we have found $\chi_{ss} > \chi_t$ where $\chi_{ss}$ refers to the steady-state result, using eqn. (14). We have argued[9] that this is the result of either (1) a very rapid transfer step that is too fast to be detected by our photon counting technique, or (2) the existance of a common precursor state for both the naphthalene monomer and the anthryl moiety. In the present case $\chi_t$ and $\chi_{ss}$ are in reasonable agreement for adsorbed anthracene (see Table 5). In general this is not the case for AnOH. However, we have found the fluorescence decay properties of these polymers to be very sensitive to the pH and the history of the solution. Consequently we do not wish to base too many conclusions on lifetime data pending later detailed studies. While it is reasonable that much of the AnOH can reside in the bulk solution after the polymer is saturated, this does not seem reasonable for the extremely hydrophobic Anth. Since the polymer hydrodynamic diameter was observed to increase upon addition of Anth we propose that anthracene aggregation occurs, which diminishes the intensity of sensitized fluorescence per anthryl moiety (see Discussion, section B).

As in the case of the excimer fluorescence, the fluorescence decay for adsorbed anthracene does not have a clearly demonstrable rise time component (see Table 6). The major lifetime component for Anth is on the order of the lifetime of directly excited Anth. Thus these results suggest a rapid sensitization of Anth followed by normal fluorescence decay without significant continued sensitization at long times. We assign the minor long lifetime component to residual excimer fluorescence.

For covalently-bound anthracene in $H_2O$ the degree of naphthalene lifetime shortening is very sensitive to pH (see Table 7). While $\chi$ changes only very slightly[10] as the pH is lowered from 10.2 to 9.1 the average lifetime of the naphthalene decreases dramatically. This implies a dramatic change in the local structure in the hydrophobic portion of the coil which is only weakly reflected in the changes of the hydrodynamic diameter of the coil (see Table 1).

## Discussion

One of the most intriguing aspects of the present polymer system is the structure of the polymer at the molecular level in aqueous or alcoholic solution. Normally one expects aggregation of hydrophobic regions of an amphiphilic copolymer but in the present case the alternating structure may prevent strong interaction of nearest neighbor naphthalene groups. Possibly an intracoil "lamellar" structure is feasible which permits facile intrachain energy migration but does not permit a high density of excimer forming sites (efs). As was discussed earlier, there seems to be no direct relationship between either the coil size, as measured by ILS, and the density of efs or the efficiency of intracoil sensitization of an anthryl trap. Additionally, the comparison of solubilized anthryl derivatives with covalently bound anthracene does not lead to any clear-cut model for the polymer structure in water. In this context perhaps the most revealing observation is

Table 6. Fluorescence Decay Parameters for P2VN-alt-MA/Adsorbed An

| [An] (M) | pH | $^1N^*$ emission[b] | | $\bar{\tau}$[c] | $^1D^*$ plus sensitized $'A^*$ emission[d] | | $\chi_t$[e] | $\chi_{ss}$ |
|---|---|---|---|---|---|---|---|---|
| 0[a] | 9.0 | 3.22/.604 | 28.78/.396 | 13.3 | | | | |
| | 7.1 | 3.61/.712 | 24.63/.288 | 9.7 | | | | |
| $2.6 \times 10^{-7}$ | 9.6 | 8.8 /.896 | 10.4 /.104 | 9.0 | 6.7/.734 | 34.7/.266 | .32 | .37 |
| | 7.1 | 1.0 /.763 | 6.5 /23.7 | 2.3 | 6.7/.848 | 58.8/.152 | .76 | .56 |
| $4.3 \times 10^{-7}$ | 8.5 | 1.88/.758 | 10.2 /.242 | 3.9 | 7.1/.835 | 50.8/.165 | .71 | .51 |
| | 7.4 | 1.03/.834 | 6.7 /.166 | 2.0 | 5.3/.857 | 70.0/.143 | .79 | .62 |
| $5.7 \times 10^{-7}$ | 9.1 | 1.0 /.654 | 8.6 /.346 | 3.6 | 6.4/.764 | 83.8/.236 | .73 | .60 |
| | 7.4 | 1.0 /.861 | 6.6 /.139 | 1.8 | 4.8/.756 | 37.0/.234 | .81 | .66 |

a. From Table 2.

b. Exc. 290 nm, emission 340 nm.

c. See eqn. (7).

d. Exc. 290 nm, emission > 400 nm.

e. See eqn. (15).

Table 7. Fluorescence Decay Parameters for P2VN-alt-MA-co-An (1.2 mol %)

| sample | $^1N^*$ emission[a] | | sensitized $^1A^*$ emission[b] | | direct $^1A^*$ emission[c] | |
|---|---|---|---|---|---|---|
| | $\tau_1/A_1$ | $\tau_2/A_2$ | $\tau_1/A_1$ | $\tau_2/A_2$ | $\tau_1/A_1$ | $\tau_2/A_2$ |
| a) $H_2O$(pH10.2) | 3.72/.732 | 34.8/.268 | 10.0/.523 | 19.1/.476 | | |
| -without An[d] | 6.41/.466 | 37.8/.534 | ----- | ----- | ----- | ----- |
| b) $H_2O$(pH9.1) | 0.7/.683 | 4.99/.317 | | | 7.4/.516 | 14.0/.484 |
| -without An[d] | 3.22/.604 | 28.8/.396 | ----- | ----- | ----- | ----- |
| c) MeOH/$MeO^-$ | 8.0/.352 | 16.3/.648 | 2.8/.449 | 11.5/.551 | 3.3/.491 | 7.6/.509 |
| -without An | 21.3/1.00 | ----- | ----- | ----- | ----- | ----- |

a. exc. 290 nm, emission 340 nm

b. exc. 290 nm, emission > 400 nm

c. exc. 345 nm, emission > 400 nm

d. from Table 2 (for ease of comparison)

the simple fact that the alternating copolymer will dissolve in water while the corresponding random copolymer will not.

From the point of view of intracoil energy migration the $Cu^{2+}$ quenching results are quite interesting since an extraordinarily high value of the quenching constant was obtained, ca. $4 \times 10^{14}$ $M^{-1}$ $s^{-1}$. We have observed apparent quenching constants for $Cu^{2+}$ - polyelectrolyte systems on the order of $10^{11}$ - $10^{12}$ $M^{-1}$ $s^{-1}$ previously[3], which is higher than diffusion controlled because of the electrostatic binding of $Cu^{2+}$ to the polyanion[6]. However, these cases have always exhibited primarily static quenching[11] with the implication that most of the fluorescence quenching occurs between static $Cu^{2+}$ - chromophore pairs rather than by random collisions between these moieties. However, in the present case the $Cu^{2+}$ quenching is primarily dynamic (see Fig. 3) yet with such a high quenching rate constant there must be preferential binding of the $Cu^{2+}$ to the polymer coil. In the next subsection (A) we will explore these results from the point of view of quasi one-dimensional energy transfer along the coil. While it is not possible to quantitatively determine the "dimensionality" and average energy transfer time between naphthalenes, we will argue that this polymer does not behave like a 1-D system and hence must have a highly convoluted structure in solution.

In subsection (B) we will discuss the results with the anthracene energy trap more extensively, from the point of view of intrachain energy migration. This system is more difficult to analyze with simple models because of the large Förster radius for naphthalene to anthracene single-step energy transfer (25Å[7]).

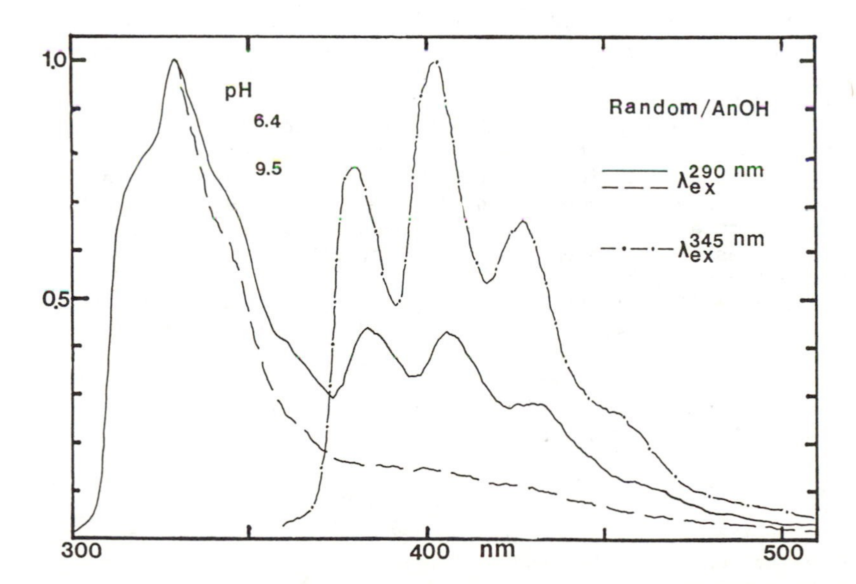


Fig. 7 Fluorescence spectra of P2VN(10 mol%)-co-MA/AnOH(ads) at indicated pH. All spectra excited at 290 nm except o-o excited at 345 nm.

A. Energy Migration and $Cu^{2+}$ Quenching Results

The simpliest model to adopt for $Cu^{2+}$ quenching is to postulate the binding of $Cu^{2+}$ to the pendent $-CO_2^-$ groups (although the preferred bidentate structure would not seem to be feasible) such that each $Cu^{2+}$ species is in contact with at least one naphthyl moiety. There are no allowed $Cu^{2+}$ optical transitions that overlap the fluorescence of naphthalene such that the quenching mechanism is expected to be short range.

One dimensional lattices have been studied extensively by Lakatos-Lindenberg, Hemenger and Pearlstein[12] and the results used recently by Frank et. al.[13] to analyze excimer formation in polymers. For a 1-D lattice with N units bounded by a "disruptive quencher" Pearlstein writes[12b]:

$$G_N(t) = (2/N(N+1))\sum^{n'} \cot^2(\alpha_k/2)\exp[-4Wt\sin^2(\alpha_k)] \tag{18}$$

where $G_N(t)$ is the survival probability of an excitation placed randomly on this lattice ($G_N(0)=1$), W is the energy transfer rate between lattice sites (assumed to be constant for all pairs) and

$$\alpha_k = (2k-1)\pi/N \tag{19}$$

$$n' = N/2 \text{ (N even)} \tag{20}$$

$$= (N+1)/2 \text{ (N odd)}$$

To correct for the finite lifetime of the excited state $G_N(t)$ must be multiplied by $e^{-t/\tau_o}$ where $\tau_o$ is the unperturbed excited state lifetime.

The term "disruptive quencher" is applied to the case in which all exciton contacts with the quencher results in completely effective quenching[12a]. It is not at all clear that this property is applicable to $Cu^{2+}$ quenching. The case of a non-disruptive quencher is much harder to analyze and does not lead to a convenient expression for the excitation decay, analogous to eqn. (18). Based on classical diffusion equations it seems plausible that the disruptive quencher model is applicable so long as the quenching rate at the quenched site is of the same order of magnitude as the transfer rate from that lattice site to neighboring sites[14].

In eqn. (18) N refers to the number of lattice sites between disruptive quenchers. Since these quenchers are placed randomly on the lattice there will be a distribution of N for a given mole fraction of quenched sites (q) and lattice length (L). Fitzgibbon and Frank[13] have derived the following expression for the probability of a given sequence length:

$$P(N,L) = (N/L)(L-N-3)q^2(1-q)^{N+1} \tag{21}$$

In the following we ignore the case of one quencher and one reflective end (i.e. the end of the polymer chain). We also ignore explicit consideration of efs although L in eqn. (21) can be taken to be the average lattice length between efs. For our purposes we replace the sum over all N by the most probable value of N which is obtained from the maximum value of P(N,L). This is

$$N^* = [2+La-(4+(La)^2)^{1/2}]/2a \quad (22)$$

where

$$a = |\ln(1-q)| \quad (23)$$

and q is the mole fraction of traps. As one would intuitively expect for small q, $N^* = L/2$.

One experimental observable that can be obtained easily is the longest lifetime component of the fluorescence decay. This corresponds to k = 1 in eqn. (18). Thus the observed longest lifetime fluorescence lifetime would correspond to

$$1/\tau_1 = 1/\tau_o + \pi^2 W/(N^*)^2 \quad (24)$$

in which it is assumed that $\sin(\pi/2N^*) = \pi/2N^*$ and hence $N^* >> \pi/2$. Using (22) it can be shown that for small q

$$1/\tau_1 = 1/\tau_o + (4\pi^2 W/L^2)(1+Lq/2+\cdots) \quad (25)$$

Thus this 1-D model predicts that the slow component of the fluorescence decay obeys a linear quenching law at low quencher concentrations. Use of the exact solution for N* leads to a rapid non-linear decrease in the value of $\tau_1$ which in principle could be fit to different values for L and W.

However, when we look at the magnitude of this component for large N* we find an approximate value of $8/\pi^2 = .811$ which is much larger than is observed for the pre-exponential factor of the long-lived component (see Table 3). This seems to be a general property of a strictly 1-D lattice treated by a master equation formalism. That is to say, the slowest component of the predicted decay function is by far the most important component.

The steady state quenching can be obtained by integrating the time dependent emission (see eqn. (16)). Using $G_{N^*}(t)e^{-t/\tau_o}$ we obtain:

$$\bar{\tau} = \tau_o \sum_{k=1}^{n'} (2\cot^2(\alpha_k/2)/N^*(N^*+1))/(1 + 4W\tau_o \sin^2(\alpha_k/2)) \quad (26)$$

The steady state quenching curve can be evaluated as a function of q by evaluating N*. For q = 0 the result depends on L which is the average separation of the efs (i.e. L is not equal to degree of polymerization in general). Note that in (26) $\tau_o$ is the lifetime of the isolated chromophore. A detailed analysis of $\bar{\tau}$ as a function of q will not be presented here except to note that $\bar{\tau}(o)/\bar{\tau}(q)$ is not linear in q over any extended range of q that yields values of $\bar{\tau}(o)/\bar{\tau}(q)$ as large as our experimental values for what we believe to be reasonable choices of W and L. We note that the exact result for infinite, disruptively quenched chains yields a quadratic quenching law at low quencher concentration[12a].

It is obviously very dangerous to attempt to draw a firm conclusion based on an approximation to an idealized model. However, we feel that our observations are not well-accomodated to a simple 1-D lattice model.

Another approach to this problem is to envoke the so-called "spectral dimension". This is based on a random walk on a disordered lattice and is defined by[15]

$$S = M^{\tilde{d}/2} + \dots \qquad (27)$$

where M is the number of steps taken by the walker and S is the average number of unique lattice sites visited. The dimension d lies in the range 1 (ideal 1-D lattice) and 2 (ideal 3-D lattice). If the average number of steps is taken to be $\langle\tau\rangle/\tau_h$, where $\langle\tau\rangle$ is the average lifetime defined by eqn. (8) and $\tau_h$ is the time/step, then the trapping probability will be on the order of

$$P \approx Sq = (\langle\tau\rangle/\tau_h)^{\tilde{d}/2} q \qquad (28)$$

If the quencher is less than 100% effective then presumably q would be multiplied by a factor less than unity. Since $I_\theta/I \approx 5$ for $[Cu^{2+}] \approx 2.75 \times 10^{-7}$M (which corresponds to $q = 2.75 \times 10^{-3}$) (see Fig. 3) and $\langle\tau\rangle \approx 6.6$ ns we obtain:

$$P \approx 4/5 \approx (6.6/\tau_h)^{\tilde{d}/2} (2.75 \times 10^{-3}) \qquad (29)$$

Based on this approach the estimated value of $\tau_h$ will be very dependent on d. The value of $\tau_h$ that corresponds to different d values is given in Fig. 8. The values of $\tau_h$ range between ca. .08 ps to ca. 25 ps as d ranges from 1 to 2. This illustrates another property of a 1-D lattice. The efficiency of sampling unique sites is much lower than for lattices of higher dimensionality,

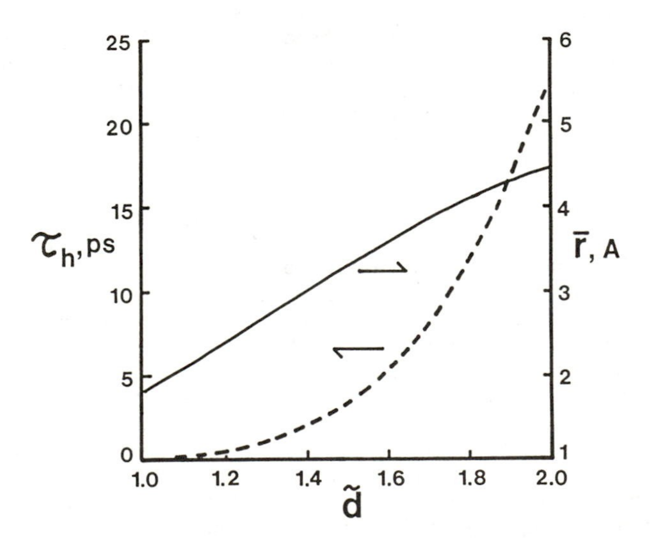


Fig. 8 Plot of $\tau_h$ and $\bar{r}$ as function of d based on experimental sensitization efficiency (see text).

hence for a given quenching efficiency a relatively fast hopping rate is required for a 1-D lattice.

Of course we do not know what value of $\tau_h$ is "reasonable". If we assume that the Förster mechanism is responsible for energy migration, then the average hopping time can be estimated from

$$\tau_h^{-1} = \tau_o^{-1}\langle(3K^2/2)(R_o/r)^6\rangle \qquad (30)$$

where $\tau_o$ is the unperturbed single state lifetime of the naphthalene and $R_o$ is the Förster radius for self-transfer. For 2-methylnaphthalene these parameters are 60 ns and 11.75Å respectively[7]. Eqn. (30) can be used to define an average distance $\bar{r}$

$$\bar{r} = (\tau_h/\tau_o)^{1/6}R_o \qquad (31)$$

in which $K^2 = 2/3$ has been assumed. This is also plotted in Fig. 8 as a function of d. These $\bar{r}$ values are quite small over the whole range of d and in fact are less than the separation expected for excimer formation. We also note that our estimated $\tau_h$ values are much smaller than the estimates of 100 - 250 ps by Fitzgibbon and Frank[13] for atactic P2VN in a good solvent, based on Monte Carlo simulations[16].

Based on these estimates we conclude that: 1) the polymer structure must be more compact than one-dimensional in aqueous solution, 2) the average separation of chromophores must be significantly less than the $R_o$ value of 11.75Å such that the proportionality to $r^{-6}$ may no longer be valid, 3) despite the proximity of the naphthalene groups there is not a high density of efs, presumably because face-to-face geometries are not favored. While conclusions 1) and 2) are based on the simplified expressions (27), (28) with a single parameter d which determines $\tau_h$, we re-emphasize that this physical system must be homogeneous with respect to energy transfer or we would observe an "unquenched component" in the naphthalene fluorescence decay as $Cu^{2+}$ is added. This is not the case as can be seen in Table 3; for all $Cu^{2+}$ additions both $\tau_1$ and $\tau_2$ are quenched.

In all the above the emphasis has been on energy migration as a mechanism to rationalize the dynamic quenching by $Cu^{2+}$. An equally valid mechanism would envoke $Cu^{2+}$ diffusion along the chain, presumably by "hopping" between $-CO_2^-$ groups. As we will discuss in the next subsection the apparent quenching rate for anthryl traps is similar to that obtained for $Cu^{2+}$. Since down chain diffusion is either impossible (for covalently-bound anthracene) or unlikely (for adsorbed anthracene) in these cases we prefer the postulate of energy migration.

B. Discussion of Quenching by Anthryl Traps

In the previous subsection the case of $Cu^{2+}$ fluorescence quenching was discussed. One of the most important features of these results was the very high second-order quenching constant. The results for anthracene were described in terms of $\chi$, the quantum efficiency of sensitization. This is related to a rate constant as follows:

$$\chi = a/(1+a)$$

where

$$a = k_{ET}[An]/k_N \tag{33}$$

where [An] is the overall concentration of the anthracene, $k_N$ is the unperturbed unimolecular decay rate of the chromophore excited state and $k_{ET}$ is the second order rate constant of interest. As can be seen from (32) as a becomes very large $\chi$ approaches unity. This also implies that as $\chi$ approaches unity the accuracy in the derived $k_{ET}$ decreases. For $k_N$ we use $\langle\tau\rangle$ for the unquenched sample. The value of $k_q$ for solubilized An at pH ~ 9 and pH ~ 7 is on the order of $1.0(\pm .1) \times 10^{14}$ and $2.6(\pm 0.5) \times 10^{14}$ $M^{-1}$ $s^{-1}$ respectively. (see Table 6 for $\chi$ values). Note that in the latter case the Stern-Volmer constant decreases with [An], as would be expected if aggregation occurs.

For covalently bound Anth $\chi$ is very high so the accuracy in estimating $k_{ET}$ is low; we estimate $k_q$ to be in the range $3\text{-}7 \times 10^{14}$ $M^{-1}$ $s^{-1}$ (depending on pH), i.e. very similar to that for $Cu^{2+}$. Given the large $R_o$ for naphthalene - anthracene energy transfer it is surprising that $k_q$ is not even larger than this.

Because of the extreme hydrophobicity of the anthracene molecule we believe that all anthracene in the solution phase is intimately associated with the polymer. Thus at first glance it is also surprising that $k_q$ is not equal to that of the covalently-bound anthracene. As was pointed out in the Results section, Anth appears to stimulate polymer aggregation, presumably by forming anthracene aggregates which are simultaneously in contact with the polymer. A heterogeneous distribution of energy acceptors might be expected to be less efficient in quenching the naphthalene excited states if a significant fraction of naphthalene groups are relatively distant from anthracene moieties. However, in such a case there should also be a significant component in the naphthalene fluorescence decay with the same lifetime as unquenched naphthalene. This is not observed. Thus we propose that our estimate of $\chi_{ss}$ is too low because 1) the fluorescence quantum yield of sensitized fluorescence is lower than directly excited anthracene because of anthracene self quenching within the aggregates, and 2) the aggregates are more effective energy traps than "isolated" anthracene groups because the former are located in naphthalene rich regions.

One of the distinctive features of the $Cu^{2+}$ quenching results was that the lifetime and intensity quenching were essentially equivalent. While the naphthalene fluorescence decay is definitely shortened by the presence of the covalent or adsorbed anthracene, there was no systematic change of $\chi_t$ with pH or mole fraction of anthracene. Part of the difficulty may be experimental - for high $\chi$ values the intensity of naphthalene fluorescence is very low such that accuracy is sacrificed in the decay function. Because of the large $R_o$ value for naphthalene - anthracene one expects a significant "static" component in the naphthalene quenching. A more detailed examination of the time dependence of energy transfer in disordered systems of the present type will be the subject of future investigations in this and related polymer systems.

## Acknowledgments

We would like to express our gratitude for the financial support of the National Science Foundation (DMR-834755), and the Robert A. Welch Foundation (F-356). The light scattering system was purchased by joint NSF funding (DMR-8418612) and the University of Texas at Austin. The Center for Fast Kinetics Research is supported jointly by the Biotechnology Branch of the Division of Research Resources of the NIH (Grant RR00886) And the University of Texas at Austin.

## Footnotes/References

1. Bai, F.; Chang, C.H.; Webber, S.E. Macromolecules **1986**, 19, 588.
2. Fox, R.B.; Price, T.R.; Cozzens, R.F.; Echols, W.H. Macromolecules **1974**, 7, 937.
3. a) Webber, S.E. Macromolecules **1986**, 19, 1658; b) Delaire, J.A.; Rodgers, M.A.J.; Webber, S.E. J. Phys. Chem. **1984**, 88, 6219.
4. Aspler, J.S.; Hoyle, C.E.; Guillet, J.E. Macromolecules **1978**, 11, 925.
5. Webber, S.E.; Avots-Avotins, P.E.; Dumie, M. Macromolecules **1981**, 14, 105.
6. See for example a) Chu, D.Y.; Thomas, J.K. Macromolecules **1984**, 17, 2142; b) Sassoon, R.E.; Rabani, J. J. Phys. Chem. **1985**, 89, 5500.
7. Berlman, I.B. "Energy Transfer Parameters of Aromatic Molecules"; Academic Press, New York **1973**. These are the same values used in ref. 13.
8. Fredrickson, G.H.; Frank, C.W. Macromolecules **1983**, 16, 572.
9. Bai, F.; Chang, C.H.; Webber, S.E. Macromolecules (in press).
10. However, this small change in $\chi$ corresponds to a large change in $I_A/I_N$ as $\chi$ 1.
11. That is to say, the quenching of the excited state lifetime is much less efficient than the quenching of the luminescence.
12. a) Lakatos-Lindenberg, K.; Hemenger, R.P.; Pearlstein, R.M. J. Chem. Phys. **1972**, 56, 4852; b) Pearlstein, R.M. ibid 2431.
13. Fitzgibbon, P.D.; Frank, C.W. Macromolecules **1982**, 15, 733.
14. Specifically if $Nk_q/W >> 1$ where $k_q$ and W are the quenching rate and hopping rate respectively. (S.E. Webber, unpublished calculations, based on the classical "surface evaporation" model (J. Crank, Mathematics of Diffusion, (Oxford Press, 1983), section 4.3.6).
15. Zumofen, G.; Blumen, A.; Klafter, J. J. Chem. Phys. **1985**, 82, 3198.
16. We note that the $\tau_h$ values of Fitzgibbon and Frank (ref. 13) correspond to $\bar{r}$ values of ca. 4.1 - 6.3Å.

RECEIVED March 13, 1987

## Chapter 30

# Polymer Models for Photosynthesis

**James E. Guillet, Yoshiyuki Takahashi [1], and Liying Gu [2]**

**Department of Chemistry, University of Toronto, Toronto M5S 1A1, Canada**

Macromolecules containing aromatic chromophores can display efficient electronic energy transfer to low-energy traps. By analogy with the biological process of photosynthesis, we have termed such molecules "antenna molecules". Synthetic polymers can thus mimic the function of the light-harvesting chlorophyll pigment layers without reproducing their exact structure. We have linked aromtic chromophroes in both organic and water-soluble polymers which provide useful antennas for solar photochemistry. For example, antennas consisting of phenyl anthracene groups are effective in increasing the absorption cross-section for tetraphenyl porphine groups by at least an order of magnitude. Sulfonated poly(2-vinylnaphthalene) polymers are also useful catalysts for singlet oxygen reactions in aqueous solution. As a model of the reaction center in natural photosynthesis we have prepared copolymers of acrylic acid containing small amounts of porphyrin (P) and anthraquinone (Q) moieties. The conformation of these polymers in aqueous solution is highly dependent on both the pH and ionic strength of the solvent, and this can be used to control the average distance between P and Q. ESR measurements confirm the formation of separated ion pairs ($P^{\dot{-}}$ and $Q^{\dot{+}}$) when the porphyrin group is irradiated in solution at -40°C.

The primary photochemical step in photosynthesis is now generally recognized to be a one-electron transfer from the singlet excited state of a chlorophyll species (Chl) to an electron acceptor. This reaction takes place within a reaction center protein that

---

[1] Current address: Oji Paper Co., Ltd., 1-10-6 Shinonome Koto-ku, Tokyo, Japan
[2] Current address: Institute of Chemistry, Academia Sinica, Beijing 100 080, People's Republic of China

0097-6156/87/0358-0412$06.00/0

spans the thylacoid membrane of the chloroplast organelle of green leaves and algae. In the simplest photosynthetic systems the electron acceptor contains the quinone moiety, such as an ubiquinone, menaquinone or plastoquinone. An essential feature of this process is that the donation of an electron must lead to a separation of the charged species $Chl^{\dagger}$ and $Q^{\bar{\cdot}}$ so that they may undergo further reactive steps in the photosynthetic sequence.

In green plants the primary charge-separation process occurs in reaction sites which contain only a small fraction of the total pigment material. The bulk of the chlorophyll in the chloroplast is photochemically inert, functioning as an "antenna pigment" by transferring light through non-radiant interactions to the reaction centers. In this way the turnover rate for reactive sites is increased, the occurrence of this energy-transfer process having the same effect as if the extinction coefficient of the reactive center were increased by a factor of over 100. In biological photochemistry, this effect is known as the "antenna effect".

It has been shown that macromolecules containing aromatic chromophores can also display efficient electronic energy transfer to low-energy traps [1]. By analogy with the biological process of photosynthesis, we have termed such molecules "antenna molecules". Macromolecules containing chromophores attached to a polymeric backbone often display very high efficiency of singlet energy transfer. The function of the connecting macromolecular chain and the plant thylacoid membrane is similar in that both serve as anchors supporting high local concentrations of the chromohores. Synthetic polymers can thus mimic the function of the light-harvesting pigment layers without reproducing their exact structure.

In earlier studies in these laboratories it has been shown that polymers containing repeating naphthalene or phenanthrene groups and small numbers (from 0.1 to 2%) of traps such as anthracene, anthraquinone, or phenyl ketone, demonstrated singlet exciton transfer from the absorbing site in the antenna to the trap [1]. The efficiency of energy transfer in the trap can be evaluated in terms of the quantity $\chi$, which is defined as the number of photons transferred to the trap divided by the number of photons absorbed by the antenna.

The efficiency can be calculated either from the emission from the trap if the trap is a fluorescing moiety, or from the quenching of emission from the antenna chromophores. An alternative way of expressing this efficiency is to calculate the number of antenna donor chromophores, n, quenched by each trap. Figure 1 shows the value of n for antennas containing naphthalene repeating units with the three traps mentioned earlier [2]. It can be seen that depending on the donor-trap combination, the value of n can range from about 50 to about 150 in these systems. This represents an increase in the absorption cross-section for the trap of one to two orders of magnitude, depending on the molar extinction coefficient of the two species at the wavelength of excitation.

In later work it was shown that water-soluble antennas could be made by copolymerizing aromatic monomers such as vinyl naphthalene and naphthylmethyl methacrylate with polyelectrolytes such as acrylic acid [3,4]. The high efficiency of these antennas in dilute aqueous base was attributed to the hypercoiling of the poly-(acrylic acid) chain to give a pseudo micellar structure such as that illustrated schematically in Figure 2. We believe that such structures are formed spontaneously in solution due to the hydrophobic interactions of the large aromatic components stabilized by the interaction of water with the hydrophillic carboxyl anions. It was later shown that other types of polyelectrolytes involving partial sulfonation of poly(vinylnaphthalene) [5,6] and copolymers of aromatic monomers with styrene sulfonate would also lead to polymers which in aqueous solution achieved this hypercoiled configuration.

It therefore seemed logical to apply these same princples to the problem of producing a model of the active site in photosynthesis. In the simplest natural photosynthetic systems the first chemical step is the transfer of an electron from a porphyrin compound to a quinonoid structure. This process is initiated by excitation transfer from antenna chlorophyll pigments to a porphyrin in the active center. Attempts have been made to mimic this process by including porphyrins and quinones in films, vesicles and micelles with varying degrees of success. An alternative approach [7] is to link a porphyrin P group covalently to a quinone Q by a chain of methylene groups. By adjusting the length of the chain it was possible to obtain compounds which are soluble in organic solvents in which photoelectron transfer from P to Q can be observed. However, a further requirement is that the radical ion species created in this first step be isolated from each other so that they do not recombine and lose the excitation energy as heat.

It was found experimentally that if the porphyrin and quinone are too close together back-transfer will occur readily and detection of the stabilized ion radical species will be difficult. However, there appears to be a distance of approximately 10 Å at which electron transfer can occur from the excited porphyrin to the quinone while the back-transfer process is sufficiently slow that radical ions can be observed by ESR and other techniques. For example, McIntosh et al. [8] have detected a charge-separated species by electron paramagnetic resonance spectroscopy of a number of covalently linked P-Q compounds.

In our recent work we decided to take advantage of the hypercoiling effects observed in poly(acrylic acid) containing large hydrophobic groups to see if we could find compositions containing porphyrins and quinones where the distance between the P and Q groups could be adjusted by changing the pH or ionic strength of the solution. Accordingly, several polymers were synthesized by copolymerization with acrylic acid with monomers of the structure

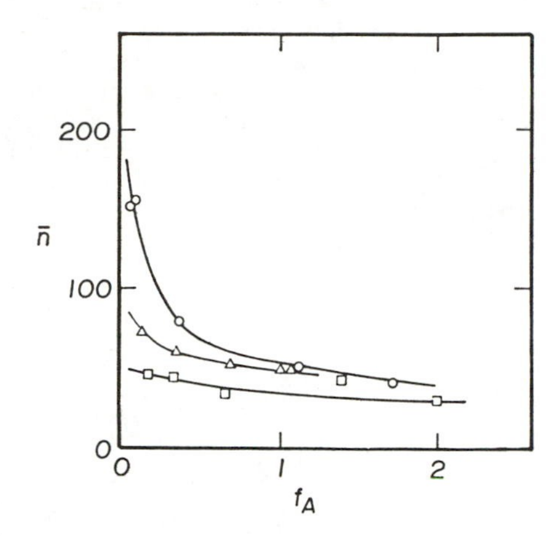


Figure 1. Number of naphthalene donors quenched per trap $\bar{n}$ as a function of trap concentration for (○) 1NMMA-2AQMMA, (△) 2VN-PVK, and (□) 2NMMA-9AMMA in 2MeTHF at 77K (~4 × $10^{-4}$ M in naphthalene). (Reprinted from Ref. 2. Copyright 1985 American Chemical Society.)

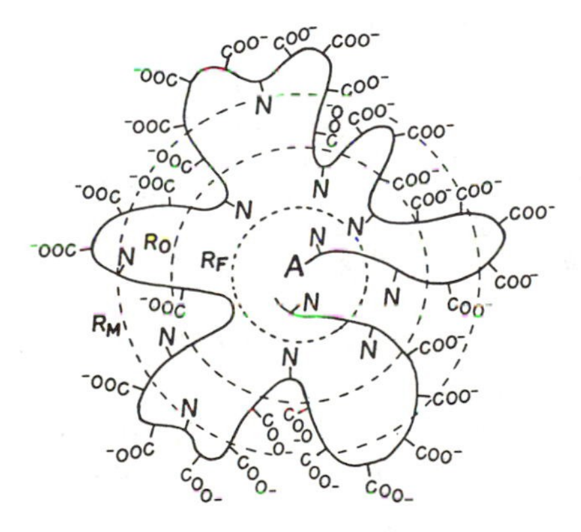


Figure 2. Proposed hypercoiled structure of anthracene end-trapped copolymers of acrylic acid and NMMA in dilute alkaline solution showing Förster radii for various energy-transfer processes.

shown below. The composition and molecular weights of the polymers synthesized are shown in Table I [9]. As can be seen, Polymer A contains about two porphyrin groups per chain, but no anthraquinone, while the second sample, B, contains one porphyrin and about three anthraquinone groups per molecule. Comparison of the emission fluorescence emission spectra of these two polymers on excitation of the porphyrin ring by irradiation with light in the range of 500 to 650 nm showed some quenching of the emission in aqueous solutions of polymer B with respect to the control sample A. In acid conditions the aqueous solution is a bright green color, while in base it is a brownish purple.

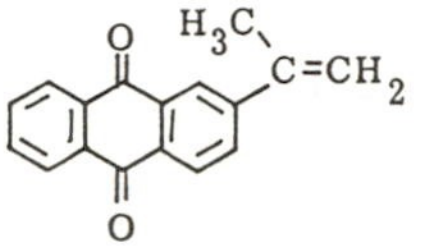

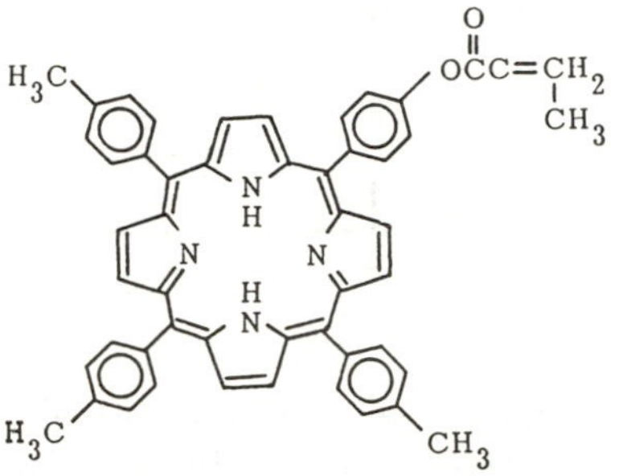

Measurements of the ESR spectra of these solutions were also made at various temperatures. In a frozen aqueous glass at -40°C the ESR signals were observed in sample B, but not in sample A. The G factor of the photo-driven signal at pH 11 was 2.0037 ± 0.0002, in excellent agreement with the assumption that it is a spin exchange average of $P^{\dagger}$ G = 2.0025 an $AQ^{\bar{\cdot}}$ G = 2.0047. The yield of $P^{\dagger}$ $Q^{\bar{\cdot}}$ radical pairs was about 3% based on double integration of the signals. Under acid conditions the resonance is symmetrical with a G factor of 2.0025 which suggests that the AQ radical ion has been protonated and that only the signal for $P^{\dagger}$ is observed. When the light is turned off, the signal decays slowly over a period of about 15 minutes at -40°C.

Table I. Properties of Acrylic Acid Copolymers

| Polymer | $\overline{M}_v$ [a] | $\overline{DP}$ | mol % P [b] | $\eta_P$ | mol % Q [c] | $\eta_Q$ |
|---|---|---|---|---|---|---|
| A | $17.5 \times 10^4$ | 2430 | 0.11 | 2.7 | | 0 |
| B | $5.5 \times 10^4$ | 764 | 0.14 | 1.0 | 0.40 | 3.0 |

*a*. From intrinsic viscosity in dioxane using K = $76 \times 10^{-5}$ and *a* = 0.50 (see: Sandler, S. R.; Karo, W. "Polymer Syntheses", Academic Press: New York, 1977; Vol. II, Chapter 9. *b*. From UV absorbance in methanol solution at 412 nm based on ε for model compound tetratolylporphine = $4.2 \times 10^5$. *c*. From UV absorbance in MeOH at 255 nm based on ε for model compound 2-methylanthraquinone = $5.0 \times 10^4$.

From these results, it was concluded that inclusion of porphyrin-quinone moities in a polyelectrolyte such as poly(acrylic acid) provides a useful model to study photoelectron transfer processes observed in artificial photosynthesis. The advantage of the use of such polymers is that experiments may be carried out in aqueous, rather than organic media and that the presence of these polyelectrolytes may stabilize some of the resultant separated ion pairs. Furthermore, variations in the efficiency of electron transfer and other important parameters in the process can be made by changing the pH or ionic strength of the medium.

Having developed a successful model for the active center it was also of interest to see if one could produce antenna structures capable of exciton transfer to the porphyrin moiety. The photophysical properties of a suitable polymer have already been reported by Hargreaves and Webber [10]. The phenyl anthracene group has the necessary overlap of its emission spectrum with the absorption spectrum of the porphyrin ring. Thus it would be expected to accept singlet excitons migrating in a chain containing repeating phenyl anthracene groups. Accordingly, we synthesized copolymers of 10-phenyl 9-anthrylmethyl methacrylate containing minor amounts of monomer I [11]. Evidence for efficient energy migration and transfer to the porphyrin ring was observed by measurement of the excitation spectrum of the porphyrin florescence emission with and without the antenna molecule. The results are shown in Figure 4. It is clear from this figure that a substantial increase in the absorption cross-section for the porphyrin has been obtained by the use of the phenylanthracene antenna.

The hypercoiled conformation assumed by polyelectrolytes containing large aromatic groups has a further interesting application. It was found that aqueous solutions of such polymers, when exposed to small amounts of water-insoluble aromatics dissolved in ether, will concentrate the large hydrophobic molecules in the center of the hypercoiled conformation. This is equivalent to having a reversible trap. When the antenna molecules in the copolymer are irradiated with light, the excitation energy moves around on the inside of the coil and if the spectroscopic properties of the antenna and trap are properly chosen, most of the energy will be transferred to the trap. For example, Figure 5 shows the emission from an aqueous solution of a copolymer of sulfonated poly(vinyl naphthalene) containing small amounts of perylene [6]. Analysis indicates that there is approximately oe perylene trap per polymer molecule. If the irradiation is carried out in the presence of air there is a very rapid removal of the perylene fluorescence by the conversion of the perylene to its endo peroxide via the intermediacy of singlet oxygen. The reaction in shown schematically below.

$$\xrightarrow{{}^{1}O_2}$$

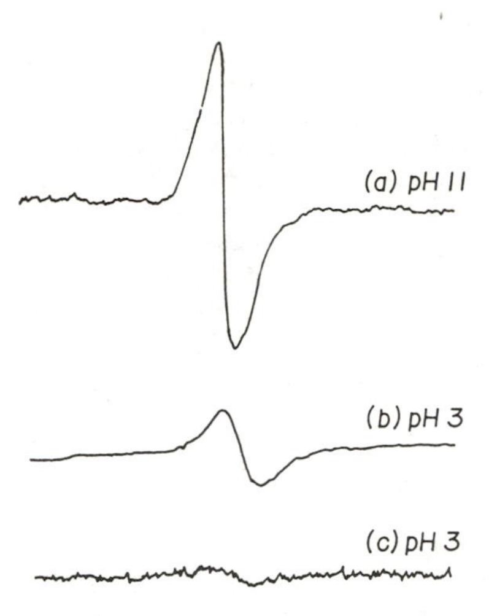


Figure 3. (a,b) EPR signal of poly(AA-POMA-AQ); (c) EPR signal of poly(AA-POMA). EPR operating conditions were 0.5-mT modulation amplitude at 100 kHz and 5-mW microwave power at 9.05 Hz; the temperature was 235 K. The spectral range displayed is 10 mT. (Reprinted from Ref. 9. Copyright 1985 American Chemical Society.)

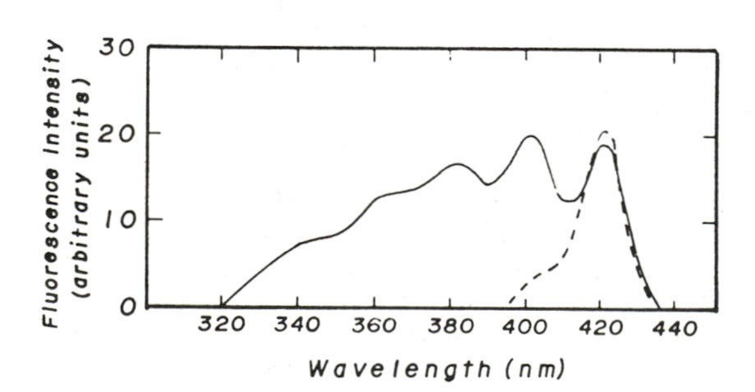


Figure 4. Modified excitation spectra of copolymer $I_H$. ——— calculated spectrum by subtraction; — — — excitation spectrum of tetratolylporphine measured under the same experimental conditions (monitored at 650 nm).

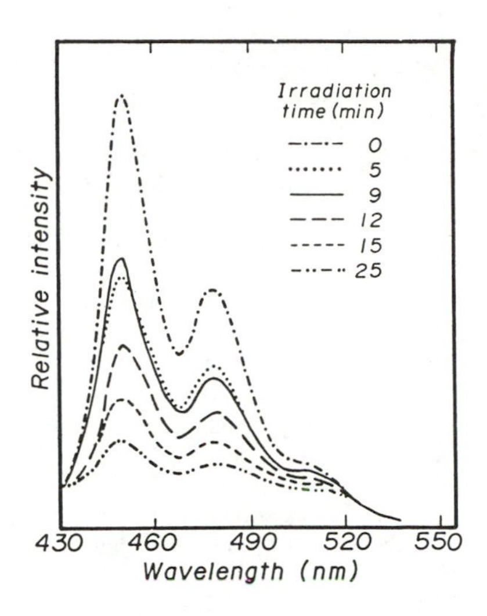


Figure 5. Perylene emission from an aqueous solution of sulfonated poly(2-vinylnaphthalene) (SP2VN) before and after irradiation in air.

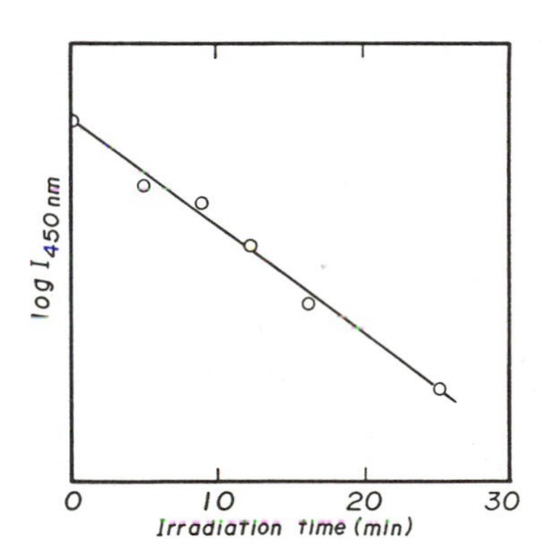


Figure 6. Rate of conversion of perylene to endoperoxide as estimated from decrease in fluorescence intensity at 450 nm.

As shown in Figure 6 the rate appears to be first order, up to relatively high degrees of conversion, and is several orders of magnitude faster than in the absence of the polymer electrolyte. After the perylene has been reacted in this way it can be removed from the aqueous solution by extraction with ether, more perylene can then be added, and the reaction continued.

We believe that these new photocatalysts may have important applications to synthetic chemistry. Because of their high catalytic efficiency and their analogy to biological catalysts which ususally contain a hydrophobic pocket in which chemical reaction can take place, we call these unusual molecules "photozymes". The conformation of these polymers can be changed radically by changing either the ionic strength or the pH of the medium. This gives considerable flexiblity in designing a system with both a high efficiency for trapping aromatic molecules, and for producing the desired product. However, potential applications for these new systems have yet to be developed.

In conclusion, it is now possible to study the photochemistry of a large number of interesting organic materials in aqueous systems using as photocatalysts polyelectrolytes containing sensitizing antenna groups. The aqueous medium provides many potential advantages. In addition to being a very cheap solvent, important variations in the photochemistry can be observed by controlling the pH and ionic strength of the solution.

## Acknowledgments

We acknowledge the financial support of this research by the Natural Sciences and Engineering Research Council of Canada. Y.T. is grateful to the Oji Paper Co. Ltd. for financial support during his studies at the University of Toronto.

## Literature Cited

1. Guillet, J. E. Polymer Photophysics and Photochemistry; Cambridge University Press: Cambridge, 1985; Chapter 9.
2. Ren, X.-X.; Guillet, J. E. Macromolecules 1985, 18, 2012.
3. Holden, D. A.; Rendall, W. A.; Guillet, J. E. Ann. N. Y. Acad Sci. 1981, 366, 11.
4. Guillet, J. E.; Rendall, W. A. Macromolecules 1986, 19, 224.
5. Guillet, J. E.; Wang, J.; Gu, L. Macromolecules 1986, 19, 2793.
6. Guillet, J. E.; Gu, L., submitted for publication.
7. (a) Kong, J. L. Y.; Loach, P. A. In Frontiers of Biological Energetics -- Electrons to Tissues; Dutton, P. L.; Leigh, J. S.; Scarpa, A., Eds.; Academic Press: New York, 1978; Vol. 1, pp 73-82. (b) Kong, J. L. Y.; Loach, P. A. J. Heterocycl. Chem. 1980, 17, 737. (c) Kong, J. L. Y.; Spears, K. G.; Loach, P. A. Photochem. Photobiol. 1982, 35, 545.

8. (a) McIntosh, A. R.; Siemiarczuk, A.; Bolton, J. R.; Stillman, M. J.; Ho, T.-F.; Weedon, A. C. J. Am. Chem. Soc. 1983,105, 7215. (b) Ho, T. F.; McIntosh, A. R.; Bolton, J. R. Nature (London) 1980, 286, 254.
9. Guillet, J. E., Takahashi, Y., McIntosh, A. R.; Bolton, J. R. Macromolecules, 1985, 18, 1788.
10 Hargreaves, J. S.; Webber, S. E. Macromolecules 1984, 17, 235.
11. Takahashi, Y.; Guillet, J. E., unpublished work.

RECEIVED March 13, 1987

# Chapter 31

# Complex Formation Between Poly(acrylic acid) and Poly(ethylene glycol) in Aqueous Solution

**Hideko Tamaru Oyama, Wing T. Tang, and Curtis W. Frank**

**Department of Chemical Engineering, Stanford University, Stanford, CA 94305**

Pyrene groups were attached to poly(ethylene glycol) (PEG) at both chain ends to allow pyrene excimer fluorescence to be used as a molecular probe of the complexation between PEG and poly(acrylic acid) (PAA). The excimer to monomer intensity ratio, $I_D/I_M$, was measured distinguishing intramolecular and intermolecular excimer formation. The decrease in $I_D/I_M$ for the intramolecular excimer showed that the addition of PAA to PEG induces a decrease of intramolecular mobility of PEG. On the other hand, the fluorescence behavior of the intermolecular excimer showed that the local concentration of PEG is increased in the vicinity of PAA as a result of hydrogen bond interaction. Both phenomena were more pronounced in the PAA-PEG complex formed from the PAA of higher molecular weight.

The objective of this study is to utilize fluorescence spectroscopy to examine complex formation between the proton-donor poly(acrylic acid) (PAA) and the proton-acceptor poly(ethylene glycol) (PEG). In the early studies, intermolecular complex formation was investigated by potentiometry, viscometry, calorimetry, turbidity, and sedimentation. (1,2) It was found from viscosity and pH measurements that the process is highly cooperative and that the complex contains stoichiometrically equivalent carboxyls of PAA and ether groups of PEG. Calorimetry demonstrated that the enthalpy and entropy of complexation are both positive in aqueous solution, suggesting that, while complexation involves hydrogen bonding between carboxyl and ether groups, hydrophobic interaction is also an important driving force to cause complexation.

However, more recently, the application of fluorescence techniques has attracted attention. Anufrieva et al. applied the polarized luminescence method to the stoichiometric hydrogen bond

0097-6156/87/0358-0422$06.00/0

complexes between PEG and poly(methacrylic acid) (PMAA) or PAA in aqueous solution. It was found that the relaxation time $\tau_w$ of anthracene-labeled PMAA and PAA became much longer upon complexation with PEG: from 77nsec to 290nsec for PMAA and 23nsec to 50nsec for PAA. (1,3) The same phenomenon was also observed in anthracene-labeled PEG, for which $\tau_w$ changed from less than 1nsec to 350nsec in the PMAA-PEG complex. They concluded from the similarity of $\tau_w$ values between those of PMAA and PEG polymers as well as the increase in $\tau_w$ that the mobility of component polymers in the complex is restricted by a fairly long continuous linear succession of bonds between monomer units of the complementary polymer chains, very much like a ladder structure.

Morawetz and coworkers employed PAA labeled with the fluorescent dansyl group (DAN-PAA) to study hydrogen bonding polymer complexes. (4-7) The characteristic feature of the dansyl group that was employed for the equilibrium study involved the fluorescence intensity, which is about an order of magnitude greater in organic solvents than in water. Using a stopped flow apparatus with fluorescence detection, the kinetics of complexation as well as that of dissociation was investigated. (5) They concluded that complex formation involved an initial diffusion controlled hydrogen bonding followed by an extensive conformational transition of the two polymer chains necessary to achieve the additional hydrogen bonding to stabilize the complex. Most recently, Turro and Arona investigated complex formation using PAA containing pyrene groups distributed randomly in the side chains. (8) They demonstrated that the intensity ratio of monomer to excimer emission could be used to monitor the extent of intermolecular association and polymer displacement reactions for terpolymer systems in aqueous solution.

The main focus of this study is to utilize excimer formation between pyrene groups attached to PEG chain ends as a molecular probe of intermolecular complex formation. The change upon complexation was monitored by UV-visible absorption, excitation, and fluorescence spectroscopies as well as by fluorescence lifetime measurements.

## Experimental

Three PAA samples of molecular weights 1850, 4600 and 890,000 were used. The first two were obtained from Polysciences and the last was synthesized by conventional radical polymerization. (9) Pyrene end-labeled poly(ethylene glycol) (PEG*) was synthesized by direct esterification between poly(ethylene glycol) and 1-pyrene butyric acid (PBA). (10) Monodisperse PEG (polydispersity 1.05) of weight-average molecular weight 4800 was purchased from Polysciences Inc. and suitably modified. Gel permeation chromatography confirmed that the polydispersity of the product was unchanged and that the high reaction temperature had not degraded the PEG. UV-visible absorption was measured to calculate the tagging percentages with methyl 1-pyrene butyrate as the model compound in THF. The product was determined to be fully labeled at both chain ends with ±5% accuracy. PEG* and PAA were dissolved separately in glass-distilled deionized water and adjusted to $1x10^{-3}$ M and $1x10^{-1}$ M per repeating unit, respectively. The composition of the complex was described by the molar ratio of the two repeating units, [PAA]/[PEG].

The UV-visible absorption spectra were measured with a Cary 210 spectrophotometer manufactured by Varian. The fluorescence spectra were taken with a spectrofluorometer that has been described previously. (11) The excitation spectra were measured by a Spex Fluorolog 212 spectrofluorometer. The monomer excitation spectrum was monitored at 376nm and the excimer excitation spectrum was at 500nm in the scanning excitation range between 300nm and 370nm. The fluorescence emission lifetimes were determined using a single photon counting apparatus from Photochemical Research Associates (PRA). Nitrogen gas was used for the flashlamp in order to have sufficient signal intensity.

## Results

Intramolecular and Intermolecular Excimer Formation of PEG*. The fluorescence spectra of $1x10^{-3}$M-PEG* aqueous solution at 303K showed emission from both the locally excited pyrene chromophore (monomer) and also from the excimer, as shown in Figure 1. The emission from a monomer entity observed between 370nm and 430nm was assumed to have the same envelope as 1-pyrene butyric acid, whose spectrum is shown by a dotted line in the figure. The broad structureless band centered at 480nm is due to the excimer.

We note at the outset that there are two types of excimer formation in the $1x10^{-3}$M-PEG* aqueous solution. The first type of excimer forming site (EFS) results intermolecularly by association between chromophores from two different polymer chains. The number of these sites is directly dependent on the local concentration of labeled chains. The second type of excimer site arises from association between aromatic rings on the same polymer chain. In the present case, this means that cyclization of the labeled PEG chain must occur.

To distinguish the two types of excimers clearly a solution of 1% of PEG* (both chain ends labeled by chromophores) and 99% of PEG (unlabeled) was also prepared, fixing the total polymer concentration to be $1x10^{-3}$M. Under this condition, only intramolecular excimer is formed because the excimer to monomer intensity ratio, $I_D/I_M$, was observed to be constant regardless of small variation of PEG* (labeled) concentration in the mixture with unlabeled PEG. We assumed that this allowed the behavior of an individual PEG chain to be examined as discussed in the following section. Furthermore, fully labeled PEG chains in $1x10^{-3}$M aqueous solution were considered to have two kinds of excimers, i.e. formed both intramolecularly and intermolecularly. As a result, the contribution of intermolecular excimer formation was assumed to be given by subtracting the $I_D/I_M$ data on the 1%PEG*-99%PEG system from those on the fully labeled PEG system. In order to do this treatment the $I_D/I_M$ values were carefully determined from the spectral area for a monomer and an excimer entity, respectively. The intermolecular excimer formation will be discussed in the third section.

Intramolecular Excimer Formation in the PEG*-PAA Complex. The fluorescence spectra of the (1%PEG* and 99%PEG) mixture were measured with addition of PAA aqueous solution. The total concentration of labeled and unlabeled PEG was fixed to be $1x10^{-3}$M per repeating unit. Figure 2 shows the results for (1%PEG* + 99%PEG). Note that even though the PAA-PEG complex is known to form a stoichiometric complex, (1,2) there is no discontinuity at the stoichiometric ratio; the $I_D/I_M$ continues to drop even in the presence of excess PAA. However, acetic acid does not cause any change over the whole composition range.

Furthermore, fluorescence lifetimes for monomer and excimer emission were investigated with complexation, as shown in Figure 3. The observed monomer emission decay curves could be fitted to a two exponential function over the whole composition range. The lifetimes of monomer decay slightly increased with addition of PAA(890K). The excimer emission decay also showed a two exponential decay curve except for pure PEG* aqueous solution, which could be analyzed by a single exponential fitting. The excimer emission decay did not show a rising time within the resolution of our PRA nanosecond fluorometer for all compositions. The increase in the lifetime upon complexation was more pronounced in the excimer emission than in the monomer emission.

Intermolecular Excimer Formation in the PEG*-PAA Complex. Next, the same experiments were repeated using the aqueous solution of fully tagged PEG materials (PEG*). Figure 4 shows the change in $I_D/I_M$ for PEG* by addition of the PAA proton donor. The initial $I_D/I_M$ value is higher than that in Figure 2 due to intermolecular excimer formation, the contribution for which was calculated to be 9.6% of the total $I_D/I_M$ value. Although acetic acid is supposed to be another type of proton donor, it had absolutely no effect over the whole region, the same as that observed in Figure 2. Upon increase of the molecular weight of PAA it appears that complex formation is facilitated.

In order to obtain the net contribution of the intermolecular excimer, the smoothed $I_D/I_M$ data shown in Figure 2 were subtracted from smoothed data in Figure 4. The results are given in Figure 5. The results show that the higher molecular weight of PAA causes stronger intermolecular aggregation and reaches the final value of $I_D/I_M$ earlier. At the higher [PAA]/[PEG], the molecular weight dependence of PAA was almost negligible.

Effects of Complexation on the Absorption and Excitation Spectra. The UV-visible absorption spectrum of PEG* has peaks between 250nm and 400nm, as shown in Figure 6. The peaks around 380nm, between 300nm and 360nm, and between 250nm and 300nm correspond to the absorptions of the $^1L_b$ band, the $^1L_a$ band, and $^1B_b$ band from $^1A$ of the pyrene ring, respectively. They are a result of excitation to the first, the second, and the third singlet excited states. The $^1L_b$ absorption was too weak to investigate the effect of complexation. However, the $^1L_a$ bands at 327.8nm and 343.8nm were observed to red-shift upon complexation by 2.3nm and 1.6nm, respectively, as shown in Figure 7. This corresponds to a decrease of energy by 193cm$^{-1}$ and 148cm$^{-1}$. The absorption bands attained constant values at [PAA]/[PEG]=2. Moreover, the excitation spectra also showed similar red-shifts of about 2.5nm as complex formation occurred.

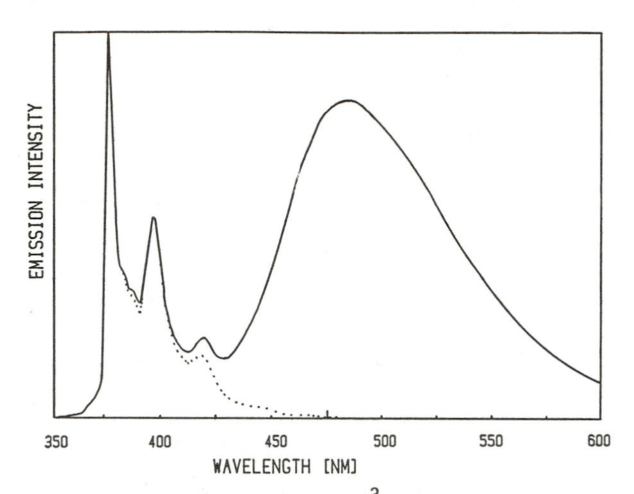


Figure 1 Emission spectrum of $1 \times 10^{-3}$M-PEG* aqueous solution at 303K. The spectrum shown by a dotted line in the figure is for 1-pyrenebutyric acid. (Reproduced from Ref. 16. Copyright 1987 American Chemical Society.)

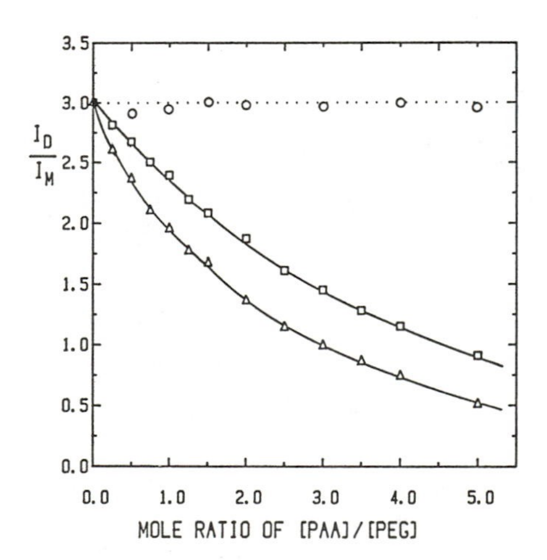


Figure 2 Change in the intramolecular excimer to monomer intensity ratio with the addition of PAA at 303K.
1) △ : PAA(890K), 2) □ : PAA(1850), 3) ○ : acetic acid
(Reproduced from Ref. 16. Copyright 1987 American Chemical Society.)

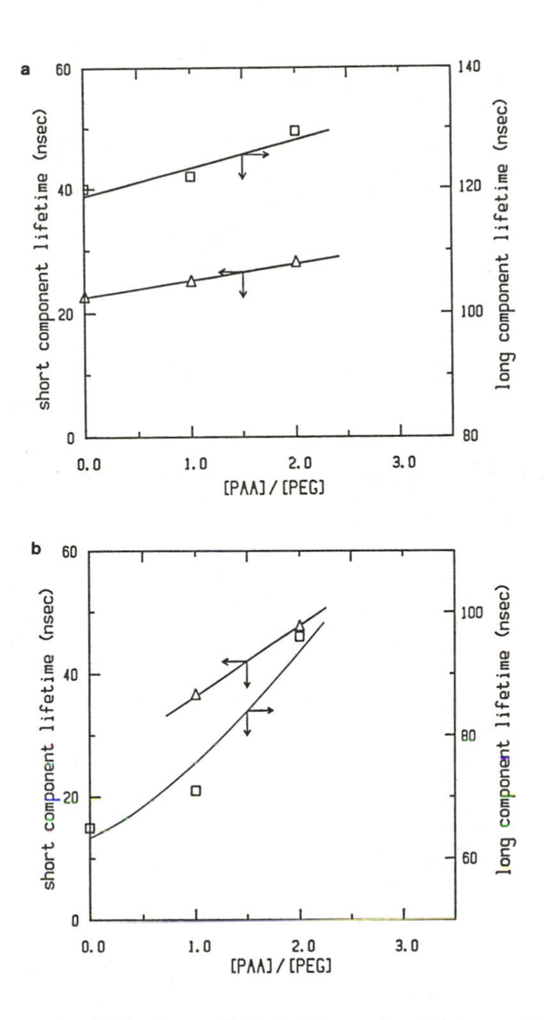


Figure 3 Change in lifetime of monomer and excimer emissions upon complexation. a) monomer emission decay, b) excimer emission decay (Reproduced from Ref. 17. Copyright 1987 American Chemical Society.)

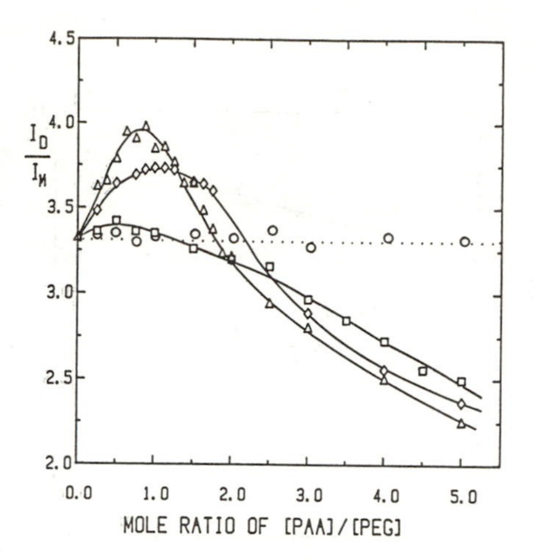


Figure 4 Change in the ratio of excimer to monomer intensity of fully labeled PEG* with the addition of PAA at 303K.
1) △ : PAA(890K), 2) ◇ : PAA(4600), 3) □ :PAA(1850),
4) ○ : acetic acid
(Reproduced from Ref. 16. Copyright 1987 American Chemical Society.)

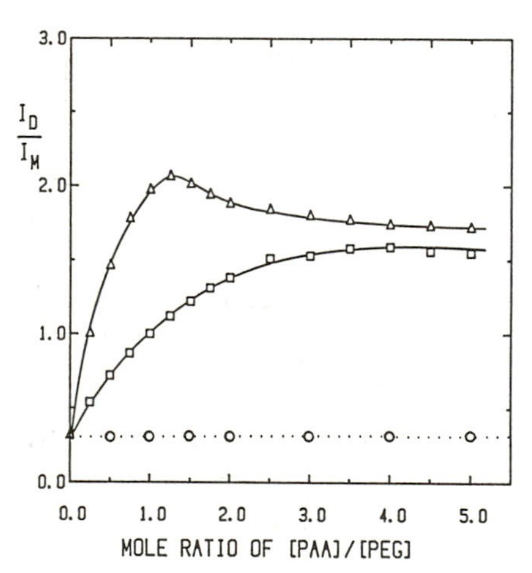


Figure 5 Change in the intermolecular excimer to monomer intensity ratio with the addition of PAA at 303K. It was calculated by the subtraction of data in Figure 2 from those in Figure 4.
1) △ : PAA(890K), 2) □ : PAA(1850), 3) ○ : acetic acid
(Reproduced from Ref. 16. Copyright 1987 American Chemical Society.)

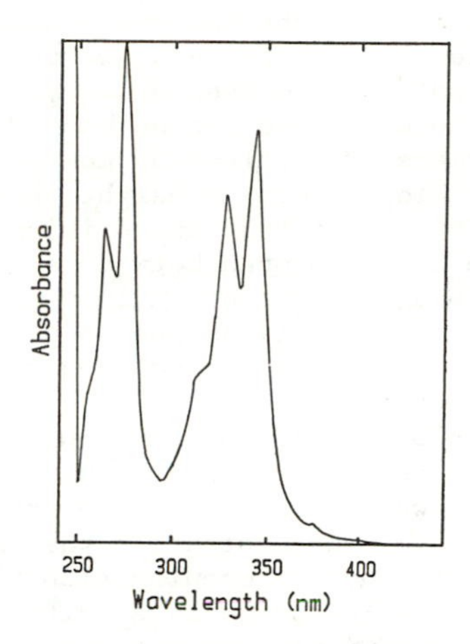


Figure 6 Absorption spectrum of PEG* in aqueous solution. (Reproduced from Ref. 17. Copyright 1987 American Chemical Society.)

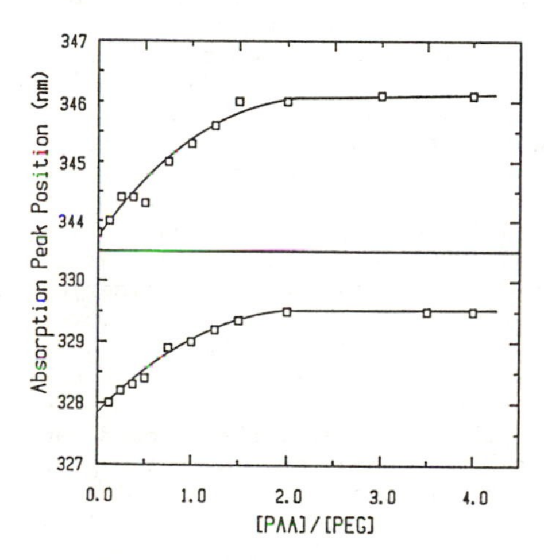


Figure 7 Absorption peak shift of PEG* upon the addition of PAA(890K)

## Discussion

From these observations about the intermolecular and intramolecular excimers formed between PEG* chain ends, we infer several important effects of complexation on the component polymers.

First, hydrogen bond interaction between PAA and PEG was observed for all polymers, regardless of molecular weight. Only the proton donor analog of low molecular weight, acetic acid, did not interact with PEG. This means that the driving force towards the formation of a single hydrogen bond between an ether and a monocarboxylic acid in aqueous solution is very small and that a stable complex can be formed only by the cooperative interaction of many such groups. Antipina et al. concluded that in order for a complex to be formed between PAA and PEG the molecular weight of the PEG had to be around 6000. (12) In addition, Ikawa et al. also reported that they did not observe any change in viscometric data when the PEG molecular weight was less than 8800 for the PAA-PEG complex, when their polymer concentration was $3x10^{-3}$M at 298K. (13) However, in the present study on complexation the pyrene excimer fluorescence technique was shown to be sensitive enough to observe the interaction between PEG* with weight-average molecular weight of 4800 and PAA with a wide range of molecular weight between 1850 and 890K.

Second, the longer chain of PAA showed a stronger interaction with PEG. The present observation of the molecular weight dependence of PAA on complexation is similar to the results of the molecular weight dependence of PEG, where Antipina observed that longer chains of PEG reduced the viscosity a greater amount for molecular weights between 6000 and 40,000 because of complexation. (12)

Third, the addition of PAA induced a decrease of intramolecular excimer formation for PEG*. Even though the PAA-PEG complex is found to have an equal mole of each repeating unit by past studies, (1,2) $I_D/I_M$ for the isolated chain decreases continuously with an increase of the [PAA]/[PEG] molar composition ratio. In addition, the transient measurement for monomer and excimer emission showed an increase of two lifetime components, which was more pronounced in the excimer emission decay. Turro and Arona also reported lifetime changes upon complexation for systems in which a pyrene group was randomly tagged to a side group of PAA. (8) They found that the monomer and excimer emission decay curves showed contributions from short-lived and long-lived components, even though the decay curves could not be satisfactorily fitted to an equation containing two exponential terms. They observed that the long-lived lifetime for excimer emission increased from 58nsec to 83nsec at a (1/1) stoichiometric ratio. In Figure 3, we also observed an increase in lifetimes upon complexation, in which the excimer lifetime showed a more pronounced increase than the monomer lifetime, as they observed. The suppression of intramolecular excimer formation and the increase in lifetime are probably due to a decrease in intramolecular mobility of the PEG chain upon complex formation.

Fourth, the addition of PAA induces an increase in the local concentration of PEG in the vicinity of the PAA chain. This was observed as an initial increase of intermolecular excimer formation, as shown in Figure 5. The hydrogen bond interaction between a carboxyl group of PAA and an ether group of PEG causes the aggregation, which results in facilitating the pyrene excimer formation between two different PEG* chains. The aggregate formed through hydrogen bonds is generally proposed to have two possible forms depending on component polymer structures and preparation conditions. The first is the so-called "zip-up" structure or "ladder" structure for which the mechanism of linear consecutive junction leads to the complex formation. The other is the "scrambled-salt" structure in which random crosslinking takes place, sometimes yielding a gel structure. (1,2) To distinguish these types, it is necessary to perform a viscosity measurement. Unfortunately, the molecular weights and polymer concentrations of our samples were too low to permit it. Another interesting feature on the complex structure is that the data in Figure 5 suggest that the composition of the complex has a stoichiometry of [PAA]/[PEG] between 2 and 3 under our experimental condition. In this case, the complex system does not have a perfect match of the two complementary polymers, presumably involving a considerable number of unbound functional groups.

Fifth, the pyrene excimer of PEG* in aqueous solution is formed by a different mechanism compared to the one appropriate for organic solvents. Winnik and his collaborators have demonstrated that the $I_D/I_M$ is inversely proportional to solvent viscosity in non-protonic solvents, as expected for a diffusion-controlled process. (14) However, in water and methanol the extent of excimer emission and the rate of intramolecular excimer formation were substantially greater than one would infer on the basis of solvent viscosity alone. They suggested that the hydrophobic chromophore is undoubtedly better solvated by the environment within the polymer coil than if the pyrene were surrounded completely by water or methanol. This situation would be expected to lead to an increased $I_D/I_M$ due to a reduced distance of separation between pyrenes.

In the present study, it was observed that the excimer decay of intramolecular excimer did not show a rising time. This means that the observed excimer formation results from the arrangement of pyrene groups which are already pre-formed in the ground state. Under such a condition the excimer is essentially formed instantaneously upon excitation without the necessity of molecular diffusion to form the coplanar sandwich structure. Considering the observation that the majority of excimer in pure PEG* aqueous solution is formed intramolecularly, PEG* chains must already exist in a cyclyzed structure before excitation in aqueous solution. This implies that the hydrophobic interaction between chromophores becomes significant in aqueous solution, as Winnik has also pointed out. A more detailed study on the hydrophobic interaction in water is reported separately. (15)

An additional piece of evidence regarding the mechanism is the red-shift in the absorption and excitation spectra upon complexation. Complexation facilitates the intermolecular excimer formation because PEG* chains are attracted to the vicinity of PAA and the local pyrene concentration increases. Thus the observed red-shift in the absorption and excitation spectra upon complexation seems to result from the pyrene-pyrene interaction which is pre-formed intermolecularly as a result of the specific structure of complex. Here also the hydrophobic attraction between chromophores must play an important role to cause such a ground state interaction.

Finally, another significant point is that the excimer was formed between pyrenes at the PEG* chain ends intermolecularly even though the system had a larger molecular weight of PEG than that of PAA, as observed in both PEG*-PAA(1850) systems. This means that in the PEG-PAA complex the hydrogen bond works very effectively to allow two or more long PEG chains to become close enough to form an excimer around a short PAA chain.

## Summary

Pyrene excimer fluorescence was used as a sensitive proximity probe in the intermolecular complex system. Upon addition of PAA solution, the intramolecular mobility of the PEG chain was suppressed resulting in decreased intramolecular excimer formation. Simultaneously, the local concentration of PEG is increased in the vicinity of PAA. The excimer formation seems to result from the arrangement of pyrene groups which is already pre-formed in the ground state. The effect of complexation is observed to be more pronounced in the PEG-PAA with a higher molecular weight of PAA. A more complete account of this work will appear elsewhere. (16,17)

## Acknowledgments

This work was supported by the Polymers Program of the National Science Foundation through DMR 84-07847 and in part by the Army Research Office through DAAG 29-82-K-0019.

## Literature Cited

1. Bekturov, E. A.; Bimendina, L. A. Advances in Polymer Science 41; Springer-Verlag: Berlin Heidelberg New York, 1981; p.99.
2. Tsuchida, E.; Abe, K. Advances in Polymer Science 45; Springer-Verlag: Berlin Heidelberg New York, 1982.
3. Anufrieva, E. V.; Pautov, V. D.; Geller, N. M.; Krakoviak, M. G. ; Papisov, I. M. Dokl. Akad. Nauk SSSR 1975, 220, 353.
4. Bednář, B.; Li, Z.; Huang, Y.; Chang, L. C.-P.; Morawetz, H. Macromolecules 1985, 18, 1829.
5. Chen, H. L.; Morawetz, H. Eur. Polym. J. 1983, 19, 923.
6. Chen, H. L.; Morawetz, H. Macromolecules 1982, 15, 1445.
7. Bednář, B.; Morawetz, H.; Shafer, J. A. Macromolecules 1984, 17, 1634.
8. Turro, N. J.; Arona, K. S. Polymer 1986, 27, 783.
9. Oyama, H. T.; Nakajima, T. J. Polym. Sci., Polym. Chem. Ed. 1983, 21, 2987.
10. Cuniberti, C.; Perico, A. Eur. Polym. J. 1977, 13, 369.
11. Oyama, H. T.; Frank, C. W. J. Polym. Sci., Polym. Phys. Ed. 1986, 24, 1813.
12. Antipina, A. D.; Baranovskii, V. Yu.; Papisov, I. M.; Kabanov, V. A. Vysokomol soyed. 1972, A14, 941. (translated in Polymer Sci. U.S.S.R. 1972, 14, 1047.)
13. Ikawa, T.; Abe, K.; Honda, K.; Tsuchida, E. J. Polym. Sci., Polym. Chem. Ed. 1975, 13, 1505.

14. Cheung, S.; Winnik, M. A.; Redpath, A. E. C. Makromol. Chem. 1982, 183, 1815.
15. Char, K.; Frank, C. W.; Gast, A. P.; Tang, W. T. Macromolecules in press.
16. Oyama, H. T.; Tang, W. T.; Frank, C. W. Macromolecules 1987, 20, 474.
17. Oyama, H. T.; Tang, W. T.; Frank, C. W. Macromolecules 1987, 20, 1839.

RECEIVED April 29, 1987

# Chapter 32

# Interaction of Cationic Species with Polyelectrolytes

**Deh-Ying Chu and J. K. Thomas**

**Department of Chemistry, University of Notre Dame, Notre Dame, IN 46556**

The interactions of various cationic species with polyacids, such as poly(methacrylic acid), PMA and poly(acrylic acid), PAA have been studied. In particular, the effect of polyacid conformation on the interaction is discussed in detail, and also the nature of the aggregation of PMA and cationic surfactants alkyltrimethylammonium bromide, $C_nTAB$. The effect of the intermediate conformation states of PMA around pH 4-6 is noted, where the photophysical properties of cationic probes bound to PMA dramatically change, effects such as a large enhancement of the fluorescence intensity of Auramine O, Au O at pH 4.5, a blue shift of the luminescence spectra of tris(2,2'-bipyridine)ruthenium(II) complex, $Ru(bpy)_3^{2+}$ at pH 5, and a great increase of the excimer yield of 1-pyrenebutyltrimethyl ammonium bromide, $C_4PN^+$ at pH 6.

Water soluble synthetic polyelectrolytes have attracted increasing attention in recent years, mainly because of their wide utility in industrial applications, and also because of their resemblance to biopolymers. Poly(methacrylic acid), PMA, a weak polyelectrolyte, exhibits a marked pH induced conformational transition. A wide variety of techniques have been employed to gain more information on the nature of the conformational transition of PMA, these techniques include: viscometry, potential titrimetry,(1-5) Raman spectrometry,(6) calorimetry,(7-9) electrical conductometry,(10) dilatometry,(11) $^1H$ NMR linewidth,(12) viscoelastic studies,(13) kinetics of chemical reactions,(14) small-angle neutron scattering,(15) pH jump,(16,17) and fluorescent probing.(18-27)

The data tend to support a two state model, i.e. at low pH, a compact globular conformation (A states) and at high pH, an extended rod-like form (B states) for the conformation transition of PMA. The conformational transition is considered to be highly cooperative and occurs in one step. Nevertheless, the

0097-6156/87/0358-0434$06.00/0

conformational transition as observed in Raman spectroscopy which indicates a multiplicity of structures,(6) exhibits progressive states rather than a cooperative change. Different arrangements of the lumophore pyrene in PMA indicate different degrees of opening of the polymer compact coil as the pH of an aqueous solution increases.(28) The covalently bound pyrene indicates a later transition than that of guest molecules included by simple solubilization. Details of the uncoiling process are, therefore, still open to question and other measurements are desirable.

In order to extend earlier work, several positively charged luminescent probes, tris(2,2'-bipyridine) ruthenium(II) complex, $Ru(bpy)_3^{2+}$, Auramine O, Au O, 1-pyrenebutyltrimethyl ammonium bromide, $C_4PN^+$ and 1-pyreneundecyltrimethylammonium iodide, $C_{11}PN^+$ have been employed to monitor the nature of the conformational transition of PMA over the pH range of 2 to 8, from A states into B states. The purpose of the present work is, to investigate the intermediate states in the uncoiling process of PMA polymer coil with increasing pH and to study any effects that are induced by polyelectrolytes on the photophysics and photochemistry of cationic probes. $Ru(bpy)_3^{2+}$, which has a large bipyridine ligand structure, Au O which is non-fluorescent in water, $C_4PN^+$ and $C_{11}PN^+$, which form excimers, all report on various features of their environment.

Polymer-surfactant systems have also been the subject of many recent studies. However, early studies have mainly focussed on the interaction between nonionic polymers and anionic surfactants, sodium dodecyl sulfate, SDS.(29-36) A great variety of experimental data via different techniques exists, but the nature of the surfactant-polymer association in these systems is relatively weak. For instance, the studies have shown that in the Poly(ethylene oxide), PEO - SDS system, there is no interaction between SDS and PEO for SDS concentrations below the CMC critical micelle concentration, of SDS. There are few reports concerning the interaction of polyelectrolytes with surfactants of opposite charge,(37-40) especially the interaction of weak polyelectrolytes with cationic surfactants.(41) However, there is no report on the effect of chainlength of a cationic surfactant on the state of aggregation and no detailed studies on the nature of aggregates. Therefore, the effect of cationic surfactants on the conformational transition of PMA has been investigated in the present study. The interaction between cationic surfactants with PMA permits a study of the effect of charge density and conformation of polyelectrolyte on the aggregation process, and also the effect of chainlength cationic surfactants on the conformational transition of PMA.

## EXPERIMENTAL

Poly(methacrylic acid), PMA, and poly(acrylic acid), PAA, used in this study were obtained from Polyscience and Aldrich Chemicals Inc., respectively. The molecular weight of PMA measured by standard viscosity methods was $1.1 \times 10^4$. The molecular weight of PAA was given as $2.5 \times 10^5$ by the Aldrich Chemical Inc. PMA samples of different molecular weight ($3.9 \times 10^4$, $1.6 \times 10^5$ and $6.4 \times 10^5$) used for studying effects of molecular weight, were synthesized by using different amounts of AIBN and monomer. Free radical polymerization were carried out as reported earlier.(42) The concentrations of polymer solutions are expressed as weight/volume ratio, i.e. grams per liter. Unless stated to the contrary, all polymer samples are used in 1 g/L.

Cationic surfactants, $C_nTAB$, such as decyltrimethylammonium bromide, $C_{10}TAB$ (Kodak), dodecyltrimethylammonium bromide, $C_{12}TAB$ (Kodak) and cetyltrimethylammonium bromide, $C_{16}TAB$ (Sigma) were purchased as indicated and then purified by recrystallization from ethanol. However, hexyltrimethylammonium bromide ($C_6TAB$) and octyltrimethylammonium bromide ($C_8TAB$) were synthesized by refluxing either 1-bromohexane or 1-bromooctane (Aldrich) with trimethylamine methanol solution and finally recrystallized twice from benzene.(5) An anionic surfactant, sodium dodecyl sulfate, SDS (BDH) was used as received.

The cationic probes used in this study are displayed in Figure 1. 1-Pyrenebutyltrimethyl ammonium bromide ($C_4PN^+$), 1-pyrene-undecyl trimethyl ammonium iodide ($C_{11}PN^+$), were used as received from Molecular Probes. Tris(2,2'-bipyridine)ruthenium(II) chloride, $Ru(bpy)_3^{2+}$ (G. Fredrick Smith) was purified by double recrystallization from the deionized water. Auramine O, Au O (Aldrich) and Pyrene (Kodak) were purified by triple and double recrystallization from ethanol, respectively.

The pH of the sample was adjusted with concentrated HCl or NaOH aqueous solution and measured with a Sargent-Welch combination electrode at room temperature, 20° C, using a Model LS pH meter. Before taking measurements, the pH meter was calibrated with standard buffer solutions of pH 4, pH 7 and pH 10.

Steady state absorption spectra and emission spectra were recorded on a Perkin-Elmer 552 UV-Vis and MPF-44B fluorescence spectrophotometer respectively. The ratio of $I_e/I_m$ is the ratio of the intensity of excimer ($\lambda$ 480 nm) to monomer fluorescence ($\lambda$ 377 nm). The ratio of $I_3/I_1$ is the ratio of the intensity of the pyrene monomer fluorescence intensity of peak 3 ($\lambda$ 384 nm) to peak 1 ($\lambda$ 373 nm).

Fluorescence decay curves were determined by a PRA LN-1000 nitrogen laser system with a response of less than $10^{-9}$ seconds.(42) The monomer fluorescence of pyrene and pyrene derivatives was monitored at 400 nm, and that of the excimer fluorescence at 480 nm.

## RESULTS AND DISCUSSION

**Studies with Au O and $Ru(bpy)_3^{2+}$.** Studies of binding of cationic dyes to polyelectrolytes has attracted interest for some time. Auramine O, Au O, a cationic diphenylmethane dye is non-fluroescent in water, but fluoresces strongly in rigid media or in the bound states of a compact PMA coil.(18-23) Other cationic dyes such as Crystal Violet (CV)(24-25), Acridine Orange (AO)[26] and Rhodamine B (RB)(27) are also found to bind strongly to PMA. These dyes are reported to bind to both the open and coiled states of the polymer, and it is concluded that binding is stronger in the A states than in the B states. However, the present photophysical studies on the binding of Au O to PMA show some additional features compared to other previous studies, i.e., the dye binding is stronger in the intermediate states rather than in the A states.

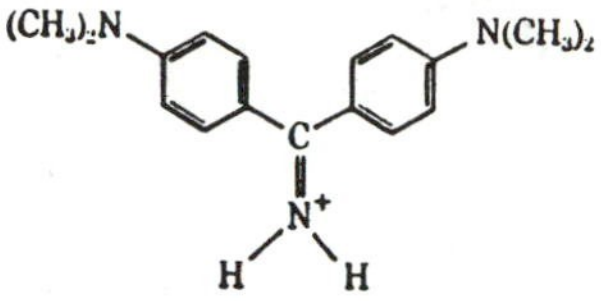

Auramine O ( Au O)

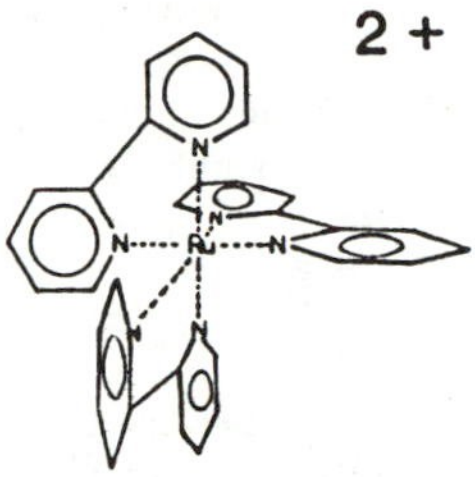

$Ru(bpy)_3^{2+}$

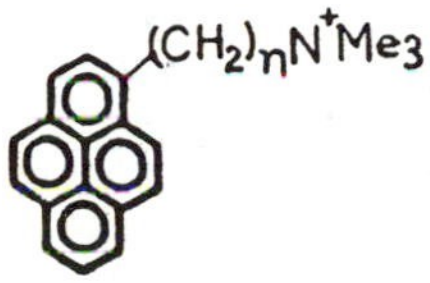

n=4 $C_4PN^+$

n=11 $C_{11}PN^+$

1. The cationic fluorescent probes used in this study.

The maximum wavelength of absorption of Au O in water is not dependent on pH. However, in aqueous solutions of PMA, the spectrum is significantly dependent on pH (shown in Figure 2A). At pH 4-5, the spectra move to longer wavelengths (the $\lambda_{max}$ of two peaks are 375 nm and 442 nm at pH 4.5), while the spectra in the solutions at pH 2 and 8 are identical to that in water, where the $\lambda_{max}$ of two peaks are 368 nm and 430 nm. Absorption spectra of Au O in glycerol, in water and in SDS are displayed in Figure 2B, for the sake of comparison. Again a bathochromic shift of maximum wavelength is observed in glycerol, $\lambda_{max}$ of 372 nm and 438 nm, while in SDS the $\lambda_{max}$ are 370 nm and 437 nm. The data indicate a special interaction between Au O and PMA polymer coils (pH 4-5) which is similar to the restriction placed by glycerol on mobility of Au O. This restriction is stronger at pH 4-5 than at other pH and also stronger than that in SDS. The restriction increases the conjugation of molecular electrons, i.e., increases the coplanarity of Au O as required for the most effective overlap of the arene $\pi$ orbitals and non-bounding electrons of nitrogen. It can be seen that in the Au O - PMA system at pH 8, electrostatic binding does not cause a bathochromic shift, while partially ionised PMA (pH 4-5), shifts the absorption spectrum about 10 nm toward the longer wavelength region. The polymer conformation in aqueous solutions at pH 4-5 effectively immobilizes Au O leading to the above spectral effects.

A wavelength of 388 nm was chosen for excitation in the studies of Au O fluorescence, as the absorbance is invariant or nearly invariant over the pH 2-8. The relative fluorescence intensity of Au O in aqueous solutions as a function of pH is shown in Figure 3A. The relative fluorescence intensity exhibits a marked enhancement on increasing the pH from 2 to 4.5, followed by a marked decrease, almost to zero, on increasing the pH above 5; it can be noted that the relative intensity of Au O fluorescence reaches a maximum at pH 4.5. The fluorescence intensity at pH 4.5 is four hundred times greater than that in water, one hundred times larger than that in SDS anionic surfactants solution ($5 \times 10^{-2}$ M) and in poly(acrylic acid) at pH 3-4, and even larger than that in glycerol. The steady state fluorescence data correspond well to the data obtained via absorption spectra studies.

A simple explanation of the observed data is as follows: anionic sites are formed on the PMA as the compact PMA coil opens with increasing pH,leading to binding of Au O to these sites. On binding to these sites, the Au O causes a restriction which tends to pull the polymer chain around the probe molecule, thus decreasing the internal rotation or other motion of bonds of the probe molecule, leading to an increase in the fluorescence yield. Structural rigidity enhances the fluorescence by inhibiting radiationless processes that compete with fluorescence, and by preventing a large Frank-Condon geometric difference between the excited singlet state $S_1$ and the ground state $S_0$. The above data indicate that the environment of Au O in PMA at pH 4.5 is much more rigid than at other pH, on SDS micellar surfaces, and even in glycerol. At higher pH, Au O is completely bound to an extended highly negatively charged polymer, and close to the aqueous

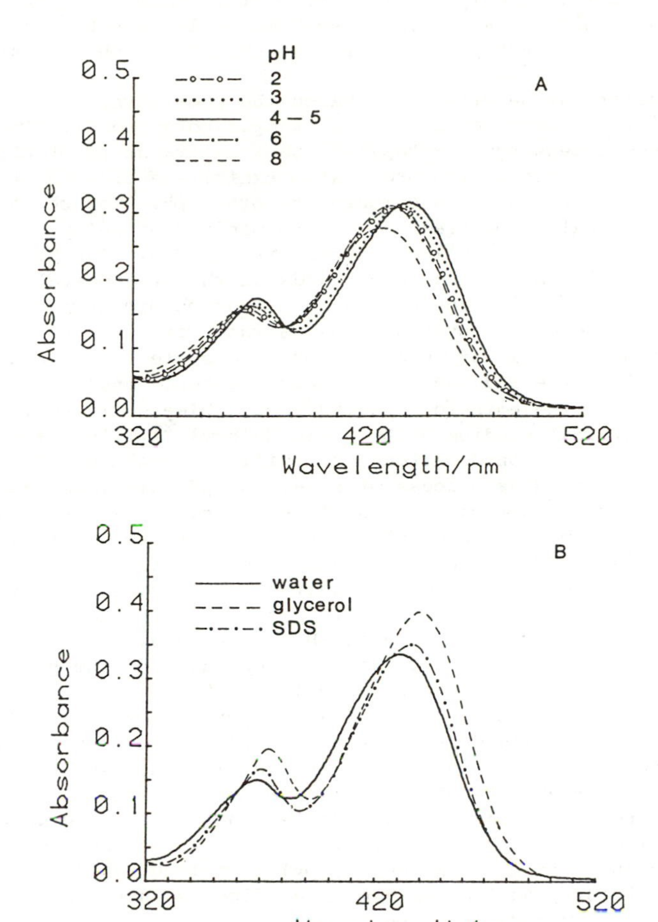


2. Absorption spectra of $2 \times 10^{-5}$ M Auramine O.
A. in aqueous solutions of PMA at various pH.
B. in water, in SDS and in glycerol.

environment, the environment of Au O is non-rigid and the polymer has little effect, the $I_r$ is close to that in water. At pH 2, no electrostatic binding exists, but Au O molecules may be partially solubilized in compact polymer coil, so that $I_r$ at pH 2 is higher than in water.

A similar unique effect of PMA on the photophysics of $Ru(bpy)_3^{2+}$ is observed at pH 5, for example both the lifetime and luminescence intensity of $Ru(bpy)_3^{2+}$ show maxima at pH of about 5. The luminescence of the probe also exhibits a blue spectral shift at this particular pH compared to other pH. The change in the photophysical properties are due to binding of $Ru(bpy)_3^{2+}$ into a partially coiled or swollen polymer PMA at pH 5. The binding is electrostatic in nature and the ligands of the organometallic complex probe are quite restricted in a hydrophobic environment, so that unlike more mobile systems such as water or a stretched polymer, complete relaxation of the excited state is not achieved. Hence, the lifetime and the yield of luminescence increase accordingly and the emission spectra show a blue shift.(42)

Photophysical studies of Au O and $Ru(bpy)_3^{2+}$ illustrate common features, i.e. the conformational transition induced by pH is a progressive continuous process over several pH units, and at pH 4-5, the compact polymer coil is partly swollen. This particular pH region allows some cationic species to bind into the swollen polymer coils of PMA. Binding is electrostatic in nature, however the mobility of cationic probe is quite restricted in a hydrophobic environment. With the probes used, this effect increases the $I_r$ of the probe which shows a maximum at pH 4.5 or pH 5. The pH at which photophysical effects are maximized is dependent on the probe properties such as size, water solubilities, etc. Au O possesses a smaller size than $Ru(bpy)_3^{2+}$, at least in one dimension, and it may be partially solubilized or intercalated into a compact polymer coil at low pH, as well as binding into a swollen polymer coil at pH 4.5. The larger size of $Ru(bpy)_3^{2+}$ may deter its solubilization in compact polymer coils, without preventing its binding into a swollen polymer at higher pH, i.e., pH 5, where the polymer is further expanded to contain the $Ru(bpy)_3^{2+}$ molecule. In accordance with this picture, Au O shows a maximum fluorescence yield in PMA at pH 4.5, and $Ru(bpy)_3^{2+}$ at pH 5.

The effect of the intermediate polymer states on the photophysics of $C_4PN^+$, a more hydrophobic cationic probe than Au O and $Ru(bpy)_3^{2+}$, is not as marked, as this probe is solubilized in polymer A states and in intermediate states. A comparison can be seen in Figure 3A, which shows the variation of $I_r$ vs. pH for 2 x $10^{-6}$ M $C_4PN^+$ in PMA. The data are similar to the case of pyrene, which shows little variation in photophysical properties on increasing pH from 2 to 4, quite unlike Au O and $Ru(bpy)_3^{2+}$.

Similar studies in aqueous solutions of poly(acrylic acid), PAA, i.e., variation of $I_r$ vs. pH are shown in Figure 3B. A smaller increase in $I_r$ with Au O fluorescence was observed around pH 3-4 than in the case of PMA, however, it is still 6-7 times larger than that in water. The data again indicate that due to the lack of methyl side groups i.e. lack of hydrophobic interaction PAA forms much looser coils at low pH than PMA, with smaller effects of the conformational transition on the photophysics of Au O.

Studies with $C_4PN^+$ and $C_{11}PN^+$. Intramolecular pyrene excimer formation was used to study the cyclization dynamics of polymers, such as pyrene end capped poly(ethylene oxide),(43) polystyrenes,(44) and other polymers.(45) However, there is no report on the effect of any conformational transition of polyelectrolytes induced by pH on the excimer formation. In this study, a cationic pyrene derivatives, the probe $C_4PN^+$ exhibits a maximum in excimer formation with pH, a feature not available with other cationic probes, while the long chain cationic probe, $C_{11}PN^+$, fails to exhibit such a pH effect on excimer formation.

The fluorescence spectra of 6 x $10^{-5}$ M $C_4PN^+$ in aqueous solutions of PMA at pH 4, 5, 6 and 7 are displayed in Figure 4. It can be seen that excimer formation is unfavorable in aqueous solutions of PMA at pH 4 and 5, where a large monomer fluorescence spectrum with fine structure ($\lambda_{max}$ 377 nm) is observed. At pH 6, a greatly enhanced excimer fluorescence spectrum is observed at longer wavelengths ($\lambda_{max}$ 480 nm), which is structureless and broad, while the monomer fluorescence and its fine structure are dramatically reduced. On increasing pH another unit, from 6 to 7, a greatly reduced excimer fluorescence is observed with a small increase in monomer fluorescence.

Little excimer formation is observed in aqueous solutions without added PMA, even up to concentrations of $C_4PN^+$ of 4 x $10^{-4}$ M. However, in aqueous solutions of PMA at pH 6, the excimer yield of 4 x $10^{-6}$ M $C_4PN^+$ is already significant, although at other pH the excimer is still not observed. The data indicate that the conformation of PMA plays an important role in excimer formation of $C_4PN^+$. Figure 5A, represents the effect of pH on the ratio of the relative intensity of excimer to monomer fluorescence, $I_e/I_m$ in aqueous solutions of PMA with various concentrations of $C_4PN^+$. The data clearly demonstrate that the maximum excimer formation occurs at pH 6, while in solutions at pH smaller than 5 or larger than 7, that the excimer yield is much lower than at pH 6. Earlier studies have shown that the uncoiling process of compact PMA coils is a continuous process over several pH units.(28,42) In solutions of pH 4-5, although the PMA coils are already swollen due to partial ionization, free pyrene cannot be solubilized by the polymer, while a cationic probe is bound into polymer coil. However, the rigid environment hinders any readjustment of the pyrene lumophores, a situation that is unfavorable for excimer formation. On increasing pH from 5 to 6, swollen PMA coils are looser and further expanded, and where more than one cationic probe is bound to a polymer coil, excimer formation is possible via readjustment of the probes in the loose polymer coil. However, in the case of polymer solutions of B states, short chain pyrene lumophores are not bound in close proximity and excimer formation is not favored. Higher concentrations of $C_4PN^+$ lead to enhanced $I_e/I_m$ as the probes are concentrated on the polymer chain.

Experimental studies using a long chain cationic pyrene derivative, $C_{11}PN^+$ are given in Figure 5B. $C_{11}PN^+$ forms excimer at low concentration, 2 x $10^{-6}$ M, in aqueous solutions of PMA at pH $\geqslant$ 7, nevertheless, a bell shaped curve with a maximum at pH 6 is not shown. This can be explained if the long carbon chains of the

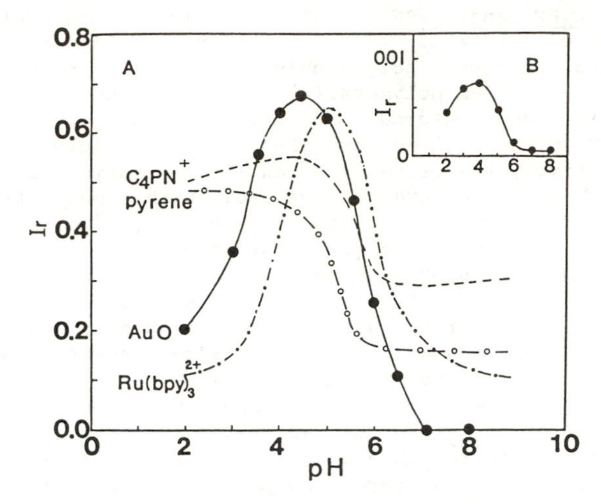


3. A. Relative fluorescence intensity of cationic probes: 2 x $10^{-5}$ M Au O ($\lambda_{excitation}$ 388 nm), 2 x $10^{-5}$ M $Ru(bpy)_3^{2+}$ ($\lambda_{excitation}$ 450 nm), 2 x $10^{-5}$ M $C_4PN^+$ ($\lambda_{excitation}$ 340 nm) and 2 x $10^{-6}$ M pyrene fluorescence ($\lambda_{excitation}$ 340 nm) as a function of pH in aqueous solutions of PMA.

B. Relative fluorescence intensity of 2 x $10^{-5}$ M Au O ($\lambda_{excitation}$ 388 nm) in PAA.

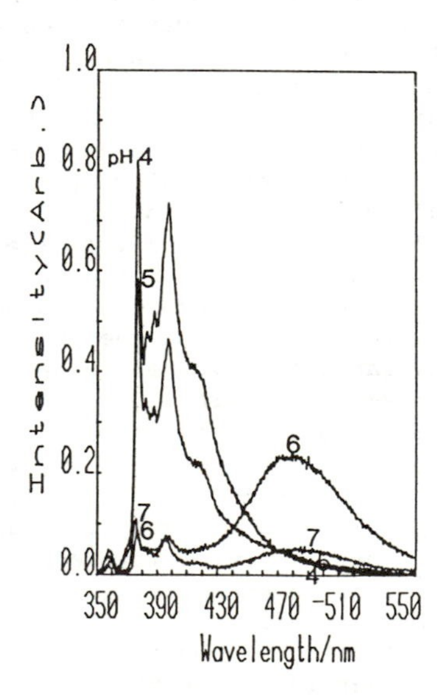


4. Fluorescence spectra of $C_4PN^+$ in aqueous solutions of PMA at pH 4, 5, 6 and 7 ($\lambda_{excitation}$ 356) nm.

$C_{11}PN^+$ cause a local clustering of the probes thus enhancing static excimer formation.

If excimers are formed by migration together of two pyrene molecules, then excimer emission intensity will increase with time, as exhibited by pyrene in micelles.(46) However, if pyrene is stacked or clustered together on the assembly then the excimer emission at 480 nm will be observed immediately after the laser pulse.(47) Figure 6A illustrates an immediate formation and accelerated decay of excimer in 1 x $10^{-4}$ M $C_4PN^+$ in aqueous solution of PMA at pH 6, showing that $C_4PN^+$ is stacked or clustered in PMA. The first order decay curves of, ln(Intensity of excimer $C_4PN^+$ fluorescence) vs. time in aqueous solutions of PMA at pH 6, 7 and 8 are given in Figure 6B. The lifetime of the excimer is larger at pH 6 than at pH 7 or pH 8. The longer lifetime of the excimer at pH 6 confirms that the environment of the probe at pH 6 is less polar than at pH 7-8 and that at pH 6 PMA is not fully open or extended but still exists as a loose coil.

Conformational Transition of PMA Induced by $C_nTAB$. Aqueous solutions of PMA at pH 8, in the absence and in the presence of $C_{10}TAB$ were examined by transmission electron microscopy. No particles were observed in simple aqueous solutions of PMA at pH 8 either via the positively charged or the negatively charged stains (uranyl acetate or phospho-tungstic acid). However, in the presence of $C_{10}TAB$ (8 x $10^{-3}$M), spherical particles of about 350 Å - 500 Å diameter can be clearly observed via the above process, the electron micrograph of aggregates of PMA - $C_{10}TAB$ is shown in Figure 7. This indicates that the PMA polymer chain is extended and stretched at pH 8 as shown in earlier studies(1-17), while addition of cationic surfactants such as, $C_{10}TAB$ refold the polymer chain to form spherical particles.

Photophysical studies both steady state and pulsed, of pyrene and its positive and negatively charged derivatives,(48) confirm that a conformational transition of PMA is induced by $C_nTAB$. The stretched PMA chain at pH 8 collapses on addition of the cationic surfactants.

Critical Aggregate Concentration - the CAC. Figure 8 shows a sharp increase both in $I_r$ and $I_3/I_1$ of pyrene fluorescence over a narrow range of $C_{10}TAB$ concentration, in aqueous solution of PMA at pH 8. It is noted that the ratio of $I_3/I_1$ in PMA - $C_{10}TAB$ solutions reaches a steady value (~ 0.70) at a $C_{10}TAB$ concentration, which is called a Critical Aggregate Concentration, CAC. At the CAC, the hydrophobic aggregates of $C_{10}TAB$ and PMA are formed which host hydrophobic molecules such as pyrene. The CAC corresponds to the midpoint of the transition in a plot of $I_r$ of pyrene fluorescence versus [$C_{10}TAB$]. The decay rate constants sharply decrease at a $C_{10}TAB$ concentration 3 x $10^{-3}$ M, which is in good agreement with the CAC determined via fluorescence ratio of $I_3/I_1$ measurement. This indicates that techniques that have been successfully used for the CMC measurements in micellar systems,(49) can also be used to investigate the aggregates of PMA and $C_{10}TAB$. The initial addition of $C_{10}TAB$ (< CAC), causes a decrease in $I_r$, the pyrene fluorescence intensity and an increase in the decay rate constants, due to

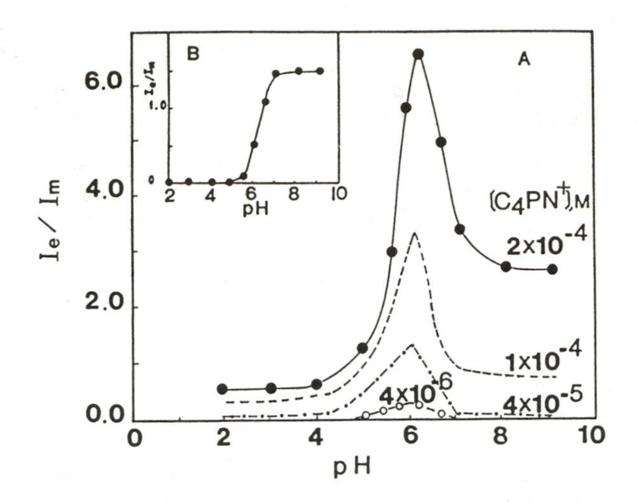


5. Dependence of the ratio of relative intensity of excimer to monomer fluorescence, $I_e/I_m$, on the pH of aqueous solutions of PMA.
A. Various concentrations of $C_4PN^+$
B. 2 x $10^{-6}$ M $C_{11}PN^+$.

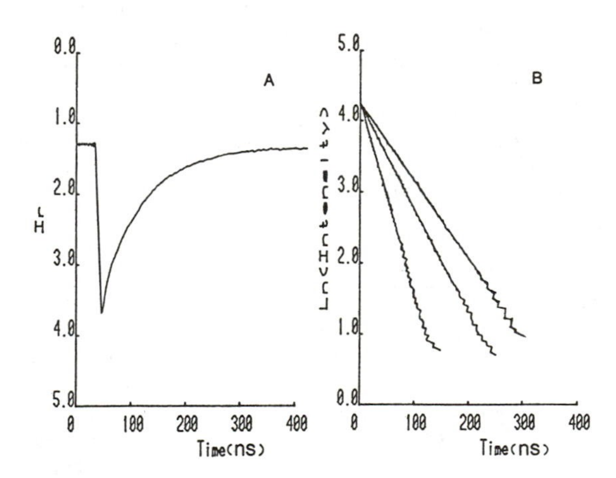


6. A. Observed decay ($I_r$ vs. t) of the $C_4PN^+$ excimer fluorescence in aqueous solutions of PMA, at pH 6.
B. First order decay (ln $I_r$ vs. t.) of the $C_4PN^+$ excimer fluorescence in aqueous solutions of PMA, at pH 6, 7 and 8.

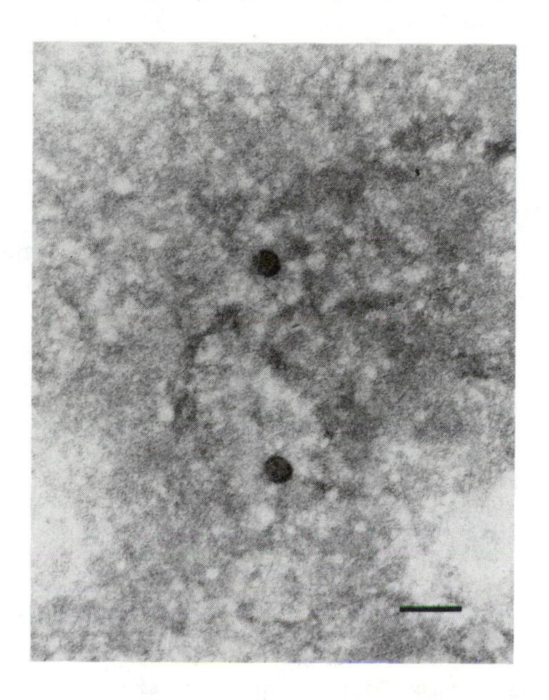

7. Electron micrograph of the PMA-$C_{10}$TAB aggregates. [$C_{10}$TAB] = 8 x $10^{-3}$ M. PMA of Molecular weight = 1.1 x $10^4$. (The length of the inserted bar is equal to 1000 Å).

quenching by free bromide ions. These data also show that pyrene is not associated with PMA or solubilized in PMA at pH 8, but solubilized in the water phase. However, at $C_{10}TAB$ concentrations above the CAC, pyrene is preferentially solubilized in hydrophobic regions formed by the aggregates of PMA - $C_{10}TAB$. Here it is protected from quenching by bromide in the aqueous phase and $I_r$ increases sharply.

Figure 9 represents a plot of log(CAC) versus the number of carbon atoms of a cationic surfactant molecule, $C_nTAB$ and log(CMC) vs. n is also plotted for the sake of comparison. The figure shows that each CAC is markedly lower than the CMC of the corresponding surfactant. This is particularly true for $C_{10}TAB$, $C_{12}TAB$ and $C_{16}TAB$, where the CACs are one to two orders of magnitude lower than the CMC! Similar experiments in $C_6TAB$ - PMA did not lead to an increase in $I_r$ and $I_3/I_1$ up to surfactant concentrations of 0.5 M, and, a short chain quaternary ammonium salt, tetrabutyl ammonium iodide TBI, does not affect the conformation of PMA. The CAC for the $C_8TAB$ surfactant is close to the CMC of the $C_8TAB$ micelle. The above experiments tend to indicate that the interaction between PMA and cationic surfactants is chainlength dependent. The shorter ($n \leq 8$) the chainlength of $C_nTAB$, the weaker the interaction between $C_nTAB$ and PMA, giving rise to a larger CAC.

The photophysical data show sharp changes over a narrow range of surfactant concentration, CAC. It is suggested that a cooperative process takes place consisting of a coiling of polymer assisted by reduced charge density and the hydrophobic interactions of the surfactant chains bound to PMA. The CAC also depends on the PMA concentrations following the stoichiometric relationship:

$$\text{CAC } (10^{-3}\text{M}) = 0.7 + 2.2 \times [\text{PMA}] \quad (1)$$

$$(0.1 \text{ g/L} < [\text{PMA}] \leq 2 \text{ g/L})$$

The data show that in surfactant systems the polymer acts as an important component in aggregate formation, rather than behaving as an inert additive.

Nature of the Aggregates. The aggregates are hydrophobic, loose structures with some residual surface charge from the anionic polymer. The fluorescence decay rate constants $k_A$ of pyrene in aggregates of PMA and $C_nTAB$, are identical to corresponding rate constants the $k_m$ in micellar alkyltrimethylammonium chloride, $C_nTAC$. It is concluded that the interior of aggregates of PMA - $C_nTAB$ is hydrophobic and similar to that of a micelle.

Several positively charged pyrene derivatives, such as $C_4PN^+$ and $C_{11}PN^+$, and negatively charged derivatives, such as 1-pyrenebutyric acid ($PyC_3H_7COOH$), 1-pyrenedecanoic acid ($PyC_9H_{19}COOH$) provide further evidence about the hydrophobic structure of the PMA - $C_{10}TAB$ aggregates. All fluorescence decay rate constants $k_0$ of the pyrene derivatives, in PMA - $C_{10}TAB$ aggregates are closer or similar to those found in hexanol or in PMA at pH 3. The longer the carbon chain of the probe, the closer the agreement of $k_0$ in hexanol and in the aggregates. This is reasonable from the point of view of the hydrophobic effects involved. The quenching data show that the surface of the PMA -

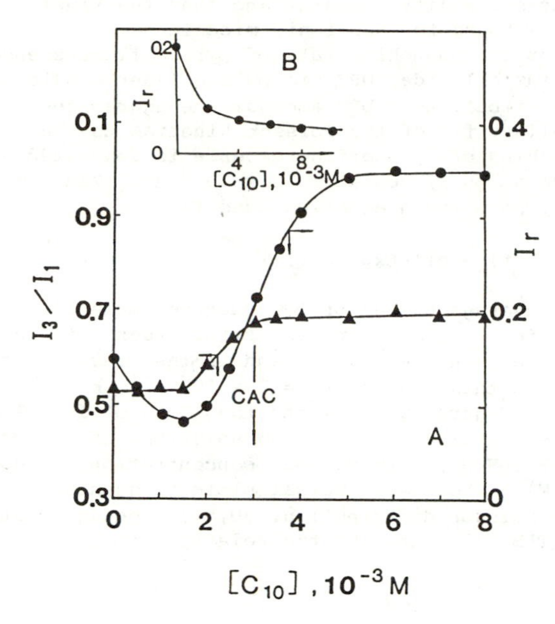


8. A. Relative fluorescence intensity, $I_r$ (o) and intensity ratio, $I_3/I_1$ (Δ) of pyrene as functions of concentrations of $C_{10}TAB$, $[C_{10}]$, in aqueous solutions of PMA at pH 8, [pyrene] = $2 \times 10^{-6}$ M.
B. Relationship between $I_r$ of pyrene fluorescence and $[C_{10}]$ in water. (Reproduced from Ref. 48. Copyright 1986 American Chemical Society.)

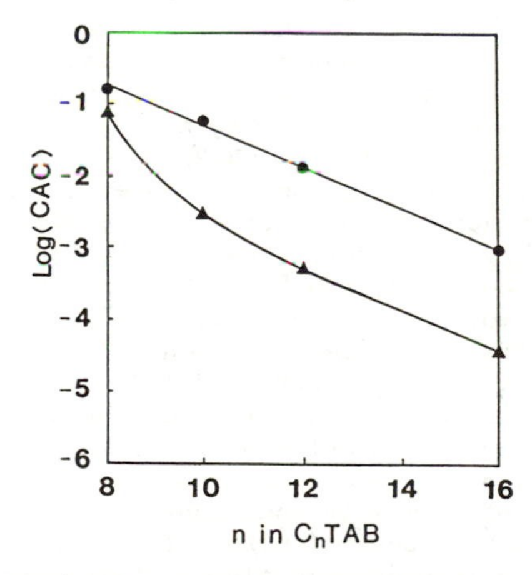


9. Plot of Log(CAC) vs. n in $C_nTAB$ (▲) and plot of log(CMC) vs. n in $C_nTAB$(●). (Reproduced from Ref. 48. Copyright 1986 American Chemical Society.)

$C_{10}TAB$ aggregates has little charge and that the kinetic data are similar to those found for non-ionic micelles.

Analysis of the quenching data of pyrene fluorescence by 1-dodecylpyridinium chloride, DPC via pulsed laser studies confirms the Poisson distribution of DPC amongst the aggregates. Figure 10A shows the excellent fit of the Poisson kinetics to the time-dependent quenching of Pyrene fluorescence in deaerated aqueous solution of PMA at pH 8, contained $8 \times 10^{-3}$ M $C_{10}TAB$ and quencher $2 \times 10^{-5}$ M DPC. The Poisson equation used is

$$I = I_0 \left\{Exp(-k_0t) - \bar{n}[1-Exp(-k_qt)]\right\} \qquad (2)$$

where $\bar{n}$ is the average number of DPC quencher molecules solubilized in each aggregate, $k_o$ and $k_q$ are the first order rate constants for the decay of pyrene in the absence and in the presence of quencher, respectively. In this figure, $k_0 = 2 \times 10^6\ s^{-1}$, $k_q = 2.0 \times 10^7\ s^{-1}$ $n_{calcd.} = 0.27$. Figure 10B presents the plots of $ln(I_r)$ of pyrene fluorescence versus time in deaerated solutions of PMA contained $8 \times 10^{-3}$ M $C_{10}TAB$ (pH 8), with various concentrations of quencher DPC ($0-1.4 \times 10^{-4}$ M). These are typical plots of quenching data according to a Poisson distribution.(50,51) The aggregation numbers of $C_{10}TAB$ calculated by the relationship

$$[C_{10}TAB]/\bar{N} = [DPC]/\bar{n}_{calcd.} \qquad (3)$$

correspond well to the results from steady state experiments ($\bar{N}$ = 105 ± 10).(48) The above experimental data confirm that the model for aggregation of PMA - $C_{10}TAB$ is not via local small and random clusters, but via discrete structures,that are much larger than pure $C_{10}TAB$ micelles (N ~ 36).(52)

It is concluded that the aggregates of PMA - $C_{10}TAB$ are large structures consisting of about one hundred $C_{10}TAB$ molecules and one coiled polymer chain. The interior of the aggregate has hydrophobic domain that is similar to that of a micelle. However, the bromide ions are only in bulk aqueous phase, and not close to the surface of aggregate. The degree of polarization of 2-methylanthracene fluorescence in PMA - $C_{10}TAB$ aggregate is four fold smaller than in PMA compact coil at pH 2, two fold smaller than in $C_{10}TAB$ micelle, indicating that the aggregate is a much looser structure than a compact PMA coil, or a $C_{10}TAB$ micelle. The data of electron microscopy show that the aggregates are spherical particles which are about 350-500 Å, rather than helix forms.

PMA samples of molecular weight $3.9 \times 10^4$, $1.6 \times 10^5$ and $6.4 \times 10^5$ emphasize the effect of molecular weight of PMA on the aggregates of PMA-$C_{10}TAB$. The curves of $I_3/I_1$ for $2 \times 10^{-6}$ M pyrene fluorescence in aqueous solutions of PMA at pH 8 show a sharp increase in above PMA samples at pH 8, indicating that in all cases the aggregates of PMA - $C_{10}TAB$ are formed above the CAC. The aggregation numbers of $C_{10}TAB$ in each aggregate, measured by analysis of the quenching data of pyrene fluorescence by DPC, are 90 ~ 100 surfactant molecules, and also contain a portion of the coiled PMA chain.

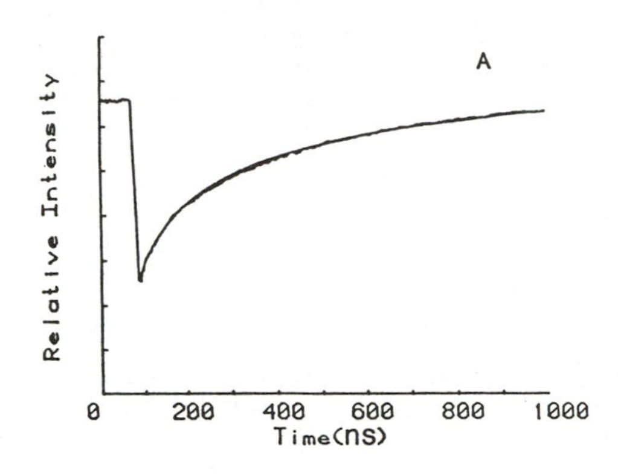


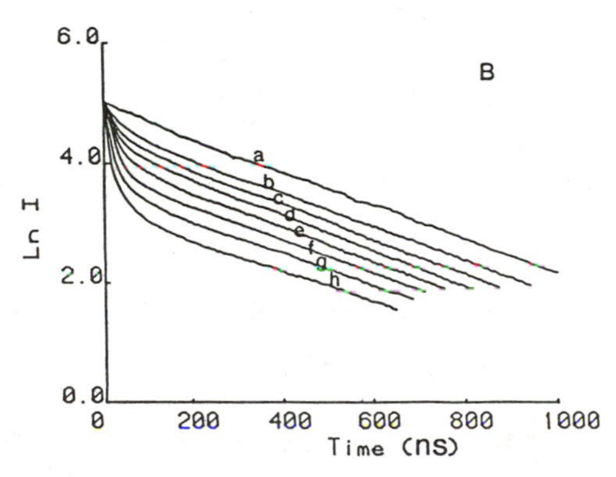


10. A. Poisson quenching fit of the time-dependent fluorescence decay of pyrene in deaerated PMA solutions at pH 8, containing 8 x $10^{-3}$ M $C_{10}$ TAB and quencher 2 x $10^{-5}$ M DPC. The smooth one is from computer fitting and the other one is real data.

B. ln(Intensity) of pyrene fluorescence vs. time in deaerated PMA solutions containing 8 x $10^{-3}$ M $C_{10}$TAB at pH 8, with various concentrations of quencher DPC, ($10^{-5}$ M): (a) 0; (b) 2.0; (c) 4.0; (d) 6.0; (e) 8.0; (f) 10.0; (g) 12.0; (h) 14.0.

CONCLUSIONS This study shows that fluorescence probing techniques are useful and powerful tools for investigation of conformational transitions of polyelectrolytes as induced by cationic surfactants, pH or other means. Studies on the interaction of cationic probes with polyelectrolytes provide useful information on the intermediates that lie between A states and B states. It is concluded that the conformational transition induced by pH is a progressive process over several pH units. Studies on the interaction of cationic surfactants with PMA at pH 8 show that the aggregates formed are large loose structures, while the interior of the aggregate has a hydrophobicity that is similar to that of a micelle.

Acknowledgment.
We thank the National Science Foundation for support of this work via Grant CHE-01226-02.

Literature Cited.

1. Katchalsky, A.; Eisenberg, H. J. Polym. Sci., 6, 145, 1951.
2. Silberberg, A.; Eliassaf, J.; Katchalsky, A. J. Polym. Sci., 7, 393, 1951.
3. Mandel, M.; Stadhouder, M. G. Makromol. Chem., 80, 141, 1964.
4. Anufrieva, E. V.; Birshtein, T. M. J. Polym. Sci. C, 16, 3519 1968.
5. Arnold, R. J. Coll. Sci., 12, 549, 1957.
6. Koenig, J. L.; Angood, A. C.; Semen, J.; Lando, J. B. J. Am. Chem. Soc., 91, 7250, 1969.
7. Crescenzi, V.; Quadrifoglio, F.; Delben, F. J. Polym. Sci., A 2, 10, 347, 1972.
8. Delben, F.; Crescenzi, V.; Quadrilfoglio, F. Eur. Polym. J., 8, 933, 1972.
9. Daoust, H.; Thanh, H. L; Ferland, P.; St-cyn, D. Can. J.Chem. 63, 1568, 1985.
10. Kern, E. E.; Anderson, D. K. J. Polym. Sci. A-1, 6, 2765, 1968.
11. Suzuki, K.; Taniguchi, Y. J. Polym. Sci. A-2, 8, 1679, 1970.
12. Kay, P. J.; Kelly, D. P.; Milgate, G. I.; Treloar, F. E. Makromol. Chemie., 177, 885, 1976.
13. Okamoto, H.; Wada, Y. J. Polym. Sci. Polym. Phys., 12, 2413, 1974.
14. Jager, J.; Engberts, J. B. F. N. J. Org. Chem., 50, 1474, 1985; J. Am. Chem. Soc., 106, 3331, 1984.
15. Moan, M.; Wolff, C.; Cotton, J. P.; Ober, R. J. Polym. Sci. Polym. Symp., 61, 1, 1977; Polym., 16, 781, 1975.
16. Irie, M. Makromol. Chem. Rapid Commun., 5, 413, 1984.
17. Bednar, B.; Marowetz, H.; Shafer, J. A. Macromolecules, 18, 1940, 1985.
18. Stork, W. J. H.; Van Boxsel, J. A. M.; DeGoeij, A. F. P. M.; Haseth, P. L. De.; Mandel, M. Biophys. Chem., 2, 127, 1974; Madel, M.; Stork, W. H. J. Biophys. Chem., 2, 137, 1974.
19. Anufrieva, E. V.; Birshtein, T. M.; Nekrasova, T. N.; Ptitsyn, O. B.; Sheveleva, T. V. J. Polym. Sci. C., 16, 3519, 1968.
20. Oster, G. J. Polym. Sci., 16, 235, 1955.
21. Baud, C. Eur. Polym. J., 13, 897, 1977.

22. Erny, B.; Muller, G. J. Polym. Sci. Polym. Chem., 17, 4011, 1979.
23. Wang, Y.; Morawetz, H. Macromolecules, 19, 1925, 1986.
24. Stork, W. H. J.; Hasseth, P. L. de.; Schippers, W. B.; Körmeling, C. M.; Mandel, M. J. Phys. Chem., 77, 1772, 1973.
25. Stork, W. H. J.; Hasseth, P. L. de.; Lippits, G. J. M.; Madel, M. J. Phys. Chem., 77, 1778, 1973.
26. Muller, G.; Fenyo, J. C. J. Polym. Sci. Polym. Chem., 16, 77, 1978.
27. Snare, M. J.; Tan, K. L.; Treloar, F. E. J. Macro. Sci. Chem. A., 17(2) 189, 1982.
28. Chu, D. Y.; Thomas, J. K. Macromolecules, 17, 2142, 1984.
29. Tokiwa, F.; Tsujii, K. Bull. Chem. Soc. Jpn., 46, 2684, 1973.
30. Schwuger, M. J. J. Colloid & Interf. Sci., 43, 491, 1983.
31. Shirahama, K. J. Colloid & Polym. Sci., 252, 978, 1974.
32. Cabane, B. J. Phys. Chem., 81, 1639, 1977.
33. Shirahama, K.; Tohdo, M.; Murahashi, M. J. Colloid & Interf. Sci., 86, 282, 1982.
34. Turro, N. J.; Baretz, B. H.; Kuo, P. L. Macromolecules, 17, 1321, 1984.
35. Zana, R.; Lianos, P.; Lang, J. J. Phys. Chem., 89, 41, 1985.
36. Kresheck, G. C.; Hargraves, W. J. Colloid & Interf. Sci., 95, 453, 1983; 105, 589, 1985.
37. Dubin, P. L.; Davis, D. D. Macromolecules, 17, 1294, 1984.
38. Leung, P. S.; Goddard, E. D. etc., Colloids Surf., 13, 47 & 63, 1985.
39. Hayakawa, K.; Kwak, C. T. J. Phys. Chem., 86, 3866, 1982.
40. Abuin, E. B.; Scaiano, J. C. J. Am. Chem. Soc., 106, 6274, 1984.
41. Hayakawa, K.; Santerre, J. P.; Kwak, C. T. Macromolecules, 16, 1642, 1983.
42. Chu, D. Y.; Thomas, J. L. J. Phys. Chem., 89, 4065, 1985.
43. Cuniberti, C.; Perico, A. Eur. Polym. J., 13, 369, 1977; 16, 887,1980; Ann. N.Y. Acad. Sci., 366, 35, 1981.
44. Winnik, M. A.l Redpeth, A. E. C.; Paton, K.; Danhelka, J. Polymer, 25, 91, 1984.
45. Tazuke, S.; Ooki, H.; Sato, K.; Macromolecules, 15, 400, 1982; Suzuki, Y.; Tazuke, S. Macromolecules, 14, 1742, 1981.
46. Thomas, J. K. Chem. Rev., 80, 283, 1980; Thomas, J. K. The Chemistry of Excitation at Interfaces, ACS Monograph, No. 181. Washington, D. C., 1984.
47. DellaGardia, R. D.; Thomas, J. K. J. Phys. Chem., 87, 3550, 1983.
48. Chu, D. Y.; Thomas, J. K. J. Am. Chem. Soc., 108, 6270, 1986.
49. Kalyanasundaram, K.; Thomas, J. K. J. Am. Chem. Soc., 99, 2039, 1977.
50. Atik, S. S.; Singer, L. A. Chem. Phys. Lett., 59, 519, 1978.
51. Hashimoto, S.; Thomas, J. K. J. Colloid & Interf. Sci., 102, 152, 1984.
52. Tabtar, H. V. J. Colloid & Interf. Sci., 14, 115, 1959.


RECEIVED April 27, 1987

# LUMINESCENT POLYMERIZATION PROBES

# Chapter 33

# Fluorescence Monitoring of Viscosity and Chemical Changes During Polymerization

F. W. Wang[1], R. E. Lowry[1], W. J. Pummer[1], B. M. Fanconi[1], and En-Shinn Wu[2]

[1]Polymers Division, National Bureau of Standards, Gaithersburg, MD 20899
[2]Department of Physics, University of Maryland, Baltimore County, Catonsville, MD 21228

Three approaches using fluorescent dyes dissolved in epoxy resins were used to determine the viscosity changes during the curing process. First, the intensity of excimer fluorescence from a dye which forms an intramolecular excimer was measured to determine the viscosity changes. In another approach, we used a dye whose fluorescence intensity increases with the increase in the local viscosity, and a second dye whose fluorescence intensity is insensitive to the local viscosity. The ratio of the fluorescence intensities of the two dyes was measured to monitor the cure of epoxy resins. In a third approach, we measured the diffusion coefficient of a fluorescent dye by a photobleaching technique to monitor the curing process. Finally, we used a fluorescence technique to monitor the formation of a polyimide polymer from poly(amide acid).

The manufacture of polymer matrix composites involves complex chemical and physical changes that must be adequately controlled to produce desirable products. Monitoring techniques and models to correlate monitoring data to improve processing are therefore key aspects to increasing production rates and product quality.

Fluorescence techniques are particularly useful to monitor the change in local viscosity because they are sensitive and can be easily adopted to *in-situ*, nondestructive monitoring. In a previous paper(1), we described an excimer-fluorescence technique to monitor the polymerization of methyl methacrylate. We show here an application of the excimer-fluorescence technique to monitor the cure of epoxy resins. In addition, we describe the cure monitoring of epoxy resins with the use of two fluorescent dyes, a dye whose fluorescence intensity increases with local viscosity, and another dye which serves as an internal standard with nearly constant fluorescence intensity. This second technique is similar to the ones used by Loutfy(2,3) and by Levy(4). However, to the best of

0097–6156/87/0358–0454$06.00/0

our knowledge, there has been no previous report of the application of a viscosity insensitive internal standard. Since the use of an internal standard eliminates the difficulties involved in making absolute measurements, it is a significant step forward in the application of fluorescence spectroscopy to cure monitoring in the factory environment. Furthermore, we report a third technique that utilizes the measurement of the diffusion coefficient of a photobleachable probe to monitor the cure of an epoxy resin. Finally, we describe a fluorescence method to monitor the formation of a polyimide polymer.

## EXPERIMENTAL

MONOMERS. Amine hardener, 4,4'-methylene-bis-(cyclohexylamine) (PACM), was distilled under reduced pressure and stored under dry argon. It was melted under dry argon before use. Epoxy resin, diglycidyl ether of bisphenol A (DGEBA), had an epoxy equivalent weight of approximately 175 and was used without further purification.

SYNTHESIS OF POLYIMIDE. The polyimide was prepared from 2,2-bis(3,4-dicarboxyphenyl) hexafluoropropane dianhydride (American Hoechst 6F)[1] and 2,2-bis[4(4-aminophenoxy)phenyl] hexafluoropropane (Morton Thiokol 4BDAF).

Both 6F and 4BDAF are soluble in dry glyme at 25°C. 4BDAF (0.5g, $9.6 \times 10^{-4}$ mol) was dissolved in 3 ml of dry glyme at 25°C with stirring in a 25 ml glass-stoppered flask. When dissolution was completed (usually within 3 minutes), 0.42g ($9.6 \times 10^{-4}$ mol) of solid 6F in small portions (0.1g each) was added to the solution of 4BDAF at 25°C. Within 5 minutes after the 6F was added, stirring was impeded by the increased solution viscosity. After 15 minutes of manually swirling the contents, the solution viscosity decreased sufficiently to allow normal stirring to proceed. The reaction was stopped after 44 hours at 25°C. This solution (26% solids) of the poly(amide acid) in glyme was used to prepare films for fluorescence spectroscopy as described below.

A few drops of the poly(amide acid) solution were spread on a clean quartz slide by drawing a wedge of the solution beneath another clean slide. Room temperature solvent evaporation and all heat treatments were carried out in air. The film was cured for 0.5 hour at each of seven temperatures ranging from 80° to 350°C, with oven warmup and cooling down times of up to 0.5 hour each. Front surface fluorescence from the same film region was measured at room temperature after each heat treatment. Film thickness averaged 13μm.

---

[1]Certain commercial materials and equipment are identified in this paper to specify adequately the experimental procedure. In no case does such identification imply recommendation or endorsement by the National Bureau of Standards, nor does it imply necessarily the best available for the purpose.

FLUORESCENT PROBES. An excimer-forming probe, 1,3-bis-(1-pyrene)propane (BPP), was obtained from a commercial source and used without further purification. A viscosity-sensitive probe, 1-(4-dimethylaminophenyl)-6-phenyl-1,3,5-hexatriene (DMA-DPH), was dissolved in DGEBA by rotating the resin at 45°C for several hours. An internal standard, 9,10-diphenylanthracene (DPA), at a concentration of $2x10^{-5}$ mol/l was added to the resin-hardener mixture.

REACTION CELL. The reaction cell was made of a silicone rubber gasket sandwiched between two pyrex cover slides. The cell was pressed against the flat face of a cylindrical heater, which was proportionally controlled to within 1°C. Two small holes were bored near the top of the rubber gasket for inserting a thermocouple and for filling the cell.

CURE MONITORING. In a typical experiment, 0.5 ml of the epoxy resin containing a probe at a concentration of $5x10^{-5}$ mol/l was put into a stoppered syringe kept at 40°C. The amine hardener was then rapidly mixed with the epoxy resin to form a mixture containing the same equivalents of the hardener and the resin. Finally, the resin mixture was injected into the reaction cell which had been heated to 60°C.

Cure monitoring with the probe BPP was carried out in the manner previously described in Reference 1. The sample was irradiated at 345 nm and the monomer and excimer fluorescence intensities at 377 nm and 488 nm were measured as a function of time. When DMA-DPH was used as a probe and DPA was used as an internal standard, they were excited at 420 nm and 345 nm, respectively.

FLUORESCENCE RECOVERY AFTER PHOTOBLEACHING (FRAP). The diffusion coefficient of the photobleachable probe, 1,1'-dihexyl-3,3,3',3'-tetramethylindocarbocyanine perchlorate [$DiIC_6(3)$], was determined by the FRAP method(5,6,7) described below. The concentration of the probe in the mixture containing the same equivalents of the epoxy resin and the amine hardener was $2x10^{-5}$ mol/l. The resin mixture, sandwiched between a microscope slide and a cover slip, was placed on a stage thermostatically controlled to within 0.5°C. The thickness of the resin mixture was about 5μm. In a FRAP experiment(6), a small area of the sample is illuminated with a weak beam of exciting light (monitoring beam). The fluorescence from this area is recorded as F. At a predetermined time, t = 0, the sample is momentarily illuminated with a strong laser beam (bleaching beam) to cause an irreversible bleaching of fluorophore. Following the bleaching, the fluorescence is again monitored by the monitoring beam. The fluorescence, F(t), is initially very weak, but gradually increases as the fresh fluorescent molecules diffuse into the bleached area, and eventually recovers to its original intensity. From the rate of the recovery of fluorescence intensity, the diffusion coefficient can be determined.

## RESULTS AND DISCUSSION

Figure 1 gives the change in the intensity ratio $F_M/F_D$ as a function of cure time for 1,3-bis-(1-pyrene)propane (BPP) dissolved in a mixture containing the same equivalents of the epoxy resin and the amine hardener. Here, $F_M$ and $F_D$ are, respectively, the fluorescence intensity of the monomer at 377 nm and that of the excimer at 488 nm. As cross-linking proceeded, the viscosity increased owing to the growth in molecular weight and this led to an increase in the intensity ratio. After 45 minutes of the cure, there was a small decrease in the intensity ratio due to photodegradation of the probe. The complication due to photodegradation can be eliminated by reducing the exposure of the resin to the exciting UV radiation. The use of an optical multichannel analyzer is one approach to reduce exposure times.

Owing to the lack of sensitivity of the BPP probe at the longer cure times, we have examined the use of other types of viscosity-sensitive probe molecules. Figure 2 gives the excitation and emission spectra of the viscosity sensitive dye DMA-DPH in diglycidyl ether of bisphenol A (DGEBA). The excitation and the emission spectra of the internal standard DPA are similar to the ones published by Berlman(8).

Figure 3 gives the fluorescence intensity, $F_P$, of DMA-DPH at 480 nm (the upper curve) and the fluorescence intensity, $F_R$, of DPA at 415 nm (the lower curve) as a function of cure time for a mixture containing the same equivalents of the epoxy resin and the amine hardener. The fluorescence intensity at the frequency of the DMA-DPH emission increased fivefold while that at the DPA peak emission increased only slightly. (The apparent increase in the DPA fluorescence was mainly due to the increase in the fluorescence of impurities in the epoxy resin. At the excitation wavelength of 345 nm, the fluorescence intensity of the resin mixture alone at 415nm increased by about 80%. However, this increase contributed to the much smaller apparent increase in the DPA fluorescence because at the beginning of the cure, the fluorescence intensity of DPA was four times larger than that of the resin alone.) In Figure 4, we show the intensity ratio $F_P/F_R$ as a function of cure time. The intensity ratio increased steadily with cure time, reaching a plateau value at 70 minutes. This intensity ratio is not sensitive to the inhomogeneity or the deformation of the sample. We can therefore use this ratio to monitor the cure of samples which shrink during polymerization or contain reinforcing fibers or particles.

Figure 5 shows the logarithm of the translational diffusion coefficient of the probe $DiIC_6(3)$ as a function of cure time when the resin was cured at 45°C, 60°C, and 75°C. In all cases, the translational diffusion coefficient increased slightly at the beginning of the cure when the sample temperature was raised from 22°C to the cure temperature because of the decrease in the resin viscosity with the increase in temperature. As cross-linking proceeded, the viscosity increased owing to the growth in molecular weight, and this was reflected by the decrease in the translational diffusion coefficient.

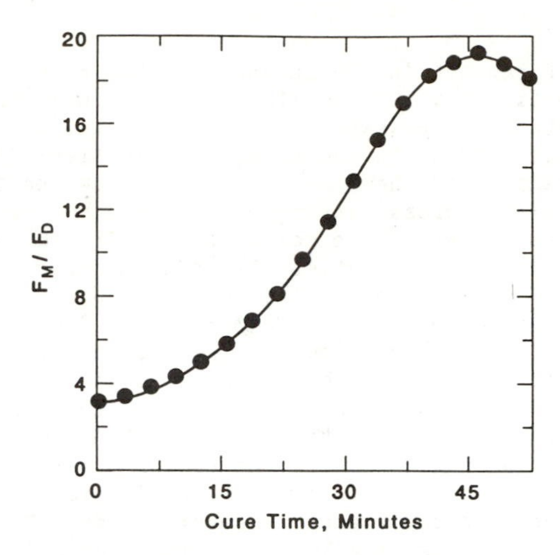


Figure 1. Ratio of monomer fluoresence intensity at 377nm ($F_M$) to excimer fluorescence intensity at 488nm ($F_D$) for 1,3-bis-(1-pyrene)propane in epoxy as a function of cure time. The excitation wavelength was 345 nm.

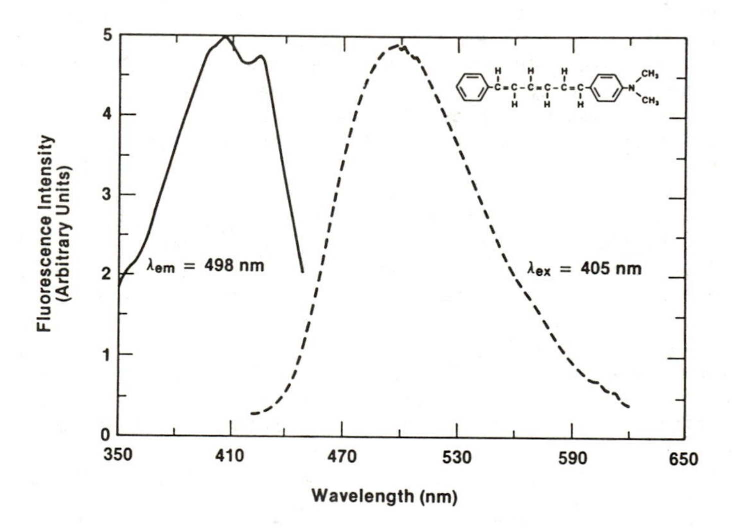


Figure 2. Excitation and emission spectra of 1-(4-dimethylaminophenyl)-6-phenyl-1,3,5-hexatriene(DMA-DPH) in diglycidyl ether of bisphenol A (DGEBA). The excitation and emission wavelengths were 405 nm and 498 nm, respectively.

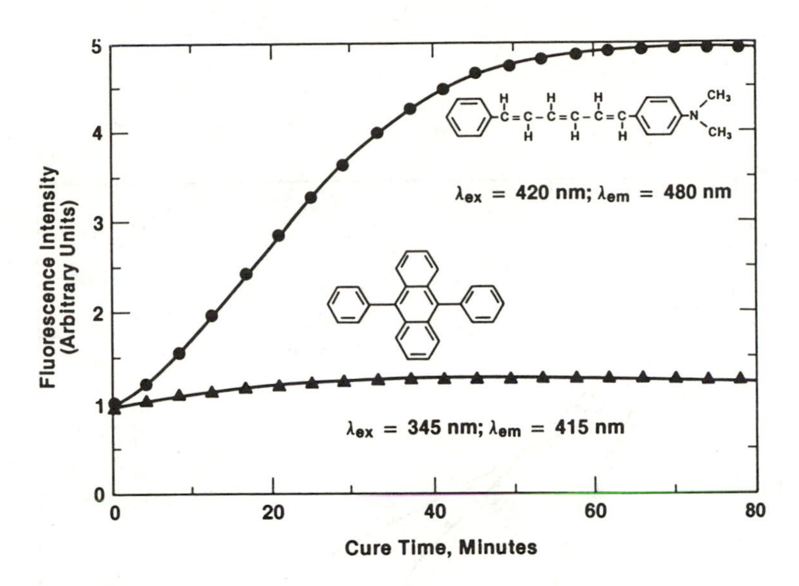


Figure 3. Fluorescence intensities of DMA-DPH at 480 nm (circles) and of DPA at 415 nm (triangles) as a function of cure time.

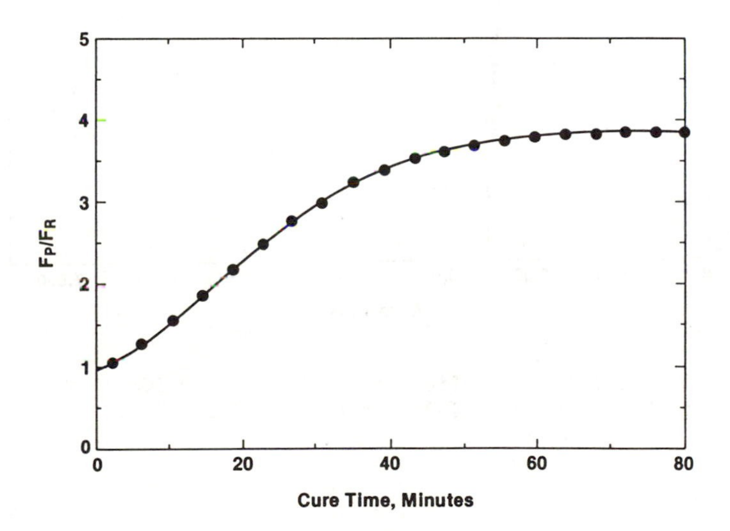


Figure 4. Ratio of the fluorescence intensity of the probe DMA-DPH ($F_P$) and the fluorescence intensity of the internal standard DPA ($F_R$) as a function of cure time.

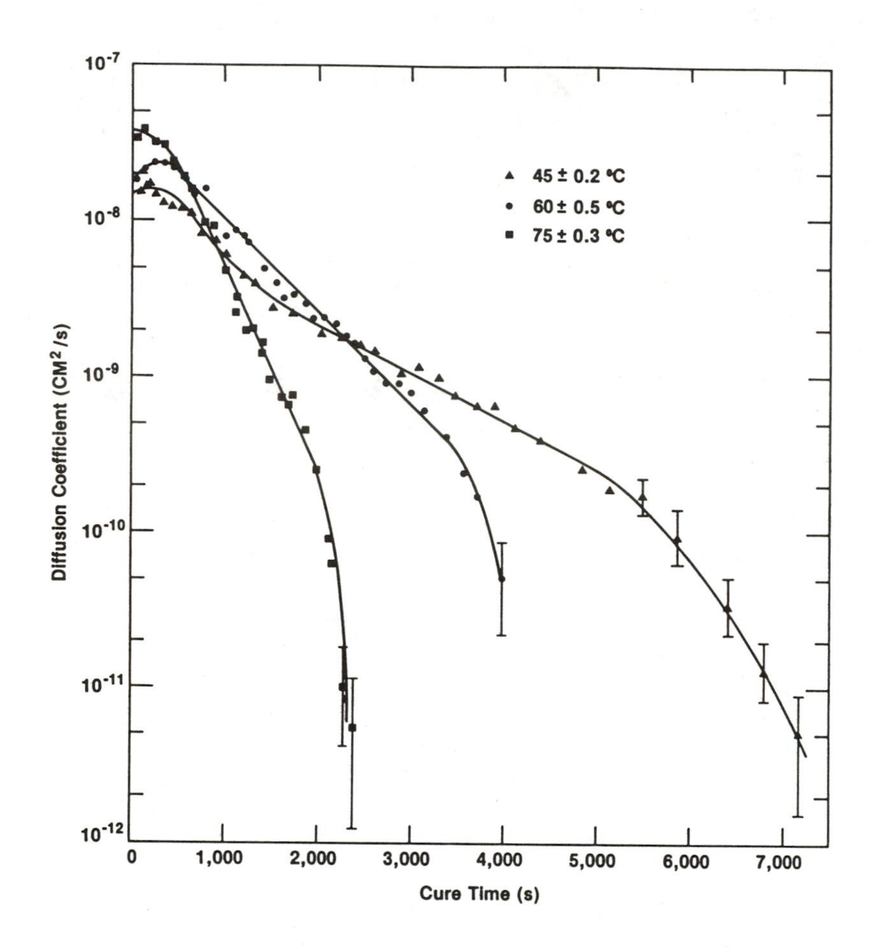


Figure 5. Diffusion coefficient of the probe $DiIC_6(3)$ as a function of cure time at 45°C, at 60°C, and at 75°C.

When the resin was cured at 60°C and 75°C, the translational diffusion coefficient decreased nearly linearly with the cure time until it decreased abruptly, at the cure time of 4000 s and 2400 s, for the resin cure at 60 °C and 75 °C, respectively. This abrupt decrease in the diffusion coefficient was presumably due to the vitrification of the resin. Diffusion measurements were also carried out on the resin mixture cured at 60°C using another fluorescent probe, 4-(N,N-dioctyl)amino-7-nitrobenz-2-oxa-1,3-diazole. The results obtained were consistent with those obtained with the use of $DiIC_6$ (3).

For the resin cured at 45°C, the behavior of the translational diffusion coefficient, D, with cure time was as follows: a slight initial increase was followed by a decrease in D until the cure time of 1700 s, a slower decrease in D between 1700 s and 5000 s, and finally a faster decrease in D after the cure time of 5500 s. However, unlike the resin cured at 60°C or 75°C, no precipitous decrease in D was observed even after 7000 s, presumably because the polymer network formed at 45°C was substantially different from that formed at 60°C or 75°C.

Figure 6 shows the uncorrected excitation and emission spectra of a film of the poly(amide acid) after thermal treatment in an air filled oven for 30 minutes at each of the temperatures ranging from 80°C to 350°C. The excitation and the emission wavelengths were 480nm and 550 nm, respectively. The fluorescence intensity at 550 nm increased steadily when the cure temperature was raised from 80°C to 350°C. The uncorrected excitation spectra were more complex. A band at 460 nm appeared at the expense of the one at 330 nm after the second heating. However, the intensity of the band at 330 nm relative to the one at 460 nm increased steadily with additional thermal treatments. In any event, the formation of this polyimide polymer can be readily monitored <u>in-situ</u> by fluorescence spectroscopy.

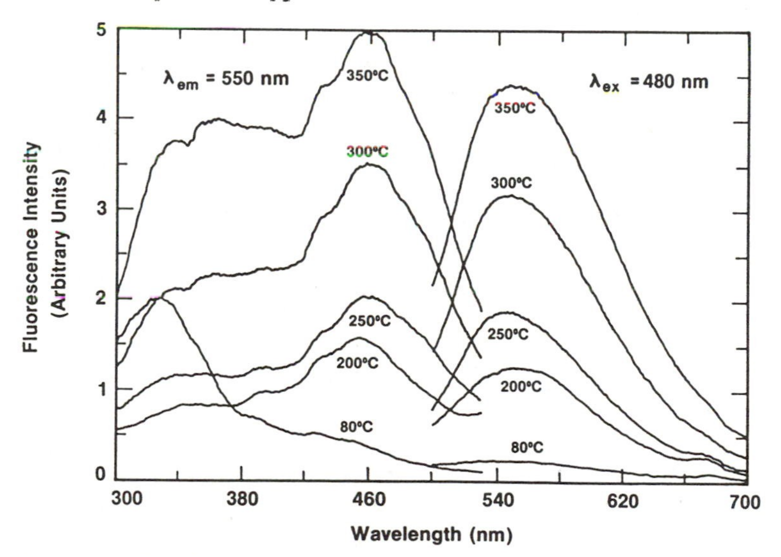


Figure 6. Excitation and emission spectra of poly(amide acid) film after heat treatments.

## CONCLUSION

We have described three fluorescence techniques for monitoring the cure of epoxy resins. The first one is based on intramolecular excimer fluorescence, the second one is based on the enhancement of fluorescence intensity with the medium viscosity, and the third one is based on the measurement of the translational diffusion coefficient of a fluorescent probe. Finally, we have demonstrated the fluorescence monitoring of the formation of a polyimide polymer.

## ACKNOWLEDGMENT

This work was partially supported by the U. S. Army Research Office through Contract MIRP ARO 111-84.

## REFERENCES

1. Wang, F. W. Lowry, R. E. and Grant, W. H. Polymer 1984, 25, 690.

2. Loutfy, R. O., Macromolecules 1981, 14, 270.

3. Loutfy, R. O., J. Polym. Sci. Polym. Phys. Ed. 1982, 20, 825.

4. Levy, R. L. and Ames, D. P. Proc. Org. Coat. Appl. Polym. Sci. 1983, 43, 116.

5. Wu, E-S., Jacobson, K., Szoka, F., and Portis, A. Biochemistry 1978, 17, 5543.

6. Axelrod, D., Koppel, D. E., Schlessinger, J., Elson, E. and Webb, W. W., Biophys. J. 1976, 16, 1055.

7. Jacobson, K., Derzko, Z., Wu, E-S, Hou, Y. and Poste, G. J. Supramol. Struct. 1977, 5, 565.

8. Berlman, I. B. "Handbook of Fluorescence Spectra of Aromatic Molecules" 2nd Edn., Academic Press, New York, 1971, p.364.

RECEIVED April 27, 1987

Chapter 34

# Application of Reactive Dye Labeling Technique for Cure Characterization of Epoxy Networks

**C. S. P. Sung**

**Institute of Materials Science, Department of Chemistry, University of Connecticut, Storrs, CT 06268**

Cure characterization of epoxy network polymers by using reactive dye labels is reported. Two types of reactive dyes, p,p'-diaminoazobenzene (DAA) and p, p'-diaminostilbene (DAS) were used to mimic the reactivities of diamine type curing agents, diamino-diphenyl sulfone (DDS) for the former and methylene dianiline (MDA) for the latter. As cure proceeds, bathochromic shifts in uv-vis absorption spectra with subsequent enhancement of fluorescence intensity are observed due to the reaction of the dyes with the epoxies. In the case of the DAA label, the analyses of the spectral shifts (about 60 nm) provide estimates of the composition of cure products, the amine reactivity ratio and the activation energy for the amine-epoxy reaction. Strong enhancement of fluorescence (about 100 times) was due to the formation of tertiary amino DAA labels. Thus, fluorescence intensity can be used for sensitive cure monitoring, especially at later stages of cure. For the DAS label, bathochromic shifts of about 20 nm are found in uv spectra due to the conversion of the primary amine groups to tertiary amine groups in DAS. Fluorescence spectra show similar bathochromic shifts with enhanced emission (about 3 times), especially after gelation. Comparison of the fluorescence intensity of DAS with the fully substituted DAS (tt-DAS), a probe molecule, as a function of cure of the matrix indicates that the chemical reaction as well as the medium viscosity contribute to the emission enhancement, particularly at later stages of cure. A calibration curve relating emission intensity with the extent of epoxide reaction by the IR method has been established to estimate cure extent from emission measurement.

0097–6156/87/0358–0463$06.00/0

Characterization of cross-linked polymers such as epoxy networks is required in order to establish structure-property relations. While a number of physicochemical techniques have been employed to provide a better characterization of cure process, none of them can continuously monitor the composition of cure reaction products beyond gelation. Therefore, our laboratory has been involved in recent years in developing new methods to track cure reaction products throughout the cure process. The approach we have taken is based on labelling with reactive dyes to mimic the cure agents and to use the photochemical and photophysical behavior of the dyes for the analysis of the cure process (1-4). Our approach differs from the usual one, which monitors the viscosity dependent behavior of fluorescencent probes (5). In contrast to such probes, our dyes react with the epoxide groups and transform their primary amine groups into secondary and tertiary amines as cure proceeds. The uv-vis absorption and fluorescence spectra of the dyes show systematic changes by such substitution of the primary amine groups, which form the basis for the analyses of the cure process. Here we discuss the use of two types of such reactive dyes, p, p'-diaminoazobenzene (DAA) and p,p'-diaminostilbene (DAS), which were used to mimic the reactivities of two diamine type curing agents, diamino diphenyl sulfone (DDS) and methylene dianiline (MDA). These two curing agents are often used for high temperature epoxy matrices for graphite reinforced composites.

## EXPERIMENTAL

Diglycidyl ether of bisphenol A (DGEBA) was recrystallized from saturated MEK solution by seeding it with purified DGEBA crystals and leaving it in the freezer (-15°C) for 1-2 weeks. DAA was purchased from Eastman Kodak and recrystallized from toluene and acetone. DDS and DGEB which were purchased from Aldrich were used without purification. DAS-dihydrochloride from Aldrich was neutralized with sodium carbonate and recrystallized from methanol to obtain the free amine. In typical cure monitoring studies, a small amount of DAA (5 ∿ 7 mg or about 0.1% by weight for UV-VIS studies and 0.01% by weight in most fluorescence studies) was added to a stoichiometric mixture of DGEBA (5.0 g) or diglycidyl ether of butanediol (DGEB 2.98 g). DDS (1.825 g) was then added and the mixture was heated with a magnetic stirrer at 120°C for 5 minutes. Two circular quartz plates were clamped together with two thin Mylar films (1.5 mil) on the edges leaving a center space for the sample. The clamped quartz plate with Mylar spacers were dipped into epoxy heated to 100°C and the sample was drawn into the center space by capillary action. UV-VIS spectra and fluorescence spectra were measured after curing in an oven for a specific time and cooling the sample to room temperature. Fluorescence was measured with excitation at 450 nm, using a Perkin-Elmer MPF-66 spectrometer with a Model 7500 Data Station. UV-VIS spectra were obtained with a Perkin-Elmer Diode Array (Model 3840) System with a Model 7500 Data Station. For the epoxy system of DGEB and MDA, 0.1% DAS was used, with the excitation at 372 nm for fluorescence spectra. (For more details on experimental conditions, refer to Ref. (3) and (4)).

## RESULTS AND DISCUSSION

### Labelling by p,p'-diaminoazobenzene (DAA)

Cure Composition from UV-VIS Studies: Fig. 1 compares UV-VIS spectra obtained as a function of cure time in DGEB-DDS and DGEBA-DDS at 160°C. In both sets of spectra, significant red shifts of DAA derivatives are observed as the cure time increases. The conversion of DDS to a tertiary amine also red-shifts the absorption of the epoxy matrix, resulting in a shift of the minimum point of the spectra from 360nm to 380nm. In the DGEBA-DDS epoxy, the matrix gels after about 50 minutes of cure and vitrifies after 150 minutes of cure at this cure temperature according to the Time-Temperature-Transformation (T-T-T) diagram (6). After vitrification, the cure reaction is supposedly quenched. As a consequence, the UV-VIS spectra in the DGEBA-DDS matrix does not show much change after vitrification (See Fig. 1(I)d corresponding to 300 min. cure time). In contrast, no vitrification occurs at 160°C in the DGEB-DDS epoxy since its maximum $T_g$ is only about 80°C. In this epoxy, the cure reaction has been pushed further as indicated by additional red shifts shown in Fig. 1(II)d. The disappearance of the epoxy ring as monitored by IR provides further support for this observation.

In order to insure that these spectral shifts are due to cure reactions and not to the changes in the matrix (e.g. polarity change as a function of cure), we ran UV-VIS spectra of fully substituted DAA (the 5th compound in Table I) in the DGEBA-DDS epoxy as a function of cure at 160°C. Any spectral shift in this case would be due to the matrix change since it cannot react any further with epoxide and in fact, only a negligible (<5 nm) spectral shift was observed here. Therefore, we can deduce that the spectral shifts obtained in DAA labelled epoxy arise from cure reactions only.

Table II summarizes the results on the composition of cure products by the deconvolution of UV-VIS spectra with a computer program by assuming $\lambda_{max}$ positions of the model cure products according to Table I and a Gaussian distribution curve for each species. The error in resolving closely overlapping peaks as in our spectra can be significant, especially when the cure is intermediate (e.g. Fig. 1b and c). We tried to fit the curve until the overall error calculated by the program was below two percent. Still, we estimate that the error in the composition of each cure product can be as large as ten percent, since corresponding to a certain cure time these values, as given in Table II, may not represent a unique solution to the particular spectrum. The last column of Table II lists the extent of amine reaction ($\xi_a$) as defined by the following equation

$$\xi_a = [A_{ps} + 2(A_{ss} + A_{pt}) + 3A_{st} + 4A_{tt}]/4 \qquad (1)$$

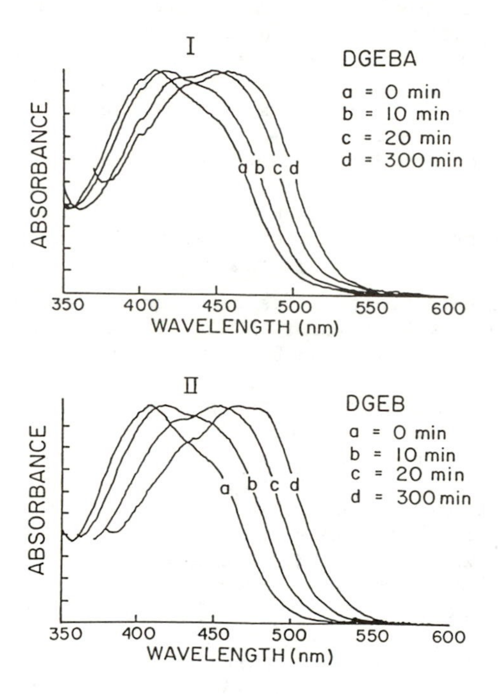


Figure 1. UV-VIS spectra of p,p'-diaminoazobenzene (DAA) in a stoichiometric mixture of DGEBA-DDS (I) and DGEB-DDS (II) as a function of cure time at 160°C. (Reproduced from Ref. 4. Copyright 1986 American Chemical Society.)

Table I

Positions of $\lambda_{max}$ for Model Compounds of DAA and GPE

| Model Comp'd.* | $\lambda_{max}$ (nm) | $\Delta\lambda$ |
|---|---|---|
| DAA (PP) | 410 | 0 |
| 1st (PS) | 420 | 10 |
| 2nd (SS)<br>3rd (PT) | 445 | 35 |
| 4th (ST) | 460 | 50 |
| 5th (TT) | 470 | 60 |

* refer to Scheme I for chemical structure. Designation in parentheses; P means primary amine, S for secondary and T for tertiary amine.

where $A_{ps}$, $A_{ss}$, $A_{pt}$, $A_{st}$ or $A_{tt}$ corresponds to the fractional amount of cure products as defined by Scheme I. $A_{pp}$ is the fraction of unreacted diamine. From Table II, we can obtain the following trends: (i) As predicted by the spectral shifts of Fig. 1, the fraction of branch points and cross-linkers increases with cure time for both epoxy matrices; and (ii) the cure reaction for DGEBA-DDS epoxy seems somewhat slower than DGEB-DDS, especially at longer cure times.

Kinetic Analyses of Epoxy Cure

Since epoxy homopolymerization may be neglected in the absence of catalysts (7), the major cure reactions can be assumed to be the reactions between epoxide and amine groups as expressed in Scheme I. This kinetic scheme defines the rate constant $k_1$ due to the conversion of the primary amine to the secondary amine while $k_2$ is due to the conversion of the secondary amine to the tertiary amine.

Based on the Scheme I, one can write a series of kinetic differential equations as described by Dusek et al (8) as follows:

$$-\frac{d[pp]}{dt} = 4k_1[pp][b] \quad (2)$$

$$-\frac{d[ps]}{dt} = 2k_1[ps][b] + k_2[ps][b] - 4k_1[pp][b] \quad (3)$$

TABLE II

Composition of Cure Products in DGEB-DDS and DGEBA-DDS Epoxy as a Function of Cure Time at 160°C

| Epoxy | Cure Time (min.) | Cure Products (%) | | | | | Extent of Amine Reaction ($\xi_a$) |
|---|---|---|---|---|---|---|---|
| | | $A_{PP}$ | $A_{PS}$ | $A_{SS}+A_{PT}$ | $A_{ST}$ | $A_{TT}$ | |
| | 0 | 78 | 20 | 2 | 0 | 0 | .059 |
| | 5 | 44 | 34 | 14 | 6 | 2 | .219 |
| | 10 | 33 | 33 | 16 | 11 | 7 | .318 |
| | 20 | 13 | 20 | 32 | 22 | 14 | .510 |
| DGEB- | 30 | 6 | 12 | 20 | 31 | 31 | .676 |
| DDS | 45 | 2 | 8 | 18 | 33 | 40 | .750 |
| | 70 | 2 | 4 | 17 | 36 | 41 | .775 |
| | 100 | 1 | 5 | 13 | 30 | 52 | .815 |
| | 150 | 1 | 5 | 11 | 21 | 62 | .848 |
| | 300 | 1 | 3 | 10 | 37 | 47 | .814 |
| | 800 | 1 | 3 | 13 | 34 | 49 | .813 |
| | 0 | 75 | 25 | 0 | 0 | 0 | .063 |
| | 5 | 61 | 20 | 9 | 6 | 5 | .183 |
| | 10 | 35 | 41 | 9 | 9 | 5 | .270 |
| | 15 | 25 | 28 | 36 | 5 | 5 | .343 |
| DGEBA- | 30 | 10 | 17 | 17 | 23 | 34 | .636 |
| DDS | 45 | 8 | 10 | 23 | 25 | 33 | .653 |
| | 60 | 9 | 15 | 19 | 25 | 33 | .646 |
| | 100 | 10 | 15 | 26 | 25 | 23 | .667 |
| | 150 | 10 | 15 | 15 | 29 | 27 | .675 |
| | 300 | 8 | 14 | 23 | 29 | 27 | .638 |
| | 1140 | 7 | 20 | 12 | 20 | 42 | .680 |

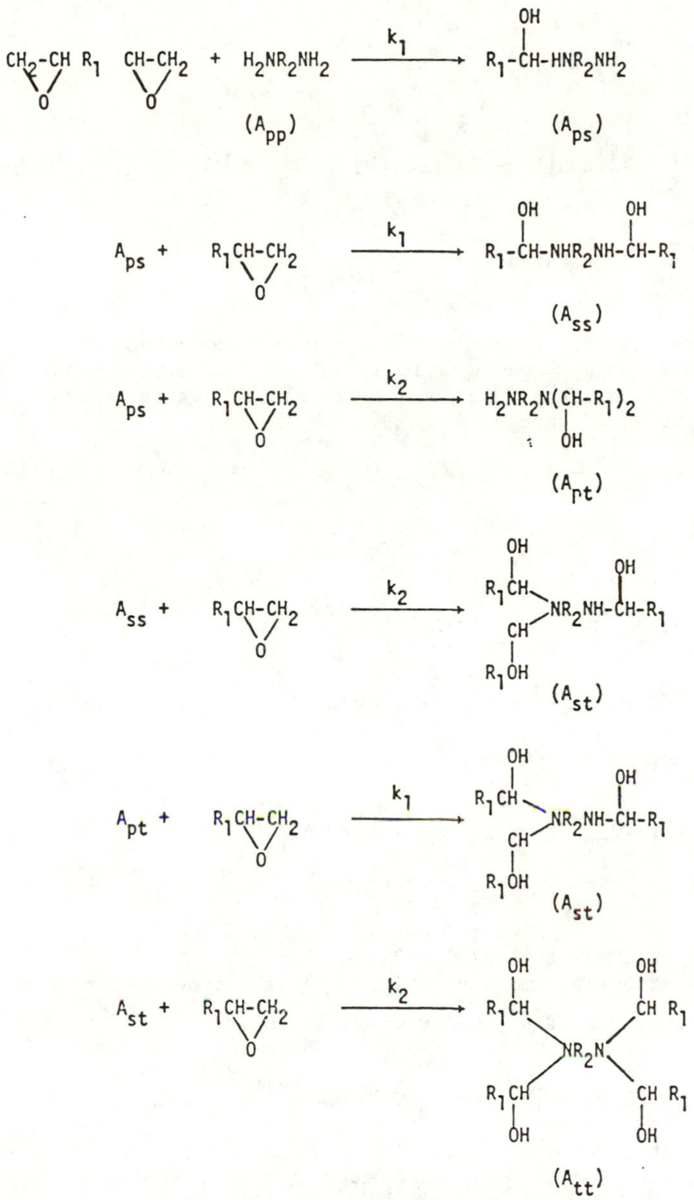

$$\underset{\diagdown O \diagup}{CH_2-CH}\ R_1\ \underset{\diagdown O \diagup}{CH-CH_2} + \underset{(A_{pp})}{H_2NR_2NH_2} \xrightarrow{k_1} \underset{(A_{ps})}{R_1-CH(OH)-HNR_2NH_2}$$

$$A_{ps} + R_1\underset{\diagdown O \diagup}{CH-CH_2} \xrightarrow{k_1} \underset{(A_{ss})}{R_1-CH(OH)-NHR_2NH-CH(OH)-R_1}$$

$$A_{ps} + R_1\underset{\diagdown O \diagup}{CH-CH_2} \xrightarrow{k_2} \underset{(A_{pt})}{H_2NR_2N(CH(OH)-R_1)_2}$$

$$A_{ss} + R_1\underset{\diagdown O \diagup}{CH-CH_2} \xrightarrow{k_2} \underset{(A_{st})}{(R_1CH(OH))(CH(OH)R_1)NR_2NH-CH(OH)-R_1}$$

$$A_{pt} + R_1\underset{\diagdown O \diagup}{CH-CH_2} \xrightarrow{k_1} \underset{(A_{st})}{(R_1CH(OH))(CH(OH)R_1)NR_2NH-CH(OH)-R_1}$$

$$A_{st} + R_1\underset{\diagdown O \diagup}{CH-CH_2} \xrightarrow{k_2} \underset{(A_{tt})}{(R_1CH(OH))_2NR_2N(CH(OH)R_1)_2}$$

Scheme I
Kinetic Scheme of Epoxy Cure Reactions

$$- \frac{d[ss]}{dt} = 2k_2[ss][b] - 2k_1[ps][b] \quad (4)$$

$$- \frac{d[pt]}{dt} = 2k_1[pt][b] - k_2[ps][b] \quad (5)$$

$$- \frac{d[st]}{dt} = k_2[st][b] - 2k_2[ss][b] - 2k_1[pt][b] \quad (6)$$

$$- \frac{d[tt]}{dt} = -k_2[st][b] \quad (7)$$

where [b] is the concentration of unreacted epoxy groups.

By solving the above equations, we can obtain the fraction of each cure species, as a function of the reactivity ratio of $k_2/k_1$ and the fraction of unreacted diamine ($A_{pp}$)

$$A_{ps} = 2p(A_{pp}{}^{q} - A_{pp}) \quad (8)$$

$$A_{ss} = p^2(-2A_{pp}{}^{q} + A_{pp} + A_{pp}{}^{r/2}) \quad (9)$$

$$A_{pt} = -2pA_{pp}{}^{q} + rpA_{pp} + 2A_{pp}{}^{1/2} \quad (10)$$

$$A_{st} = p^2[(r+2)A_{pp}{}^{q} - rA_{pp} - (2-r)A_{pp}{}^{1/2} - 2A_{pp}{}^{r/2} + (2-r)A_{pp}{}^{r/4}] \quad (11)$$

$$A_{tt} = p^2[-rA_{pp}{}^{q} + (r^2/4)A_{pp} + (r/p)A_{pp}{}^{1/2} + A_{pp}{}^{r/2} - (2-r)A_{pp}{}^{r/4} + (\frac{r}{2} -1)^2] \quad (12)$$

where $r = k_2/k_1$, $p = 1/(1-r/2)$ and $q = (1+r/2)2$
The overall extent of amine reaction ($\xi_a$) as defined before can now be written in terms of $A_{pp}$ and r only, as follows:

$$\xi_a = 1 - [1/(2-r)]\,[(1-r)A_{pp}{}^{1/2} + A_{pp}{}^{r/4}] \quad (13)$$

Now we attempt to determine the r value that best fits with the experimental data. As demonstrated by the branching theory of Bidstrup and Macosko (9), and of Dusek (10,11), the r value has a strong effect on many structural parameters of the network. From the UV-VIS spectra obtained at three cure temperatures (140°, 160° and 180°C), the fraction of each cure product was deconvoluted and plotted with the maximum error bars as a function of the extent of amine reaction in Fig. 2, for DGEB-DDS.

In the literature, many values of r have been reported (12,13). While the majority of the reported r values are close to one, much smaller r values in the range of 0.1 to 0.2 have also been reported(14,15) and, in fact, an r value of 0.1 has sometimes been used by theoreticians(9). One of our primary objectives in this work is then to clarify the confusion concerning the reported r values.

In order to find the best-fit value of r, we show two theoretical curves according to Equations 8 through 12 corresponding to an r of 0.1 (dotted line) or 1 (solid line). A comparison of the experimental data with these two sets of curves, especially the $A_{ss}$ profile, clearly eliminates the possibility of r being close to 0.1. Rather, an r value of unity seems to fit the data quite well, especially for the DGEB-DDS epoxy. Thus, the reaction rates of primary amine-epoxy and secondary amine-epoxy are practically indistinguishable. This is also the conclusion reached by Prime (16) after a careful review of the literature and especially in view of the thermal analyses data.

Assuming r = 0.1, we have calculated $k_1$ from the kinetic equation (2). After the integration of Eqn. (2) and proper substitution, we get;

$$\frac{\xi_a}{\xi_a - 1} = -4k_1 t \qquad (14)$$

Fig. 3 shows the plot of $\xi_a/(\xi_a-1)$ versus the cure time for the DGEBA-DDS epoxy at three cure temperatures. At cure times beyond gelation, the reaction rate constant which is proportional to the slope of these curves is clearly reduced. A similar decrease in the reaction rate has been observed by thermal analyses (17) and IR (18). By drawing a straight line through the first few data points for the slope, we estimate the initial $k_1$ to be $5.4 \times 10^{-3}$, $1.3 \times 10^{-2}$ and $3.1 \times 10^{-2}$ $min^{-1}$ at 140°, 160° and 180°, respectively. From the Arrhenius plot combining the data for both epoxy matrices, an activation energy of 15.7 Kcal/mole and a pre-exponential factor of $1.2 \times 10^{6}$ $min^{-1}$ were estimated. These estimates from our studies are similar to the values reported from the use of DSC and other techniques which measure the overall extent of reaction. (16)

## Fluorescence Studies

When excited around 320 nm, both epoxy matrices are highly fluorescent with an emission maximum around 370 ∿ 380 nm. When the fluorescence intensity is calibrated for thickness fluctuations, the fluorescence intensity is constant, independent of the extent of cure. Therefore, the inherent fluorescence of the epoxy matrix itself is not useful to monitor cure reactions.

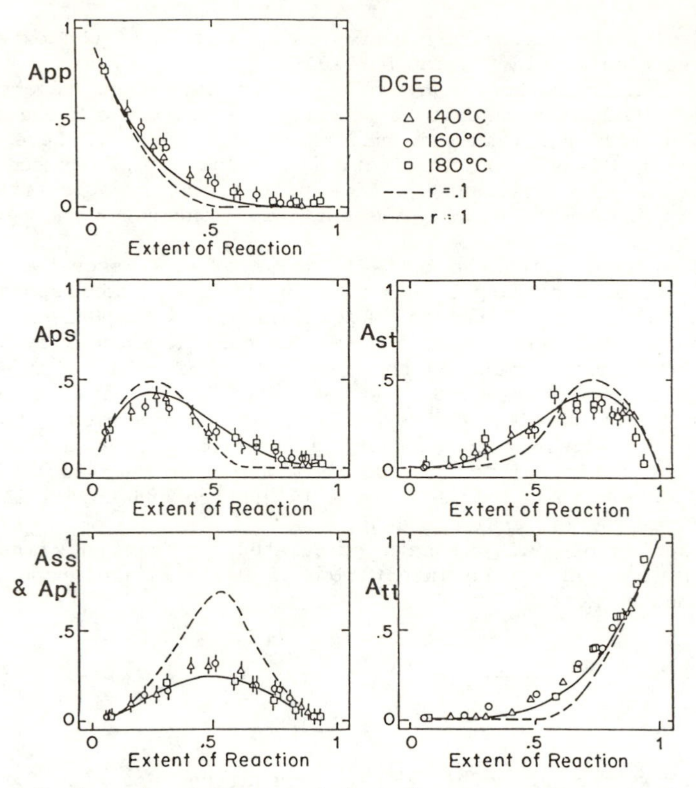


Figure 2. Experimental composition of cure species as a function of extent of amine reaction for DGEB-DDS. Refer to Scheme I for definitions of cure species. Two dotted lines show predictions for reactivity of ratio of either 0.1 or 1. (Reproduced from Ref. 4. Copyright 1986 American Chemical Society.)

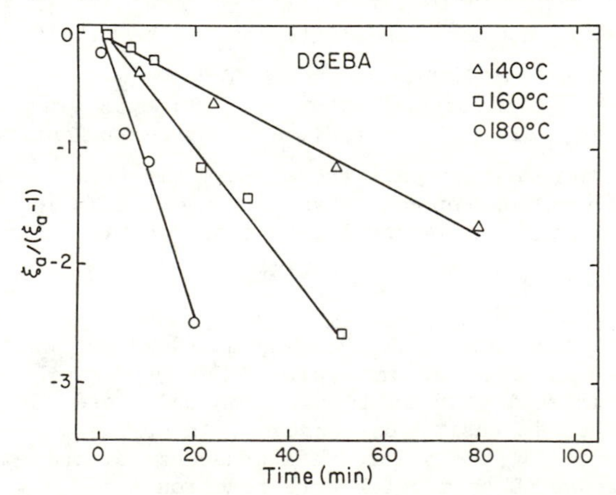


Figure 3. Plot of $\xi_a/(\xi_a-1)$ versus cure time to calculate rate constant ($k_1$) for epoxy-primary amine reaction. (Reproduced from Ref. 4. Copyright 1986 American Chemical Society.)

However, when the DAA label is excited, for example at 456 nm, the fluorescence intensity exhibited strongly cure dependent behavior. Fig. 4 shows such fluorescence spectra for DGEBA-DDS-DAA and DGEB-DDS-DAA respectively in the spectral range of 450 nm to 800 nm. In both epoxy matrices, at zero cure time, hardly any fluorescence is observed. But with increasing cure time, fluorescence increases with a broad emission peak around 560 nm. At long cure times, the emission peak seems to have red-shifted slightly (by 5 ∿ 10 nm). This small red shift is in sharp contrast to much larger red shifts observed in UV-VIS spectra (ref. Fig. 1.I and II). Changes of polarity in the solvent medium are known to cause large shifts in emission spectra. (19) Therefore, we can conclude that polarity did not change much as the epoxy cured. This trend was also suggested by small shifts in UV-VIS spectra by the fully substituted DAA.

In order to quantify the fluorescence intensity changes, we plot the relative fluorescence intensity at 560 nm as a function of cure time at three cure temperatures for both epoxies in Fig. 5. The relative fluorescence is calculated after dividing by the label's UV-VIS peak area in order to calibrate for thickness fluctuation. The label's peak area remains constant as long as the thickness does not change due to the same extinction coefficients observed for cure products of DAA. The fluorescence intensity for the DGEB epoxy is about the same as for the DGEBA epoxy up to gelation time. However, fluorescence increases continuously beyond gelation for DGEB but levels off in DGEBA epoxy. This is due to the vitrification which only occurs in the DGEBA epoxy. In the DGEBA epoxy, we have shown (2) that the increase in fluorescence derives from the cure products alone rather than from the viscosity changes. Thus the total fluorescence intensity can be written as $I_F = c\Sigma F_i A_i$, where $F_i$ is the relative fluorescence quantum yield for each cure product, $A_i$ is their concentration, and c is the experimental constant. The relative fluorescence quantum yield for the cross-linker model compound was found to be independent of cure. Using the concentration values obtained by deconvolution of UV-VIS spectra (e.g. Table II), we found that a predicted $I_F$ agrees well with the experimental points. Much greater values of $I_F$ for DGEB are thus a direct consequence of further cure reactions as indicated by both UV-VIS results and by IR monitoring of epoxy ring disappearance.

In order to correlate fluorescence intensity at 565 nm with the overall extent of amine reaction, we plotted $I_F$ versus $\xi_a$ in Fig. 6, for both epoxies cured at three isothermal cure temperatures. In this figure, $\xi_a$ is estimated from the deconvolution of UV-visible spectra. All the data fall on a single smooth curve whose slope is much sharper at later stages of cure especially after gelation. In other words, fluorescence intensity constitutes a sensitive gauge for monitoring cure beyond gelation, because it derives mostly from tertiary amine products.

Labelling by p,p'-diaminostilbene (DAS)

Figure 7 shows uv-vis spectra of DGEB-MDA-DAS (0.1%) following cure at 140°C. Before cure, the DAS absorption maximum lies at 352 nm.

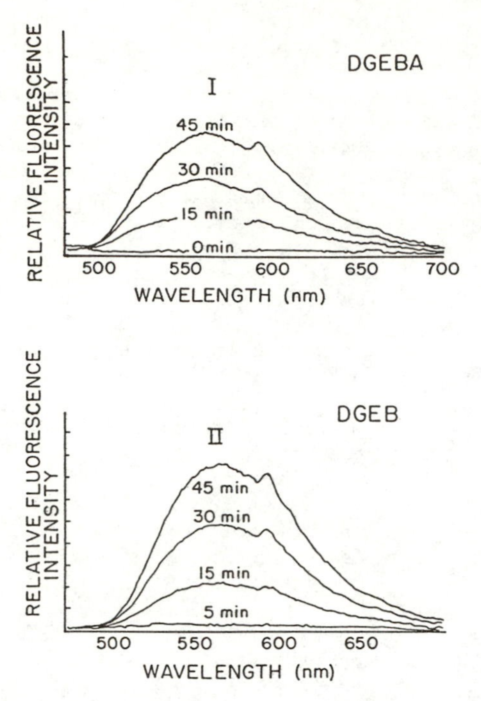


Figure 4. Fluorescence spectra of p,p'-diaminoazobenzene in a stoichiometric mixture of DGEBA-DDS (I) and DGEB-DDS (II) as a function of cure time at 160°C (excitation at 456 nm). Reproduced from Ref. 4. Copyright 1986 American Chemical Society.)

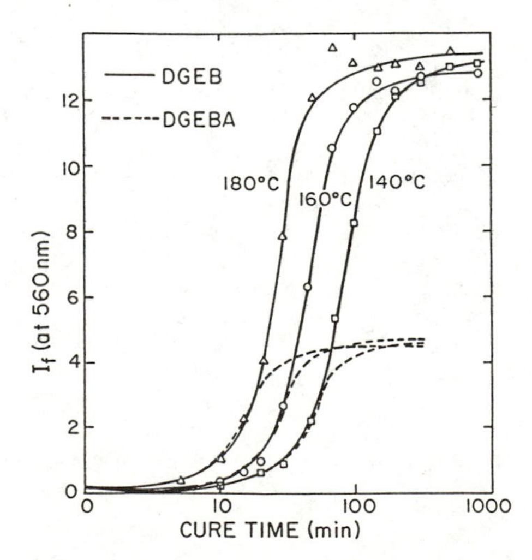


Figure 5. Relative fluorescence intensity at 565 nm as a function of cure time at three cure temperatures for DGEBA-DDS and DGEB-DDS epoxies. (Reproduced from Ref. 4. Copyright 1986 American Chemical Society.)

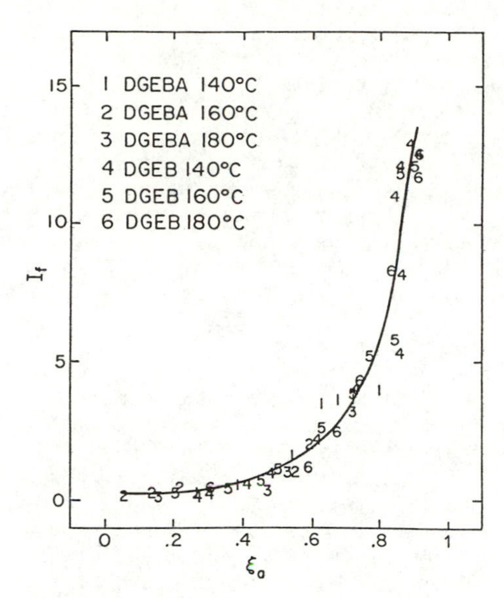


Figure 6. Correlation of relative fluorescence at 565 nm with the extent of amine reaction by UV-VIS. (Reproduced from Ref. 4. Copyright 1986 American Chemical Society.)

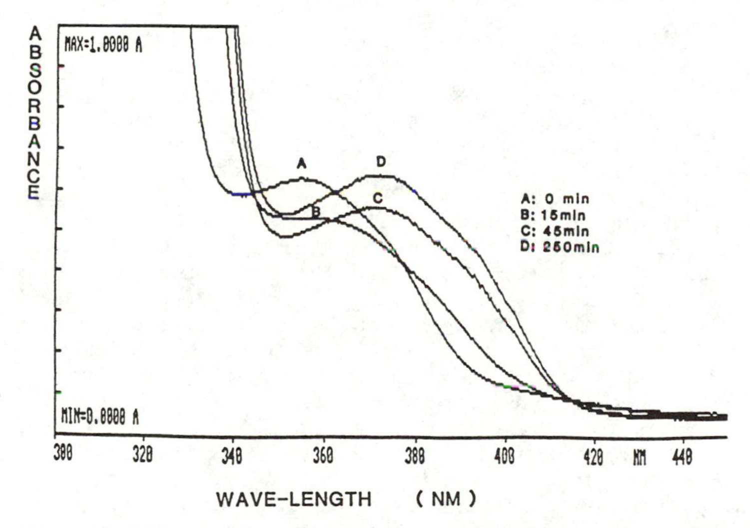


Figure 7. UV-VIS absorption spectra of DAS in DGEB-MDA epoxy as a function of cure time at 140°C. (Reproduced with permission from Ref. 3. Copyright 1987 Butterworth & Co. [Publishers] Ltd.)

DAS in DGEB shows another absorption peak at 327 nm, which is hidden under the MDA absorption in the DGEB-MDA epoxy. As cure proceeds, we observe red shifts of the peak at 352 nm. After 300 minutes of cure at 140°C, the DAS absorption peak is shifted to 371 nm. Further curing even at high temperatures does not shift the position of this peak beyond 372 nm, which is consistent with the model reactions.(3) Fig. 8 shows fluorescence spectra as a function of cure time at 140°C. At zero cure time, the emission spectrum is broad with a maximum near 418 nm. As cure increases, an enhancement of emission was noted, as well as the splitting into two emission peaks at 418 nm and 430 nm. This splitting is due to the red-shifted emission of a tertiary-amine containing DAS (tt-DAS) as a result of cure reactions. When the emission intensity at 418 nm is plotted as a function of cure time at two cure temperatures (140° and 120°C, respectively), we obtain s-shaped curves as illustrated in Fig. 9. At these cure temperatures, the gelation for the DGEBA-MDA epoxy is known to take place after 8 min. and 16 min. respectively, according to a T-T-T diagram. (6) We may assume that the gelation takes place at similar times in the DGEB-MDA epoxy. We note in Fig. 9 that after gelation, fluorescence emission increases sharply at the cure temperature. Also, it is noted that the overall increase in emission intensity is about three and a half times greater at longer cure times, as compared to zero cure time.

In order to delineate the effect of viscosity on the fluorescence intensity, we cured the DGEB-MDA epoxy containing a small amount of tt-DAS as a probe molecule. In this experiment, any fluorescence change is due to the viscosity effect of the medium since tt-DAS can no longer react in the matrix. Fig. 10 compares the emission intensity of DAS with tt-DAS in the DGEB-MDA matrix as a function of cure time at 140°C. Emission intensity in Fig. 10 is calibrated to reflect emission per mole of DAS or tt-DAS. At zero cure time, the emission from tt-DAS is 2.4 times greater than DAS. While DAS shows an increase in emission after gelation, the emission from tt-DAS is constant well past gelation up to the cure time of 100 minutes. At this cure time, about 75% of the epoxide in the matrix has reacted, as determined by IR spectra. After 100 minutes the emission from tt-DAS increases sharply, at least three times. From these trends, we may conclude that the enhanced fluorescence of DAS near gelation is due largely to the chemical reactions taking place; only at later stages is it due to the medium viscosity effect. Even though we cannot separate in a quantitative way the contributions from chemical reactions and the medium viscosity effect at a given cure time when DAS was used, a plot of emission intensity in the DAS labelled epoxy as a function of the overall extent of epoxide reaction can be used to estimate the extent of cure. Such a plot is shown in Fig. 11, for a cure at 140°C, where a sharp increase in emission is shown after gelation.

In *trans*-stilbene, fluorescence and rotation around the CH = CH bond are known to be the main photophysical processes. Assuming that the photophysical behavior of DAS and tt-DAS is similar to that of *trans*-stilbene, the increase in fluorescence of tt-DAS,

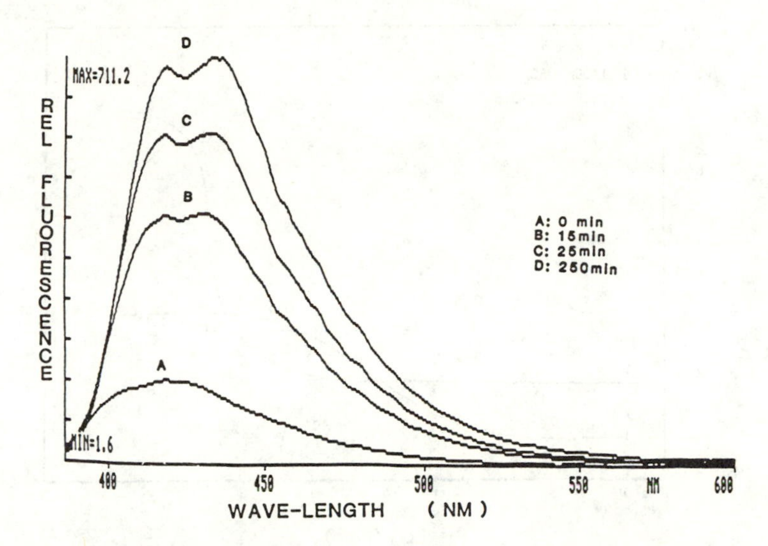


Figure 8. Fluorescence spectra of DAS in DGEB-MDA epoxy as a function of cure time at 140°C. (Reproduced with permission from Ref. 3. Copyright 1987 Butterworth & Co. [Publishers] Ltd.)

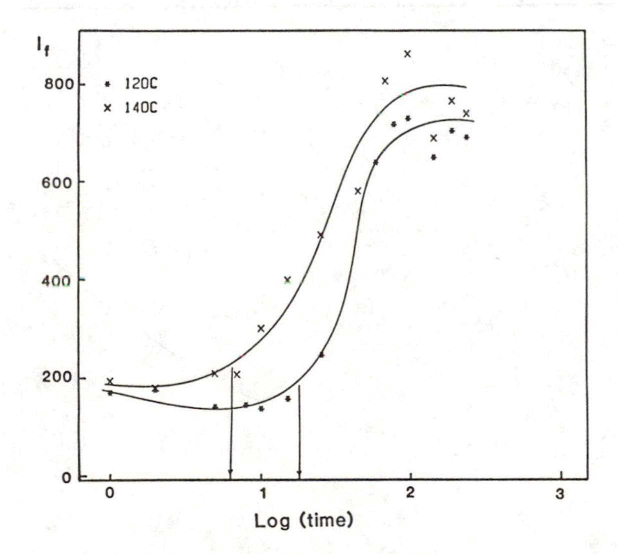


Figure 9. Fluorescence intensity at 418 nm as a function of cure time at 140°C and 120°C, respectively, in DAS (0.1%) labelled DGEB-MDA epoxy. (Reproduced with permission from Ref. 3. Copyright 1987 Butterworth & Co. [Publishers] Ltd.)

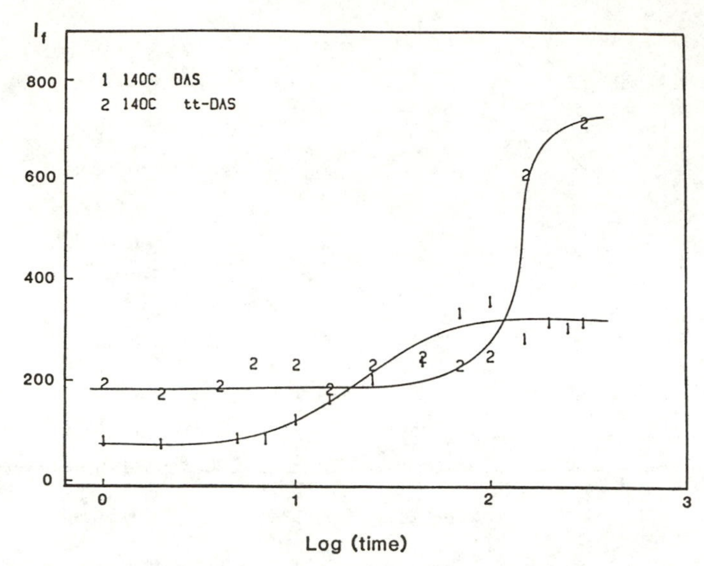


Figure 10. Fluorescence intensity at 418 nm as a function of cure time at 140°C, comparing DAS and tt-DAS in DGEB-MDA epoxy. (Reproduced with permission from Ref. 3. Copyright 1987 Butterworth & Co. [Publishers] Ltd.)

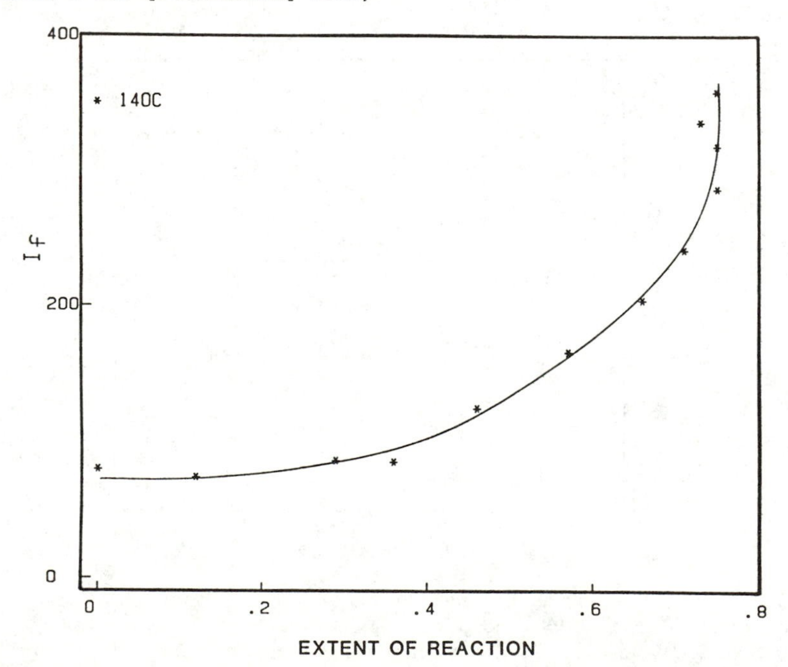


Figure 11. Correlation of fluorescence intensity at 418 nm with the extent of epoxy reaction ($\xi_b$)by IR method at 140°C cure. (Reproduced with permission from Ref. 3. Copyright 1987 Butterworth & Co. [Publishers] Ltd.)

especially in cured epoxy, can be explained by a higher activation barrier to the rotation around the CH = CH bond. (3)

Acknowledgment

The support of this research by the Army Research Office (Contract No. DAAG 29-85-K-0055) and the National Science Foundation, Polymers Program (Grant No. DMR 82-05897) is acknowledged. With great pleasure, I acknowledge several capable collaborators (Dr. I. J. Chin, Ms. E. Pyun, Mr. R. Mathisen, Dr. H. L. Sun.)

References

1. Chin, I. J.; Sung, C. S. P. Macromolecules, 1984, 17, 2603.
2. Sung, C. S. P.; Chin, I.-J.; Yu, W. C. Macromolecules, 1985, 18, 1510.
3. Sung, C. S. P.; Mathisen, R. Polymer, 1987, in press.
4. Sung, C. S. P.; Pyun, E.; Sun, H. L., Macromolecules, 1986 19, 2922.
5. Wang, F. W.; Lowry, R. E.; Fanconi, B. M. ACS Polym. Mater. Sci. & Eng. Proceedings, 1985, 53-2, 180.
6. Enns, J. B. and Gillham, J. K., Polymer Characterization Ed. by Craver, C. D., Advances in Chemistry Series #203, ACS, 1983.
7. Byrne, C. A.; Hagnauer, G. L. and Schneider, N. S., Polym. Comp., 1983, 4, 206.
8. Dusek, K.; Ilavsky, M. and Lunak, S., J. Polym. Sci., Symp. No. 53, 1975, 29.
9. Bidstrup, S. A. and Macosko, C. W., Proceedings of ANTEC '84, SPE 1984, 278.
10. Lunak, S. and Dusek, K., J. Polym. Sci., Symp. No. 53, 1975, 45.
11. Miller, D. R. and Macosko, C. W., Macromolecules, 1980, 13, 1063.
12. See references cited in Table 1 of Charlesworth J., J. Polym. Sci., Polym. Chem., 1980, 18, 621.
13. Zukas, W. X.; Schneider, N. S. and MacKnight, W. J., Polym. Materials Sci. & Eng. Preprint, 1983, 49-2, 588.
14. Bell, J. P., J. Polym. Sci., A-2, 1970, 8, 417.
15. Morgan, R. J.; Happe, J. A. and Mones, E. T., Proc. 28th SAMPE Symp., April 12-14, 1983, 596.
16. Prime, R. B., Thermal Characterization of Polymeric Materials, Ed. Turi, E. A., Academic Press: 1981, Fig. 29, p. 479.
17. We can calculate $k_1$ by expressing [b] in terms of [pp], followed by integration. However, it does not give as reliable values as based on $[\xi_a]$.
18. Riccardi, C. C.; Adabbo, H. E. and Williams, R. J. J., J. Appl. Polym. Sci., 1984, 29, 2481.
19. For a review, Lakowicz, J. R., Principles of Fluorescence Spectroscopy, Plenum: NY, 1983; Chap. 7.

RECEIVED September 10, 1987

# PHOTOPHYSICS OF SILICON-BASED POLYMERS

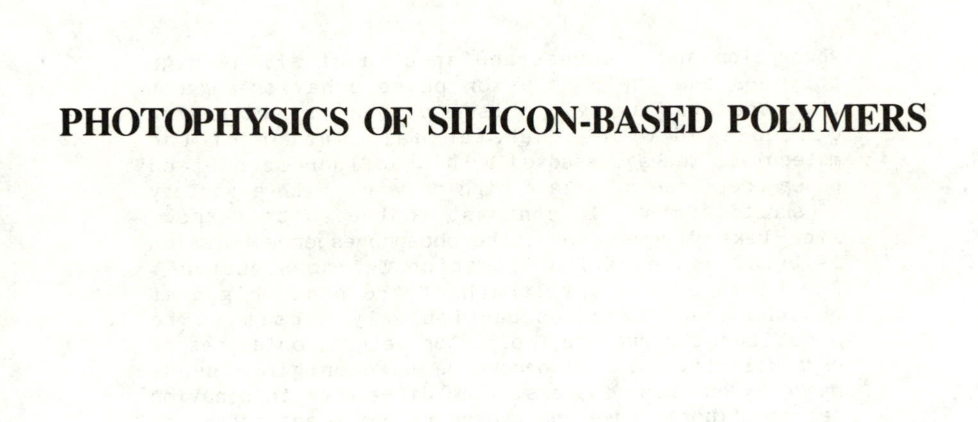

## Chapter 35

# Emission Spectra of Polysilylenes

**L. A. Harrah and J. M. Zeigler**

**Sandia National Laboratories, Albuquerque, NM 87185**

Absorption and fluorescence spectra of silane high polymers and their solution phase behavior suggest that the states giving rise to these spectra are delocalized over a substantial portion of the molecule. We have studied both the fluorescence and phosphorescence of silane high polymers with a variety of substituents. In contrast to the quite narrow, line-like fluorescence, the phosphorescence emission is broad and exhibits vibrational fine structure. Furthermore, the wavelength of the band origin of phosphorescence is not particularly sensitive to substituent structure, molecular weight, or degree of crystallinity. The fluorescence band origin depends markedly on these factors. The difference in behavior between fluorescence and phosphorescence suggests that the states giving rise to phosphorescence are quite localized while those responsible for fluorescence are delocalized along the chains. The emission yields and structure suggest that the immediate precursor to photochemistry is the triplet.

Polysilylenes, a new class of polymers synthesized (1,17) by Na metal-mediated Wurtz coupling of dichlorosilanes, are receiving considerable attention because of their extreme photosensitivity and interesting electronic properties. They are being studied as photoresists, and as exceedingly $O_2$-reactive ion etch resistant imaging and contrast enhancement layers in multilayer lithography (2-6). Highly efficient photoconductivity with nondispersive transport of the charge carriers through relatively thick films has been reported for alkyl as well as aryl polysilylenes (7). The photosensitivity of these materials depends markedly on the nature and structure of the substituents and, for some alkyl substituted polymers, results in photovolatilization of the polymer under deep UV exposure (2-3). Photoscission of the polysilylene chain, respon-

0097-6156/87/0358-0482$06.00/0

sible for initiating the photovolatilization process, also occurs in the non-photovolatilizing polysilylenes and allows their use as solvent-developed resists (4-6).

While these technologically important applications are the primary driving force for the current investigations, the silylene polymers also display a wealth of phenomena of fundamental scientific interest. Dilute solutions of many of the alkyl substituted polymers exhibit thermochromism (9) which may be associated with a random coil to rod-like conformational transition on cooling (8-10). This thermochromism has been treated using a rotational isomeric state model (9,10) and a theoretical investigation (11) has suggested that the driving force for both the thermochromism and the sharp, first-order-like transition observed with some polymer-solvent combinations, results from a complex solvent-polymer polarizability interaction induced by the change in chain electronic structure with conformational rearrangement caused by rotation about the silicon-silicon backbone bonds. Similar phenomena occur in neat thin films of the alkyl polymers (12,13) and may also be related to electronic-conformational coupling. Evidence has been presented that indicates that solution phase conformational transitions also occur (14) that are not manifested in thermochromism for polysilylenes having physically large substituents.

The absorption spectra of silylene polymers and telomers have been extensively reported and it has been shown that the position of the absorption maximum shifts to the red with increasing degree of polymerization (6,15-19). We and others have reported the existence of narrow, line like fluorescence with no observed vibrational structure for a number of medium and high molecular weight polymers (2,9,16,20,22). The narrow, line-like fluorescence and the chain length dependence of the absorption spectra both indicate substantial delocalization of the electronic states involved in the transitions.

The delocalization of the electronic states involved in the photoexcitation of these polymers suggest that they should not undergo photodissociation from the singlet manifold. We have examined the luminescence from these polymers, both in solution and as neat thin films, in an attempt to identify the electronic precursor to their interesting photochemistry and to provide guidance in interpreting their other spectroscopic, electronic, and physical properties.

We have studied the thermochromism of fluorescence and show this behavior to be consistent with the rotational isomeric state model previously proposed to explain solution thermochromism in absorption (9,10). Weak, structured phosphorescence is observed from all polymers studied. The contrast between the structured phosphorescence and the narrow fluorescence is interpreted as evidence that the triplet state is the immediate precursor to photochemistry. Finally, the change in the fluorescence character in the aryl series on going from phenyl substitution to naphthyl substitution suggests a change in the nature of the transition from one involving mixed side chain-backbone states in the phenyl case to one which is primarily side chain-like for naphthyl-substituted polysilylenes.

## Experimental

The polymers used in this study were prepared in the manner described in references 1 and 17 by a Wurtz-type reductive coupling of the appropriate dichlorosilane(s) with sodium metal dispersion in toluene or a mixed solvent (1a,b). Fluorescence, phosphorescence and absorption spectra were taken on a spectrophotometer of our own design (17). For these studies the excitation source for fluorescence was a 200 watt Hg•Xe arc lamp filtered to reduce the intensity at the exciting line using neutral density filters with optical densities from 2.0 to 4.0. Phosphorescence excitation used a nitrogen laser with a pulse length of 0.8 nsec. The beam from this laser was expanded to reduce the energy density and normally illuminated a 2 $cm^2$ area of sample. The phosphorescence spectra were obtained by integrating the light emitted at each wavelength over 10 pulses to reduce the noise introduced by the pulse to pulse variations in laser output. For each pulse the phototube output was gated on at 200 μsec after the .8 nsec laser pulse. Data was then collected for an interval of 10 msec for the early time emission and for up to 1 sec to gather long time spectra. Decays were derived from the data at a single wavelength by again delaying 200 μsec after the laser pulse and then gathering data in 500 μsec intervals for up to 1 sec. Up to 2000 laser pulses were used to assure adequate counting statistics.

Film spectra were taken using thin films of polymer spun-cast on 1.6 mm thick fused silica flats, 25 x 25 mm square. All film samples were cast from xylene which had been previously dried over 4A molecular sieves. For the solution spectra, toluene was used after distillation from NaK alloy under Ar. The alkane solvents were obtained from a variety of sources and used without further purification.

Fluorescence quantum yields were estimated using the method of Renschler and Harrah (24) with p-terphenyl as a standard and assuming its quantum yield to be 0.93 (25). Emission spectra shown are corrected for variation in monochromator throughput and phototube response by the methods described in reference 17.

Absorption spectra were run using the same collection optics as for emission measurements. The source for absorption measurement was a deuterium arc lamp operating at 30 watts and projected onto a slit through a 1.0 OD filter, again, to avoid photolysis during the scan. The source slit was projected onto the sample with a pair of fused silica lenses arranged at the achromatizing distance for about 3200 A. The slit image was adjusted to illuminate the sample in the same area as the excitation slit image for emission measurement. The reference for absorption measurement was a scan of the light source with sample cell and solvent in the normal configuration. For films, the reference was a bare fused silica flat.

## Thermochromism in Fluid Solution

We have previously suggested (9a,10) a rotational isomeric state model to explain the solution thermochromism exhibited by the unbranched alkyl substituted polysilylenes. This model treats the absorption spectrum as a superposition of the spectra of isolated chain segments separated by conformational defects. Figure 1 compares spectra calculated using this model with observed absorption spectra for poly(n-hexyl methyl silylene).

The proposed model assumes that the electronic states of these molecules are delocalized along an undefected portion of the polymer chain and that chain defects effectively isolate the absorbing segments. The transition energies for each segment are given by the expression

$$E_T = A + 4\beta \sin \frac{\pi}{2(n-1)} \quad (1)$$

used by Boberski and Allred (15) to explain the chain length dependence in the permethyl silylene telomers. In this expression, A is a constant, β is the Huckel transfer integral (the value derived for the permethyl telomers by Boberski and Allred is used for Figure 1) and n is the number of defect-free bonds in the segment. The probability of occurrence of a defect-free segment of n bonds is calculated using a modified expression derived by Levinson (26) to treat energy transfer in linear chains with traps,

$$P(n) = n(f_d^{\,2})(1 - f_d)^{n-1} \quad (2)$$

where n is again the number of undefected bonds in a sequence.

To derive an absorption spectrum, several additional assumptions are made: the chains are sufficiently long to avoid end effects; defects isolate the electronic states of the segment; adjacent defects do not interfere; and each component has an intrinsic homogeneous width. The agreement shown between the calculated and observed spectra is surprisingly good considering the simplicity of the model.

A consequence of the model is that the peak of the absorption spectrum can be used to calculate the fraction of defect structures

$$f_d = \frac{E_T - E_0}{2\beta} \quad (3)$$

where $E_T$ is the energy of the absorption maximum at temperature T, $E_0$ is the low temperature limit (A in expression 1) and β has the previous significance. From expression 3, the energy of a defect structure can be derived. Figure 2 illustrates the Arrhenius behavior of $f_d$ for the n-propyl methyl polymer.

Previous studies have shown that energy transfer occurs over distances of 50-100 Å in these polymers (22,27). If this transfer occurs in solution following absorption, one would expect emission not from the total distribution of isolated segments, but only from the segment having the lowest transition energy. Our model predicts that the fluorescence should occur from a distribution of longest segments and should peak near the fluorescence transition energy for the most probable segment length. The energy of the fluorescence maximum should obey the expression

$$E_{fT} = E_{f0} + 4\beta \sin \frac{\pi}{2(N_p+1)} \quad (4)$$

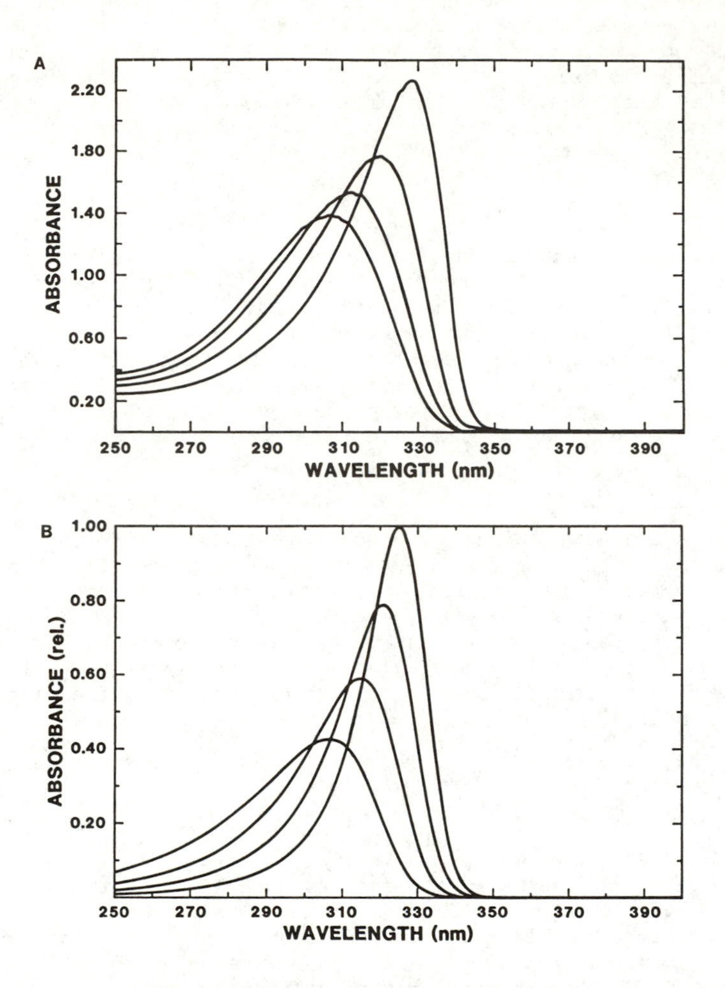


Figure 1. A. Absorption spectra of poly(n-hexyl methyl silylene) at four temperatures
B. RISM simulation of spectra.

where $E_{f0}$ is the low temperature limit for the fluorescence maximum, $E_{fT}$ is the fluorescence maximum at temperature T and $N_p$ is the most probable segment length given by

$$N_p = \frac{1}{\ln(1-f_d)} \qquad (5)$$

$f_d$ has its previous significance and can be calculated from the defect energy determined from the absorption spectrum.

Figure 3 shows a plot of the fluorescence maximum for poly(n-propyl methyl silylene) in hexane vs. sin $\pi/2(N_p + 1)$ using a defect energy of 1650 calories/mole. Although the agreement here seems good, the uncertainty in location of the fluorescence maximum is large as the band broadens near room temperature and may disguise a real nonlinearity. Nevertheless, this result lends credibility to the rotational isomeric state model for thermochromism in these polysilylenes.

For most of the unbranched alkyl substituted polymers, in solvents of high refractive index, the smooth transition described by the above model changes to an abrupt, first order like transition (8). This has been described in terms of electronic-conformational coupling (11). The thermochromism for the n-propyl methyl polymer is summarized in Figure 4 for two solvents.

Branched alkyl and aromatic substituted polymers do not show the red shift with decreasing temperature. Cyclohexyl methyl, isopropyl methyl, phenyl methyl, and trimethylsilyl methyl silylene homopolymers exhibit only a small but discernable blue shift with decreasing temperature. This behavior is interpreted to suggest that, in these hindered systems, there is little energy difference between the various chain conformations although light scattering measurements do imply some weak transition-like behavior (14). This behavior may signify that branching makes one of the conformations, presumably the *trans*, so high in energy that only *gauche* conformers are accessible.

## Thermochromism - Thin Films

In all of the polysilylenes studied, the fluorescence from neat thin films on fused silica substrates exhibits a blue shift upon cooling. In cases where our studies have spanned the glass transition of the polymer, no change in behavior is seen (Figure 4). In the polymers which have substantial crystallinity, an abrupt shift in behavior occurs at the crystalline melting point; above this temperature the films behave in much the same fashion as the fluid solutions. These phenomena have been extensively studied (9,12,13) and will not be treated here.

The blue shift probably does not involve changes in conformational structure (no change upon passing through Tg) and most likely reflects only the influence of the increased density on the transition. In these nonpolar materials, one expects a red shift on density increase (28) and perhaps a partially compensating shift to the blue due to preferential narrowing in the long wavelength tail (reference 28, pg. 42). The observed blue shift suggests that the narrowing effect has more than offset the normally expected density-determined red shift. We have examined a film of poly(phenyl methyl

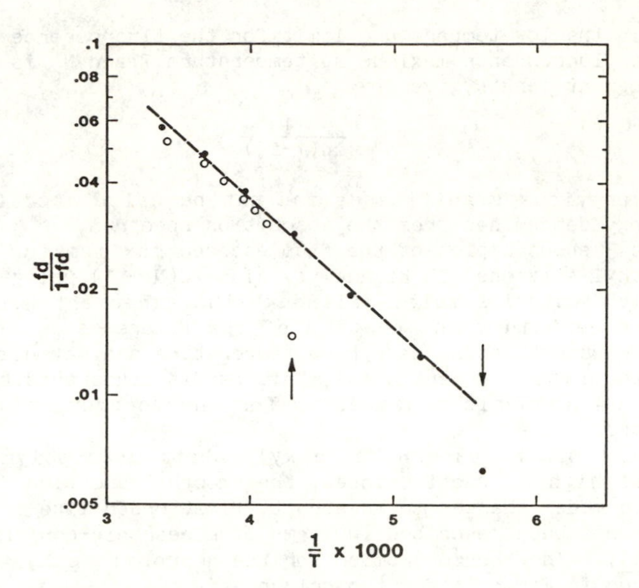


Figure 2. Thermal activation of chain defects in poly(n-propyl methyl silylene).

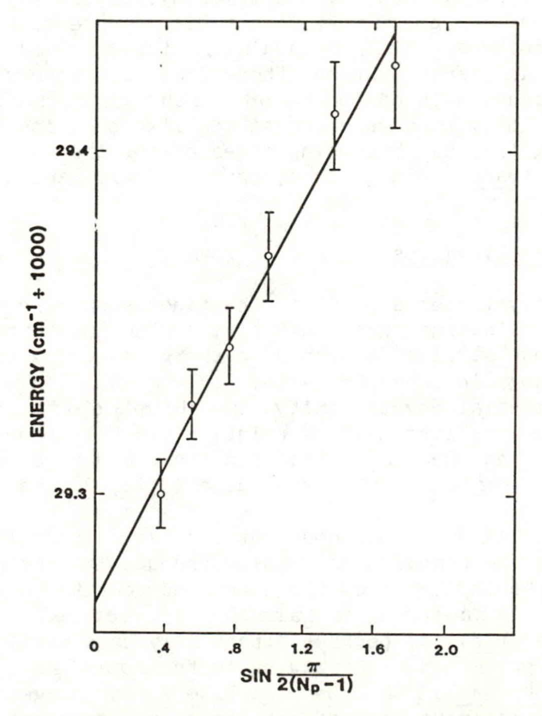


Figure 3. Dependence of fluorescence maximum of most probable chain length for poly(n-propyl methyl silylene).

silylene) cast onto a diamond anvil at high pressure and have observed a blue shift in the absorption band with increasing pressure. This result is consistent with the conclusion that the thermochromism below the crystalline melting point in these polymers is primarily related to the density increase. Figure 5 summarizes the data on film thermochromism. We have normalized these data to the room temperature fluorescence maximum for presentation convenience.

## Phosphorescence

Figure 6 displays the phosphorescence spectra of three alkyl polysilylenes obtained from pulsed laser excitation of thin film samples at low temperature. For these spectra, the data was integrated over the period from 200 μsec to 0.01 sec following the laser pulse. All of the alkyl polysilylenes studied show a substantial amount of emission in the region of fluorescence which has essentially the same shape as the normal fluorescence. We interpret this emission as delayed fluorescence resulting from triplet-triplet annihilation. The delayed fluorescence provides a convenient comparison between the fluorescence shape and width and that of the phosphorescence.

The three polymers, poly(n-propyl methyl silylene), poly(n-hexyl methyl silylene), and poly(n-dodecyl methyl silylene), exhibit spectra with a surprising similarity in structure. Clearly, the phosphorescent emission exhibits vibrational structure and is quite broad (>7500 $cm^{-1}$) while the fluorescence is narrow (<600 $cm^{-1}$) and structureless. The narrow fluorescent emission and absence of discernable vibrational structure indicate, as does the chain length dependence of absorption (6,17), that the electronic states involved are delocalized over substantial portions of the polymer chain. The breadth and structured nature of the phosphorescence indicates that the triplet is quite localized. Some vibrational progressions are apparent in these spectra and the splittings may reflect involvement of a silicon-carbon stretching vibration. Takeda and Matsumoto (29) have performed a band structure calculation on $(SiH_2)_n$ which suggests that the LUMO in that polymer has both silicon-silicon and silicon-hydrogen antibonding character. The existence of silicon-carbon vibrational splittings in the phosphorescence spectra may indicate that the calculational result is correct and the silicon substituent antibonding character of the LUMO is general.

It is not clear, if the singlet excitations are delocalized, why the triplets, presumably with the same origin and the terminus orbitals, should be localized. Recent calculations by Bigelow (30) suggest that the origin and terminus orbitals for the singlet and triplet states may not, in fact, be the same, but no experimental evidence to confirm this result has yet been brought forth. Nevertheless, the localization of the triplet does suggest its involvement in the photodissociation process. Further evidence is available from the quantum yields for the two emissions. For the alkyl polysilylenes the quantum yields for fluorescence are all, with the exception of the phenethyl substituted polymer, in excess of 0.4 (*vide* Table I). While we have not measured the phosphorescence yields, a rough comparison with poly(vinyl naphthalene)

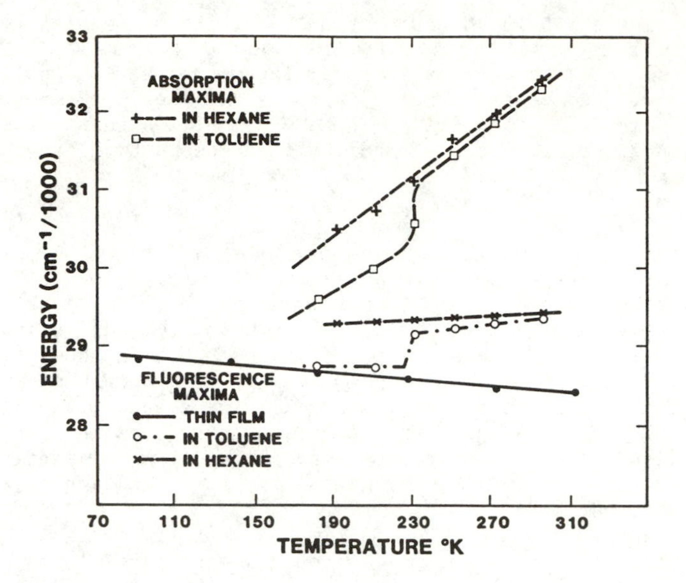


Figure 4. Thermochromism in poly(n-propyl methyl silylene).

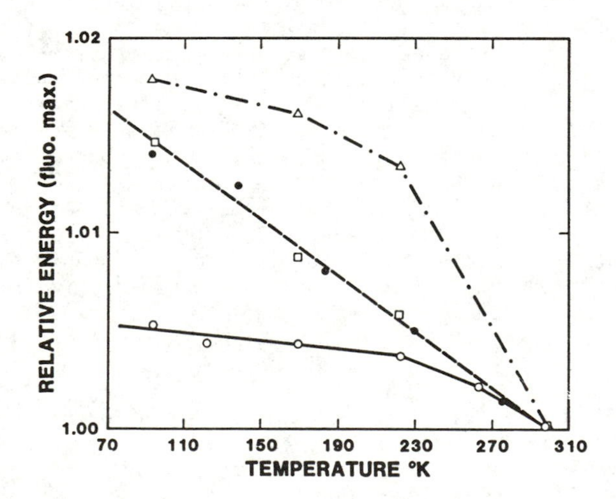


Figure 5. Thin film thermochromism for several polysilylenes.

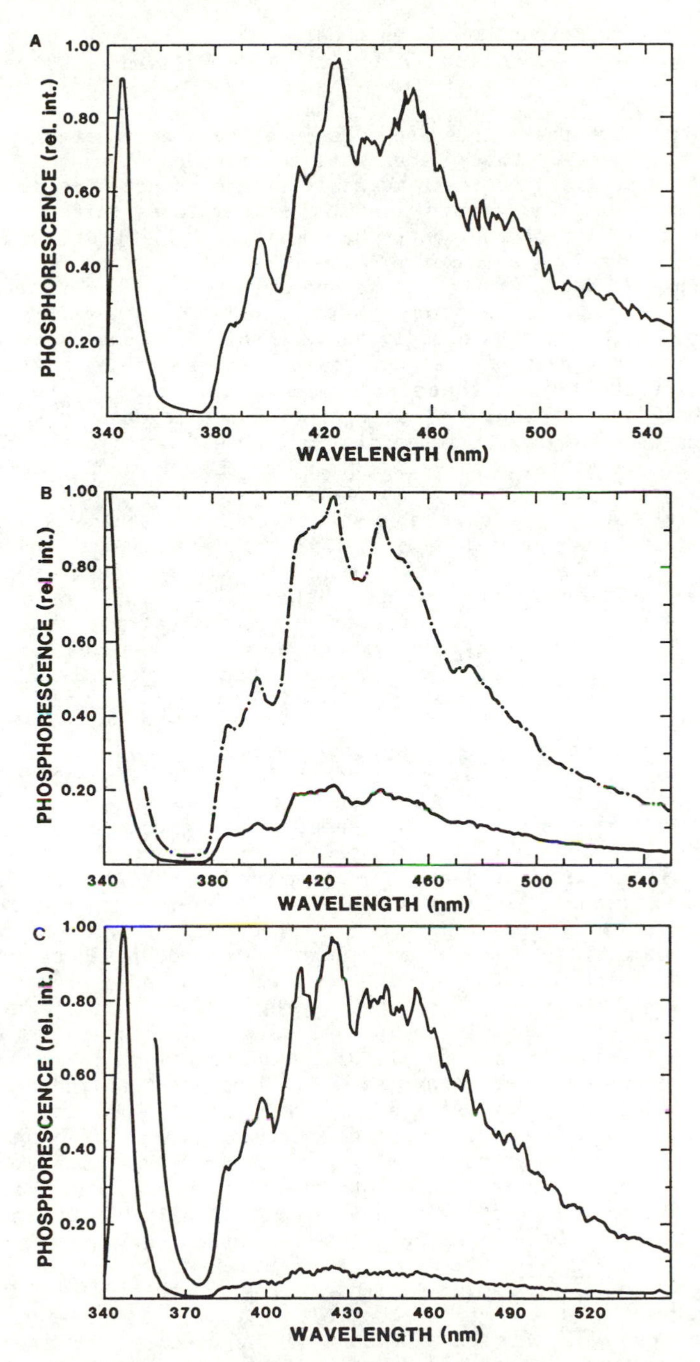


Figure 6. Laser excited delayed emission from
A. Poly(n-propyl methyl silylene)
B. Poly(n-hexyl methyl silylene)
C. Poly(n-dodecyl methyl silylene)

phosphorescence intensity places these yields at less than $10^{-4}$. No phosphorescence is detected in these films at temperatures above 250°K.

Even though the triplet appears localized, at least on a vibrational time scale, the apparent delayed fluorescence does indicate substantial mobility on the time scale of phosphorescence decay. We have examined these decays and found them to be quite nonlinear. With the exception of the β-naphthyl polymer, the first half lives are about 1-2 x $10^{-3}$ seconds.

Figure 7 shows the phosphorescence spectra during the 0.01 to 0.1 time periods for n-propyl methyl and n-hexyl methyl films. Comparison with Figure 6 clearly shows a shift to longer wavelengths and a loss in vibrational structure. Again, as in the earlier time period, the spectra of these polymers are similar. The delayed fluorescence is absent and this observation together with the red shift suggest trapping at impurities or, more probably, at photoinduced scission sites. Table II summarizes some spectral data on the phosphorescence of thin films studied.

Figure 8 is the phosphorescence spectrum taken from a glassy solution of poly(n-propyl methyl silylene) in methyl cyclopentane at 89°K. This emission is similar in width to the film emission, as are the solution spectra of the other polymers. Again delayed fluorescence is evident but the sharp vibrational fine structure is lost. The solution and film spectra are not expected to be comparable since they represent conformational equilibria (at room temperature for film and the Tg of 3-methylpentane for the solutions).

The phosphorescence spectra of the two aryl polysilylenes studied are shown in Figure 9 and their fluorescence at room temperature in Figure 10. Although phosphorescence quantum yields for these two polymers were not measured, estimates based on comparison with the alkyl intensities indicate that these polymers emit with substantially greater yield. Todesco and Kamat (21) have measured the phosphorescence yield of a copolymer of α-naphthyl methyl and dimethyl silylene units to be 0.39. Our naphthyl polymer gives clearly the most intense phosphorescence and probably has a quantum yield near the 0.39 value for the copolymer. We estimate the phenyl polymer yield to be ≈ 1/10 of the naphthyl polymer.

The fluorescence of the phenyl polymer is similar in shape to the fluorescence from the alkyl polymers and the similar shape of the phosphorescence spectrum, as well, suggests that the origins of the electronic spectrum are also much the same. The apparent increased quantum yield for phosphorescence in poly(phenyl methyl silylene) probably reflects a mixing of the ring electronic levels with the levels of the chain. Both the fluorescence and phosphorescence of the naphthyl derivative are substantially altered relative to the phenyl polymer. Fluorescence resembles that of poly(β-vinyl naphthalene) (17,29) which is attributed to excimer emission. Phosphorescence is similar to naphthalene itself. These observations suggest that the replacement of an alkyl with phenyl moiety does not change the basic nature of the electronic state but may incorporate some π character. Upon a naphthyl substitution both the fluorescence and phosphorescence become primarily $\pi$-$\pi^*$ like.

Table I. Solution and Film Fluorescence Data for Polysilylenes

| Polymer Substituents | Solution (Room Temp.) | | | Film (Low Temp.) | |
|---|---|---|---|---|---|
| | $\lambda_f$ | $\phi_f$ | width ($cm^{-1}$) | $\lambda_f$ | width ($cm^{-1}$) |
| phenyl, methyl | 353 | $.08-.25^+$ | 1043 | 353 | 706 |
| p-anisyl, methyl | 362 | $0.2-0.5^+$ | 1220 | | |
| β-naphthyl, methyl | 437 | - | 4167 | | |
| n-propyl, methyl | 341 | .76 | | 345 | 447 |
| n-hexyl, methyl | 337 | .61 | 1497 | 343 | 553 |
| n-dodecyl, methyl | 337 | .67 | 1500 | 347 | 532 |
| di n-hexyl | 342 | .42 | 1275 | 371 | 472 |
| β phenethyl, methyl | 338 | .14 | | | |
| Co i-propyl, methyl; n-propyl, methyl | 340 | .69 | 1143 | 342 | 592 |

+ excitation wavelength dependent
* highly crystalline

Table II. Polysilylene Phosphorescence Origins and Vibrational Splittings

| Polymer Substituents | λ Origin | Vibrational Splittings |
|---|---|---|
| phenyl, methyl | 390 | 1667 |
| β-naphthyl, methyl | 467 | 1450 |
| n-propyl, methyl | 380 | 722/1667 |
| n-hexyl, methyl | 380 | 717/1694 |
| n-dodecyl, methyl | 380 | 801/1589 |
| di n-hexyl | 375 | 848/1613 |
| co i-propyl, methyl; n-propyl, methyl | 380 | 894 |

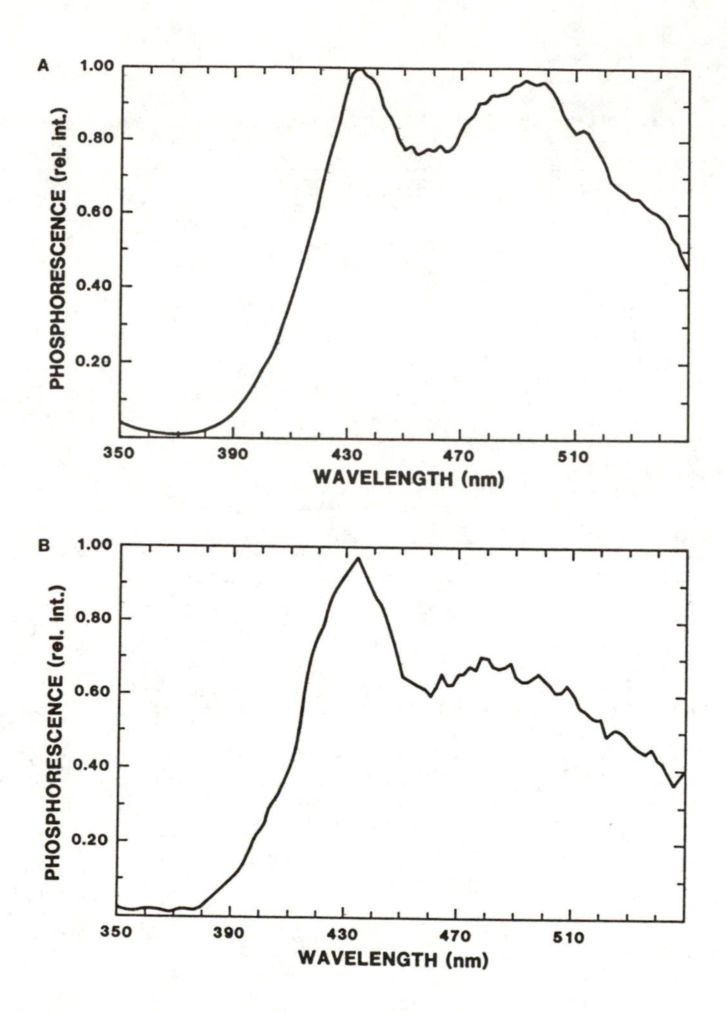


Figure 7. Laser excited delayed emission at times greater than 0.01 seconds from

A. Poly(n-propyl methyl silylene)

B. Poly(n-hexyl methyl silylene)

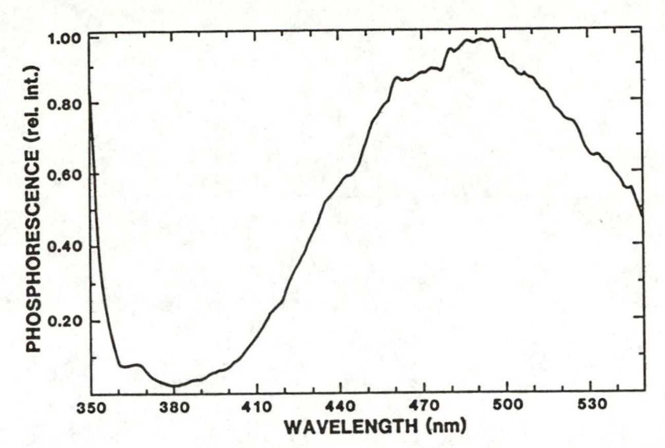


Figure 8. Laser excited delayed emission from poly(n-propyl methyl silylene) in methylcyclopentane.

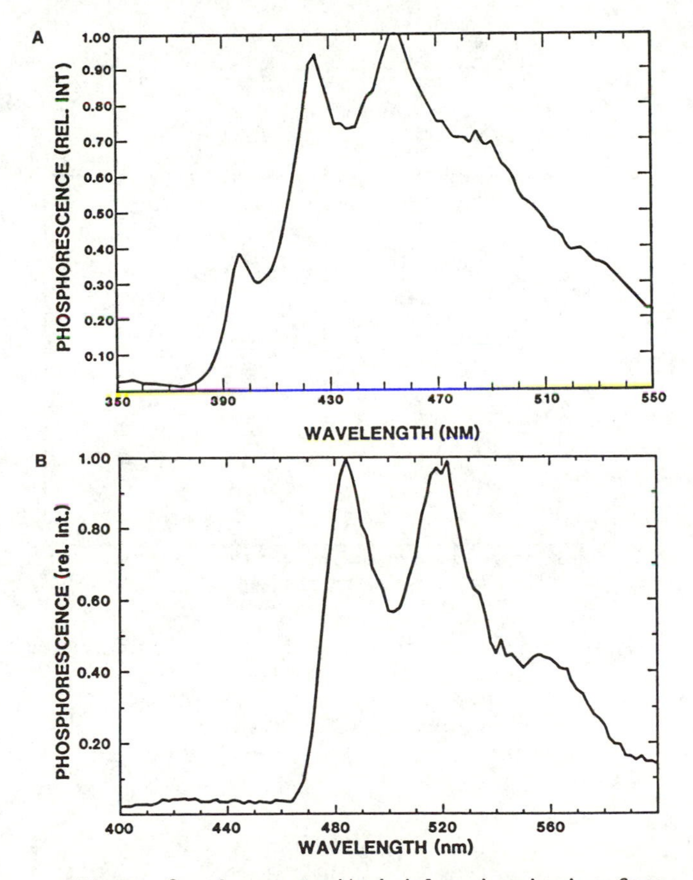


Figure 9. Laser excited delayed emission from
A. Poly(phenyl methyl silylene)
B. Poly (β-naphthyl methyl silylene)

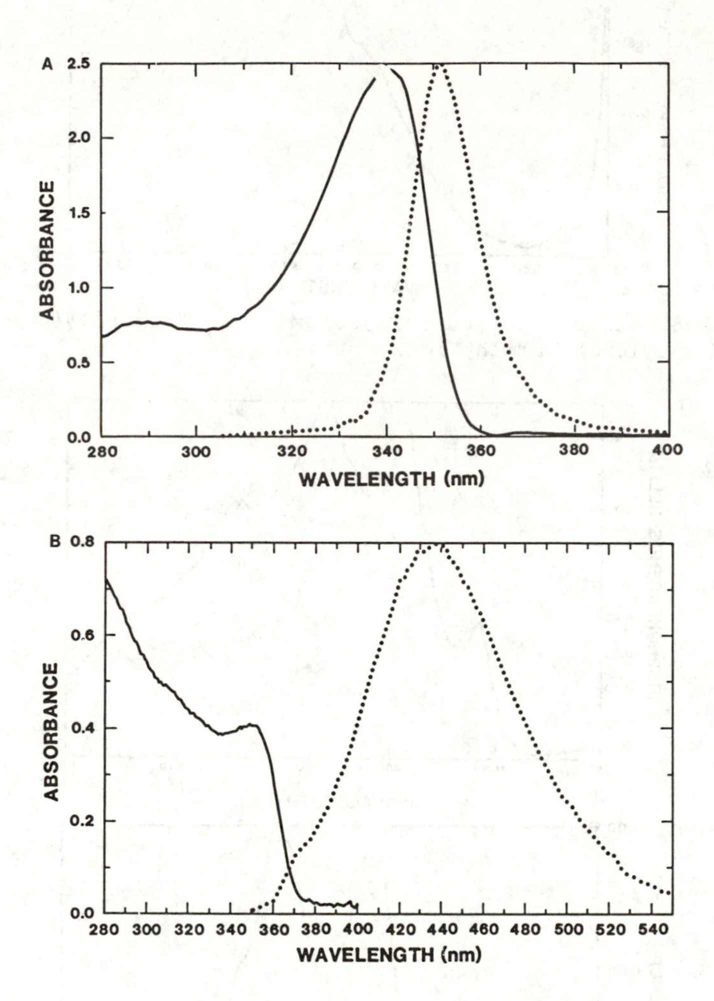


Figure 10. Room temperature absorption and emission spectra in hexane

A. Poly(phenyl methyl silylene)

B. Poly($\underline{\beta}$-naphthyl methyl silylene)

Summary

We have examined the emission spectra of a variety of polysilylenes as thin films and solutions. The solution fluorescence thermochromism provides evidence to support the rotational isomeric state model used to interpret the absorption spectrum. The structured character and low yield of phosphorescence in the alkyl polysilylenes suggest that the triplet is the immediate precursor to photochemical scission. The change in character of both fluorescence and phosphorescence on progressing from phenyl to naphthyl in the aryl series indicates that the transitions in the naphthyl polymers are principally $\pi$-$\pi^*$.

Acknowledgment

This work performed at Sandia National Laboratories was supported by the U.S. Department of Energy under contract number DE-AC04-76DP00789.

Literature Cited

1. (a) Zeigler, J. M. Polymer Preprints 1986. 27(1), 109.
   (b) Zeigler, J. M.; Harrah, L. A.; and Johnson, A. W. Polymer Preprints 1987, 28(1), 0000.
   (c) West, R. J. Organometal. Chem. 1986, 99, 300.
2. Zeigler, J. M.; Harrah, L. A.; and Johnson, A. W. SPIE Advances in Resist Process and Technology II 1985, 539, 166.
3. Johnson, A. W.; Zeigler, J. M.; and Harrah, L. A. SPIE Optical Microlithography IV 1985.
4. Hofer, D. C.; Miller, R. D.; and Wilson, C. G. SPIE Advances in Resist Technology 1984, 469, 16.
5. Hofer, D. C.; Miller, R. D.; Wilson, C. G.; and Neureuther, A. SPIE Advances in Resist Technology 1984, 469, 108.
6. (a) Miller, R. D.; Hofer, D. C.; Wilson, C. G.; Trefonas, P., III; and West, R. Polymer Preprints 1984, 25, 307.
   (b) Miller, R. D.; Hofer, D. C.; Wilson, C. G.; Trefonas, P., III ACS Symposium Series 1984, No. 266, 293.
7. Kepler, R. G.; Zeigler, J. M.; Harrah, L. A.; and Kurtz, S. R. Phys. Rev. B, in press; Kepler, R. G.; Zeigler, J. M.; Harrah, L. A.; and Kurtz, S. R. Bull. Am. Phys. Soc. 1983, 28, 362.
8. Harrah, L. A. and Zeigler, J. M. J. Poly. Sci., Poly. Lett. Ed. 1985, 23, 209.
9. (a) Harrah, L. A. and Zeigler, J. M. Proc. 1984 International Congress of Pacific Basin Societies, 1984, Honolulu, Hawaii; Abstract 09F0016.
   (b) Trefonas, P., III; Damewood,J. R., Jr.; West, R.; and Miller, R. D. Organometallics 1985, 4, 1318.
10. Harrah, L. A. and Zeigler, J. M. Bull. Am. Phys. Soc. 1985, 30, 540.
11. (a) Schweizer, K. S. Chem. Phys. Lett. 1986, 125, 118.
    (b) Schweizer, K. S. Polymer Preprints 1986, 27, 354.
    (c) Schweizer, K. S. J. Chem. Phys. 1986, 85, 1156.
    (d) Schweizer, K. S. J. Chem. Phys., 85 1986, 1176.
12. Miller, R. D.; Hofer, D. C.; Rabolt, J. F.; and Fickes, G. N. J. Am. Chem. Soc. 1985, 107, 2172.

13. (a) Lovinger, A. J.; Schilling, F. C.; Bovey, F. A.; and Zeigler, J. M. Macromolecules 1986, 19, 2663.
(b) Schilling, F. C.; Bovey, F. A.; Lovinger, A. J.; and Zeigler, J. M. Macormolecules 1986, 19, 2657.
(c) Damewood, J. R., Jr. Macromolecules 1985, 18, 1795.
14. Zeigler, J. M.; Adolf, D. B.; and Harrah, L. A. Proc. 20th Organosilicon Symp., 1986, Abs. #P-2.25, Tarrytown, NY
15. Boberski, W. G.; and Allred, A. L. J. Organometal. Chem. 1975, 88, 65.
16. Harrah, L. A. and Zeigler, J. M. Bull. Am. Phys. Soc. 1983, 28(3), 362.
17. Harrah, L. A. and Zeigler, J. M. Macromolecules, in press.
18. Pitt, C. G. and Bock, H. J. Chem. Soc., Chem. Comm. 1972, 29.
19. Pitt, C. G.; Carey, R. N.; and Toren, E. C., Jr. J. Am. Chem. Soc. 1972, 94, 3806.
20. Kagawa, T.; Fujino, M.; Takeda, K.; and Matsumoto, N. Solid State Comm. 1986, 57(8), 635.
21. Todesco R. V. and Kamat, P. V. Macromolecules 1986, 19, 196.
22. Johnson G. E. and McGrane, K. M. Polymer Preprints 1986, 27, 352.
23. Frank, C. W. and Harrah, L. A. J. Chem. Phys. 1974, 61, 1526.
24. Renschler, C. L. and Harrah, L. A. Anal. Chem. 1985, 55, 798.
25. Berlman, I. B. "Handbook of Fluorescence Spectra of Aromatic Molecules," 2nd Ed., Academic: New York, 1971.
26. Levinson, N. J. Soc. Ind. Appl. Math. 1962, 10, 442.
27. Harrah, L. A. and Zeigler, J. M. J. Poly. Sci., Poly. Lett. Ed., in press.
28. Becker, R. S. "Theory and Interpretation of Fluorescence and Phosphorescence," Wiley Interscience, New York, 1969.
29. Takeda, K. and Matsumoto, N. Phys. Rev. B. 1984, 30, 5871.
30. Bigelow, R. W. Chem. Phys. Lett. 1986, 126, 63.

RECEIVED March 13, 1987

# Chapter 36

# Spectroscopic and Photophysical Properties of Poly(organosilylenes)

**G. E. Johnson and K. M. McGrane**

**Xerox Corporation, Webster Research Center, Webster, NY 14580**

The results of an experimental study concerning the nature of the excited electronic states of poly(organosilylenes), a class of polymers in which the backbone consists of covalently bonded silicon atoms, are presented. A comparison of the absorption spectra of dilute solutions of alkyl and phenyl substituted silicon backbone polymers at room temperature and 77°K reveals pendant group dependent thermochromic effects which are attributed to temperature dependent conformational changes. The very narrow bandwidth fluorescence spectrum of these materials at 77°K is shown to be consistent with the formation of an exciton band in which electronic excitation is delocalized. Based on these results, as well as photoselection or polarized luminescence measurements, a model is developed which describes individual chains of these sigma bonded silicon backbone polymers as consisting of a distribution of variable length all-trans sequences each with its own effective conjugation length. Energy migration along the polymer backbone is found to occur by a mechanism in which energy is transferred from shorter to longer sequences.

Linear macromolecules in which the main chain is composed of covalently bonded silicon atoms constitute a class of materials which has recently attracted a great deal of renewed attention (1,2). This resurgence of interest has been stimulated in large part by advances in synthetic methods. High molecular weight materials with a variety of alkyl and/or aryl pendant groups are now available which can be cast into films or spun into fibers, are formable and in general are tractable, (3-6) a feature which distinguishes them from the first intractable silicon-based polymers synthesized over sixty years ago (7). These materials have been found to possess a number of remarkable properties which have led to a variety of technological

0097-6156/87/0358-0499$06.00/0

applications (8). Perhaps the most notable property of the organosilicon polymers is their sensitivity to ultraviolet light; consequently considerable efforts are currently being devoted to the development of these materials as dry or self developable resists for photolithographic applications by a number of industrial laboratories (9-12). Polymers with a silicon backbone have also been found to transport photoinjected or photogenerated positive charge carriers with remarkably high mobilities ($>10^{-4} cm^2 \cdot volt^{-1} \cdot sec^{-1}$) and thus constitute a novel and potentially important class of transport materials (13-15).

In addition to the interest this class of materials has attracted recently regarding their development for technological applications they have, for a number of years, received considerable attention from a more basic scientific viewpoint (1). Structurally the linear polysilylenes are analogs of the saturated alkanes, however, they have the distinct advantage, at least to the experimental spectroscopist, that their lowest lying electronic transition is strongly red-shifted into the easily accessible near ultraviolet region of the spectrum (16). Thus whereas both the carbon and silicon catenates exhibit a shift of their lowest lying electronic transitions to longer wavelengths with increasing chain length, the alkane absorption remains confined to the far ultraviolet while that of the silicon catenates appears to approach asymptotically, a limiting value at wavelengths longer than 300nm (16,17). In essence, one has at hand a class of materials which are sigma bonded like the alkanes yet which are amenable to detailed investigation utilizing more conventional experimental techniques than the vacuum ultraviolet methods generally required for the alkanes and other saturated hydrocarbons.

In many respects the organosilicon polymers exhibit properties which are more characteristic of an unsaturated aromatic than of a completely saturated sigma bonded compound. For example they exhibit remarkably low ionization potentials with values approximately 3 eV lower than an alkane of comparable length (18). Because of the low ionization energy of the sigma electrons, the organosilicon catenates have been found to form charge transfer complexes with certain strong organic electronic acceptors and thus constitute one of the few examples of a $\sigma$ electron donor- $\pi$ acceptor complex (19,20). A number of cyclic polysilanes have been found to undergo one-electron oxidation and reduction to form radical cations and anions (18,19,21,22). Electron spin resonance studies of the radical ions have shown the delocalized nature of the unpaired electrons in both the cation and anion (1,21,22).

While it is only recently that high molecular weight silicon backbone polymers have become available for detailed study, compounds containing the silicon-silicon bond, including oligomeric polysilylenes, have attracted considerable interest for many years. Thus there exists a number of papers of fundamental importance with particular relevance to the polysilylenes. Paramount among these are the contributions of Pitt and coworkers on the optical spectroscopy (18,23,24) and that by Bock and coworkers on the photoelectron spectroscopy of permethylated linear and cyclic silanes and the homologous series of linear silanes (25-27). A most useful survey of the field in general and the group IV catenates in particular as it existed up to 1977 is contained in the book titled "Homoatomic Rings, Chains and Macromolecules of Main-Group Elements" (28). A more recent review devoted to silicon backbone polymers alone has also appeared (29).

In this report, the results of an investigation which probes the nature of the excited electronic states of high molecular weight poly(organosilylenes) is presented. A model is developed, based on a variety of experimental observations, which describe individual polymer chains in terms of a distribution of variable length sequences of monomer units in an all-trans conformation. The fluorescence emanating from these units is exciton-like, suggesting that energy is delocalized on a very rapid time scale within these units.

## EXPERIMENTAL METHODS

MATERIALS. All the silicon backbone polymers investigated here were synthesized in-house by a Wurtz type reductive condensation of the appropriately substituted dichlorosilane with highly dispersed molten sodium metal. Because of the rapid photodegradation experienced by the polymers on exposure to ultraviolet light, the reactions were carried out in low level yellow light. The polymers were obtained, following appropriate work up of the reaction mixture, by precipitation from solution by a nonsolvent. The polymer is filtered, washed, reprecipitated and filtered and finally dried in vacuum. The molecular weight and molecular weight distribution of the polymers was determined by gel permeation chromatography and are based on polystyrene standards. The polymers, along with molecular weight data for those materials that were characterized, are listed in Table I.

Table I. Molecular Weights of the Poly(organosilylenes)

| Polymer | $\bar{M}_w$ | $\bar{M}_n$ | $\bar{M}_w/\bar{M}_n$ |
|---|---|---|---|
| poly(methylphenylsilylene)[a] (PMPS) | 623,950<br>4,870 | 223,230<br>3,400 | 2.8<br>1.43 |
| poly(methylcyclohexylsilylene (PMHS) | ------- | ------- | ---- |
| poly(methyl n-octylsilylene (PMOS) | 532,500 | 237,400 | 2.24 |
| poly(methyl n-propylsilylene (PMPrS) | 47,300 | 12,970 | 3.65 |

a. Bimodal distribution

Solvents utilized in this investigation, along with their source and method of purification, are listed below.

Benzene; J.T. Baker Chemical Co. Phillipsburg, NJ; Baker 'Analyzed' reagent grade or Burdick and Jackson Laboratories, Inc., Muskegan, Michigan. "Distilled in Glass" grade-both used as received.

2-methylbutane (isopentane); Aldrich Chemical Company, Inc., Milwaukee, Wis., spectrophotometric grade, Gold Label, used as received.

2-methyltetrahydrofuran (2MTHF); Aldrich Chemical Company, Inc., Milwaukee, Wis.; stabilized with 1% BHT. The solvent was freed of stabilizer by shaking in a separatory funnel with ~1 N NaOH until the aqueous phase remained clear. The stabilizer free 2MTHF was dried over molecular sieves (Davison M-518, Type 4A) overnight and then distilled. The middle third of the distilled 2MTHF was collected and stored over molecular sieves. The purified material was found to form a clear, stain-free amorphous glass at 77°K, devoid of interfering emission at the exciting wavelengths, used in these experiments.

3-methylpentane (3MP); Aldrich Chemical Company, Inc., Milwaukee, Wis. The solvent was shaken in a separatory funnel with sulfuric acid to remove possible aromatic impurities. After removal of the sulfuric acid layer the 3MP was shaken with dilute NaOH solution, separated and dried over molecular sieves (Davison M-518 Type 4A). The 3MP was then distilled and stored over molecular sieves for use in absorption and 77°K absorption and emission spectroscopic measurements.

INSTRUMENTATION. Absorption spectra were measured on a Cary 17D spectrophotometer. For room temperature spectra, the solutions were contained in 1 cm path length Hellma fused quartz cells. Spectra at 77°K were recorded on rigid glass solutions contained in 1 cm path length spectrosil grade quartz tubes mounted in a small fused quartz dewar which, when carefully cleaned and care exercized to prevent ice particles from entering the dewar, maintained the liquid nitrogen in a completely bubble free condition throughout the course of a measurement. These precautions yielded very clean spectra, free from the noise generally introduced by light scattering from the bubbles of boiling liquid nitrogen. A stream of dry nitrogen gas directed into the Cary 17D sample compartment prevented the formation of condensation on the dewar. The low temperature spectra were recorded against 1 cm of solvent at room temperature in the reference compartment of the Cary 17D.

Emission spectra were measured on a fluorometer constructed from the following components. Excitation was provided by a 200-w Hg-Xe lamp (Oriel) mounted on the entrance slit of a 0.25m Bausch and Lomb grating monochromator. Emission was viewed at right angle to the excitation with a RCA C31034 photomultiplier tube housed in a thermoelectrically cooled Pacific Precision Instruments Model 3407 PMT housing. The PMT was mounted on the exit slit of a McPherson Model 218, 0.3m, f/5.3 scanning monochromator. The instrument grating was blazed for 500nm with 1200 grooves/mm giving a reciprocal linear dispersion of 2.65nm/mm. The photomultiplier photocurrent was measured with a Keithley 610B electrometer and the output monitored with a Soltec Model 3312 recorder. Optics for collimating and focusing the excitation onto the sample as well as for collecting and properly focusing the emission onto the entrance slit of the viewing monochromator were of fused quartz. When polarization of emission spectra were determined using the method of photoselection a quartz Glan-Thomson polarizing prism (Karl Lambrecht) was inserted in the exciting and viewing optical paths. Fluorescence lifetimes were measured by the method of time correlated single photon counting on a system constructed with Ortec, Inc. components.

## RESULTS AND DISCUSSION

Presentation and discussion of the experimental results will proceed in the following manner. First the absorption spectra of the polysilylenes will be described and a comparative analysis of the spectra in room temperature fluid solvent media and rigid low temperature glasses at 77°K made. This will be followed by a description of the rather remarkable emission properties of these materials with emphasis on results obtained at 77°K. Included as part of the emission spectroscopic properties are the results of photoselection or polarization of emission measurements obtained in a rigid glass at 77°K. Based on these results a model is developed which describes individual chains of these silicon polymers in terms of a distribution of all-trans sequences with variable effective conjugation lengths.

DILUTE SOLUTION ABSORPTION SPECTRA. Figure 1 shows the absorption spectra of a dilute solution of PMPS in 2MTHF at room temperature and at 77°K. At room temperature the spectrum is characterized by a structureless band with a maximum absorbance, at 337nm with an extinction coefficient, $\epsilon$, at $\lambda$(max) of $8.7\times10^{3}$ liter-mole r.u.$^{-1}$cm$^{-1}$. Extinction coefficients are calculated based on the molecular weight of the silylene repeat unit (r.u.) which in the case of methylphenylsilylene, for example, is 120g/mole. The strong band at $\lambda > 300$nm is followed by a second, weaker structureless band with a peak intensity at 270nm. Absorption by the 2MTHF solvent prevents the observation of higher lying absorption bands.

Cooling the solution to 77°K to form the rigid glassy state leads to changes in the absorption spectrum, most notable of which is the increase in intensity of the lowest energy absorption band. The band also exhibits a slight red shift to yield a $\lambda$(max) of 339nm while experiencing a decrease in band width. The extinction coefficient at $\lambda$(max) is $1.27\times10^{4}$ liter-mole r.u.$^{-1}\cdot$cm$^{-1}$. a value nearly 50% greater than that at room temperature.

The same series of measurements was carried out on dilute solutions of PMHS in 2MTHF, however, since this material was not characterized as to its molecular weight and MWD and exhibited limited solubility in 2MTHF, no attempt was made to determine quantitative values of extinction coefficients. The following qualitative observations are noted. This polymer exhibits a structureless low energy band with a λ(max) at 320nm followed by a weak shoulder on the high energy side. Cooling to 77°K sharpens the low energy band slightly but the λ(max) remains at 320nm. The band does not show the increase in intensity exhibited by PMPS and, in fact, the absorbance at λ(max) is slightly diminished at 77°K compared to room temperature.

The most dramatic behavior is exhibited by PMOS and PMPrS. Figure 2 shows the absorption spectra of dilute solutions of PMPrS in 3MP at room temperature and 77°K. In fluid 3MP at room temperature PMPrS exhibits a moderately intense, structureless band with λ(max) at 306nm and $\epsilon=6.03\times10^{3}$ liter-mole $r.u.^{-1}\cdot cm^{-1}$. Upon cooling the solution to the rigid glassy state at 77°K the lowest lying absorption band shifts $2740cm^{-1}$ to the red ( λ(max)=334nm), sharpens dramatically, and becomes significantly more intense with ε increasing to $1.88\times10^{4}$ liter-mole $r.u.^{-1}\cdot cm^{-1}$. At room temperature the full width at half maximum (FWHM) of the lowest lying band is $5240cm^{-1}$ while at 77°K the FWHM decreases to $1250cm^{-1}$. Similar features are exhibited by PMOS.

Based upon the absorption spectra of the four poly(organosilylenes) investigated here it appears that the basic nature of the lowest excited electronic state remains relatively unperturbed by a variation in the nature of the pendant groups attached to the silicon backbone. There are obviously spectral shifts since the (max) of the lowest lying absorption band in the n-propyl and n-octyl derivatives each occurs at 306nm while that of the cyclohexyl occurs at 320nm and the phenyl at 337nm, however, the spectral bandshape remains practically unchanged. The lowest lying transition appears to be a property of the silicon backbone and has in fact been assigned as a transition from the highest occupied delocalized silicon molecular orbital to either the delocalized antibonding sigma orbital ( σ*) or silicon 3dπ type orbital (23-27). It should be noted here, however, that Robin has presented persuasive arguments against any significant 3dπ orbital participation in the lowest energy electronic excitation of the silicon catenates (30). The most dramatic effect is the large red shift and narrowing of the lowest absorption band of the poly(methyl n-alkylsilylenes) on cooling to 77°K. This thermochromic effect has been attributed to a conformational change of the polymer (31-33). In this regard it should be recalled, as noted earlier, that the position of the lowest lying transition in this class of materials is dependent on molecular weight. This was predicted early on through simple HMO calculations (16), corroborated by measurements on oligomeric organosilanes and recently extended to high molecular weight poly(alkylsilylenes) by Trefonas, et. al. (17). The λ(max) of the lowest lying band of the poly(alkylsilylenes) increases progressively with an increase

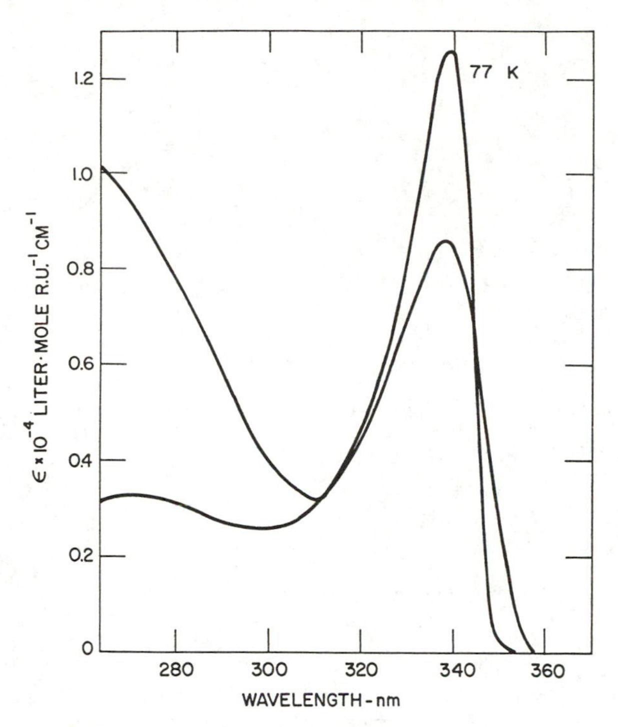


Figure 1. Dilute solution absorption spectra of poly(methylphenylsilylene) in 2-methyltetrahydrofuran at room temperature and 77 K.

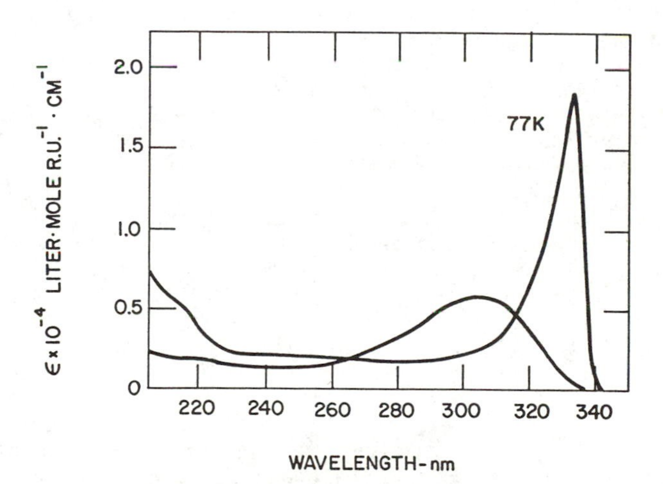


Figure 2. Dilute solution absorption spectra of poly(methyl-n-propylsilylene) in 3-methylpentane at room temperature and 77 K.

in the chain length and achieves a limiting value as the number of monomer units approaches 40 (17). Similar molecular weight effects have been reported by Harrah and Zeigler for poly(methylphenylsilylene) (34). The materials investigated here exceed this DP and hence, should be in the molecular weight regime when the λ(max) has arrived at its limiting value, at least in fluid solution. The fact that the poly(methyl n-alkysilylenes) investigated here experience a significant red shift and spectral narrowing upon cooling to 77°K is thus a consequence of a transition to a more thermodynamically stable conformational state with decreasing temperature. In contrast, the two polymers with the phenyl and cyclohexyl pendant groups do not exhibit this effect and their spectra remain relatively invariant with a decrease in temperature to 77°K. This apparently is a consequence of the steric constraints imposed by these relatively bulky substituents. A recent investigation by Cotts (35) suggests that the poly(organosilylenes) exist as somewhat expanded, slightly stiffened random coils in solution. Measurements of the radius of gyration over a range of temperatures indicated very little change (35) and this appears consistent with West and Miller's suggestion (33) that the thermochromism is due to an increase in population of the trans conformation in the silicon backbone upon cooling.

EMISSION SPECTRA. Emission spectra were measured for each of the silicon polymers both at room temperature and in rigid glass matrices at 77°K. Room temperature measurements in fluid solvent will not be reported since the results are subject to considerable variation. This is a consequence of the very rapid photodegradation these materials experience under ultraviolet excitation. Contrary to the extremely time dependent fluorescence signals observed in fluid solution, in rigid organic glasses at 77°K the photochemical degradation is virtually eliminated and fluorescence spectra can be recorded simply and reproducibly. Figure 3 shows the emission spectrum of a dilute solution of PMPS in 2MTHF at 77°K along with the lowest energy absorption band. The spectrum is characterized by a very sharp fluorescence band with a λ(max) at 347.7nm followed by a weaker, very broad structureless emission band which extends out beyond 500nm. The fluorescence is intense and, although an absolute quantum yield determination was not made, based on the level of emission of this material compared to other compounds of known quantum yield whose emission spectra have been measured on the same experimental set-up, the fluorescence quantum yield of PMPS in dilute 2MTHF solution at 77°K is estimated to be on the order of 0.6-0.7. The fluorescence lifetime was found to be on the order of 1 nsec or less. (It is not possible to give a precise value since the fluorescence decay was too fast to be resolved on the time correlated single photon counting system with the existing light pulser. However, subsequent to submission of this manuscript an article has appeared reporting the fluorescence lifetime of poly(di-n-hexylsilylene) to be 130 psec (36)). Clearly the fluorescence quantum yield and lifetime values are indicative of a highly allowed transition. Most striking is the sharpness of this band compared to the lowest lying absorption band. Behavior of this type signifies that the electronic excitation is delocalized along the polymer chain and can be described as a mobile Frenkel exciton (37,38). The absence of significant vibrational structure signifies that the Coulombic interaction between electronic transition dipole moments on neighboring silicon-silicon bonds is strong and that the rate of energy migration or delocalization is fast relative to the typical period of nuclear vibrations (39).

Figure 4 presents the emission spectrum of a dilute solution of PMPrS in 3MP at 77°K. As was the case with the phenyl derivative the emission spectrum is characterized by a very sharp fluorescence band peaking slightly to the red of the λ(max) of the lowest lying absorption band. PMPrS also displays a broad unstructured luminescence band extending well beyond 500nm. The intensity of this band in the case of PMPrS is more than three orders of magnitude less intense at its λ(max) (417nm) than the sharp fluorescence band. The intensity of this low energy emission band is even less in the case of PMOS and, is virtually identical to that of PMPrS with respect to its spectral location and band shape. These poly(organosilylenes) thus exhibit a variety of luminescence properties which are most unusual and in fact are considerably different than those exhibited by their structural analogs, the saturated alkanes. Lipsky and coworkers were the first to observe fluorescence from saturated hydrocarbons (40) and established, through a series of extensive investigations, correlations between molecular structure and the emission characteristics (41). For the n-alkanes the fluorescence was broad, unstructured, and virtually independent of the degree of catenation (at least from n=5 to n=17). The fluorescence λ(max) displayed an unusually large Stoke's shift. An increase in the degree of catenation led to an increase in the fluorescence quantum yield and lifetime. Typically fluorescence quantum yields are low ($<10^{-2}$) while fluorescence lifetimes are on the order of 1-10 nsecs. By way of contrast, as noted above, the most striking characteristic of the poly(organosilylenes) is their high quantum yield for fluorescence and the extreme narrowness of this band. In dilute solutions at 77°K the fluorescence band width (FWHM) is typically on the order of 500 $cm^{-1}$ furthermore, the λ(max) depends on the degree of catenation. The effect of molecular weight on the fluorescence spectrum was not investigated in a quantitative manner, however, it was demonstrated that λ (max) shifted to shorter wavelength in three PMPS fractions of decreasing molecular weight and shifted to shorter wavelengths in all materials which were subjected to ultraviolet photolysis and then excited at shorter wavelengths. From the results of the absorption and emission spectroscopic measurements two results of particular importance have emerged; one is the dramatic red shift and sharpening of the lowest lying transition of the conformationally mobile n-propyl and n-octyl derivatives on passage from room temperature to 77°K, the other is the

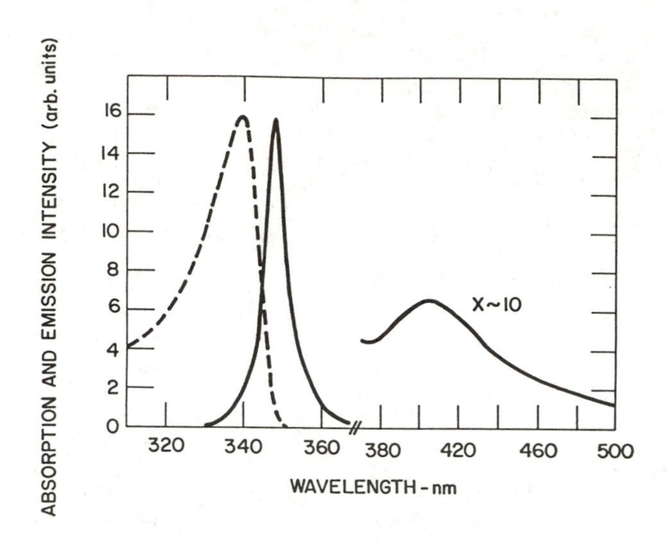


Figure 3. Absorption and emission spectra of poly(methylphenylsilylene) in 2-methyltetrahydrofuran at 77 K. The fluorescence spectrum was measured with a resolution of 0.265nm. The long wavelength emission was measured with a resolution of 1.32nm.

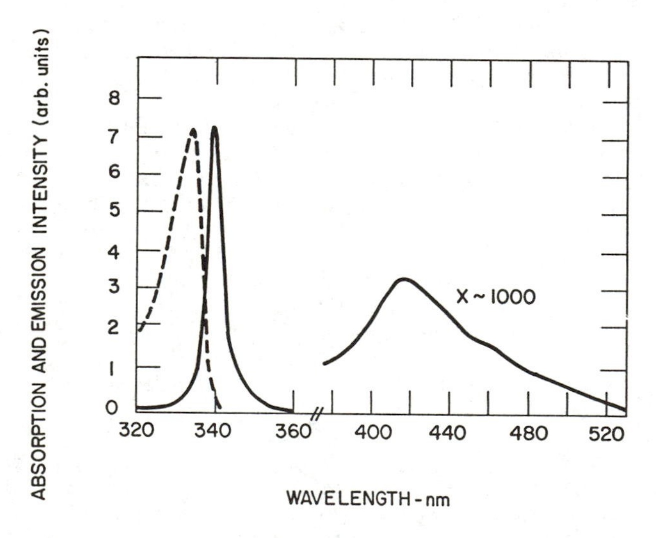


Figure 4. Absorption and emission spectra of poly(methyl-n-propylsilylene) in 3-methylpentane at 77 K. The fluorescence spectrum was measured with a resolution of 0.265nm. The long wavelength emission was measured with a resolution of 1.32nm.

very narrow band width of the fluorescence of each of the polymers. As will be discussed further below, each of these observations are consistent with an interpretation which calls upon the formation of an exciton band.

PHOTOSELECTION. An experimental technique which has proven to be of great utility in studies concerned with the assignment of electronic transitions in small molecules, particularly those that luminesce, is that termed photoselection (42). The principle underlying this method depends on the fact that molecules with appropriate symmetry properties absorb light preferentially along certain molecular axes or in or out of molecular planes. Even in a sample of randomly oriented molecules only those whose absorbing transition dipole moments are aligned with the oscillating electric vector of the incident exciting light are excited and an ensemble of excited molecules is photoselected; no effort to actively orient molecules, as for example in a single crystal or stretched polymer film, is made. It is the anisotropic nature of light itself along the polarization of the molecular transition dipole moment which is responsible for the selective excitation of only certain molecules out of the completely randomly oriented sample. When the absorbing molecules are rigidly fixed in space, such as in a solid amorphous matrix (typically a glassy organic solvent at 77°K), then the emission emanating from the sample is found to be polarized with the degree of polarization depending on the conditions of excitation and the relative orientation of absorbing and emitting transition dipole moments. Relationships giving the polarization ratio or degree of polarization of emission to be expected under various experimental conditions of excitation with respect to intrinsic molecular properties have been derived (42).

Here photoselection techniques are used to investigate the poly(organosilylenes), however, not with the strict intent to make state assignments but rather in a more qualitative manner to prove or disprove delocalization or migration of electronic excitation along the main chain of the polymer. It will be recalled that the very sharp fluorescence band, characteristic of all these materials, was suggestive of an emission from an exciton band and that the electronic excitation migrated very rapidly along the silicon backbone. If this is so one expects emission that is completely unpolarized since there should be no correlation between the orientation of the transition dipole moment in absorption and emission. For example, photoselection experiments on a dilute solution of poly(N-vinylcarbazole) in 2MTHF at 77°K yielded exactly this result; the emission was found to be completely unpolarized consistent with effective migration among carbazole pendant groups (43).

The results of a photoselection experiment on a dilute solution of PMPrS in a rigid glass consisting of a 7:3 volume ratio of 3MP and isoP at 77°K are shown in Figure 5. The polarization of the fluorescence, excited at four different wavelengths throughout the lowest absorption band, is shown as a function of the viewing wavelength. The results are not in accord with the intuitive expectation that the fluorescence would be unpolarized as a result of extensive energy migration along the polymer backbone. (A polarization ratio of 1.0 indicates unpolarized emission.) An additional significant result is that the polarization ratio increases as the excitation approaches the red edge of the lowest absorption band. (The large polarization ratios at the shorter viewing wavelengths under 334nm and 338nm excitation are not real; they are distorted due to contributions from scattered and reflected excitation from the quartz dewar and sample tube as the viewing and exciting wavelengths approach one another.) This indicates that as the excitation wavelength approaches the sharp edge of the lowest absorption band a condition is being achieved in which the absorbing and emitting transition dipole moments become parallel to one another. This is like the case of photoselection experiment on a dilute, rigid solution of small molecules in which the transition moment linking the ground and lowest lying excited state lies along a single molecular axis. In this case the absorption and fluorescence are necessarily along the same molecular axis and polarization ratios approaching 3 are observed.

In the case of polymers the situation is considerably more complex . Because of the polydisperse nature of polymers in general, in principle it is not entirely correct to describe solutions or films in terms of a single chain with its own unique properties. This is particularly so in the case of the silicon polymers where it is known that the location of the lowest absorption band is molecular weight dependent (17,31). In addition, the shift of λ (max) and the change in band shape with temperature suggests that a conformational dependency is also operational. The PMPrS used here has a rather broad MWD of 3.65 a fact which could be considered as being responsible for the rather broad structureless nature of the lowest absorption band observed in fluid solution at room temperature. In essence the band could be interpreted as reflecting the envelope of the individual absorption bands of chains of varying length within the distribution. This does not appear likely, however, since the vast majority of chains within the distribution will have molecular weights in the region where the absorption λ(max) has reached its limiting value. A more appropriate description, and one which is in accord with the temperature dependence of the absorption spectra, is based on a distribution of "effective conjugation lengths" within a single chain. (For a discussion of the concept of "effective conjugation length" and how it pertains in particular to poly(diacetylenes) see (44). In the ground state these regions of variable "effective conjugation length" (ECL) absorb independently and lead to the broad structureless absorption observed at room temperature; as temperature is lowered conformational changes occur which extend the ECL and narrow its distribution thus yielding the red shift and band narrowing. The polarization of fluorescence results can be interpreted, in fact, as indicating the migration of energy from regions of shorter to longer ECL with a subsequent loss of correlation between the absorbing and emitting transition dipole moments. Since the polarization ratio is always greater than one, the loss of correlation is not complete.

In order to proceed it is now necessary to consider the nature of the lowest excited state of these polymers. One description which appears to be particularly appropriate to these materials is that given by the molecular exciton theory (37,38). This of course is suggested by the nature of the fluorescence spectrum itself and in addition this approach has proven to be quite successful in the interpretation of the electronic states of the alkanes, the structural analogs of the poly(organosilylenes) (45,46). The basic assumption in this theory is that the absorption bands shift due to an interaction of silicon-silicon bond excitations each considered as an independent system. The absorption due to each bond is assigned as a $\sigma^* \leftarrow \sigma$ transition of the Si-Si bond electrons and thus the transition moment is polarized along the bond. (See Figure 6) The Coulombic interaction of the transition dipole moments on individual bonds can be shown to form an exciton band which, for the case at hand (i.e. oblique transition dipoles where the angle between them is given by the Si-Si-Si bond angle), leads to two allowed transitions (37,38). One is strongly red shifted (with respect to a disilane) and corresponds to an in-phase arrangement of transition dipoles while the other is blue shifted and corresponds to an out-of-plane arrangement of transition dipoles. Within a planar zig-zag, all-trans sequence along the polymer, the lowest transition will be polarized along the backbone while the higher energy transition will be polarized perpendicular to the backbone. Thus within the molecular exciton model the polymer should have a strongly allowed transition at an energy far below that of a single Si-Si bond and this transition should be polarized along the trans sequence contained in the polymer backbone. The fluorescence will necessarily be along the same axis and a theoretical polarization ratio of 3 is expected. The high energy transition will lie at energies to the blue of that of a single Si-Si bond and will be spectroscopically inaccessible. In any event, within this simple model, these states cannot lead to absorption polarized in a direction perpendicular to the fluorescence emission in a spectral region which overlaps the lower energy absorption. The change in the polarization ratio as the excitation energy increases from the high energy end to the "red edge" of the lowest absorption band is consequently attributed to energy migration along individual chains. Within any individual chain there is likely to be a distribution of all-trans sequences of varying length. If each of these sequences were strictly electronically isolated from one another then the polarization ratio would be independent of the excitation wavelength and approach a value of 3. This is not the case and apparently energy initially absorbed by the shorter sequences can transfer to longer sequences, perhaps by a Förster type nonradiative resonance mechanism (47) with a subsequent randomization of absorbing and emitting transition

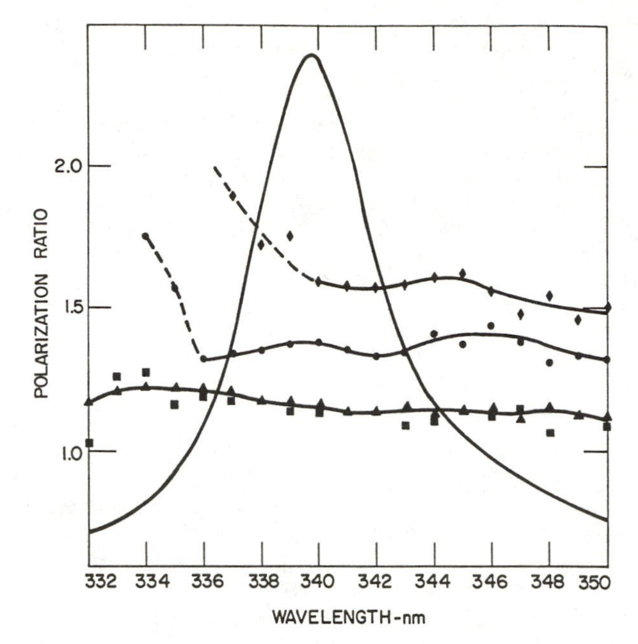


Figure 5. Polarization of fluorescence spectra of a dilute solution of poly(methyl n-propylsilylene) in 7:3 3-methylpentane: isopentane at 77 K. Excitation wavelengths: ■ , 297 nm: ▲ , 313nm; ● , 334nm; ◆ , 338nm.

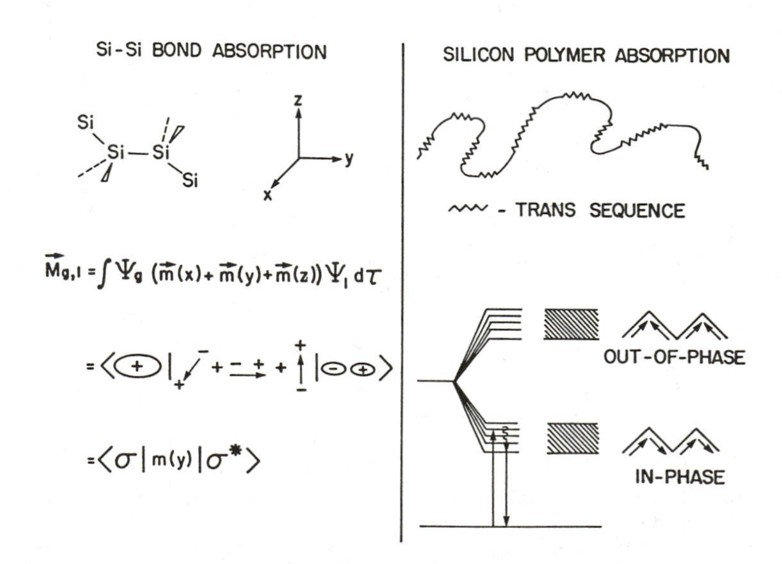


Figure 6. Pictorial description of the silicon-silicon bond transition dipole moment and the formation of an exciton band in all-trans sequences of poly(organosilylenes).

moments and reduction of the polarization ratio. The above picture is obviously much too simple to describe the fine details of complex systems such as these, however, it appears to offer at least a rationalization of some of the more essential features.

## CONCLUSIONS

The preceeding results, which have been concerned with the properties of isolated polymer chains, appear consistent with an interpretation which describes the lowest lying absorption band of these silicon main-chain polymers as arising from highly allowed transitions linking the ground state with collective states delocalized over variable length all-trans sequences of monomer units. Polarization of emission experiments have indicated that this collective excitation, which is delocalized on a very rapid time scale over the individual sequences, can migrate along the chain from shorter (higher energy) to longer sequences on which it is ultimately trapped. In essence, each isolated silicon polymer chain can be considered to contain a collection of ultraviolet absorbing chromophores consisting of variable length all-trans sequences each with its own ECL. Chain defects, such as gauche kinks, are likely responsible for disrupting the conjugation along the backbone (48). It is this variation in ECL which leads to the broadness of the lowest energy absorption band. Energy initially localized within the shorter sequences can migrate along the chain due to dipole-dipole coupling of transition dipole moments (i.e. there is overlap between the emission of short sequences and the absorption of longer sequences). Trapping of electronic excitation within the longest sequences is responsible for the sharp, exciton-like fluorescence band. It is important to note that this description, which pertains strictly to isolated polymer chains in dilute solution, very likely holds for the solid state as well. A variety of experiments on solid films of these silicon polymers indicate that interchain interactions are weak and that the polymer chains maintain their individual identity. (Johnson, G.E., Xerox Corporation, Webster Research Center, unpublished data). The spectroscopic and photophysical properties of these poly(organosilylenes) in the solid state are, in a very real sense, much like those expected for small molecules molecularly dispersed in a host polymer matrix. Perhaps in this regard it is not unreasonable to draw an analogy between the spectral characteristics of a distribution of variable length all-trans sequences with that of a distribution of small molecules inhomogeneously broadened due to variable environmental perturbations.

## ACKNOWLEDGMENT

Both authors wish to express their gratitude to Dr. Milan Stolka. K.M. McGrane acknowledges the valuable direction provided by Dr. Stolka to the synthetic aspects of this work. G.E. Johnson is grateful to Dr. Stolka for introducing him to this class of polymers and for his interest throughout the course of this investigation.

LITERATURE CITED

1. West, R., Carberry, E; Science 1975, 189, 179.
2. Allcock, H.R. Chem. Eng. News, March 18, 1985, p.22.
3. Trefonas III, P; West, R.; Miller, R.D.; Hofer, D. J. Polym. Sci. Polym. Lett. Ed. 1983, 21, 819.
4. Wesson, J.P.; Williams, T.C. J. Polym. Sci. Polym. Chem. Ed. 1980, 18 959.
5. Zhang, X.; West, R. J. Polym. Sci. Polym. Chem. Ed. 1984 22, 159.
6. Zhang, X.; West, R. J. Polym. Sci. Polym. Chem. Ed. 1984, 22, 225.
7. Kipping, F.S. J. Chem. Soc. 1924, 125, 2291.
8. See for example, High Technology, July 1985, p.70.
9. Zeigler, J.M.; Harrah, L.A.; Johnson, A.W. SPIE Adv. Resist. Tech. Proc. II, 1985, 539 166.
10. Hofer, D.C.; Miller, R.D.; Willson, C.G. Proc. SPIE 1984, 469, 16.
11. Hofer, D.C.; Miller, R.D.; Willson, C.G.; Neureuther, A.R. Proc. SPIE 1984, 469, 108.
12. Miller, R.D.; Hofer, D.; McKean, D.R.; Willson, C.G.; West R.; Trefonas III, P.T. in Materials for Microlithography, Thompson, L.F.; Willson, C.G. and Frechet, J.M.J., Eds. ACS Symposium Series 1984 266, 293.
13. Yuh, H.-J.; Abkowitz, M.; McGrane, K.; Stolka, M. Bull. Am. Phys. Soc. 1986, 31, 381.
14. Kepler, R.G.; Zeigler, J.M.; Harrah, L.A. Bull. Am. Phys. Soc. 1984,29, 509.
15. Kepler, R.G.; Zeigler, J.M.; Harrah, L.A.; Kurtz, S.R. Bull Am. Phys. Soc. 1983, 28, 362.
16. Pitt, C.G.; Jones, L.L.; Ramsey, B.G. J. Am. Chem. Soc. 1967, 89, 5471.
17. Trefonas III, P.T.; West, R.; Miller, R.D.; Hofer, D. J. Polym. Sci. Polym. Lett. Ed. 1983, 21, 823.
18. Pitt, C.G.; Bursey, M.M.; Rogerson, P.F. J. Am. Chem. Soc. 1970, 92, 519.
19. Traven, V.F.; West, R. J. Am. Chem. Soc. 1973, 95, 6824.
20. Sakurai, H.; Kira, M.; Uchida, T. J. Am. Chem. Soc. 1973, 95, 6826.
21. Bock, H.; Kaim, W.; Kira, M.; West, R. J. Am. Chem. Soc. 1979, 101, 7667.
22. Carberry, E.; West, R.; Glass, G.E. J. Am. Chem. Soc. 1969, 91, 5446.
23. Pitt, C.G. J. Am. Chem. Soc. 1969, 91, 6613.
24. Pitt, C.G.; Carey, R.N.; Toren, Jr., E.C. J. Am. Chem. Soc. 1972, 94, 3806.
25. Bock, H.; Ensslin, W. Angew. Chem. Int. Ed. 1971, 10, 404.
26. Bock, H.; Ramsey, B.G. Angew. Chem. Int. Ed. 1973, 12, 734.
27. Bock, H. Angew. Chem. Int. Ed. 1977, 16, 613.

28. Homoatomic Rings. Chains and Macromolecules of Main-Group Elements Rheingold, A.L., Ed.; Elsevier Scientific Publishing Company, 1977.
29. West, R. J. Organomet. Chem. 1986, 300, 327.
30. Robin, M.B. Higher-Lying Excited States of Polyatomic Molecules Academic Press, New York 1974; Chapter IIIG.
31. Miller, R.D.; Hofer, D.; Rabolt, J.; Fickes, G.N. J. Am. Chem. Soc. 1985, 107, 2172.
32. Harrah, L.A.; Zeigler, J.M. J. Polym. Sci. Polym. Lett. Ed. 1985, 23, 209.
33. Trefonas III, P.; Damewood, Jr., J.R.; West, R.; Miller, R.D. Organometallics, 1985 4, 1318.
34. Harrah, L.A.; Zeigler, J.M. Bull. Am. Phys. Soc. 1983, 28, 362.
35. Cotts, P.M. Am. Chem. Soc. Polym. Mater. Sci. Eng. 1985, 53, 336.
36. Klingensmith, K.A.; Downing, J.W.; Miller, R.D.; Michl, J. J. Am. Chem. Soc. 1986, 108, 7438.
37. McRae, E.G. and Kasha, M. Physical Processes in Radiation Biology, (Academic Press, New York, 1964, p. 23.
38. Kasha, M.; Rawls, H.R.; El-Bayoumi, M.A. Pure Appl. Chem. 1965, 11, 371.
39. Kasha, M. Radiation Research 1963, 20, 55.
40. Hirayama, F.; Lipsky, S. J. Chem. Phys. 1969, 51, 3616.
41. Rothman, W.; Hirayama, F.; Lipsky, S. J. Chem. Phys. 1973, 58, 1300.
42. Albrecht, A.C. J. Mol. Spectroscopy 1961, 6, 84.
43. Johnson, G.E. Macromolecules 1980, 13, 145.
44. Wenz, G.; Miller, M.A.; Schmidt, M.; Wegner, G. Macromolecules, 1984, 17, 837.
45. Raymonda, J.W.; Simpson, W.T. J. Chem. Phys. 1967, 47, 430.
46. Partridge, R.M. J. Chem. Phys. 1968, 49, 3656.
47. Förster, Th. Discussions Faraday Soc. 1959, 27. 7.
48. Johnson, G.E., McGrane, K.M. Polym. Preprints, 1986, 27(2), 351.


RECEIVED March 13, 1987

# Author Index



# Affiliation Index

# Subject Index

G

H



I



J



K

L



M



N

*Production and indexing by Colleen P. Stamm*
*Jacket design by Carla L. Clemens*

*Elements typeset by Hot Type Ltd., Washington, DC*
*Printed and bound by Maple Press, York, PA*